Differential Geometry and Integrable Systems

CONTEMPORARY
MATHEMATICS

308

Differential Geometry and Integrable Systems

A Conference on
Integrable Systems in Differential Geometry
University of Tokyo, Japan
July 17–21, 2000

Martin Guest
Reiko Miyaoka
Yoshihiro Ohnita
Editors

American Mathematical Society
Providence, Rhode Island

This volume contains the proceedings of a conference on Integrable Systems in Differential Geometry held at the University of Tokyo, Japan on July 17–21, 2000.

2000 *Mathematics Subject Classification.* Primary 35Q53, 37K05, 37K10, 53A05, 53A07, 53A10, 58D25, 58D27, 58E20, 70H06.

Library of Congress Cataloging-in-Publication Data

Differential geometry and integrable systems : a conference on integrable systems in differential geometry, July 2000, University of Tokyo / Martin Guest, Reiko Miyaoka, Yoshihiro Ohnita, editors.

p. cm. — (Contenporary mathematics, ISSN 0271-4132 ; 308)

Includes bibliographical references.

ISBN 0-8218-2938-6 (alk. paper)

1. Geometry, Differential—Congresses. 2. Hamiltonian systems—Congresses. I. Guest, Martin A. II. Miyaoka, Reiko, 1951– III. Ohnita, Yoshihiro, 1958– IV. Contemporary mathematics (American Mathematical Society) ; v. 308.

QA641 .D3847 2002
516.3′6—dc21 2002028162

∞ The paper used in this book is acid-free and falls within the guidelines
established to ensure permanence and durability.
Visit the AMS home page at `http://www.ams.org/`

10 9 8 7 6 5 4 3 2 1 07 06 05 04 03 02

Contents

Preface

Geometry is perhaps the oldest branch of mathematics, and the concept of integrability goes back to the early days of the calculus of several variables. It is not surprising that the tides of interaction have ebbed and flowed throughout this long period of coexistence. What may be surprising is the lack of a systematic theory relating the two subjects: each has developed in its own way, with its own problems, techniques and priorities. There is good reason to believe that this state of affairs is changing, however, and the main reason for this change — and consequent optimism for a more systematic approach — is the development of new mathematical tools over the last century and especially over the last thirty years.

These new tools include infinite dimensional Lie theory, the theory of moduli spaces of geometric structures, algebraic topology, symplectic geometry — and computers. The computer experiments of the 1950's, which led to the discovery of soliton theory, were the catalyst for rapid theoretical developments, employing a dazzling array of techniques across the whole spectrum of mathematics. The renewal of strong links between mathematics and quantum physics since the 1970's came at a fortuitous time, because many "integrable systems" have their origins in physics, and differential geometry plays a fundamental role here as well. Another phenomenon which has proved favourable to the subject is the return to mathematical fashion of concrete examples and explicit calculations, following several decades of preoccupation with abstract principles.

In view of these developments, an international conference was held in Japan, at the University of Tokyo, during the period 17-21 July 2000, with the title "Integrable Systems in Differential Geometry". Choosing a suitable title was not an easy task. The words "geometry" and "integrable systems" were essential, but the precise combination implies a deliberate choice of emphasis, and we were reluctant to impose our own artificial constraints on such a rapidly emerging subject. In the end we decided that differential geometry should be the underlying theme, partly because of the large number of differential geometers in Japan who were likely to attend, and partly because of the research areas of those who had already expressed interest in taking part in such a conference. But we intended to cover relations between integrable systems and geometry in the widest sense, and we were gratified that many speakers and participants represented areas such as physics, algebraic geometry, differential equations, and topology. We believe that a successful aspect of the conference was the juxtaposition of speakers and topics from a variety of subjects, some very much at the heart of recent connections between geometry and integrable systems, others in areas where there is potential for interaction in the future. We hope that the conference, and in particular these proceedings, will encourage differential geometers to embrace new ideas and techniques from integrable systems, and researchers in integrable systems to recognise the role of geometry as a source of intuition and concrete examples.

For practical purposes the proceedings are divided into three volumes. "Surveys on Geometry and Integrable Systems", published by the Mathematical Society of Japan in the series *Advanced Studies in Pure Mathematics*, contains mainly expository articles which relate geometry and integrable systems. "Differential Geometry and Integrable Systems", published by the American Mathematical Society in the series *Contemporary Mathematics*, contains both research and expository articles whose main focus is differential geometry. "Integrable Systems, Topology, and Physics", in the same series, contains articles related to geometry in a wider sense: topology, algebraic geometry, symplectic geometry, and mathematical physics. We have provided brief topical introductions for each of the three volumes, but there are a number of excellent general articles on geometry and integrable systems where the reader can find more details, such as the survey article [**4**] and the introductions to [**1**] and [**5**]. The article by Hitchin in [**5**] gives an overview of some major links between differential geometry, algebraic geometry and integrable systems, and the collections [**2**], [**3**] contain many more examples. References are listed at the end of the Introduction.

The conference was organised under the auspices of the Mathematical Society of Japan, as its 9-th International Research Institute, and financial support was provided by the Mathematical Society of Japan, the Japan Association for Mathematical Sciences, and from individual research grants of the Japan Society for Promotion of Science. The members of the Organising Committee were Takushiro Ochiai, Chairman (University of Tokyo), Martin Guest (Tokyo Metropolitan University), Mitsuhiro Itoh (University of Tsukuba), Reiko Miyaoka (Sophia University), Tohru Morimoto (Nara Women's University), Yoshihiro Ohnita (Tokyo Metropolitan University), Simon Salamon (Oxford University), Takeshi Sasaki (Kobe University), Chuu-Lian Terng (Northeastern University), and Gudlaugur Thorbergsson (Cologne University). On behalf of the entire Organising Committee we would like to thank several institutions and many individuals whose cooperation was essential for the success of the conference. In addition to the above funding agencies we thank the University of Tokyo for providing facilities at the Graduate School of Mathematical Sciences building in Komaba. Our conference secretaries Junko Furihata, Shiho Imai, and Yukari Takigawa, as well as the student assistants, all worked hard to ensure the smooth running of day to day activities and the associated administrative procedures. All articles were refereed, and we thank those mathematicians who served as anonymous referees for their important contributions. Carolyn Dong provided invaluable assistance with the preparation of TeX files. Finally we thank all speakers and participants for making the conference possible.

Martin Guest, Reiko Miyaoka and Yoshihiro Ohnita
9-th MSJ-IRI Proceedings Editors
Tokyo, June 2002

Introduction

The articles in this volume have a differential geometric emphasis. Many examples in the theory of integrable systems have their origins in differential geometry, where, conversely, questions of integrability have always been fundamental.

Variational principles give rise to natural geometric objects such as geodesics, minimal surfaces, harmonic maps, and Yang-Mills connections, so these have the same foundations as the various Euler-Lagrange equations of mathematical physics. Such equations often have a Hamiltonian formulation via the Legendre transformation, and this is the starting point of symplectic geometry and the study of symmetries of differential equations, a major theme of the theory of integrable systems. While some yield directly to integration through the Liouville-Arnold theorem, most partial differential equations remain very hard to solve explicitly by analytical methods, and the opportunity to exploit symmetries of naturally-occurring equations is one of the attractions of the integrable systems approach. Sometimes this approach is spectacularly successful, the KdV equation being a well known case, and this has certainly attracted the attention of geometers.

These symmetry groups are in general infinite dimensional, and they are usually not immediately visible in the initial geometrical formulation of the problem. For this reason, new techniques are required beyond the familiar methods of (finite dimensional) Lie group theory in differential geometry. Since infinite dimensional Lie theory has been developed most effectively at the infinitesimal level — that is, for Lie algebras — the subject has at times an algebraic character. But all these symmetries tend to be related, and the hidden, infinite dimensional groups can usually be traced back to more evident, finite dimensional ones. Thus, classical Lie theory and the theory of symmetric spaces in differential geometry feature in many of the articles in this volume.

The greatest impact of the theory of integrable systems on differential geometry has so far been in surface theory, especially in the study of surfaces with special properties in Euclidean space, and several articles deal with this subject. Other articles are concerned with integrability in the geometry of higher dimensional manifolds, and this topic seems likely to receive a lot more attention as the results from surface theory become more widely known and understood. The theory of harmonic maps (and their generalizations such as p-harmonic maps and harmonic morphisms) into symmetric spaces combines aspects of both these directions, as does classical submanifold theory.

Infinite dimensional symmetry groups arise naturally when these kinds of problems can be reformulated as zero curvature equations, i.e. equations which express the flatness of a certain connection, for then the gauge group comes into play. An alternative point of view regards the zero curvature equation as a family of Lax equations, or a single Lax equation on a function space. The resulting infinite dimensional "symmetry groups" have a strong influence on the geometry of the original problem, and provide a variety of approaches to revealing new information. For example one can study orbits of the action, or try to characterize solutions of a certain type in terms of properties of their orbits, or one can regard the group action simply as a way of producing new solutions from old ones.

Differential geometers may have been reluctant to venture into this field in the past, perhaps because it seems to demand a background in analysis or algebra or even physics. But this much is certain: the methods of integrable systems theory have already been so productive in geometry that they cannot in future be ignored. The discovery of Wente tori, the work of Hitchin and of Uhlenbeck on harmonic maps, the new insights into submanifold theory — all of which are related to the presence of previously unimaginable symmetry groups — have established a bond between integrable systems and differential geometry which is likely to endure.

References

[1] A.I. Bobenko and R. Seiler (eds.), *Discrete Integrable Systems*, Oxford Lecture Series in Math. and its Appl. 16, Oxford Univ. Press, 1999.

[2] A. Fordy and J.C. Wood (eds.), *Harmonic Maps and Integrable Systems*, Vieweg, 1994.

[3] N.J. Hitchin, G.B. Segal and R.S. Ward, *Integrable Systems*, Oxford Graduate Texts in Math. 4, Oxford Univ. Press, 1999.

[4] S. Novikov, *Role of integrable models in the development of mathematics*, Mathematical Research Today and Tomorrow. Viewpoints of Seven Fields Medalists. (C. Casacuberta and M. Castellet., eds.), Lecture Notes in Mathematics 1525, Springer, 1992.

[5] C.-L. Terng and K. Uhlenbeck (eds.), *Surveys in Differential Geometry: Integrable Systems*, volume IV, International Press, 1998.

List of speakers and titles

All speakers are listed below in alphabetical order. The locations of articles published in the conference proceedings are indicated as follows:

SGIS: "Surveys on Geometry and Integrable Systems", *Advanced Studies in Pure Mathematics*, Mathematical Society of Japan

DGIS: "Differential Geometry and Integrable Systems", *Contemporary Mathematics*, American Mathematical Society

ISTP: "Integrable Systems, Topology, and Physics", *Contemporary Mathematics*, American Mathematical Society

The title refers to the published article when this differs from the title of the talk.

N. Ando [DGIS]
The Index of an Isolated Umbilical Point on a Surface
J. Berndt [—]
Ricci Curvature and Staeckel Systems
A. Bobenko [SGIS]
Exploring Surfaces through Methods from the Theory of Integrable Systems: the Bonnet Problem
J. Bolton [DGIS]
The Toda Equations and Equiharmonic Maps of Surfaces into Flag Manifolds
J.-M. Burel and E. Loubeau [DGIS]
p-Harmonic Morphisms: the $1 < p < 2$ *Case and Some Non-trivial Examples*
F. Burstall, F. Pedit and U. Pinkall [DGIS]
Schwarzian Derivatives and Flows of Surfaces
L. Casian and Y. Kodama [ISTP]
Twisted Tomei Manifolds and the Toda Lattices
J.-H. Chang [ISTP]
Quantization of Benney Hierarchies
V. De Smedt and S. Salamon [DGIS]
Anti-Self-Dual Metrics on Lie Groups
J. Dorfmeister [SGIS]
Generalized Weierstrass Representations of Surfaces
J. Dorfmeister, J. Inoguchi and M. Toda [DGIS]
Weierstraß-Type Representation of Timelike Surfaces with Constant Mean Curvature
B. Dubrovin [—]
Geometry of Bihamiltonian Evolutionary Equations

N. Ejiri [DGIS]
A Differential-Geometric Schottky Problem, and Minimal Surfaces in Tori
E.V. Ferapontov [DGIS]
Surfaces in 3-space Possessing Nontrivial Deformations which Preserve the Shape Operator
A. Fujioka and J. Inoguchi [SGIS]
Timelike Surfaces with Harmonic Inverse Mean Curvature
K. Fukaya [ISTP]
Floer Homology for Families — a Progress Report
R. Goto [ISTP]
Rozansky-Witten Invariants of Log Symplectic Manifolds
C.-H. Gu [SGIS]
Darboux Transformations and Generalized Self-Dual Yang-Mills Flows
M.A. Guest* [ISTP]
An Update on Harmonic Maps of Finite Uniton Number, via the Zero Curvature Equation
E. Heintze [—]
From Submanifold Geometry to Kac-Moody Algebras
F. Hélein and P. Romon [SGIS]
From CMC Surfaces to Hamiltonian Stationary Lagrangian Surfaces
F. Hélein and P. Romon [DGIS]
Hamiltonian Stationary Lagrangian Surfaces in Hermitian Symmetric Spaces
T. Hoffmann [—]
Discrete Hashimoto Surfaces and a Discrete Smoke Ring Flow
J. Hong [—]
Fano Manifolds with Contact Structures
H. Hu [DGIS]
Line Congruences and Integrable Systems
J. Hurtubise [—]
Generalised Hitchin Systems
S. Ianus [—]
Recent Results on Semi-Riemannian Submersions
M. Igarashi [—]
On Compact Non-Kaehlerian Hermite-Liouville Surfaces
J. Inoguchi (see J. Dorfmeister; see A. Fujioka)
R. Inoue [ISTP]
The Lattice Toda Field Theory for Simple Lie Algebras
U. Jeromin [—]
Infinitesimal Geometry: a Link between Smooth and Discrete Theories?
X. Jiao [DGIS]
Factorizations of Harmonic Maps of Surfaces into Lie Groups by Singular Dressing Actions
H. Jin and X. Mo [DGIS]
On Submersive p-Harmonic Morphisms and their Stability
D. Joyce [SGIS]
Special Lagrangian 3-folds and Integrable Systems
K. Kiyohara [DGIS]
On Kähler-Liouville Manifolds
Y. Kodama (see L. Casian)

M. Kokubu, M. Umehara and K. Yamada [DGIS]
Minimal Surfaces that Attain Equality in the Chern-Osserman Inequality
H. Konno [ISTP]
On the Cohomology Ring of the HyperKähler Analogue of the Polygon Spaces
X. Liu [SGIS]
Quantum Product, Topological Recursion Relations, and the Virasoro Conjecture
E. Loubeau (see J.-M. Burel)
A.-L. Mare* [ISTP]
On the Theorem of Kim Concerning $QH^*(G/B)$
S. Matsutani [—]
The Frenet-Serret and Generalized Weierstrass Relations from the Dirac Equations
V.S. Matveev [DGIS]
Low Dimensional Manifolds Admitting Metrics with the Same Geodesics
C. McCune [—]
Rational Minimal Surfaces through the UP-Iteration
I. McIntosh [SGIS]
Harmonic Tori and their Spectral Data
T. Miwa [—]
Combinatorics of Coinvariants
R. Miyaoka [SGIS]
Isoparametric Geometry and Related Fields
X. Mo (see H. Jin)
Y. Nagatomo [ISTP]
Geometry of the Twistor Equation and its Applications
A. Nakayashiki [ISTP]
On the Cohomology of Theta Divisors of Hyperelliptic Jacobians
M. Noumi and Y. Yamada [—]
Birational Realization of Weyl Groups and Discrete Integrable Systems
Y. Ohnita and S. Udagawa [DGIS]
Harmonic Maps of Finite Type into Generalized Flag Manifolds, and Twistor Fibrations
Y. Ohyama [ISTP]
Isomonodromy Deformations and Twistor Theory
K. Okamoto [—]
The Painleve-Garnier Systems
K. Ono [ISTP]
Simple Singularities and Symplectic Fillings
L. Ornea [—]
Complex Structures on some Stiefel Manifolds
D. Osipova [—]
Symmetric Submanifolds of Symmetric Spaces
T. Otofuji [ISTP]
Quantum Cohomology of Infinite Dimensional Flag Manifolds
R. Palais [—]
An Efficient Method for Solving Initial Value Problems for Weakly Nonlinear Wave Equations
J. Park [DGIS]
Submanifolds Associated to Grassmannian Systems

H. Pedersen [SGIS]
Kahler Ricci Solitons
F. Pedit (see F. Burstall)
U. Pinkall (see F. Burstall)
P. Romon (see F. Helein)
W. Rossman, M. Umehara and K. Yamada [SGIS]
Period Problems for Mean Curvature One Surfaces in H^3 (with Applications to Surfaces of Low Total Curvature)
S. Saito, N. Suzuki and H. Yamaguchi [ISTP]
Discrete Conjugate Nets of Strings
Y. Sakane and T. Yamada [DGIS]
Harmonic Cohomology Groups of Compact Symplectic Nilmanifolds
S. Salamon (see V. De Smedt)
B.A. Shipman [ISTP]
Nongeneric Flows in the Full Kostant-Toda Lattice
B.A. Springborn [DGIS]
Bonnet Pairs in the 3-Sphere
I.A.B. Strachan [ISTP]
Frobenius Manifolds and Bi-Hamiltonian Structures on Discriminant Hypersurfaces
N. Suzuki (see S. Saito)
R. Szoke [—]
Involutive Structures on the Tangent Bundle of Symmetric Spaces
M. Tanaka [DGIS]
Subspaces in the Category of Symmetric Spaces
T. Taniguchi [ISTP]
Periodicity Conditions for Harmonic Maps Associated to Spectral Data
H. Tasaki [DGIS]
Integral Geometry of Submanifolds of Real Dimension Two and Codimension Two in Complex Projective Spaces
Y. Terashima [ISTP]
Higher Dimensional Parallel Transports for Deligne Cocycles
C.-L. Terng [SGIS]
Geometries and Symmetries of Soliton Equations and Integrable Elliptic Equations
M. Toda (see J. Dorfmeister)
P. Topalov [SGIS]
A-integrability of Geodesic Flows and Geodesic Equivalence
S. Udagawa (see Y. Ohnita)
M. Umehara (see M. Kokubu; see W. Rossman)
A.P. Veselov [—]
Geometry of the Singularities of Integrable Schrödinger Operators
H.-Y. Wang [ISTP]
Geometric Nonlinear Schrödinger Equations
J.C. Wood [DGIS]
Jacobi Fields along Harmonic Maps
D. Wu [—]
Global Poisson Group Actions of the ZS-AKNS Flows
H. Wu [DGIS]
Denseness of Plain Constant Mean Curvature Surfaces in Dressing Orbits
K. Yamada (see M. Kokubu; see W. Rossman)

T. Yamada (see Y. Sakane)
Y. Yamada (see M. Noumi)
H. Yamaguchi (see S. Saito)

*invited speaker who either did not attend the conference or did not give a talk

Contemporary Mathematics
Volume **308**, 2002

The Index of an Isolated Umbilical Point on a Surface

Naoya Ando

1. Introduction

Let F be a smooth function defined on a neighborhood of $(0,0)$ satisfying

$$F(0,0) = \frac{\partial F}{\partial x}(0,0) = \frac{\partial F}{\partial y}(0,0) = 0,$$

such that the origin $o := (0,0,0)$ of $\mathbf{R}^3$ is an isolated umbilical point of its graph. Let r_0 be a positive number such that the graph has no umbilical point on $\{0 < x^2 + y^2 < r_0^2\}$ and ϕ_F a continuous function on $(0, r_0) \times \mathbf{R}$ such that for each $(r, \theta) \in (0, r_0) \times \mathbf{R}$,

$$\cos \phi_F(r,\theta) \frac{\partial}{\partial x} + \sin \phi_F(r,\theta) \frac{\partial}{\partial y}$$

is in a principal direction at $(r\cos\theta, r\sin\theta)$. Then the *index* $\mathrm{ind}_o(F)$ of o on the graph of F is represented as follows:

$$\mathrm{ind}_o(F) = \frac{\phi_F(r, \theta + 2\pi) - \phi_F(r,\theta)}{2\pi}. \tag{1.1}$$

The purpose of this paper is to survey results we have obtained on the index of an isolated umbilical point.

For $l \in \mathbf{N} \cup \{\infty\}$, let $\mathcal{C}_o^{(\infty,l)}$ be the set of smooth functions defined on a connected neighborhood of $(0,0)$ in $\mathbf{R}^2$ such that $(\partial^{m+n} F / \partial x^m \partial y^n)(0,0) = 0$ for each $F \in \mathcal{C}_o^{(\infty,l)}$ and for all non-negative integers m, n satisfying $0 \le m + n < l$. We have: $\mathcal{C}_o^{(\infty,l)} \supset \mathcal{C}_o^{(\infty,l+1)} \supset \mathcal{C}_o^{(\infty,\infty)} \neq \{0\}$, where $l \in \mathbf{N}$. Let F be an element of $\mathcal{C}_o^{(\infty,2)}$ such that o is an umbilical point of its graph, that is, there exists a real number a_F satisfying

$$F(x,y) = a_F(x^2 + y^2)/2 + o(x^2 + y^2). \tag{1.2}$$

2000 *Mathematics Subject Classification.* Primary 53A05; Secondary 53A99, 53B25.

The author is grateful to the referee for his many helpful comments and suggestions, and to Professor Guest for his advice on the usage of English. The author is a Research Fellow of the Japan Society for the Promotion of Science.

Let σ_F be an element of $\mathcal{C}_o^{(\infty,2)}$ defined by

$$\sigma_F := \begin{cases} 0 & \text{if } a_F = 0, \\ \dfrac{1}{a_F} - \dfrac{|a_F|}{a_F}\sqrt{\dfrac{1}{a_F^2} - (x^2+y^2)} & \text{if } a_F \neq 0. \end{cases}$$

If $a_F \neq 0$, then the graph of σ_F is part of a sphere of radius $1/|a_F|$. We obtain $F - \sigma_F \in \mathcal{C}_o^{(\infty,3)}$. Let $\mathcal{C}_o^{\langle\infty,2\rangle}$ be the subset of $\mathcal{C}_o^{(\infty,2)}$ such that each $F \in \mathcal{C}_o^{\langle\infty,2\rangle}$ satisfies (1.2) for some $a_F \in \mathbf{R}$ and $F - \sigma_F \notin \mathcal{C}_o^{(\infty,\infty)}$. For an integer $k \geq 3$, let $\mathcal{P}^k$ be the set of homogeneous polynomials of degree k. Then for each $F \in \mathcal{C}_o^{\langle\infty,2\rangle}$, there exists an integer $k_F \geq 3$ and a nonzero element $g_F \in \mathcal{P}^{k_F}$ satisfying $F - \sigma_F - g_F \in \mathcal{C}_o^{(\infty,k_F+1)}$. For $g \in \mathcal{P}^k$ and $\theta \in \mathbf{R}$, let $\mathrm{Hess}_g(\theta)$ be Hessian of g at $(\cos\theta, \sin\theta)$ and η_g a continuous function on $\mathbf{R}$ such that for any $\theta \in \mathbf{R}$, ${}^t(\cos\eta_g(\theta), \sin\eta_g(\theta))$ is an eigenvector of $\mathrm{Hess}_g(\theta)$, and let S_g denote the set of numbers at each of which Hess_g is a scalar matrix. In Section 2, we shall prove the following:

PROPOSITION 1.1 ([A6, Propositions 1.1, 1.2]). *Let F be an element of $\mathcal{C}_o^{\langle\infty,2\rangle}$ such that o is an isolated umbilical point of its graph.*

1. *For any $\theta_0 \in \mathbf{R} \setminus S_{g_F}$, there exists a number $\phi_{F,o}(\theta_0)$ satisfying*
$$\lim_{r\to 0} \phi_F(r,\theta_0) = \phi_{F,o}(\theta_0);$$
2. *for any $\theta_0 \in \mathbf{R}$,*
   (a) *there exist numbers $\phi_{F,o}(\theta_0+0)$, $\phi_{F,o}(\theta_0-0)$ satisfying*
   $$\lim_{\theta\to\theta_0\pm 0} \phi_{F,o}(\theta) = \phi_{F,o}(\theta_0 \pm 0),$$
   (b) *$\Gamma_{F,o}(\theta_0) := \phi_{F,o}(\theta_0+0) - \phi_{F,o}(\theta_0-0)$ is an element of $\{n\pi/2\}_{n\in\mathbf{Z}}$;*
3. $\mathrm{ind}_o(F) = \dfrac{\eta_{g_F}(\theta+2\pi) - \eta_{g_F}(\theta)}{2\pi} + \dfrac{1}{2\pi}\sum_{\theta_0\in S_{g_F}\cap[\theta,\theta+2\pi)} \Gamma_{F,o}(\theta_0)$.

PROPOSITION 1.2 ([A6, Proposition 3.1]). *Let F be an element of $\mathcal{C}_o^{\langle\infty,2\rangle}$ satisfying $S_{g_F} = \emptyset$. Then on each of the graphs of F and g_F, o is an isolated umbilical point.*

By (3) of Proposition 1.1 together with Proposition 1.2, we see in particular that if $F \in \mathcal{C}_o^{\langle\infty,2\rangle}$ satisfies $S_{g_F} = \emptyset$, then:

$$\mathrm{ind}_o(F) = \mathrm{ind}_o(g_F) = \frac{\eta_{g_F}(\theta+2\pi) - \eta_{g_F}(\theta)}{2\pi}. \tag{1.3}$$

In Section 3, we shall describe a way of computing $\eta_g(\theta+2\pi) - \eta_g(\theta)$ for any $g \in \mathcal{P}^k$. In addition, we shall prove the following:

THEOREM 1.3 ([A4], [A1, Theorem 1.2]). *For $k \geq 3$, let g be an element of $\mathcal{P}^k$ such that o is an isolated umbilical point of its graph. Then*

1. $\Gamma_{g,o}(\theta_0) = -\pi/2$ *for any $\theta_0 \in S_g$;*
2. $\mathrm{ind}_o(g) \in \{1 - k/2 + i\}_{i=0}^{[k/2]}$.

REMARK 1.4. We have obtained (1) of Theorem 1.3 in the proof of (1) of Theorem 1.3 in [A4]; (2) of Theorem 1.3 corresponds to Theorem 1.2 in [A1].

In Section 4, we shall prove the following:

THEOREM 1.5 ([A6, Theorem 1.3]). *Let F be an element of $\mathcal{C}_o^{\langle\infty,2\rangle}$ such that on each of the graphs of F and g_F, o is an isolated umbilical point. Then*

1. $-\pi/2 \leq \Gamma_{F,o}(\theta_0) \leq \pi/2$ *for any* $\theta_0 \in S_{g_F}$;
2. $\mathrm{ind}_o(g_F) \leq \mathrm{ind}_o(F) \leq 1$.

Generally, it has been expected that the index of an isolated umbilical point is not more than one. We call this conjecture the *index conjecture*. There are two conjectures related to this: Carathéodory's conjecture and Loewner's conjecture. Carathéodory's conjecture asserts that there exist at least two umbilical points on a compact, strictly convex surface in $\mathbf{R}^3$. If the index conjecture is true, then we see from the Hopf-Poincaré theorem that there exist at least two umbilical points on a compact, orientable surface of genus zero, and this immediately gives the expected answer to Carathéodory's conjecture. Let F be a real-valued, smooth function of two real variables x, y, and set $\partial_{\bar{z}} := (\partial/\partial x + \sqrt{-1}\partial/\partial y)/2$. Then Loewner's conjecture for $n \in \mathbf{N}$ asserts that if a vector field

$$\mathrm{Re}(\partial_{\bar{z}}^n F)\frac{\partial}{\partial x} + \mathrm{Im}(\partial_{\bar{z}}^n F)\frac{\partial}{\partial y}$$

has an isolated zero point z_0, then the index of z_0 with respect to this vector field is not more than n ([K], [T]). It is known that Loewner's conjecture for $n = 2$ is equivalent to the index conjecture ([SX]). In Section 4, we shall prove the following:

THEOREM 1.6 ([A6, Theorem 1.5]). *Let F be an element of $\mathcal{C}_o^{\langle\infty,2\rangle}$ such that o is an isolated umbilical point of its graph. Suppose $\Gamma_{F,o}(\theta_0) \leq \pi$ for any $\theta_0 \in S_{g_F}$. Then* $\mathrm{ind}_o(F) \leq 1$.

Let S be a Weingarten surface in $\mathbf{R}^3$ and w_S a function of two variables satisfying $w_S(K_S, H_S) \equiv 0$ on S, where K_S and H_S are the Gaussian curvature and the mean curvature of S, respectively. In this paper, we suppose that w_S is smooth, and, according to [HaW], we call S *special* if w_S satisfies

$$H_S\frac{\partial w_S}{\partial X}(K_S, H_S) + \frac{1}{2}\frac{\partial w_S}{\partial Y}(K_S, H_S) \neq 0 \tag{1.4}$$

at each umbilical point of S.

THEOREM 1.7 ([A6, Theorem 1.6]). *Let F be an element of $\mathcal{C}_o^{\langle\infty,2\rangle}$ whose graph is special Weingarten. Then :*

1. *On each of the graphs of F and g_F, o is an isolated umbilical point;*
2. $\mathrm{ind}_o(F) = \mathrm{ind}_o(g_F) = 1 - k_F/2$.

In Section 4, we shall prove that g_F is a spherical harmonic function (Proposition 4.2). This implies $S_{g_F} = \emptyset$. Therefore from Proposition 1.2, we obtain (1) of Theorem 1.7. In Section 3, we shall show that for a spherical harmonic function g of degree $k \geq 3$, $\mathrm{ind}_o(g) = 1 - k/2$ (Example 3.6). Using this, together with (1.3), we obtain (2) of Theorem 1.7.

REMARK 1.8. Since $k_F \geq 3$, we see that if F is as in Theorem 1.7, then $\mathrm{ind}_o(F) < 0$ holds. Hartman-Wintner proved that if S is connected, non-totally umbilical and special Weingarten, then each umbilical point is isolated and its index is negative ([HaW]).

REMARK 1.9. Let S be a surface with constant mean curvature in $\mathbf{R}^3$. Then S is special Weingarten. It is known that an umbilical point of S is just a zero point of the Hopf differential Ω_S on S. Since Ω_S is holomorphic with respect to isothermal coordinates, we see that if S is connected and non-totally umbilical, then each umbilical point p_0 is isolated. It is known that the index of p_0 is equal to $-\mathrm{ord}_{p_0}(S)/2$, where $\mathrm{ord}_{p_0}(S)$ is the order of Ω_S at p_0 ([Ho]).

REMARK 1.10. In [A6], we have proved Proposition 1.1, Proposition 1.2, Theorem 1.5, Theorem 1.6 and Theorem 1.7 under the additional condition that F is real-analytic.

2. Proof of Propositions 1.1 and 1.2

Let f be a smooth function of two variables x, y and G_f the graph of f. We set

$$p_f := \frac{\partial f}{\partial x}, \quad q_f := \frac{\partial f}{\partial y}, \quad r_f := \frac{\partial^2 f}{\partial x^2}, \quad s_f := \frac{\partial^2 f}{\partial x \partial y}, \quad t_f := \frac{\partial^2 f}{\partial y^2}$$

and

$$\mathrm{grad}_f := \begin{pmatrix} p_f \\ q_f \end{pmatrix}, \quad \mathrm{grad}_f^{\perp} := \begin{pmatrix} -q_f \\ p_f \end{pmatrix}, \quad \mathrm{Hess}_f := \begin{pmatrix} r_f & s_f \\ s_f & t_f \end{pmatrix}.$$

For $\phi \in \mathbf{R}$, we set $u_\phi := {}^t(\cos\phi, \sin\phi)$. Let $\langle\ ,\ \rangle$ be the scalar product in $\mathbf{R}^2$. Then by Proposition 2.2 in [A6] together with Lemma 2.4 in [A6], we see that at a point of G_f, a tangent vector $(\cos\phi)\partial/\partial x + (\sin\phi)\partial/\partial y$ is in a principal direction if and only if the following holds:

$$\langle \mathrm{Hess}_f u_\phi,\ u_{\phi+\pi/2}\rangle + \langle \mathrm{grad}_f,\ u_\phi\rangle\langle \mathrm{grad}_f^{\perp},\ \mathrm{Hess}_f u_\phi\rangle = 0. \tag{2.1}$$

Let F be an element of $\mathcal{C}_o^{\langle\infty,2\rangle}$ satisfying (1.2) for some $a_F \in \mathbf{R}$ and $\tilde{\Phi}_F$ a smooth function on $(0, r_0) \times \mathbf{R}^2$ for some $r_0 > 0$ such that $\tilde{\Phi}_F(r,\theta,\phi)$ denotes the left hand side of (2.1) at $(r\cos\theta, r\sin\theta)$ for any $(r,\theta,\phi) \in (0,r_0)\times\mathbf{R}^2$. Then by $\tilde{\Phi}_{\sigma_F} \equiv 0$, we see that for each $F \in \mathcal{C}_o^{\langle\infty,2\rangle}$,

$$\lim_{r\to 0} \frac{\tilde{\Phi}_F(r,\theta,\phi)}{r^{k_F-2}} = \langle \mathrm{Hess}_{g_F}(\theta)u_\phi, u_{\phi+\pi/2}\rangle. \tag{2.2}$$

PROOF OF PROPOSITION 1.1. By (2.2), we obtain (1). Let θ_1, θ_2 be numbers satisfying $\theta_1 < \theta_2$ and $S_{g_F} \cap (\theta_1, \theta_2) = \emptyset$. Then we see that there exists an element $z(\theta_1,\theta_2)$ of $\{n\pi/2\}_{n\in\mathbf{Z}}$ satisfying

$$\phi_{F,o}(\theta) = \eta_{g_F}(\theta) + z(\theta_1,\theta_2) \tag{2.3}$$

for any $\theta \in (\theta_1,\theta_2)$. Since S_{g_F} is an empty or discrete set in $\mathbf{R}$, we obtain (2). By (1.1) together with (1), we see that for $\theta \in \mathbf{R}\setminus S_{g_F}$,

$$\mathrm{ind}_o(F) = \frac{\phi_{F,o}(\theta+2\pi) - \phi_{F,o}(\theta)}{2\pi}. \tag{2.4}$$

By (2.3), we obtain

$$\phi_{F,o}(\theta+2\pi) - \phi_{F,o}(\theta) = \eta_{g_F}(\theta+2\pi) - \eta_{g_F}(\theta) + \sum_{\theta_0 \in S_{g_F}\cap(\theta,\theta+2\pi)} \Gamma_{F,o}(\theta_0). \tag{2.5}$$

By (2.4) together with (2.5), we obtain (3). □

PROOF OF PROPOSITION 1.2. By (2.2), we obtain Proposition 1.2. □

3. Homogeneous polynomials

For an integer $k \geq 3$. Let g be an element of $\mathcal{P}^k$, and for $\theta \in \mathbf{R}$, set $\tilde{g}(\theta) := g(\cos\theta, \sin\theta)$. A number $\theta_0 \in\mathbf{R}$ is called a *root* of g if $(d\tilde{g}/d\theta)(\theta_0) = 0$. The set of roots of g is denoted by R_g. Let $R(\mathrm{Hess}_g)$ be the set of numbers such that each $\theta_0 \in R(\mathrm{Hess}_g)$ satisfies $\theta_0 \in \{\eta_g(\theta_0) + n\pi/2\}_{n\in\mathbf{Z}}$. For $\theta \in\mathbf{R}$, we set $\mathrm{grad}_g(\theta) := \mathrm{grad}_g(\cos\theta, \sin\theta)$. Then:

$$(k-1)\mathrm{grad}_g(\theta) = \mathrm{Hess}_g(\theta)u_\theta. \tag{3.1}$$

Therefore we obtain $S_g \subset R_g$. By (3.1), we obtain

$$\langle \mathrm{Hess}_g(\theta)u_\theta,\ u_{\theta+\pi/2}\rangle = (k-1)\frac{d\tilde{g}}{d\theta}(\theta). \tag{3.2}$$

Therefore we obtain $R(\mathrm{Hess}_g) \subset R_g$.

Suppose $R_g =\mathbf{R}$. Then k is even and g is represented by $(x^2+y^2)^{k/2}$ up to a constant. By direct computations, we obtain $S_g = \emptyset$. Then Proposition 1.2 says that o is an isolated umbilical point of G_g. By (3.2), we see that $R(\mathrm{Hess}_g) = \mathbf{R}$, i.e., there exists a number $z_0 \in \{n\pi/2\}_{n\in\mathbf{Z}}$ satisfying $\eta_g(\theta) = \theta + z_0$ for any $\theta \in \mathbf{R}$. Therefore by (1.3), we obtain $\mathrm{ind}_o(g) = 1$. In the following, suppose $R_g \neq\mathbf{R}$. Then for any $\theta_0 \in R_g$, there exists a positive integer μ satisfying $(d^{\mu+1}\tilde{g}/d\theta^{\mu+1})(\theta_0) \neq 0$. The minimum of such integers is denoted by $\mu_g(\theta_0)$. A root $\theta_0 \in R_g$ is said to be

1. *related* if θ_0 satisfies $\tilde{g}(\theta_0) = 0$ or if $\mu_g(\theta_0)$ is odd;
2. *non-related* if θ_0 satisfies $\tilde{g}(\theta_0) \neq 0$ and if $\mu_g(\theta_0)$ is even.

Suppose that $\theta_0 \in R_g$ is related. Then it is said that the *critical sign* of θ_0 is positive (respectively, negative) if the following holds:

$$\tilde{g}(\theta_0)\frac{d^{\mu_g(\theta_0)+1}\tilde{g}}{d\theta^{\mu_g(\theta_0)+1}}(\theta_0) \leq 0 \text{ (respectively, } > 0).$$

The critical sign of θ_0 is denoted by $\text{c-sign}_g(\theta_0)$.

LEMMA 3.1 ([A6, Lemma 4.1]). *The set $R_g \setminus R(\mathrm{Hess}_g)$ consists of the numbers at each of which Hess_g is a nonzero scalar matrix. In addition, an element $\theta_0 \in R_g \setminus R(\mathrm{Hess}_g)$ is related and satisfies* $\text{c-sign}_g(\theta_0) = -$.

It is said that the *sign* of $\theta_0 \in R(\mathrm{Hess}_g)$ is positive (respectively, negative) if there exists a neighborhood U_{θ_0} of θ_0 in $\mathbf{R}$ satisfying

$$\{\theta - \eta_g(\theta) - (\theta_0 - \eta_g(\theta_0))\}(\theta - \theta_0) > 0 \text{ (respectively, } < 0)$$

for any $\theta \in U_{\theta_0} \setminus \{\theta_0\}$.

PROPOSITION 3.2 ([A6, Proposition 4.3]). *For $\theta_0 \in R(\mathrm{Hess}_g)$, θ_0 is related if and only if the sign of θ_0 is positive or negative.*

If $\theta_0 \in R(\mathrm{Hess}_g)$ is related, then the sign of θ_0 is denoted by $\mathrm{sign}_g(\theta_0)$.

PROPOSITION 3.3 ([A6, Proposition 4.4]). *Let θ_0 be a related root of g satisfying* $\text{c-sign}_g(\theta_0) = +$. *Then $\theta_0 \in R(\mathrm{Hess}_g)$ and* $\mathrm{sign}_g(\theta_0) = +$.

REMARK 3.4. In [A2], we have proved that if θ_0 is a related element of $R(\mathrm{Hess}_g)$ satisfying c-sign$_g(\theta_0) = -$, then $\mathrm{sign}_g(\theta_0) = +$ (respectively, $-$) holds if and only if there does not exist (respectively, exists) an umbilical point other than o on $L(\theta_0) := \{(\rho\cos\theta_0, \rho\sin\theta_0)\}_{\rho\in\mathbf{R}}$.

Let $n_{g,+}$ (respectively, $n_{g,-}$) denote the number of the related elements of $R(\mathrm{Hess}_g)$ in $[\theta, \theta+\pi)$ with positive (respectively, negative) sign.

PROPOSITION 3.5 ([A6, Proposition 4.2]). *For any* $\theta \in \mathbf{R}$,

$$\frac{\eta_g(\theta+2\pi)-\eta_g(\theta)}{2\pi} = 1 - \frac{n_{g,+}-n_{g,-}}{2}.$$

EXAMPLE 3.6. Let g be a spherical harmonic function of degree $k \geq 3$. Then we see that any root $\theta_0 \in R_g$ is related and satisfies c-sign$_g(\theta_0) = +$. Therefore from Proposition 3.3, we obtain $\theta_0 \in R(\mathrm{Hess}_g)$ and $\mathrm{sign}_g(\theta_0) = +$ for any $\theta_0 \in R_g$. Since the number of roots of g in $[\theta, \theta+\pi)$ is equal to k, we obtain

$$\frac{\eta_g(\theta+2\pi)-\eta_g(\theta)}{2\pi} = 1 - \frac{k}{2}.$$

Since $S_g = \emptyset$, from (1.3) we obtain $\mathrm{ind}_o(g) = 1 - k/2$.

For an integer $k \geq 3$, let $\mathcal{P}^k_o$ be the set of the elements of $\mathcal{P}^k$ such that on each of their graphs, o is an isolated umbilical point.

PROOF OF (1) OF THEOREM 1.3. Let g be an element of $\mathcal{P}^k_o$. Then we see that g may be represented as follows:

$$g(x,y) = g_0(x,y) \times \prod_{i=1}^{i_0} (\alpha_i x + \beta_i y)^{\varepsilon_i},$$

where

1. $\varepsilon_i = 0, 1$ or 2,
2. (α_i, β_i) $(i = 1, \ldots, i_0)$ are nonzero vectors satisfying $\alpha_i\beta_j - \alpha_j\beta_i \neq 0$ if $i \neq j$,
3. g_0 is a homogeneous polynomial of even degree satisfying $g_0(x,y) \neq 0$ for each $(x,y) \in \mathbf{R}^2 \setminus \{(0,0)\}$.

Therefore we obtain $S_g = R_g \setminus R(\mathrm{Hess}_g)$.

By (3.1), we see that at $(\cos\theta, \sin\theta)$, the second term of the left hand side of (2.1) is equal to the following:

$$\frac{\det(\mathrm{Hess}_g(\theta))}{k-1}\langle \mathrm{grad}_g(\theta), u_\phi\rangle \sin(\phi - \theta).$$

Let θ_0 be an element of S_g. Then we may suppose $\phi_g(r, \theta_0) = \theta_0$ and we obtain $(\partial\tilde{\Phi}_g/\partial\phi)(r, \theta_0, \theta_0) \neq 0$ for any $r \in (0, r_0)$. Therefore we see that ϕ_g is smooth at (r, θ_0) and we obtain

$$\frac{\partial(\theta - \phi_g)}{\partial\theta}(r, \theta_0) = \frac{d^2\tilde{g}}{d\theta^2}(\theta_0) \Big/ k^3 r^{2k-2}\tilde{g}(\theta_0)^3.$$

Then by Lemma 3.1, we see that there exists a neighborhood U_{θ_0} of θ_0 in $\mathbf{R}$ satisfying

$$\{\theta - \phi_g(r,\theta)\}(\theta - \theta_0) > 0 \tag{3.3}$$

for any $r \in (0, r_0)$ and any $\theta \in U_{\theta_0} \setminus \{\theta_0\}$. Noticing that $\theta_0 \in R_g \setminus R(\mathrm{Hess}_g)$, we may suppose $\phi_g(r,\theta) \notin \{\eta_g(\theta) + n\pi/2\}_{n \in \mathbf{Z}}$ for any $r \in (0, r_0)$ and any $\theta \in U_{\theta_0}$. Then by (3.3), we obtain $\Gamma_{g,o}(\theta_0) = -\pi/2$. Hence we have proved (1) of Theorem 1.3. □

REMARK 3.7. In [A1], we have proved $g \in \mathcal{P}_o^k$ for an integer $k \geq 3$ and an element g of $\mathcal{P}^k$ satisfying $1 \sim 3$ in the proof of (1) of Theorem 1.3.

PROOF OF (2) OF THEOREM 1.3. We set $n_{g,s} := \sharp\{S_g \cap [\theta, \theta + \pi)\}$. Then from (3) of Proposition 1.1, (1) of Theorem 1.3 and Proposition 3.5, we obtain

$$\mathrm{ind}_o(g) = 1 - \frac{n_{g,s} + n_{g,+} - n_{g,-}}{2}.$$

Therefore we shall prove

$$n_{g,s} + n_{g,+} - n_{g,-} \in \{k - 2i\}_{i=0}^{[k/2]}. \tag{3.4}$$

Let n_g be the number of the related elements of $R(\mathrm{Hess}_g)$ in $[\theta, \theta + \pi)$ and $\{\theta_i\}_{i=1}^{n_g+1}$ a set of related elements of $R(\mathrm{Hess}_g)$ satisfying $\theta_i < \theta_{i+1}$ for $i \in \{1, \ldots, n_g\}$ and $\theta_{n_g+1} = \theta_1 + \pi$. Let $n_{g,+}^C$ (respectively, $n_{g,-}^C$) be the number of the related elements of $R(\mathrm{Hess}_g)$ in $[\theta, \theta + \pi)$ with positive (respectively, negative) critical sign. From Proposition 3.3, we obtain $n_{g,+}^C \leq n_{g,+}$. In addition, since

$$n_g = n_{g,+} + n_{g,-} = n_{g,+}^C + n_{g,-}^C,$$

we obtain $n_{g,-} \leq n_{g,-}^C$. We see that for each $i \in \{1, \ldots, n_g\}$, either $\text{c-sign}_g(\theta_i)$ or $\text{c-sign}_g(\theta_{i+1})$ is positive. Therefore we obtain $n_{g,-}^C \leq n_{g,+}^C$, and $n_{g,-} \leq n_{g,+}$. Therefore to prove (3.4), we have only to show that $k - (n_{g,s} + n_g)$ is even. Let $\tilde{n}_g$ be the number of the non-related roots of g in $(\theta_1, \theta_1 + \pi)$ and $\{\tilde{\theta}_j\}_{j=1}^{\tilde{n}_g}$ the set of the non-related roots in $(\theta_1, \theta_1 + \pi)$. Then $d\tilde{g}/d\theta$ can be represented as follows:

$$\frac{d\tilde{g}}{d\theta}(\theta) = \tilde{\gamma}(\theta) \prod_{\theta_s \in S_g \cap (\theta_1, \theta_1 + \pi)} \sin(\theta - \theta_s) \prod_{i=1}^{n_g} \sin^{\mu_g(\theta_i)}(\theta - \theta_i) \prod_{j=1}^{\tilde{n}_g} \sin^{\mu_g(\tilde{\theta}_j)}(\theta - \tilde{\theta}_j),$$

where $\tilde{\gamma}(\theta) \neq 0$ for any $\theta \in \mathbf{R}$. We see that $k - \left(n_{g,s} + \sum_{i=1}^{n_g} \mu_g(\theta_i)\right)$ is even. Therefore we see that $k - (n_{g,s} + n_g)$ is also even. Hence we have proved (2) of Theorem 1.3. □

REMARK 3.8. For each $j \in \{1 - k/2 + i\}_{i=0}^{[k/2]}$, let g_j be an element of $\mathcal{P}^k$ satisfying $\tilde{g}_j(\theta) = \cos 2(1-j)\theta$. Then $g_j \in \mathcal{P}_o^k$ and $\mathrm{ind}_o(g_j) = j$. Let $\mathcal{P}_o^{k,j}$ be the set of the elements of $\mathcal{P}_o^k$ such that on each of their graphs, the index of o is equal to j. Then

1. $\mathcal{P}_o^k = \bigsqcup_{j \in \{1-k/2+i\}_{i=0}^{[k/2]}} \mathcal{P}_o^{k,j}$,
2. $\mathcal{P}_o^{k,j} \neq \emptyset$ for any $j \in \{1 - k/2 + i\}_{i=0}^{[k/2]}$.

In particular, we obtain $\mathcal{P}_o^3 = \mathcal{P}_o^{3,1/2} \sqcup \mathcal{P}_o^{3,-1/2}$, and $\mathcal{P}_o^{3,1/2}, \mathcal{P}_o^{3,-1/2} \neq \emptyset$. We set

$$\mathcal{S}^{3,1} := \{((\cos\tau_1)x + (\sin\tau_1)y)(x^2 + y^2) \ ; \ \tau_1 \in [0, 2\pi)\},$$
$$\mathcal{S}^{3,3} := \{(\cos\tau_3)(x^3 - 3xy^2) + (\sin\tau_3)(-y^3 + 3x^2y) \ ; \ \tau_3 \in [0, 2\pi)\}.$$

For $\tau \in [0, \pi/2]$ and $(h_1, h_3) \in \mathcal{S}^{3,1} \times \mathcal{S}^{3,3}$, we set $g_{\tau,h_1,h_3} := (\cos\tau)h_1 + (\sin\tau)h_3$. We see that for any $g \in \mathcal{P}^3$, there exist numbers $c \in \mathbf{R}$, $\tau \in [0, \pi/2]$ and a pair $(h_1, h_3) \in \mathcal{S}^{3,1} \times \mathcal{S}^{3,3}$ satisfying $g = cg_{\tau,h_1,h_3}$. In [A3], we have shown

$$g_{\tau,h_1,h_3} \in \begin{cases} \mathcal{P}_o^{3,1/2} & \text{for } \tau \in [0, \arctan 1/3), \\ \mathcal{P}_o^{3,-1/2} & \text{for } \tau \in (\arctan 1/3, \pi/2], \end{cases}$$

and $g_{\arctan 1/3,h_1,h_3} \in \mathcal{P}_o^{3,-1/2}$ or $\notin \mathcal{P}_o^3$ for any $(h_1, h_3) \in \mathcal{S}^{3,1} \times \mathcal{S}^{3,3}$. Let $\langle\langle\ ,\ \rangle\rangle$ be an inner product on $\mathcal{P}^3$ defined by

$$\langle\langle g_1, g_2 \rangle\rangle := \frac{1}{\pi} \int_0^{2\pi} \tilde{g}_1(\theta)\tilde{g}_2(\theta) d\theta$$

for g_1, $g_2 \in \mathcal{P}^3$. Then for any $\tau \in [0, \pi/2]$ and any pair $(h_1, h_3) \in \mathcal{S}^{3,1} \times \mathcal{S}^{3,3}$, the norm of g_{τ,h_1,h_3} with respect to the inner product $\langle\langle\ ,\ \rangle\rangle$ is equal to one. We see that the set $\mathcal{P}_o^3$ is divided into $\mathcal{P}_o^{3,1/2}$ and $\mathcal{P}_o^{3,-1/2}$ by the cone obtained from the rectangular torus determined by $\tau = \arctan 1/3$ in the unit hypersphere

$$\left\{ g_{\tau,h_1,h_3} \in \mathcal{P}^3 \ ;\ \tau \in [0, \pi/2],\ h_1 \in \mathcal{S}^{3,1},\ h_3 \in \mathcal{S}^{3,3} \right\}$$

in $\mathcal{P}^3$. If $k \geq 4$, then we have no information about $\mathcal{P}_o^{k,j}$ other than $\mathcal{P}_o^{k,j} \neq \emptyset$ for $j \in \{1 - k/2 + i\}_{i=0}^{[k/2]}$ (for example, we do not even know whether $\mathcal{P}_o^{k,j}$ is connected with respect to the natural topology of $\mathcal{P}^k$).

REMARK 3.9. There may exist umbilical points other than o on G_g for some $g \in \mathcal{P}_o^k$. In [A2], we have proved the following: if for $\theta_0 \in \mathbf{R}$, there exist other umbilical points than o on $L(\theta_0)$, then $\theta_0 \in R_g$ holds and the set of such umbilical points is $\{\pm p_{\theta_0}\}$, where $p_{\theta_0} := (r_{\theta_0} \cos\theta_0, r_{\theta_0} \sin\theta_0)$ for some $r_{\theta_0} > 0$; if $R_g = \mathbf{R}$, then the set of the umbilical points on G_g is $\{o\} \cup \left\{x^2 + y^2 = c_0^2\right\}$ for some $c_0 > 0$. Let $\mathcal{P}_F^k$ be the subset of $\mathcal{P}_o^k$ such that the set of the umbilical points on the graph of each $g \in \mathcal{P}_F^k$ is finite. Then $\mathcal{P}_F^k = \left\{g \in \mathcal{P}_o^k \ ;\ R_g \neq \mathbf{R}\right\}$. In [A4], we have presented a way of describing the index of each isolated umbilical point of G_g for $g \in \mathcal{P}_F^k$. Then we have presented another proof of (1) of Theorem 1.3, different from that above. In addition, we have proved that if for $\theta_0 \in R_g$, there exist umbilical points other than o on $L(\theta_0)$, then the following holds:

$$\mathrm{ind}_{p_{\theta_0}}(g) = \begin{cases} 1/2, & \text{if } \theta_0 \text{ is related and satisfies c-sign}_g(\theta_0) = +, \\ 0, & \text{if } \theta_0 \text{ is non-related}, \\ -1/2, & \text{if } \theta_0 \text{ is related and satisfies c-sign}_g(\theta_0) = -. \end{cases}$$

REMARK 3.10. To prove (2) of Theorem 1.3, we first treated $g \in \mathcal{P}_o^k$ satisfying $R_g = \mathbf{R}$. In addition, we have described one way of computing $\mathrm{ind}_o(g)$ for $g \in \mathcal{P}_o^k$ satisfying $R_g \neq \mathbf{R}$. Let f be a smooth function of two variables such that at each point of G_f, there exists one principal direction parallel to the xy-plane. Then we see that $-q_f\partial/\partial x + p_f\partial/\partial y$ is in a principal direction at each point of G_f. Therefore, using (2.1), we obtain $\varpi_f \equiv 0$, where

$$\varpi_f := \left\langle \mathrm{Hess}_f \mathrm{grad}_f, \mathrm{grad}_f^\perp \right\rangle.$$

Conversely, if a function f satisfies $\varpi_f \equiv 0$, then at each point of G_f, there exists one principal direction parallel to the xy-plane. By (3.1), we see that for $g \in \mathcal{P}^k$, $\varpi_g \equiv 0$ is equivalent to $\det(\mathrm{Hess}_g)d\tilde{g}/d\theta \equiv 0$. If $\det(\mathrm{Hess}_g) \equiv 0$, then we see that there exists a vector $(\alpha, \beta) \in \mathbf{R}^2$ satisfying $g = (\alpha x + \beta y)^k$, which implies $g \notin \mathcal{P}_o^k$. Therefore we see that for $g \in \mathcal{P}_o^k$, $\varpi_g \equiv 0$ (respectively, $\neq 0$) is equivalent to $R_g = \mathbf{R}$

(respectively, $\neq \mathbf{R}$). In [A5], we have proved that if F is a real-analytic element of $\mathcal{C}_o^{\langle\infty,2\rangle}$ satisfying $\varpi_F \equiv 0$ and the condition that o is an isolated umbilical point of G_F, then G_F is part of a surface of revolution such that o lies on the axis of rotation. Then $\mathrm{ind}_o(F) = 1$ holds. On the other hand, it may be very difficult to compute $\mathrm{ind}_o(F)$ for $F \in \mathcal{C}_o^{\langle\infty,2\rangle}$ satisfying $\varpi_F \not\equiv 0$ and the condition that o is an isolated umbilical point of G_F, even if F is real-analytic. Theorem 1.5, which we shall show in the next section is our partial result for the situation $\varpi_F \not\equiv 0$.

4. Proof of the other theorems

Let F be an element of $\mathcal{C}_o^{\langle\infty,2\rangle}$ such that on each of G_F and G_{g_F}, o is an isolated umbilical point. Then we obtain $S_{g_F} = R_{g_F} \setminus R(\mathrm{Hess}_{g_F})$. We set $\tilde{\varpi}_F(r,\theta) := \varpi_F(r\cos\theta, r\sin\theta)$ and

$$k := \begin{cases} 3k_F - 4, & \text{if } a_F = 0, \\ k_F, & \text{if } a_F \neq 0. \end{cases}$$

Then, using $\varpi_{\sigma_F} \equiv 0$, we see that $\tilde{\varpi}_F/r^k$ may be continuously extended to $\{r = 0\}$. By the implicit function theorem, we obtain

LEMMA 4.1 ([A6, Lemma 5.1]). *For $\theta_0 \in S_{g_F}$, there exist a neighborhood $V_{(0,\theta_0)}$ of $(0,\theta_0)$ in $\mathbf{R}^2$ and a smooth curve $C_{(0,\theta_0)}$ in $V_{(0,\theta_0)}$ through $(0,\theta_0)$ satisfying*

1. $C_{(0,\theta_0)} = \{(r,\theta) \in V_{(0,\theta_0)}\ ;\ \tilde{\varpi}_F(r,\theta)/r^k = 0\}$;
2. *$C_{(0,\theta_0)}$ is not tangent to the θ-axis at $(0,\theta_0)$.*

PROOF OF (1) OF THEOREM 1.5. Suppose $\theta_0 \in S_{g_F}$. Let ψ_F be a continuous function on $(0,r_0) \times \mathbf{R}$ such that for any $(r,\theta) \in (0,r_0) \times \mathbf{R}$, $\mathrm{grad}_F(r\cos\theta, r\sin\theta)$ is represented by $u_{\psi_F(r,\theta)}$ up to a constant. By (2.1), we may suppose $\phi_F = \psi_F$ on $\{(0,r_0) \times \mathbf{R}\} \cap C_{(0,\theta_0)}$. In addition, since $\theta_0 \in R_{g_F} \setminus R(\mathrm{Hess}_{g_F})$, we may suppose that $V_{(0,\theta_0)}$ is a neighborhood of $(0,\theta_0)$ such that ψ_F may be continuously extended to the closure $\overline{\{(0,r_0) \times \mathbf{R}\} \cap V_{(0,\theta_0)}}$ of $\{(0,r_0) \times \mathbf{R}\} \cap V_{(0,\theta_0)}$ in $\mathbf{R}^2$. Let U_{θ_0} be a neighborhood of θ_0 in $\mathbf{R}$ satisfying $\{0\} \times U_{\theta_0} \subset V_{(0,\theta_0)}$. We set

$$\psi_{F,o}(\theta) := \psi_F(0,\theta), \qquad \chi_{F,o}(\theta) := \phi_{F,o}(\theta) - \psi_{F,o}(\theta)$$

for $\theta \in U_{\theta_0} \setminus \{\theta_0\}$. Then we see that if $(\theta - \theta_0)\chi_{F,o}(\theta) > 0$ (respectively, < 0) holds for any $\theta \in U_{\theta_0} \setminus \{\theta_0\}$, then $\Gamma_{F,o}(\theta_0) = \pi/2$ (respectively, $-\pi/2$) holds and that if there exists a nonzero number c satisfying $c\chi_{F,o}(\theta) > 0$ for any $\theta \in U_{\theta_0} \setminus \{\theta_0\}$, then $\Gamma_{F,o}(\theta_0) = 0$ holds. Hence we have proved (1) of Theorem 1.5. □

PROOF OF (2) OF THEOREM 1.5. By (3) of Proposition 1.1, (1) of Theorem 1.3 and (1) of Theorem 1.5, we obtain $\mathrm{ind}_o(g_F) \leq \mathrm{ind}_o(F)$. Let $n_{g_F,s}$ be as in the proof of (2) of Theorem 1.3 and $\{\theta_{s,i}\}_{i=1}^{n_{g_F,s}+1}$ a subset of S_{g_F} satisfying $\theta_{s,i} < \theta_{s,i+1}$ and $S_{g_F} \cap (\theta_{s,i}, \theta_{s,i+1}) = \emptyset$ for $i \in \{1,\dots,n_{g_F,s}\}$. Then by Lemma 3.1, we see that for each $i \in \{1,\dots,n_{g_F,s}\}$, the number of the related elements of R_{g_F} in $(\theta_{s,i}, \theta_{s,i+1})$ with positive critical sign is more than the number of the related elements of R_{g_F} in $(\theta_{s,i}, \theta_{s,i+1})$ with negative critical sign. Therefore by Proposition 3.3 together with Proposition 3.5, we obtain

$$\frac{\eta_{g_F}(\theta + 2\pi) - \eta_{g_F}(\theta)}{2\pi} \leq 1 - n_{g_F,s}/2. \tag{4.1}$$

By (3) of Proposition 1.1, (1) of Theorem 1.5 and (4.1), we obtain $\mathrm{ind}_o(F) \leq 1$. Hence we have proved (2) of Theorem 1.5. □

PROOF OF THEOREM 1.6. Let $S^{(0)}_{g_F}$ be the set of the elements of S_{g_F} at each of which Hess_{g_F} is a zero matrix and set $S^{(1)}_{g_F} := S_{g_F} \setminus S^{(0)}_{g_F}$. For $\theta \in \mathbf{R}$ and $i \in \{0,1\}$, set $n^{(i)}_{g_F,s} := \sharp\{S^{(i)}_{g_F} \cap [\theta, \theta+\pi)\}$. Then by Lemma 3.1, Proposition 3.3 and Proposition 3.5, we obtain

$$\frac{\eta_{g_F}(\theta+2\pi) - \eta_{g_F}(\theta)}{2\pi} \leq 1 - n^{(0)}_{g_F,s} - n^{(1)}_{g_F,s}/2. \tag{4.2}$$

If $\Gamma_{F,o}(\theta_0) \leq \pi$ holds for $\theta_0 \in S^{(0)}_{g_F}$, then by (3) of Proposition 1.1, (1) of Theorem 1.5 and (4.2), we obtain $\mathrm{ind}_o(F) \leq 1$. Hence we have proved Theorem 1.6. □

In order to complete the proof of Theorem 1.7, we have only to prove

PROPOSITION 4.2 ([A6, Proposition 6.1]). *Let F be an element of $\mathcal{C}_o^{\langle\infty,2\rangle}$ the graph of which is special Weingarten. Then g_F is spherical harmonic.*

For $F \in \mathcal{C}_o^{\langle\infty,2\rangle}$, we set $K_F := K_{\mathsf{G}_F}$ and $H_F := H_{\mathsf{G}_F}$. Then we obtain $K_F(0,0) = a_F^2$ and $H_F(0,0) = a_F$, where a_F is in (1.2). Since $K_{\sigma_F} \equiv a_F^2$ and $H_{\sigma_F} \equiv a_F$, we obtain

LEMMA 4.3 ([A6, Lemma 6.2]). *For $F \in \mathcal{C}_o^{\langle\infty,2\rangle}$,*

1. (a) *if $a_F = 0$, then*

$$K_F \in \mathcal{C}_o^{(\infty,2k_F-4)}, \quad K_F - \det(\mathrm{Hess}_{g_F}) \in \mathcal{C}_o^{(\infty,2k_F-3)},$$

   (b) *if $a_F \neq 0$, then*

$$K_F - a_F^2 \in \mathcal{C}_o^{(\infty,k_F-2)}, \quad K_F - a_F^2 - a_F \mathrm{tr}(\mathrm{Hess}_{g_F}) \in \mathcal{C}_o^{(\infty,k_F-1)};$$

2. $H_F - a_F \in \mathcal{C}_o^{(\infty,k_F-2)}$ *and* $H_F - a_F - \mathrm{tr}(\mathrm{Hess}_{g_F})/2 \in \mathcal{C}_o^{(\infty,k_F-1)}$.

Let w be an element of $\mathcal{C}_o^{(\infty,1)}$ satisfying

$$a_F \frac{\partial w}{\partial X}(0,0) + \frac{1}{2}\frac{\partial w}{\partial Y}(0,0) \neq 0,$$

and set $\Delta_{F,w}(x,y) := w\big(K_F(x,y) - a_F^2, H_F(x,y) - a_F\big)$. By Lemma 4.3, we obtain

LEMMA 4.4 ([A6, Lemma 6.3]).

$$\Delta_{F,w} \in \mathcal{C}_o^{(\infty,k_F-2)},$$
$$\Delta_{F,w} - \left(a_F \frac{\partial w}{\partial X}(0,0) + \frac{1}{2}\frac{\partial w}{\partial Y}(0,0)\right)\mathrm{tr}(\mathrm{Hess}_{g_F}) \in \mathcal{C}_o^{(\infty,k_F-1)}.$$

By (1.4) together with Lemma 4.4, we obtain Proposition 4.2.

References

[A1] N. Ando, *An isolated umbilical point of the graph of a homogeneous polynomial.* Geom. Dedicata **82** (2000), 115–137.

[A2] N. Ando, *The behavior of the principal distributions around an isolated umbilical point.* J. Math. Soc. Japan **53** (2001), 237–260.

[A3] N. Ando, *Umbilical points of the graphs of homogeneous polynomials of degree 3.* J. Math. Sci. Univ. Tokyo **8** (2001), 71–87.

[A4] N. Ando, *The behavior of the principal distributions on the graph of a homogeneous polynomial.* to appear in Tohoku Math. J.

[A5] N. Ando, *A class of real-analytic surfaces in the 3-Euclidean space.* to appear in Tsukuba J. Math.

[A6] N. Ando, *The behavior of the principal distributions on a real-analytic surface.* preprint.

[HaW] P. Hartman and A. Wintner, *Umbilical points and W-surfaces.* Amer. J. Math. **76** (1954), 502–508.

[Ho] H. Hopf, *Lectures on differential geometry in the large.* Lecture Notes in Math. vol.1000, Springer-Verlag, Berlin-NewYork, 1989.

[K] T. Klotz, *On Bol's proof of Carathéodory's conjecture.* Comm. Pure Appl. Math. **12** (1959), 277–311.

[SX] B. Smyth and F. Xavier, *Real solvability of the equation* $\partial_{\bar{z}}^2\omega = \rho g$ *and the topology of isolated umbilics.* J. Geom. Anal. **8** (1998), 655–671.

[T] C.J. Titus, *A proof of a conjecture of Loewner and of the conjecture of Carathéodory on umbilic points.* Acta Math. **131** (1973), 43–77.

Department of Mathematics, Graduate School of Science, Tokyo Metropolitan University, Minami-Ohsawa 1-1 Hachioji-shi, Tokyo 192-0397, JAPAN

E-mail address: `naoya@comp.metro-u.ac.jp`

Contemporary Mathematics
Volume **308**, 2002

The Toda Equations and Equiharmonic Maps of Surfaces into Flag Manifolds

John Bolton

1. Introduction

The 2-dimensional Toda equations provide important examples of integrable systems and, as is well known, they are closely related to certain types of harmonic maps of Riemann surfaces into Lie groups and their associated k-symmetric spaces. An excellent introduction to this area may be found in [**12**], which also contains many references to the relevant literature.

In this expository article we give a brief account of the relationship between solutions of the 2-dimensional elliptic open and affine G-Toda equations for a compact simple Lie group G and certain smooth maps, called τ-adapted maps (or primitive maps in [**8**]), from a Riemann surface into the full flag manifold G/T, where T is a maximal torus of G. We then provide an elementary proof of a theorem of Black, namely that all such maps are harmonic with respect to any G-invariant metric on G/T.

In §2 we give a geometrical characterization of τ-adapted maps by means of a simple symmetry property. At each point $x \in G/T$ there is a natural symmetry τ_x of order m, (where $m-1$ is the height of the highest root of G), which is an isometry for any G-invariant metric on G/T and has x as an isolated fixed point. A smooth map $\psi : S \to G/T$ is τ-adapted if and only if for each $p \in S$ the derivative of $\tau_{\psi(p)}$ maps $d\psi(T_pS)$ into itself and induces a rotation of T_pS through $2\pi/m$.

On the other hand, if H is a closed subgroup of G and if W is a Riemannian manifold then a map $\tilde{\psi} : W \to G/H$ is said to be **equiharmonic** [**1**] if it is harmonic with respect to any G-invariant metric on G/H. Black [**1**] has obtained necessary conditions for equiharmonicity, and the restriction of his result to the case in which W is a Riemann surface and $T \subseteq H$ with $\tilde{\psi}$ factoring through G/T may be stated as follows.

THEOREM 1. *(Black [**1**]) Let $\psi : S \to G/T$ be a τ-adapted map from a Riemann surface S to the full flag manifold G/T, where T is a maximal torus of a compact*

2000 *Mathematics Subject Classification.* Primary 53C43.

simple Lie group G. If H is a closed subgroup of G containing T then $\tilde{\psi} = \pi_H \psi : S \to G/H$ is equiharmonic, where $\pi_H : G/T \to G/H$ is the canonical projection.

However, the proof in [**1**] covers a more general situation and is quite complicated. We give a short and reasonably self-contained proof of the above important special case, starting from the most direct geometrical characterization for harmonicity of a smooth map $\phi : W \to M$ between Riemannian manifolds, namely that the tension field $\tau(\phi)$ of ϕ is zero (see, for example, Page 389 of [**10**]). In fact, if W is a surface then this condition depends only on the conformal class of the metric on W, so that harmonicity may be defined for a smooth map defined on a Riemann surface. Indeed, in this case, the harmonic map condition may be given in terms of a local complex coordinate z on W and the Riemannian connection ∇ on M by

$$\nabla_{\partial/\partial \bar{z}}\big(d\phi(\partial/\partial z)\big) = 0. \tag{1.1}$$

This condition comes from the fact that, for a Riemann surface equipped with an Hermitian metric, $\tau(\phi)$ is a scalar multiple of the left hand side of (1.1).

In a concluding remark we note that this method may be used to prove an analogous result for τ-adapted maps of surfaces into any flag manifold of a semisimple compact Lie group with its canonical r-symmetric space structure [**9**].

2. τ-adapted maps

We first recall [**16**] that if H is a closed subgroup of a Lie group G then $M = G/H$ may be given the structure of a smooth manifold such that $\pi : G \to M$ is a principal H-bundle. Left multiplication induces a transitive action of G on M, and M is a *reductive homogeneous space* if the Lie algebra $\mathbf{g}$ of G has a decomposition $\mathbf{g} = \mathbf{h} \oplus \mathbf{m}$ for some $\mathrm{Ad}_G H$-invariant summand $\mathbf{m}$, where $\mathbf{h}$ is the Lie algebra of H. In this case we may use the derivative $d\pi_e$ of π at the identity e of G to identify $\mathbf{m}$ with the tangent space $T_{eH}M$ of M at eH. If we let $L_g : M \to M$ denote left translation by $g \in G$ then, for $h \in H$, the derivative of L_h at eH is then identified with $\mathrm{Ad}(h) : \mathbf{m} \to \mathbf{m}$.

For the rest of the paper we assume that G is a compact simple Lie group with maximal torus T. Let Δ be the set of roots and, for each root $\alpha \in \Delta$, let $\mathbf{g}^\alpha \subset \mathbf{g}^{\mathbb{C}}$ be the corresponding root space.

A choice of simple roots $\{\alpha_1, \dots, \alpha_\ell\}$ determines an $\mathrm{Ad}_G(T)$-invariant decomposition of $\mathbf{g}^{\mathbb{C}}$ into m direct summands, where $m-1$ is the height of the highest root. In fact,

$$\mathbf{g}^{\mathbb{C}} = \bigoplus_{k=0}^{m-1} \mathcal{M}_k,$$

where $\mathcal{M}_0 = \mathbf{t}^{\mathbb{C}}$ and for $k = 1, \dots, m-1$, $\mathcal{M}_k$ is the direct sum of the root spaces corresponding to roots of weight congruent to k modulo m. It is clear that $\overline{\mathcal{M}_k} = \mathcal{M}_{m-k}$ and, using addition modulo m,

$$[\mathcal{M}_j, \mathcal{M}_k] \subseteq \mathcal{M}_{j+k}. \tag{2.1}$$

In particular, each $\mathcal{M}_k$ is $\mathrm{Ad}_G T$-invariant, so we may identify $T^{\mathbb{C}}_{eT}G/T$ with $\bigoplus_{k=1}^{m-1} \mathcal{M}_k$, and use the action of G on G/T to obtain a corresponding G-invariant decomposition

$$T^{\mathbb{C}}G/T = \bigoplus_{k=1}^{m-1} [\mathcal{M}_k] \tag{2.2}$$

such that $\mathcal{M}_k$ is the fibre of $[\mathcal{M}_k]$ at eT. We will be particularly interested in the space $\mathcal{M}_1$, which is the direct sum of the root spaces corresponding to roots $\alpha_0, \ldots, \alpha_\ell$, where $-\alpha_0 = m_1\alpha_1 + \ldots + m_\ell\alpha_\ell$ is the highest root of $\mathbf{g}$

Let S be a Riemann surface and let $T^{1,0}S$ denote the holomorphic tangent bundle of S. A smooth map $\psi : S \to G/T$ is said to be *τ-adapted* [**6**] if $d\psi(T^{1,0}S) \subseteq [\mathcal{M}_1]$. As we see below, this is a special case of *primitive maps* as discussed in [**8**]. To avoid some vacuous statements and trivial counterexamples involving maps into S^2 we will assume from now on that G has rank $\ell > 1$, thus ensuring that $\mathcal{M}_1 \cap \overline{\mathcal{M}_1} = \{0\}$.

We now give a geometrical characterization of τ-adapted maps in terms of the canonical m-symmetric space structure on G/T. A choice of simple roots for $\mathbf{g}$ determines an inner automorphism $\tau : G \to G$ of order m which induces, for each $x \in G/T$, a diffeomorphism $\tau_x : G/T \to G/T$ of order m having x as an isolated fixed point [**3, 8, 9**]. Furthermore, each τ_x is an isometry with respect to any G-invariant metric so that G/T inherits the structure of a Riemannian m-symmetric space. In fact, the symmetry τ is an automorphism of smallest order r that exhibits G/T as an r-symmetric space [**13**].

It turns out that, for each $k = 0, \ldots, m-1$, $\mathcal{M}_k$ is the ζ^k-eigenspace in $\mathbf{g}^{\mathbb{C}}$ of the derivative of τ, where $\zeta = e^{2\pi i/m}$. Thus for each $x \in G/T$, the fibre of $[\mathcal{M}_k]$ over x is the ζ^k-eigenspace of the derivative of the symmetry τ_x. It is in this context that our definition of τ-adapted map is a special case of the concept of primitive map, which is defined for maps of Riemann surfaces into r-symmetric spaces in [**8**].

The following geometrical characterization of τ-adapted maps in terms of the m-symmetric space structure on G/T is immediate from the definition.

PROPOSITION 1. *Let S be a Riemann surface and let $\psi : S \to G/T$ be a smooth map. Then ψ is τ-adapted if and only if for each $p \in S$ the derivative of $\tau_{\psi(p)}$ maps $d\psi(T_pS)$ into itself and induces a rotation of T_pS through $2\pi/m$.*

As noted above, if G/T is given a G-invariant metric, then each symmetry τ_x is an isometry. It now follows easily that, in this case, a map ψ is τ-adapted if and only if ψ is weakly conformal and, for each $p \in S$, the derivative of $\tau_{\psi(p)}$ maps $d\psi(T_pS)$ into itself by rotation through $2\pi/m$.

3. The Toda equations and τ-adapted maps

In this section we use the notation of §2 to describe the way in which τ-adapted maps satisfying a natural non-degeneracy condition correspond locally to solutions of the 2-dimensional elliptic open and affine Toda equations [**3, 15**],

$$2\frac{\partial^2\Omega}{\partial z\partial\bar{z}} + \sum_{k\in I} m_k e^{2\alpha_k(\Omega)}\alpha_k^{\#} = 0, \tag{3.1}$$

for the Lie algebra $\mathbf{g}$ of G. Here, $\Omega : U \to i\mathbf{t}$ is a smooth map defined on an open subset U of $\mathbb{C}$, and $(\ )^{\#} : (\mathbf{t}^{\mathbb{C}})^* \to \mathbf{t}^{\mathbb{C}}$ is the map defined using the Killing form B by

$$B(\lambda^{\#}, H) = \lambda(H), \quad \forall\ H \in \mathbf{t}^{\mathbb{C}},\ \lambda \in (\mathbf{t}^{\mathbb{C}})^*.$$

The **open Toda system** is obtained if $I = \{1, \dots, \ell\}$ and the **affine Toda system** if $I = \{0, \dots, \ell\}$ and $m_0 := 1$.

The Toda equations arise as the integrability conditions for the local existence of a certain type of lift $F : S \to G$ (the Toda frame) of a τ-adapted map $\psi : S \to G/T$. Suppose then that $\psi : S \to G/T$ is τ-adapted and let $F : U \to G$ be a local lift of ψ defined on the domain U of a local complex coordinate z of S. If G is a matrix group, then $F^{-1}(\partial F/\partial z)$ is the left translate to T_eG of $dF(\partial/\partial z)$ so, as is standard practice, we use this notation in our situation also. Then

$$F^{-1}\frac{\partial F}{\partial z} = A_0 + A_1, \tag{3.2}$$

for some $A_j : U \to \mathcal{M}_j$, $j = 0, 1$. Moreover, since F is real we also have

$$F^{-1}\frac{\partial F}{\partial \bar{z}} = \bar{A}_0 + \bar{A}_1. \tag{3.3}$$

The integrability conditions for (3.2) and (3.3) are determined by the requirement that $\partial^2 F/\partial z\partial\bar{z} = \partial^2 F/\partial\bar{z}\partial z$. Since $\mathcal{M}_1 \cap \overline{\mathcal{M}_1} = \{0\}$, these are easily seen to be given by the Maurer-Cartan equations,

$$\frac{\partial A_0}{\partial\bar{z}} - \frac{\partial\bar{A}_0}{\partial z} = [A_1, \bar{A}_1], \qquad \frac{\partial A_1}{\partial\bar{z}} = [A_1, \bar{A}_0]. \tag{3.4}$$

For each $\alpha \in \Delta$ we now fix a vector $X_\alpha \in \mathbf{g}^\alpha$ such that

$$X_{-\alpha} = -\bar{X}_\alpha, \quad [X_\alpha, X_{-\alpha}] = \alpha^{\#}. \tag{3.5}$$

A map $F : U \to G$ satisfying (3.2) is a **Toda frame** if there exists a complex coordinate z on U and a smooth map $\Omega : U \to i\mathbf{t}$ such that

$$A_0 = \frac{\partial\Omega}{\partial z}, \quad A_1 = \sum_{k\in I} \sqrt{m_k}\, e^{\alpha_k(\Omega)} X_{\alpha_k}. \tag{3.6}$$

If $I = \{1, \dots, \ell\}$ then F is called an **open Toda frame**, and if $I = \{0, \dots, \ell\}$ then F is an **affine Toda frame**.

Let $\pi : G \to G/T$ be the canonical projection. If F is a Toda frame it is clear that the corresponding map $\psi = \pi F : U \to G/T$ is τ-adapted and $d\psi(\partial/\partial z)$ has non-zero component in $[\mathbf{g}^{\alpha_k}]$ for each $k \in I$. The converse is also true.

PROPOSITION 2. *Suppose that $\psi : U \to G/T$ is a τ-adapted map of a simply connected domain U in $\mathbb{C}$ and that $d\psi(\partial/\partial z)$ has non-zero component in $[\mathbf{g}^{\alpha_k}]$ for each $k \in I$. Then ψ has a lift $F : U \to G$, where F is an open Toda frame if $I = \{1, \dots, \ell\}$ and an affine Toda frame if $I = \{0, \dots, \ell\}$. In the open case, for any complex coordinate z on U there is a lift F with corresponding A_0, A_1 satisfying (3.6) for some smooth map $\Omega : U \to i\mathbf{t}$, while in the affine case the* *complex coordinate z is uniquely determined up to a translation and a rotation by a d-th root of unity where d is an integer determined by* $\mathbf{g}$. *In both cases the Toda frame is uniquely determined by the complex coordinate up to multiplication by elements of the centre $Z(G)$ of G.*

The proof in the affine case is given in [**3**]. The proof in the open case is similar.

The link with the Toda equations is provided by the following result which is easily proved using (3.4), (3.5) and (3.6). In fact, for A_0, A_1 of the form of (3.6), the second equation of (3.4) always holds while (3.5) may be used to show that (3.1) is equivalent to the first equation of (3.4).

PROPOSITION 3. *Let $\Omega : U \to i\mathbf{t}$ be a smooth map. The Toda equations (3.1) are the Maurer-Cartan equations for the existence of a Toda frame $F : U \to G$ with corresponding A_0, A_1 satisfying (3.6).*

Hence, locally at least, τ-adapted maps satisfying the non-degeneracy condition noted in Proposition 2 correspond essentially uniquely to Toda frames, and the integrability conditions for the latter are given by the Toda equations (3.1). Geometrical interpretations of τ-adapted maps and the corresponding solutions of the Toda equations are discussed in [**2**], [**3**] and [**4**] for $G = SU(n+1)$, $SO(2n+1)$ and G_2, and in [**5**] for the other classical simple Lie groups.

In the case of an open Toda system, the corresponding τ-adapted maps are holomorphic with respect to the G-invariant complex structure on G/T for which $T^{1,0}_{eT}(G/T)$ is the direct sum of the positive root spaces. When $G = SU(n+1)$ these fit naturally into the classical picture [**11**] of holomorphic curves in $\mathbb{C}P^n$. If $\phi : S \to \mathbb{C}P^n$ is a linearly full holomorphic curve then, identifying $SU(n+1)/T^n$ with the space of full flags in $\mathbb{C}^{n+1}$, the map $\psi : S \to SU(n+1)/T^n$ given by the Frenet frame of ϕ (or equivalently the full flag in $\mathbb{C}^{n+1}$ given by the osculating spaces of ϕ) is holomorphic and τ-adapted. A Toda lift of ψ is obtained by applying Gram-Schmidt orthonormalisation to the sequence of derivatives of a suitably chosen holomorphic curve $f(z)$ with $\phi(z) = [f(z)]$. If Ω is the corresponding solution of the open Toda system then, applying α_j to (3.1) and writing $\omega_j = \alpha_j(\Omega)$, we may use the fact that each $m_j = 1$ to obtain

$$2\frac{\partial^2 \omega_j}{\partial z \partial \bar{z}} - e^{2\omega_{j-1}} + 2e^{2\omega_j} - e^{2\omega_{j+1}} = 0, \quad j = 1, \dots, n,$$

where $\omega_0 = \omega_{n+1} = -\infty$, which are the classical unintegrated Plücker formulae for ϕ. When S is compact, integration over S leads to the Plücker formulae relating the degrees and ramification indices of the associated curves of ϕ. An extension of the Plücker formulae to the open Toda system of an arbitrary semi-simple Lie algebra $\mathbf{g}$ has been obtained by Positsel'skii [**14**] and Razumov-Saveliev [**15**].

The τ-adapted maps corresponding to affine Toda systems are not holomorphic. Nevertheless, a natural generalization of the Plücker formulae for arbitrary compact simple G in the case of both open and affine Toda systems is obtained in [**6, 7**], where they are used to derive global properties of τ-adapted maps.

4. Homogeneous geometry

In preparation for using (1.1) to show that τ-adapted maps are equiharmonic, we consider G-invariant connections on the reductive homogeneous space $M = G/H$. We use the notation of previous sections and, in addition, we let $(X)^{\mathbf{h}}$ (respectively $(X)^{\mathbf{m}}$) denote the $\mathbf{h}^{\mathbb{C}}$-component (respectively $\mathbf{m}^{\mathbb{C}}$-component) of $X \in \mathbf{g}^{\mathbb{C}}$ under the decomposition $\mathbf{g}^{\mathbb{C}} = \mathbf{h}^{\mathbb{C}} \oplus \mathbf{m}^{\mathbb{C}}$. If F is a local section of $\pi : G \to M$ defined on an open set $U \subset M$ we may use the identification of $\mathbf{m}$ with $T_{eH}M$ described at the beginning of §2 to define a connection D on U by setting

$$DX = dL_F\left(d\tilde{X} + [(F^{-1}dF)^{\mathbf{h}}, \tilde{X}]\right) \tag{4.1}$$

for a vector field X on U, where $\tilde{X} : U \to \mathbf{m}$ is given by $\tilde{X} = dL_{F^{-1}}(X)$.

It may be readily checked that this connection is independent of the choice of local section F, so that (4.1) may be used to define a connection globally on M. This is a G-invariant connection, and is in fact the *canonical connection* on M as defined, for example, in [**9**].

An arbitrary connection ∇ on M may be determined by reference to the *difference tensor* λ, which is given by $\lambda(X,Y) = \nabla_Y X - D_Y X$. It is clear that ∇ is G-invariant if and only if λ is G-invariant. Such a tensor λ is determined by its values at eH where it gives an $\mathrm{Ad}_G H$-invariant bilinear map $\mathbf{m} \times \mathbf{m} \to \mathbf{m}$.

5. Proof of the theorem

If H is a closed subgroup of G containing T, then an $\mathrm{Ad}_G H$-invariant decomposition $\mathbf{g} = \mathbf{h} \oplus \mathbf{m}$ is, a fortiori, $\mathrm{Ad}_G T$-invariant so that $\mathbf{m}^{\mathbb{C}}$ is a direct sum of root spaces while $\mathbf{h}^{\mathbb{C}}$ is the direct sum of $\mathbf{t}^{\mathbb{C}}$ and the remaining root spaces.

LEMMA 1. *Let $\lambda : \mathbf{m} \times \mathbf{m} \to \mathbf{m}$ be an $Ad_G H$-invariant bilinear map. If $\mathbf{g}^\alpha$ and $\mathbf{g}^\beta$ are root spaces which are subspaces of $\mathbf{m}^{\mathbb{C}}$ and if $X \in \mathbf{g}^\alpha$, $Y \in \mathbf{g}^\beta$ then $\lambda(X,Y) \in \mathbf{g}^{\alpha+\beta} \cap \mathbf{m}^{\mathbb{C}}$.*

PROOF. Our assumption on λ implies that if $Z \in \mathbf{t}$ then

$$\lambda([Z,X],Y) + \lambda(X,[Z,Y]) = [Z,\lambda(X,Y)],$$

and the result follows. □

We now let $\psi : S \to G/T$ be a τ-adapted map with local lift F into G.

LEMMA 2. *Let $\pi_H : G/T \to G/H$ be the canonical projection and let $\tilde{\psi} = \pi_H \psi : S \to G/H$. If $F^{-1}(\partial F/\partial z)$ is given by (3.2) then*

$$d\tilde{\psi}(\partial/\partial z) = dL_F((A_1)^{\mathbf{m}}), \quad d\tilde{\psi}(\partial/\partial \bar{z}) = dL_F((\bar{A}_1)^{\mathbf{m}}).$$

PROOF. We just prove the first equality. Let $\pi : G \to G/H$ be the canonical projection. Then

$$\begin{aligned} dL_F((A_1)^{\mathbf{m}}) &= dL_F\, d\pi(A_0 + A_1) \\ &= dL_F\, d\pi F^{-1} dF(\partial/\partial z) \\ &= dL_F dL_{F^{-1}} d\pi\, dF(\partial/\partial z) \\ &= d\tilde{\psi}(\partial/\partial z). \end{aligned}$$

□

We now use condition (1.1) to show that $\tilde{\psi}$ is harmonic. Let ∇ be the Riemannian connection of a G-invariant metric on G/H, and let λ be the difference tensor of ∇ and the canonical connection D described in Section 4. Then

$$\nabla_{\partial/\partial\bar{z}}\big(d\tilde{\psi}(\partial/\partial z)\big) = D_{\partial/\partial\bar{z}}\big(d\tilde{\psi}(\partial/\partial z)\big) + \lambda\big(d\tilde{\psi}(\partial/\partial\bar{z}), d\tilde{\psi}(\partial/\partial z)\big).$$

However, by Lemma 2,

$$\begin{aligned} \lambda\big(d\tilde{\psi}(\partial/\partial\bar{z}), d\tilde{\psi}(\partial/\partial z)\big) &= \lambda\big(dL_F((\bar{A}_1)^{\mathbf{m}}), dL_F((A_1)^{\mathbf{m}})\big) \\ &= dL_F \lambda\big((\bar{A}_1)^{\mathbf{m}}, (A_1)^{\mathbf{m}}\big) \end{aligned}$$

which is equal to zero since Lemma 1, together with the fact that $-\alpha_0$ is the highest root, shows that

$$\lambda\big((\bar{A}_1)^{\mathbf{m}}, (A_1)^{\mathbf{m}}\big) \in \mathbf{t}^{\mathbb{C}} \cap \mathbf{m}^{\mathbb{C}} = \{0\}.$$

Further, Lemma 2 shows that $dL_{F^{-1}}(d\tilde{\psi}(\partial/\partial z)) = (A_1)^{\mathbf{m}}$ so, by (3.3) and (4.1),

$$\begin{aligned} D_{\partial/\partial\bar{z}}(d\tilde{\psi}(\partial/\partial z)) &= dL_F \left(\frac{\partial (A_1)^{\mathbf{m}}}{\partial \bar{z}} + [\bar{A}_0 + (\bar{A}_1)^{\mathbf{h}}, (A_1)^{\mathbf{m}}] \right) \\ &= dL_F \left([(\bar{A}_1)^{\mathbf{h}}, (A_1)^{\mathbf{m}}] \right), \quad \text{(using (3.4))}, \end{aligned}$$

which is equal to zero since (2.1) shows that

$$[(\bar{A}_1)^{\mathbf{h}}, (A_1)^{\mathbf{m}}] \in \mathcal{M}_0 \cap \mathbf{m}^{\mathbb{C}} = \mathbf{t}^{\mathbb{C}} \cap \mathbf{m}^{\mathbb{C}} = \{0\}.$$

Thus $\nabla_{\partial/\partial\bar{z}}\big(d\tilde{\psi}(\partial/\partial z)\big) = 0$ and the theorem follows from condition (1.1) for a harmonic map.

Remark. In fact, if K is the centraliser of a torus of a compact semi-simple Lie group G, then G/K is called a *flag manifold* [**9**]. Such manifolds have a natural r-symmetric structure for some positive integer $r \geq 2$ induced by an inner automorphism $\tau_K : G \to G$ of order r. This leads to a G-invariant decomposition of $T^{\mathbb{C}}G/K$ similar to that in (2.2) in which, for $k = 1, \dots, r-1$, $[\mathcal{M}_k]$ is the ζ^k-eigenbundle of the derivative of the symmetry at each point. A smooth map $\psi : S \to G/K$ is said to be τ_K*-adapted* (or *primitive* in [**8**]) if $d\psi(T^{1,0}S) \subseteq [\mathcal{M}_1]$. Such maps have a geometrical characterization similar to that given in Proposition 1 and, for $r > 2$, a result analogous to Theorem 1 may be proved using the methods of the present paper.

References

[1] M. Black. *Harmonic Maps into Homogeneous Spaces.* Longman, 1991.

[2] J. Bolton, G. Jensen, M. Rigoli and L.M. Woodward. On conformal minimal immersions of S^2 into $\mathbb{C}P^n$. *Math. Ann.*, 279 (1988) 599-620.

[3] J. Bolton, F. Pedit and L.M. Woodward. Minimal surfaces and the affine Toda field model. *J. reine angew. Math.*, 459 (1995) 119-150.

[4] J. Bolton and L.M. Woodward. The affine Toda equations and minimal surfaces. In *Harmonic Maps and Integrable Systems*, eds. A.P. Fordy and J.C. Wood. Vieweg 1993, 59-82.

[5] J. Bolton and L.M. Woodward. Minimal surfaces and the Toda equations for the classical groups. In *Geometry and Topology of Submanifolds VIII*, World Scientific, 1996, 22-30.

[6] J. Bolton and L.M. Woodward. Some geometrical aspects of the 2-dimensional Toda equations. In *Geometry, Topology and Physics*, Proceedings of the First Brazil-USA Workshop, Campinas, Brazil, 1996, Eds. B.N. Apanasov et al., de Gruyter, 1997, 69-81.

[7] J. Bolton and L.M. Woodward. Pluecker formulae for primitive maps into flag manifolds. Preprint.

[8] F.E. Burstall and F. Pedit. Harmonic maps via Adler-Kostant-Symes theory. In *Harmonic Maps and Integrable Systems*, eds. A.P. Fordy and J.C. Wood. Vieweg 1993, 221-272.

[9] F.E. Burstall and J.H. Rawnsley. *Twistor Theory for Riemannian Symmetric Spaces with Applications to Harmonic Maps of Riemann Surfaces*, Lecture Notes in Math. 1424. Springer, 1990.

[10] J. Eells and L. Lemaire. Another report on harmonic maps. *Bull. London Math. Soc.*, 20 (1988) 385-524.

[11] P.A. Griffiths and J. Harris. *Principles of Algebraic Geometry.* Wiley, 1978.

[12] M.A. Guest. *Harmonic Maps, Loop Groups, and Integrable Systems.* London Mathematical Society Student Texts 38, Cambridge University Press, 1997.

[13] A. Jimenez. Addendum to "Existence of Hermitian n-symmetric spaces and non-commutative naturally reductive spaces". *Math Z.* 197 (1988) 455-456.

[14] L.E. Positsel'skii. Local Plücker formulas for a semisimple Lie group. *Funct. Anal. and its Appl.* 25 (1991) 291–292.

[15] A.V. Razumov and M.V. Saveliev. Differential geometry of Toda systems. *Comm. Anal. Geom.*, 2 (1994) 461-511.

[16] F.W. Warner. *Foundations of Differentiable Manifolds and Lie Groups.* Scott, Foresman and Company, 1971.

Department of Mathematical Sciences, University of Durham, Durham DH1 3LE, UK

E-mail address: john.bolton@dur.ac.uk

Contemporary Mathematics
Volume **308**, 2002

p-Harmonic Morphisms: the $1 < p < 2$ Case and Some Non-trivial Examples

Jean-Marie Burel and Eric Loubeau

1. Introduction

Harmonic maps, defined as the extrema of the energy functional, find a natural generalisation in p-harmonic maps, the critical points of the p-energy functional. Though several key properties remain valid, crucial dissimilarities in their regularity make the analysis of p-harmonic maps a delicate matter.

A particular subclass of harmonic maps, the *harmonic morphisms*, has been found to enjoy notable geometrical attributes. Harmonic morphisms were defined by potential theorists as preserving local harmonic functions, and characterised as horizontally weakly conformal harmonic maps by B. Fuglede [**F78**] and T. Ishihara [**I**] (see [**F99**] for an overview of this theory).

The temptation of defining p-harmonic morphisms as preserving local p-harmonic functions and recovering a Fuglede-Ishihara characterisation is the driving force behind our work (Theorem 4.3). The proof of the case $p > 2$ is in [**L2**] and here we modify it for the case $1 < p < 2$. From this, several well-known properties of harmonic morphisms are extended to p-harmonic morphisms, in particular, a version of the Baird-Eells Theorem for $p > 1$ (Theorem 4.10). Finally, in the case of maps between spheres, we present a couple of examples in detail. Manifolds and metrics will usually be assumed smooth.

2. p-Harmonic maps

In 1956 Eells, in [**E**], showed the space of $C^k(0 \leq k < +\infty)$ maps between Riemannian manifolds to be an infinite dimensional manifold. Inspired by Morse theory, he initiated the study of $C^k(M, N)$ by looking at the critical points of functionals, in connected components, i.e. within homotopy classes. Among many possible choices, the most natural seems to be the p-energy.

2000 *Mathematics Subject Classification.* Primary 58E20.

The second author was supported by the Region Bretagne.

Let $\phi : (M, g) \to (N, h)$ be a C^2-mapping between Riemannian manifolds. Define, for a real number p, the p-energy of ϕ as:

$$E_p(\phi) = \frac{1}{p} \int_M |d\phi_x|^p \, v_g.$$

(or on any compact subset $K \subset M$). Other potential candidates were the exponential energy:

$$\mathbb{E}(\phi) = \int_M \exp e(\phi) \, v_g,$$

where $e(\phi) = \frac{1}{2}|d\phi_x|^2$, or the biharmonic energy:

$$E^2(\phi) = \frac{1}{2} \int_M |\tau_2(\phi)|^2 \, v_g,$$

(see later for the definition of $\tau_2(\phi)$). The most successful work has been on the p-harmonic energy, so we will concentrate on this. First, we must notice that for $p < 1$, E_p is not a norm and $W^{1,p}$ is not a Banach space. Besides, for $p = 1$, while $W^{1,1}$ becomes a Banach space, it is impossible to derive a Euler-Lagrange equation corresponding to critical points of the 1-energy. Therefore, from now on, we will assume $p > 1$.

DEFINITION 2.1. A C^2-map is called a *p-harmonic map* if it is a critical point, in its homotopy class, of E_p (or $E_p(K)$ for all compact subsets K).

The Euler-Lagrange equation associated to extrema of E_p is the p-tension field:

$$\tau_p(\phi) = |d\phi|^{p-4} \left[|d\phi|^2 \operatorname{trace} \nabla d\phi + \tfrac{p-2}{2} \, d\phi \operatorname{grad}(|d\phi|^2) \right].$$

THEOREM 2.2. *A map is p-harmonic if and only if $\tau_p(\phi) = 0$.*

The case $p = 2$, harmonic maps, generalises harmonic functions and geodesics. In fact, p-harmonic maps can be transformed into harmonic maps:

LEMMA 2.3. [**B-G**] *Let $\phi : (M^m, g) \to (N, h)$ $(m \geq 3)$ be a C^1-map with no degenerate point. Then ϕ is p-harmonic if and only if $\phi : (M^m, \tilde{g}) \to (N, h)$ is harmonic with $\tilde{g} = |d\phi|^{\frac{2(p-2)}{m-2}} g$.*

Assuming N to be isometrically embedded in $\mathbb{R}^q$, by Nash's Theorem for example, there exists a divergence form for τ_p:

$$\tau_p(\phi) = \operatorname{div}\left(|d\phi|^{p-2} d\phi \right) - |d\phi|^{p-2} \Pi(d\phi, d\phi),$$

where Π is the second fundamental form of the isometric embedding of N into $\mathbb{R}^q$.

If $1 < p < 2$, then the possible existence of degenerate points, i.e. such that $|d\phi(x)| = 0$, implies that τ_p might exist everywhere only in a weak sense. Moreover, it is possible to make sense of weakly p-harmonic maps, by considering $W^{1,p}$ maps and requiring either that they satisfy $\tau_p = 0$ in a distributional sense or, since we cannot consider homotopy classes any more, that they are extrema of the p-energy functional for all variations of the type:

$$u_t(x) = \pi(u(x) + t\eta(x)) \quad \text{for all } t \in]-\epsilon, +\epsilon[,$$

where $\eta \in C^\infty(M, \mathbb{R}^q)$ and π is the orthogonal projection of $\mathbb{R}^q$ unto N. In general, the regularity of weakly p-harmonic maps is problematical, a fact due essentially to the non-ellipticity of the operator τ_p at degenerate points. Away from these points, C^1 weakly p-harmonic maps are C^∞ (at least for $p \geq 2$). As a matter of fact, Tolksdorf [**T**] showed that weakly p-harmonic maps need not even be $C^{1,1}$.

As for many aspects of the theory, the case $p = 2$ is clearer: a continuous weakly harmonic map is smooth (cf. [**Mor, L-U, H**]), but if we remove the continuity condition, all sorts of things can happen, even discontinuity at each point ([**R92**]).

Only by imposing some extra conditions, can we expect better regularity. For example, Takeuchi [**T94**] showed that weakly m-harmonic maps from $\Omega \subset \mathbb{R}^m$ into $\mathbb{S}^{m-1}$ are everywhere Hölder continuous. Earlier, Duzaar had shown that minimising p-harmonic maps ($p \geq 2$) into the closed m-hemisphere have Hölder continuous first derivatives for $\alpha \in]0,1[$, if $p < m < p + 2 + 2\sqrt{p}$.

The central question in the theory of p-harmonic maps is existence. In contrast to the situation for regularity, whether p is 2 or not does not play a major role, as Duzaar and Fuchs [**D-F**] showed, by extending the crucial existence theorem of Eells and Sampson [**E-S**]:

THEOREM 2.4. [**D-F**] *Let (M, g) be a compact manifold and (N, h) a compact manifold with non-positive Riemannian sectional curvature. Then, for $p \geq 2$, any continuous map from M to N has, in its homotopy class, a p-harmonic representant, of minimal p-energy.*

When the domain manifold has dimension m, the m-energy functional becomes conformally invariant: if $\tilde{g} = f^2 g$, then $v_{\tilde{g}} = f^m v_g$, but $|d\phi|^p_{\tilde{g}} = \frac{1}{f^p}|d\phi|^p_g$. This conformal invariance implies a greater freedom of existence:

THEOREM 2.5. [**J**] *Let M and N be compact Riemannian manifolds. Let $m = \dim M$ and assume that $\pi_m(N) = \{0\}$. Then any $\phi \in C^1(M, N)$ admits in its homotopy class an m-harmonic map of minimal m-energy.*

Moreover, each homotopy class of a compact Riemannian manifold possesses a p-harmonic representative:

THEOREM 2.6. [**We**] *If N is a compact Riemannian manifold, then for any positive integer i, each class in $\pi_i(N)$ can be represented by a $C^{1,\alpha}$ p-harmonic map from $\mathbb{S}^i$ into N minimising p-energy in its homotopy class for any $p > i$.*

The proof of this result is rather short, if one knows that for $p > i$, the p-energy satisfies the Palais-Smale condition (C). This implies the existence of a minimum in each component of $L^p_1(\mathbb{S}^i, N)$, whose p-stability implies $C^{1,\alpha}$-regularity ([**H-L**]).

The core of the research carried out on p-harmonic maps has been on the case $p = 2$ (cf. [**E-L78, E-L83, E-L88**]). We now give an overview of some of the results specific to $p = 2$: Though we have the Eells-Sampson Theorem for existence, its curvature conditions cannot be dispensed with, as is shown, for example, by:

THEOREM 2.7. [**E-W**] *Any harmonic non $\pm$holomorphic map from the two-torus to the two-sphere is null homotopic. In particular, there is no harmonic map of degree one from $\mathbb{T}^2$ to $\mathbb{S}^2$.*

However, if we allow conformal changes of metrics then, surprisingly:

THEOREM 2.8. [**E-F**] *Let (M, g) and (N, h) be compact Riemannian manifolds, with* $\dim M \geq 3$. *Let $\phi : M \to N$, then there exists a metric $\tilde{g}$ conformal to g and a map $\tilde{\phi}$, in the homotopy class of ϕ, such that $\tilde{\phi} : (M, \tilde{g}) \to (N, h)$ is harmonic.*

In the non-compact case, though we have no general existence result, it seems from the next result that their energy would have to be infinite:

THEOREM 2.9. [**S-Y**] *If M is a non-compact Riemannian manifold with* $\mathrm{Ricci}^M \geq 0$ *and* $\mathrm{Riem}^N \leq 0$, *then any harmonic map $\phi : M \to N$ of finite energy is constant.*

This was later extended to $p > 2$ by Takeuchi:

THEOREM 2.10. [**T91**] *Let M be a complete non-compact Riemannian manifold with non-negative Ricci curvature and N a Riemannian manifold with non-positive sectional curvature. Then any p-harmonic map $(p \geq 2)$ $\phi : M \to N$ such that $\int_M |d\phi|^{2p-2}\, dv$ is finite, has to be constant.*

A key to the importance of harmonic maps has been the link with a major branch of differential geometry, minimal submanifolds; if ϕ is an immersion, then its mean curvature is, up to a constant, the trace of the second fundamental form $\nabla d\phi$. Therefore:

THEOREM 2.11. [**E-S**] *A Riemannian immersion is harmonic if and only if it is minimal.*

The framework of Hermitian geometry is particularly propitious for the study of harmonic maps, as was immediately observed by Eells and Sampson.

THEOREM 2.12. [**E-S**] *If ϕ is a $\pm$holomorphic map between Kähler manifolds then it is harmonic.*

Finally, the Hopf map from $\mathbb{S}^3$ to $\mathbb{S}^2$ is known to be an unstable harmonic map, but it is also a p-harmonic map for any $p > 1$. However for $p \geq 4$, it actually minimises the p-energy in its homotopy class ([**R98**]). In contrast with a theorem of Xin [**X**] which says that any harmonic map to or from a sphere (of dimension at least three) has to be unstable, Rivière shows that infinitely many homotopy classes from $\mathbb{S}^3$ to $\mathbb{S}^2$ admit minimising 3-harmonic maps.

3. Harmonic morphisms

Harmonic morphisms were introduced in potential theory by Constantinescu and Cornea in 1965 (cf. [**C-C**]) to generalise a well-known property of holomorphic maps. Indeed, holomorphic maps between Riemann surfaces possess the characteristic feature of pulling back (local) holomorphic functions to holomorphic functions. Since holomorphic functions on a Riemann surface are harmonic, in the context of differential geometry, harmonic morphisms were defined as mappings between Riemannian manifolds, which pull back (local) harmonic functions to (local) harmonic functions. More precisely:

DEFINITION 3.1. Let $\phi : (M, g) \to (N, h)$ be a continuous mapping between Riemannian manifolds. Then ϕ is called a *harmonic morphism* if, for any harmonic function $f : U \subset N \to \mathbb{R}$, its pull-back by ϕ, $f \circ \phi : \phi^{-1}(U) \subset M \to \mathbb{R}$, is harmonic as well.

An immediate consequence of this definition is that the composition of two harmonic morphisms is a harmonic morphism. Moreover, since the Laplacian is (twice) the generator of Brownian motion, there exists a translation of this definition in probabilistic terms: a map is a harmonic morphism if its image of Brownian motion is again Brownian motion (up to time-scaling). Though neither definition is very easy to work with, if one knows that the Brownian motion has to be dense, one can infer that the dimension of the domain is at least that of the codomain. This fact will be confirmed later on.

The first result of this section is to characterise harmonic morphisms in terms which are easier to handle. They will evince a particularly geometric flavour, typified by:

DEFINITION 3.2. A C^1-map $\phi : (M^m, g) \to (N^n, h)$ is called *horizontally weakly conformal* if, at each point $x \in M$, either $d\phi_x \equiv 0$ or the linear map

$$d\phi_x\big|_{(\ker d\phi_x)^\perp} : (\ker d\phi_x)^\perp \to T_{\phi(x)}N$$

is a surjective and conformal map. The conformal factor is usually called the *dilation* and denoted by $\lambda(x)(> 0)$. If ϕ is non-constant this necessitates $\dim M \geq \dim N$.

We extend the function λ to points where $d\phi \equiv 0$, called *critical points*, by giving it the value 0. For vectors, horizontal weak conformality signifies:

$$h\left(d\phi(X), d\phi(Y)\right) = \lambda^2(x) g(X, Y) \quad \text{for all } X, Y \in H_x,$$

where $V_x = \ker d\phi_x$ is the vertical space and $H_x = (\ker d\phi_x)^\perp$ the horizontal space. This property can equivalently be expressed by:

$$g\left(\nabla^M(f \circ \phi), \nabla^M(g \circ \phi)\right) = \lambda^2 h\left(\nabla^N f, \nabla^N g\right)$$

for any C^1-functions f, g on N. In fact, the map ϕ is horizontally weakly conformal if and only if the dual of $d\phi$, $(d\phi)^*$, is conformal (cf. [**E-L**]), and this, in turn, leads to a rewording of the definition in local coordinates $(x^i)_{i=1,\ldots,m}$ around x on M and $(y^\alpha)_{\alpha=1,\ldots,n}$ around $\phi(x)$ on N: A map $\phi : (M, g) \to (N, h)$ is called *horizontally weakly conformal* if

$$g^{ij}(x)\frac{\partial \phi^\alpha}{\partial x^i}(x)\frac{\partial \phi^\beta}{\partial x^j}(x) = \lambda^2(x) h^{\alpha\beta}(\phi(x)) \quad \text{for all} \alpha, \beta = 1, \ldots, n.$$

This last feature gives us the means to link, for a horizontally weakly conformal map, the energy density and the dilation:

$$e_\phi = \tfrac{1}{2}|d\phi|^2 = \tfrac{1}{2} g^{ij}\frac{\partial \phi^\alpha}{\partial x^i}\frac{\partial \phi^\beta}{\partial x^j} h_{\alpha\beta} = \tfrac{1}{2}\lambda^2 h^{\alpha\beta} h_{\alpha\beta} = \tfrac{n}{2}\lambda^2.$$

The following theorem, independently proved by B. Fuglede [**F78**] and T. Ishihara [**I**], positions harmonic morphisms within the theory of harmonic maps:

THEOREM 3.3. [**F78, I**] *A mapping between Riemannian manifolds is a harmonic morphism if and only if it is a horizontally weakly conformal harmonic map.*

PROOF. There are distinct strategies in each paper: Fuglede's depends on a knowledge of Potential Theory and some properties specific to the Laplace-Beltrami operator on Riemannian manifolds. On the other hand, Ishihara applies a more direct, and in our opinion more adaptable, line of action, decomposable into two steps:

1) Prove the existence, on the target (in fact any) manifold, of local harmonic functions with stipulated first and second derivatives (in local coordinates) at a given point q, around which we will work.

2) Write out the chain rule for $\Delta(f \circ \phi)$, f being a local harmonic function on N and ϕ a harmonic morphism from M to N. Still working at the point q, insert in the specified derivatives of f at q and, juggling between the admissible choices for the derivatives at q and the expression of $\Delta(f \circ \phi)$, extract (two) sufficient requirements on ϕ.

The first stage is achieved by:

LEMMA 3.4. [**I**] *Let (N^n, h) be a Riemannian manifold. Then for any point $q \in N$ and any set of constants $\{C_{\alpha\beta}\}$ $(C_{\alpha\beta} = C_{\beta\alpha})$, $\{C_\alpha\}$, $\alpha, \beta \in \{1, \dots, n\}$, satisfying $\sum_\alpha C_{\alpha\alpha} = 0$, there exists a harmonic function f defined on a neighbourhood $U \subseteq N$ of q such that in a system of normal local coordinates (y^α) centred on q:*

$$\frac{\partial^2 f}{\partial y^\alpha \partial y^\beta}(q) = C_{\alpha\beta}, \quad \frac{\partial f}{\partial y^\alpha}(q) = C_\alpha.$$

This was shown directly by Ishihara in [**I**], but can be seen as part of a wider result (see next section) and was probably known to PDE specialists a long time ago. Armed with this existence lemma, the remaining course of action is fairly straightforward: Let $\phi : M \to N$ be a harmonic morphism and f a harmonic function on N, then, using normal coordinates for simplicity's sake:

$$\begin{aligned} 0 = {}^M\Delta(f \circ \phi) &= df(\tau(\phi)) + \operatorname{trace}\nabla df(d\phi, d\phi) \\ &= \frac{\partial f}{\partial y^\alpha}\tau^\alpha(\phi) + g^{ij}\frac{\partial \phi^\alpha}{\partial x^i}\frac{\partial \phi^\beta}{\partial x^j}\frac{\partial^2 f}{\partial y^\alpha \partial y^\beta}. \end{aligned} \tag{3.1}$$

If we choose a harmonic function f such that all its second derivatives vanish at the point $\phi(q)$ and

$$\frac{\partial f}{\partial y^\alpha}(\phi(q)) = \delta_{\alpha\alpha_0},$$

then equation (3.1) reads: $0 = {}^M\Delta(f \circ \phi) = \tau^{\alpha_0}(\phi)$, therefore ϕ is harmonic. Next, take f harmonic such that all its first derivatives vanish at $\phi(q)$ and

$$\frac{\partial^2 f}{\partial y^\alpha \partial y^\beta}(\phi(q)) = C_{\alpha\beta} \quad (C_{\alpha\beta} = C_{\beta\alpha}),$$

with $C_{\alpha\beta} = 0$ if $\alpha \neq \beta$ and $\sum_\alpha C_{\alpha\alpha} = 0$ (this condition is forced by the harmonicity of f). Equation (3.1) becomes:

$$0 = {}^M\Delta(f \circ \phi) = \sum_\alpha C_{\alpha\alpha}\frac{\partial \phi^\alpha}{\partial x^i}\frac{\partial \phi^\alpha}{\partial x^j}g^{ij}.$$

Choosing different values for the $C_{\alpha\alpha}$'s shows that:

$$g^{ij}\frac{\partial \phi^\alpha}{\partial x^i}\frac{\partial \phi^\alpha}{\partial x^j} = g^{ij}\frac{\partial \phi^1}{\partial x^i}\frac{\partial \phi^1}{\partial x^j}, \quad g^{ij}\frac{\partial \phi^\alpha}{\partial x^i}\frac{\partial \phi^\beta}{\partial x^j} = 0 \quad (\alpha \neq \beta),$$

or in other words:

$$g^{ij}\frac{\partial\phi^\alpha}{\partial x^i}\frac{\partial\phi^\beta}{\partial x^j}=\lambda^2h^{\alpha\beta},$$

i.e. ϕ is horizontally weakly conformal. The converse, i.e. that a horizontally weakly conformal harmonic map is a harmonic morphism, is easily deduced from equation (3.1). □

REMARK 3.5. It is well-known that compositions of harmonic maps (or even harmonic functions) are not in general harmonic, in contrast to the case of harmonic morphisms.

REMARK 3.6. One consequence of Theorem 3.3 is that harmonic morphisms are submersive almost everywhere.

REMARK 3.7. This method was used by Fuglede in [**F96**] to extend the characterisation to harmonic morphisms between semi-Riemannian manifolds. The two obstacles encountered were a proper generalisation of the notion of horizontal weak conformality to the situation of semi-Riemannian geometry, solved by requiring that, if $\ker d\phi_x$ is degenerate then it must contain its orthogonal complement, and second the counterpart of Lemma 3.4. This last point was solved by Hörmander in the appendix of [**F96**] making use of, not ellipticity (in this context the Laplacian being no longer so), but semi-linearity.

Looking at equation (3.1) in detail reveals that:

THEOREM 3.8. *A continuous map $\phi:(M,g)\to(N,h)$ is a harmonic morphism if and only if there exists a (unique and continuous) function λ such that:*

$${}^M\Delta(f\circ\phi)=\lambda^2({}^N\Delta f)\circ\phi$$

for all $f\in C^2(U\subset N,\mathbb{R})$.

In fact this formula is also valid for maps and the tension field:

THEOREM 3.9. *Let $\phi:(M,g)\to(N,h)$ be a harmonic morphism and $\psi:(N,h)\to(P,k)$ a harmonic map. Then $\psi\circ\phi:(M,g)\to(P,k)$ is a harmonic map.*

When $\dim M<\dim N$, harmonic morphisms from M to N are necessarily constant, and in the case $\dim M=\dim N$, while not trivial, they are of limited interest:

THEOREM 3.10. [**F78**] *A harmonic morphism between surfaces is a $\pm$holomorphic (conformal) map, the converse being also true. A harmonic morphism between Riemannian manifolds of the same dimension (greater than or equal to three) is a homothetic map, i.e. a conformal map of constant dilation.*

REMARK 3.11. An often overlooked property of harmonic morphisms is their openness. Established by Fuglede with the help of the Harnack Inequality ([**F78**]), this implies, in particular, that if the domain space M is compact and $\phi:M\to N$ is a non-constant harmonic morphism then N has to be compact.

Though a lot has been written on harmonic morphisms (cf. [**G**]), a general existence theory is still lacking. However, for the lowest significant dimensions, the following classification was obtained by Baird and Wood:

THEOREM 3.12. [**B-W88, B-W91**] *Let* $M = \mathbb{R}^3, \mathbb{S}^3$ *or* $\mathbb{H}^3$ *with the standard metric, and let* N *be a Riemann surface. Then every non-constant harmonic morphism* $\phi : M \to N$ *has the form* $\phi = \sigma \circ \psi \circ i$*, where* $i : M \to M$ *is an isometry,* $\sigma : \tilde{N} \to N$ *a weakly conformal map between Riemann surfaces, and* ψ *is a projection or the Hopf map.*

As for harmonic maps, adding a complex structure enriches the problem:

THEOREM 3.13. [**F78**] *A* $\pm$*holomorphic map from a Kähler manifold to a Riemann surface is a harmonic morphism.*

This is merely Theorem 2.12 plus the observation that holomorphic maps into Riemann surfaces are easily seen to be horizontally weakly conformal.

The nearest to a converse of this theorem is:

THEOREM 3.14. [**Wo, V**] *Let* $\phi : M \to N$ *be a non-constant harmonic morphism from an orientable Einstein four-manifold to a Riemann surface. Then* ϕ *is holomorphic with respect to some (integrable) Hermitian structure.*

In contrast to this, Baird and Wood gave in [**B-W95**] examples of harmonic morphisms from $\mathbb{R}^n$ ($n \geq 4$) to $\mathbb{C}$ which are not holomorphic with respect to any Kähler structure.

Drawing on a well-known principle of physics, Baird and Eells introduced in [**B-E**] the *stress-energy tensor* for harmonic maps:

$$S_\phi = e_\phi \, g - \phi^* \, h,$$

where e_ϕ is the energy density, and with it gave a link between harmonic morphisms and minimal submanifolds:

THEOREM 3.15. [**B-E**] *Let* $\phi : (M^m, g) \to (N^n, h)$ *be a horizontally conformal submersion. Then:*

1. *if* $n = 2$*, then* ϕ *is a harmonic morphism if and only if the fibres are minimal.*
2. *if* $n \neq 2$*, two of the following properties imply the other:*
   (a) ϕ *is a harmonic morphism.*
   (b) *the fibres are minimal submanifolds.*
   (c) ∇e_ϕ *is a vertical field.*

Related to this is the following fact:

THEOREM 3.16. [**F82**] *If* ϕ *is a harmonic morphism with* ∇e_ϕ *vertical then* ϕ *has no critical point.*

Theorem 3.15 calls for a specific study of harmonic morphisms into surfaces and, since they must have minimal fibres, a legitimate question is of the stability of the fibres and the stability of the map. Montaldo shows that the former implies the latter:

THEOREM 3.17. [**Mon**] *Let* $\phi : (M, g) \to (N^2, h)$ *be a submersive harmonic morphism to a surface. If its fibres, necessarily minimal, are stable for the volume functional, then* ϕ *is stable as a harmonic map.*

In fact, when mapping into the two-sphere, we have a sort of converse:

THEOREM 3.18. [**C**] *A stable harmonic map from a compact Riemannian manifold into* $\mathbb{S}^2$ *is a harmonic morphism.*

Moreover, harmonic morphisms preserve also the Jacobi operator (along harmonic maps) and its kernel, the Jacobi fields (cf. [**M-W**]). In contrast to the lack of existence results, Mustafa has employed a Bochner technique for harmonic morphisms to show non-existence in several cases:

THEOREM 3.19. [**M98**] *There exists no non-constant harmonic morphism from a compact Riemannian manifold of non-negative Ricci curvature to a compact Riemann surface of genus greater than or equal to two.*

For the non-compact case:

THEOREM 3.20. [**M99**] *Let* $m \geq n$. *There exists no non-constant harmonic morphism from* $\mathbb{R}^m$ *to* $\mathbb{H}^n$. *If* ϕ *from* $\mathbb{H}^m$ *to* $\mathbb{R}^n$ *is a harmonic morphism of dilation* λ *then* $\lambda^2 \leq \frac{m-1}{n-1}$. *If* ϕ *from* $\mathbb{S}^m$ *to* $\mathbb{S}^n$ *is a submersive harmonic morphism of dilation* λ *then* $\lambda^2 > \frac{m-1}{n-1}$.

Polynomial harmonic morphisms between Euclidean spaces have recently been the focus of several papers and quadratic harmonic morphisms have been classified by Ou and Wood in [**O-W**] using Clifford systems, while Ababou, Baird and Brossard recently proved a striking result on mappings between Euclidean spaces:

THEOREM 3.21. [**A-B-B**] *Let* ϕ *be a harmonic morphism from* $\mathbb{R}^m \setminus K$, *where* K *is a polar set, into* $\mathbb{R}^n$ $(m \geq n \geq 3)$. *Then* ϕ *is a polynomial map of degree less than or equal than* $\frac{m-2}{n-2}$.

A very new addition to the theory is the discrete analogue of harmonic morphisms. Introduced by Urakawa (cf. [**U, Ts**]), and extended to metric graphs by Anand [**A**], based on the discrete Laplacian of Dodziuk [**D**] and an adapted notion of horizontal conformality, harmonic morphisms between graphs proved to be precisely horizontally conformal maps (a notable difference with the Fuglede-Ishihara characterisation is the absence of harmonicity).

4. p-Harmonic morphisms

It is easy to imagine a generalisation of harmonic morphisms for p-harmonic maps:

DEFINITION 4.1. A C^2-map $\phi : (M,g) \to (N,h)$ is called a *p-harmonic morphism* if it pulls back local p-harmonic functions on M to p-harmonic functions on N, i.e.

$$^N\Delta_p(f) = 0 \Rightarrow {}^M\Delta_p(f \circ \phi) = 0.$$

REMARK 4.2. One of the consequences of the Fuglede-Ishihara characterisation is that harmonic morphisms, only assumed to be continuous in their definition, are in fact smooth. But p-harmonic maps do not enjoy similar regularity properties, and extra care must be taken with the regularity hypothesis on p-harmonic morphisms. If one is ready to work in a distributional sense, then the C^2 condition of Definition 4.1 can be weakened to $C^{1,1}$.

An inevitable question regarding p-harmonic morphisms is whether they can be characterised in a similar manner to harmonic morphisms:

THEOREM 4.3. *A map is a p-harmonic morphism if and only if it is a horizontally weakly conformal p-harmonic map.*

The strategy employed to prove this characterisation of p-harmonic morphisms, is the method of Ishihara. First, the decisive existence lemma of [**I**] is extended to p-harmonic functions:

LEMMA 4.4. *Let (N^n, h) be a Riemannian manifold. Then for any point $q \in N$ and any set of constants $\{C_{\alpha\beta}\}$ $(C_{\alpha\beta} = C_{\beta\alpha})$, $\{C_\alpha\}$ $(\sum_\alpha C_\alpha^2 \neq 0)$ $\alpha, \beta \in \{1, \dots, n\}$, satisfying*

$$\left(\sum_\alpha C_\alpha^2\right)\left(\sum_\alpha C_{\alpha\alpha}\right) + (p-2)\sum_{\beta,\delta=1}^{n} C_\beta C_\delta C_{\beta\delta} = 0, \tag{4.1}$$

there exists a C^∞ p-harmonic function f defined on a neighbourhood $U \subseteq N$ of q such that in a system of normal local coordinates (y^α) centred on q:

$$\frac{\partial^2 f}{\partial y^\alpha \partial y^\beta}(q) = C_{\alpha\beta}, \ \frac{\partial f}{\partial y^\alpha}(q) = C_\alpha.$$

This, in fact, is part of a wider local solvability result due to Alinhac and Gerard:

THEOREM 4.5. [**A-G**, Proposition 2.4] *Let $F = F(x, u^{(\alpha)})$, $(\mid \alpha \mid \leq 2)$, be a smooth function of $n + N$ real variables $(N = (n+1)(n+2)/2)$. Let $p_0 \in \mathbb{R}^N$ be such that:*

i) $F(0, p_0) = 0$.
ii) *The operator $P = \sum \frac{\partial F}{\partial u^{(\alpha)}}(0, p_0)\partial^\alpha$ is elliptic, i.e.*

$$p_2(\xi) = \sum_{|\alpha|=2} \frac{\partial F}{\partial u^{(\alpha)}}(0, p_0)\xi^\alpha \neq 0,$$

for $\xi \in \mathbb{R}^n, \xi \neq 0$.

Then there exists in a neighbourhood of $x = 0$, a smooth function u solution to:

$$F(x, \partial^\alpha u) = 0 \quad (\partial^\alpha u(0)))_{|\alpha|\leq 2} = p_0.$$

For $p > 2$, the rest of the proof of Theorem 4.3 is in [**L2**] and we will adapt it here to the case $1 < p < 2$. Our strategy was to start by showing that if ϕ is both a p-harmonic morphism and a p-harmonic map then it must be horizontally weakly conformal ([**L2**, Proposition 4]), and then, that a p-harmonic morphism is always a p-harmonic map ([**L2**, Proposition 5]). The first step was achieved via three lemmas, the proofs of the first two, as well as the proof of [**L2**, Proposition 5], remaining valid for the case $1 < p < 2$. Only Lemma 3 (of [**L2**]) needs some adjustment:

LEMMA 4.6. $|d\phi^{\alpha_0}|^2 = |d\phi^{\beta_0}|^2$ *for all α_0, β_0.*

PROOF. Let $q \in M$ be a point such that $d\phi(q)$ does not vanish, and let us choose normal local coordinates on (M, g) and (N, h) centred around q and $\phi(q)$. We put:

$$\frac{|d\phi^\alpha|^2}{|d\phi|^2} = K_\alpha, \quad \frac{< d\phi^\alpha, d\phi^\beta >}{|d\phi|^2} = K_{\alpha\beta},$$

all the differentials being at the point q. Of course $K_{\alpha\alpha} = K_\alpha$ and if $d\phi^\alpha \equiv 0$ for a specific α, then $K_\alpha = K_{\alpha\beta} = 0$ for all β. Choose a local p-harmonic function f such that:

$$f_\alpha(q) = 1 \quad \text{for all } \alpha \quad f_{\alpha\beta}(q) = \delta^{\beta\alpha_0}_{\alpha\alpha_0} - \delta^{\beta\beta_0}_{\alpha\beta_0} \quad \alpha_0 \neq \beta_0$$

These coefficients do satisfy condition (4.1), therefore, taking into account lemmas 1 and 2 of [**L2**]:

$$\begin{aligned} 0 = \tau(f \circ \phi) &= \mid d\phi \mid^2 \mid d(f\circ\phi) \mid^{p-4} \left\{ \left(\sum_\alpha K_\alpha \right) (K_{\alpha_0} - K_{\beta_0}) + (p-2)\left(K^2_{\alpha_0} - K^2_{\beta_0}\right) \right\} \\ &= \mid d\phi \mid^2 \mid d(f\circ\phi) \mid^{p-4} \left\{ (K_{\alpha_0} - K_{\beta_0})\left(1 + (p-2)\left(K_{\alpha_0} + K_{\beta_0}\right)\right) \right\}. \end{aligned}$$

But $1 + (p-2)(K_{\alpha_0} + K_{\beta_0}) < 0$ since $K_{\alpha_0} + K_{\beta_0} \leq 1$ and $\frac{1}{(p-2)} < -1$, and necessarily $K_{\alpha_0} = K_{\beta_0}$. □

As with harmonic morphisms, this characterisation means that the only way of being a p-harmonic morphism is to commute, up to a factor, with the p-Laplacian:

THEOREM 4.7. *A C^2-map $\phi : (M, g) \to (N, h)$ is a p-harmonic morphism if and only if there exists a (unique and continuous) function λ such that:*

$$ {}^M\Delta_p(f \circ \phi) = \lambda^p({}^N\Delta_p f) \circ \phi \tag{4.2}$$

for all $f \in C^2(U \subset N, \mathbb{R})$.

Since equation (4.2) remains valid for maps, we have:

THEOREM 4.8. *Let $\phi : (M, g) \to (N, h)$ be a p-harmonic morphism. Then ϕ pulls back p-harmonic* maps *to p-harmonic* maps.

A key to producing examples is the comparison of p-harmonic morphisms for different values of p:

THEOREM 4.9. *Let $p, q > 1, p \neq q$. A p-harmonic morphism ϕ is also a q-harmonic morphism if and only if it is horizontally homothetic, i.e. $d\phi(\nabla\lambda^2) = 0$.*

A non-trivial class of examples (first brought to our attention by S. Montaldo) is a variation on the radial projection: $\psi : \mathbb{R}^m \setminus \{0\} \to \mathbb{R}^m \setminus \{0\} \quad x \mapsto x/|x|^l$, which, if $l \neq 1$, is a p-harmonic map if and only if $p = \frac{m-l}{l-1} + 2$ and is horizontally weakly conformal if and only if $l = 2$. On the other hand, when $l = 1$, the map becomes the radial projection which is horizontally homothetic, with dilation $\lambda = \frac{1}{|x|}$ and a (2-)harmonic morphism, therefore a p-harmonic morphism for any $p > 1$.

A central result for harmonic morphisms is the Baird-Eells Theorem (Theorem 3.15) linking harmonic morphisms and minimal submanifolds. The generalisation of the Fuglede-Ishihara characterisation, enables an extension (cf. [**L1**] for details):

THEOREM 4.10. *Let $\phi : (M^m, g) \to (N^n, h)$ be a submersive p-harmonic morphism. Then:*

1. *if $n = p$, the fibres are minimal.*
2. *if $n \neq p$, the following conditions are equivalent:*
   (a) *the fibres are minimal submanifolds.*
   (b) *∇e_p is a vertical field.*
   (c) *the horizontal distribution has mean curvature $\frac{1}{p}\frac{\nabla e_p}{e_p}$.*

5. Some non-trivial examples

The study of conformal deformations of the metric permits us to construct examples of minimal submanifolds. For this, we use results on p-harmonicity. With Proposition 4.9, we can produce examples of p-harmonic morphisms.

What can we say when the dilation is basic? On a manifold M endowed with a Riemannian metric, can we find a real number p (not necessarily an integer) making the map p-harmonic with respect to g or a metric conformal to g?

Let $\phi : (M^m, g) \to (N^n, h)$ be a smooth map with no critical points, i.e $|d\phi_x| \neq 0$ for all $x \in M$, and $\tilde{g}$ be a metric conformally related to g, i.e. $\tilde{g} = a^2 g$ where $a : M \to \mathbb{R}^+$ is a smooth function.

LEMMA 5.1. *Let $\phi : (M^m, g) \to (N^n, h)$ be a smooth map with no critical points and horizontally weakly conformal with dilation λ. Then ϕ is p-harmonic with respect to a conformal (singular for some values of p) metric $\tilde{g}$ if and only if*

$$\tau(\phi) + (m-p)d\phi(\nabla^g \ln a) + (p-2)d\phi(\nabla^g \ln \lambda) = 0 \tag{5.1}$$

with $a : M \to \mathbb{R}^+$ a smooth function.

PROOF. If $\tau(\phi)$ denotes the tension field of ϕ, then the tension field $\tilde{\tau}(\phi)$ with respect to the metric $\tilde{g}$ is given by

$$\tilde{\tau}(\phi) = \tfrac{1}{a^2}\left[\tau(\phi) + d\phi(\nabla^g \ln a^{m-2})\right]. \tag{5.2}$$

The p-tension field $\tilde{\tau}_p(\phi)$, with respect to $\tilde{g}$, is :

$$\tilde{\tau}_p(\phi) = |d\phi|^{p-2}\left[\tilde{\tau}(\phi) + \tfrac{(p-2)}{2} d\phi(\nabla^{\tilde{g}} \ln |d\phi|^2)\right], \tag{5.3}$$

where $|.|$ denotes the norm associated to $\tilde{g}$. The dilation is expressed by $\tilde{\lambda}^2 = \frac{\lambda^2}{a^2}$, since $\nabla^{\tilde{g}} = \frac{1}{a^2}\nabla^g$ for functions. Combining (5.2) and (5.3), ϕ is p-harmonic if and only if $\tau(\phi) + (m-p)d\phi(\nabla^g \ln a) + (p-2)d\phi(\nabla^g \ln \lambda) = 0$. □

The following propositions are direct consequences:

THEOREM 5.2. *Let $\phi : \mathbb{S}^3 \to \mathbb{S}^2$ be defined by*

$$(\cos s e^{ia}, \sin s e^{ib}) \mapsto (\cos \alpha(s), \sin \alpha(s) e^{i(ka+lb)}),$$

where the function $\alpha(s)$ is chosen such that ϕ is horizontally weakly conformal, where $s \in [0, \pi/2], a, b \in [0, 2\pi[$. There does not exist a p which makes ϕ p-harmonic with respect to the Euclidean metric on $\mathbb{S}^3$.

Nevertheless, we have:

THEOREM 5.3. *The map* $\phi : \mathbb{S}^3 \to \mathbb{S}^2$ *defined by:*

$$(\cos s e^{ia}, \sin s e^{ib}) \mapsto (\cos \alpha(s), \sin \alpha(s) e^{i(ka+lb)})$$

is p-harmonic with respect to the conformally Euclidean metric:

$$\tilde{g} = (\frac{\sin \alpha}{\sin^2 2s})^{2(p-2)/(p-3)} (k^2 \sin^2 s + l^2 \cos^2 s)^{(p-1)/(p-3)} g_{\mathbb{S}^3}.$$

Note that $\tilde{g}$ *is singular for certain values of* p.

PROOF. On the two-sphere, in the following system of coordinates $0 \leq t \leq \pi$, $0 \leq \gamma \leq 2\pi$, the Riemannian metric is $dt^2 + \sin^2 t d\gamma^2$, and an orthonormal frame of $T_{(t,\gamma)}\mathbb{S}^2$, $(0 < t < \pi)$ is $\left(\frac{1}{\sin t}\partial_\gamma, \partial_t\right)$.

For $k \neq l$, the function α, which renders the map ϕ horizontally weakly conformal is:

$$\alpha(s) = 2 \arctan \left(\frac{l-u}{l+u}\right)^{l/2} \left(\frac{k+u}{k-u}\right)^{k/2}, \tag{5.4}$$

with $u(s) = \sqrt{k^2 \sin^2 s + l^2 \cos^2 s}$. This relation is obtained by integrating the quantity

$$\dot{\alpha} = \frac{\sin \alpha}{\sin s \cos s} \sqrt{k^2 \sin^2 s + l^2 \cos^2 s}.$$

Then the tension field $\tau(\phi)$ is:

$$\tau(\phi) = \left[\ddot{\alpha} + \frac{\cos 2s}{\sin s \cos s}\dot{\alpha} - \frac{k^2 \sin^2 s + l^2 \cos^2 s}{\cos s \sin s} \sin \alpha \cos \alpha\right] \partial_t.$$

Equation (5.1) becomes:

$$\frac{\dot{a}}{a} = \frac{2 \cos 2s}{(p-3) \sin 2s} + \frac{(p-2)\dot{\alpha} \cos \alpha}{(p-3) \sin \alpha} + \frac{p-1}{p-3} \left[\frac{(k^2 - l^2) \cos s \sin s}{k^2 \sin^2 s + l^2 \cos^2 s} - 2\frac{\cos 2s}{\sin 2s}\right],$$

One solution is:

$$a^2 = \left(\frac{\sin \alpha}{\sin^2 2s}\right)^{2(p-2)/(p-3)} (k^2 \sin^2 s + l^2 \cos^2 s)^{(p-1)/(p-3)}. \tag{5.5}$$

By (5.4), we have

$$\sin \alpha = \frac{2(-1)^{l/2}(k^2 - l^2)^{\frac{l+k}{2}} \cos^k s \sin^l s}{(l+u)^l (k-u)^k + (l-u)^l (k+u)^k}.$$

Using the latter, we have that, for $p \neq 2$, the function $a^2(s)$ given by (5.5) vanishes for $s = 0$ or $\pi/2$. When $k = 2$ (respectively $l = 2$) the function $a(s)$ vanishes only for $s = 0$ (respectively for $s = \pi/2$). □

REMARK 5.4. When $p = 2$, we recover the well-known result ([**E-R**]) that the map is harmonic with respect to the metric:

$$\tilde{g} = \frac{1}{k^2 \sin^2 s + l^2 \cos^2 s} g.$$

REMARK 5.5. Let

$$A(s) = u^{\frac{2(p-1)}{p-3}} \left[\frac{(-1)^{l/2}(k^2 - l^2)^{\frac{l+k}{2}}}{2(l+u)^l(k-u)^k + (l-u)^l(k+u)^k} \right]^{\frac{2(p-2)}{p-3}}.$$

For $k = 2$ and $l = \frac{p-1}{p-2}$, the metric which makes the map p-harmonic is:

$$\tilde{g} = A(s) \times \frac{1}{\sin^2 s} g_{\mathbb{S}^3}.$$

Let $t = \ln \tan(s/2)$. The metric $\tilde{g}$ is conformal to the product metric for $\mathbb{H}^2 \times \mathbb{R}$, where $\mathbb{H}^2$ is the hyperbolic plane of constant negative curvature -1.

By this technique, we obtain:

THEOREM 5.6. *Consider the map $\phi : \mathbb{S}^4 \to \mathbb{S}^3$:*

$$(\cos s, \sin sx) \mapsto \big(\cos \alpha(s), \sin \alpha(s) H(x)\big)$$

with $s \in [0, \pi]$, H the Hopf map from $\mathbb{S}^3$ to $\mathbb{S}^2$, and $\alpha(s) = 2\arctan(\tan^2 s/2)$. Then ϕ is p-harmonic with respect to the conformally Euclidean metric (singular for some values of p):

$$\tilde{g} = \frac{(\sin s)^{2(p-2)/(p-4)}}{(1 - 1/2\sin^2 s)^{2(p-3)/(p-4)}} g_{\mathbb{S}^4}.$$

PROOF. This map is horizontally weakly conformal if and only if

$$\dot{\alpha}(s) = 2\frac{\sin \alpha}{\sin s}.$$

A solution is given by $\alpha(s) = 2\arctan(\tan^2 s/2)$. We want to determine a metric which can render ϕ p-harmonic. The tension field is found to be:

$$\tau(\phi) = \left[\ddot{\alpha} + 3\cot s\dot{\alpha} - 8\frac{\sin \alpha \cos \alpha}{\sin^2 s} \right] \partial_t.$$

A direct computation shows that $\ddot{\alpha} = 2\frac{\sin \alpha}{\sin^2 s}(2\cos \alpha - \cos s)$ and $d\phi(\nabla^g \lambda) = \ddot{\alpha}\dot{\alpha}\partial_t$. By (5.1), we obtain the following equation:

$$(4-p)\frac{\dot{a}}{a}\sin s + (4-p)\cos s + 2(p-3)\cos \alpha = 0,$$

i.e.

$$\frac{\dot{a}}{a} = -\frac{\cos s}{\sin s} + 2(p-3)/(p-4)\frac{\cos \alpha}{\sin s}. \tag{5.6}$$

Since $\cos \alpha = \frac{\cos s}{1 - 1/2\sin^2 s}$, we set $u = \sin s$ and so $du = \cos s\, ds$. We have to integrate the following quantity

$$\frac{du}{u(1 - 1/2u^2)} = \frac{du}{u} + \frac{1/2u\, du}{1 - 1/2u^2}.$$

Equation (5.6) has the solution:

$$a^2(s) = \frac{(\sin s)^{2(p-2)/(p-4)}}{(1 - 1/2\sin^2 s)^{2(p-3)/(p-4)}}.$$

For $p \neq 2$, the function $a(s)$ vanishes for $s = 0$ or π. □

REMARK 5.7. When $p = 2$, $|k| = |l| = 1$, we have that ϕ is harmonic (and so a harmonic morphism) with respect to the conformal metric:

$$\tilde{g} = \frac{2}{2 - \sin^2 s} g_{\mathbb{S}^4}.$$

REMARK 5.8. When $p = 3$, the metric which makes the map 3-harmonic is expressed by:

$$\tilde{g} = \frac{1}{\sin^2 s} g_{\mathbb{S}^4}.$$

To understand better the metric, we set $u = \ln(\tan s/2)$. Then $\tilde{g} = g_{\mathbb{S}^3} + g_{\mathbb{R}}$.

References

[A-B-B] R. Ababou, P. Baird and J. Brossard, *Polynômes semi-conformes et morphismes harmoniques.* Math. Zeit. **231** (1999), 589–604.

[A-G] S. Alinhac and P. Gerard, *Opérateurs pseudo-différentiels et théorème de Nash-Moser.* Savoirs Actuels. InterEditions \ Editions du C.N.R.S., 1991.

[A] C.K. Anand, *Harmonic morphisms of metric graphs.* In Harmonic Morphisms, Harmonic Maps and Related Topics, CRC Press (1999), 109–112.

[B-E] P. Baird and J. Eells, *A conservation law for harmonic maps.* In E. Looijenga, D. Siersma, and F. Takens, editors, Geometry Symposium. Utrecht 1980, volume 894 of Lecture Notes in Math., Springer, Berlin, Heidelberg, New York, 1981, pp. 1–25.

[B-G] P. Baird and S. Gudmundsson, *p-Harmonic maps and minimal submanifolds.* Math. Ann. **294** (1992), 611–624.

[B-W88] P. Baird and J.C. Wood, *Bernstein theorems for harmonic morphisms from* $\mathbb{R}^3$ *and* $\mathbb{S}^3$. Math. Ann. **280** (1988), 579–603.

[B-W91] ———, *Harmonic morphisms and conformal foliations by geodesics of three-dimensional space forms.* J. Australian Math. Soc. **51** (1991), 118–153.

[B-W95] ———, *Hermitian structures and harmonic morphisms in higher dimensional Euclidean spaces.* Int. J. Math. **6** (1995), 161–192.

[C] J.Y. Chen, *Stable harmonic maps into* $\mathbb{S}^2$. In Report of the first MSJ-IRI, Tohoku University, Sendai, Japan, (1993), 431–435.

[C-C] C. Constantinescu and A. Cornea, *Compactifications of harmonic spaces.* Nagoya Math. J. **25** (1965), 1–57.

[D] J. Dodziuk, *Difference equation, isoperimetric inequality and transcience of certain random walks.* Trans. Amer. Math. Soc. **284** (1984), 787–794.

[D-F] F. Duzaar and M. Fuchs, *Existence and regularity of functions which minimize certain energies in homotopy classes of mappings.* Asymptotic Anal. **5** (1991), 129–144.

[E] J. Eells, *On the geometry of function spaces.* In Proc. Symp. Int. Topologia Algebraica, Mexico City, 1958, 303-308.

[E-F] J. Eells and M.J. Ferreira, *on representing homotopy classes by harmonic maps.* Bull. London Math. Soc. **23** (1991), 160–162.

[E-L] J. Eells and L. Lemaire, *On the construction of harmonic and holomorphic maps between surfaces.* Math. Ann. **252** (1980), 27–52.

[E-L78] ———, *A report on harmonic maps.* Bull. London Math. Soc. **10** (1978), 1–68.

[E-L83] ———, *Selected topics in harmonic maps.* Regional Conference Series in Mathematics, 1983, American Mathematical Society.

[E-L88] ———, *Another report on harmonic maps.* Bull. London Math. Soc. **20** (1988), 385–524.

[E-R] J. Eells and A. Ratto, *Harmonic maps between spheres and ellipsoids.* Internat. J. Math. **1** (1990), 1–27.

[E-S] J. Eells and J.H. Sampson, *Harmonic mappings of Riemannian manifolds.* Amer. J. Math. **86** (1964), 109–160.

[E-W] J. Eells and J.C. Wood, *Restrictions on harmonic maps of surfaces.* Topology **15** (1976), 263–266.

[F78] B. Fuglede, *Harmonic morphisms between Riemannian manifolds.* Ann. Inst. Fourier (Grenoble) **28** (1978), 107–144.

[F82] ———, *A criterion of non-vanishing differential of a smooth map.* Bull. London Math. Soc. **14** (1982), 98–102.

[F96] ———, *Harmonic morphisms between semi-Riemannian manifolds.* Ann. Acad. Sci. Fenn. Ser. A I Math. **21** (1996) 31–50.

[F99] ———, *The beginnings of harmonic morphisms.* In: Harmonic Morphisms, Harmonic Maps and Related Topics. Proceedings of a conference held in Brest, France 7th-11th of July 1997, CRC Press, 1999, 3–12.

[G] S. Gudmundsson, *The Bibliography of Harmonic Morphisms.* `http://www.maths.lth.se/matematiklu/personal/sigma/harmonic/publications.html`

[H-L] R. Hardt and F.H. Lin, *Mappings minimizing the L^p-norm of the gradient.* Comm. Pure Appl. Math. **40** (1987), 555–588.

[H] F. Hélein, *Régularité des applications faiblement harmoniques entre une surface et une variété riemannienne.* C. R. A. S. Paris Ser. I **312** (1991), 591–596.

[I] T. Ishihara, *A mapping of Riemannian manifolds which preserves harmonic functions.* J. Math. Kyoto Univ. **19** (1979), 215–229.

[J] J. Jost, *A conformally invariant variational problem for mappings between Riemannian manifolds.* Preprint of the Centre for Math. Analysis, Australian National Univ., 1984.

[L-U] O. Ladyzenkaya and N. Ural'ceva, *Linear and quasilinear elliptic equations.* Academic Press, 1968.

[Li] A. Lichnerowicz, *Applications harmoniques et variétés kählériennes.* Symp. Math. III (Bologna 1970), 341–402.

[L1] E. Loubeau, *The Fuglede-Ishihara and Baird-Eells theorems for $p > 1$.* To appear in Proceedings of the International Congress on Differential Geometry in memory of Alfred Gray, Bilbao, Spain.

[L2] ———, *On p-harmonic morphisms.* Differential Geometry and its Applications **12** (2000), 219–229.

[Mon] S. Montaldo, *Stability of harmonic morphisms to a surface.* Int. J. Math. **9** (1998), 865–875.

[M-W] S. Montaldo and J.C. Wood, *Harmonic morphisms and the Jacobi operator.* Preprint, 2000.

[Mor] C.B. Morrey, *Multiple integrals in the calculus of variations.* Grundlehren, Band 130, Springer, 1966.

[M98] M.T. Mustafa, *A Bochner technique for harmonic morphisms.* J. London Math. Soc. **57** (1998), 746–756.

[M99] ———, *Restrictions on harmonic morphisms.* Conformal Geom. Dynamics **3** (1999), 102–115.

[O-W] Y.L. Ou and J.C. Wood, *On the classification of quadratic harmonic morphisms between Euclidean spaces.* Algebras, Groups and Geometries **13** (1996), 41–53.

[R92] T. Rivière, *Applications harmoniques de $\mathbb{B}^3$ dans $\mathbb{S}^2$ partout discontinues.* C. R. Acad. Sci. Paris **314** (1992), 719–723.

[R98] ———, *Minimizing fibrations and p-harmonic maps in homotopy classes from $\mathbb{S}^3$ into $\mathbb{S}^2$.* Comm. Anal. Geom. **6** (1998), 427–483.

[S-Y] R. Schoen and S.T. Yau, *Harmonic maps and the topology of stable hypersurfaces and manifolds of non-negative Ricci curvature.* Comm. Math. Helv. **51** (1976), 333–341.

[S] Y.T. Siu, *The complex-analyticity of harmonic maps and the strong rigidity of compact Kähler manifolds.* Ann. Math. **112** (1980), 73–111.

[T91] H. Takeuchi, *Stability and Liouville theorems of p-harmonic maps.* Japan J. Math. (N.S.) **17** (1991), 317–332.

[T94] ———, *Some conformal properties of p-harmonic maps and regularity for sphere-valued p-harmonic maps.* J. Math. Soc. Japan **46** (1994), 217–234.

[T] P. Tolksdorf, *Regularity for a more general class of quasilinear elliptic equations.* J. Diff. Equ. **51** (1984), 126–150.

[Ts] T. Tsuruta, *Harmonic morphisms on weighted graphs.* J. Math. Sci. Univ. Tokyo **7** (2000), 297–310.

[U] H. Urakawa, *A discrete analogue of the harmonic morphism.* In Harmonic Morphisms, Harmonic Maps and Related Topics, CRC Press (1999), 97–108.

[V] M. Ville, *Harmonic morphisms from Einstein 4-manifolds to Riemann surfaces.* Preprint, Ecole Polytechnique (1999).

[We] S.W. Wei, *Representing homotopy groups and spaces of maps by p-harmonic maps.* Indiana Univ. Math. J. **47** (1998), 625–670.

[Wo] J.C. Wood, *Harmonic Morphisms and Hermitian structures on Einstein 4-manifolds.* Int. J. Math. **3** (1992), 415–439.

[X] Y.L. Xin, *Some results on stable harmonic maps.* Duke Math. J. **47** (1980), 609–613.

UNIVERSITÉ DE BRETAGNE OCCIDENTALE, UFR SCIENCES ET TECHNIQUES, DEPARTEMENT DE MATHEMATIQUES, 6, AVENUE VICTOR LE GORGEU, BP 809, 29285 BREST CEDEX, FRANCE

Current address: Centre for Mathematical Sciences, Mathematics, Faculty of Science, Lund University, Box 118, 221 00 Lund, SWEDEN

E-mail address: `Jean-Marie.Burel@math.lu.se`

UNIVERSITÉ DE BRETAGNE OCCIDENTALE, UFR SCIENCES ET TECHNIQUES, DEPARTEMENT DE MATHEMATIQUES, 6, AVENUE VICTOR LE GORGEU, BP 809, 29285 BREST CEDEX, FRANCE

E-mail address: `loubeau@univ-brest.fr`

Contemporary Mathematics
Volume **308**, 2002

Schwarzian Derivatives and Flows of Surfaces

Francis Burstall, Franz Pedit, and Ulrich Pinkall

1. Introduction

Over the last decades it has been widely recognized that many completely integrable PDE from mathematical physics arise naturally in geometry. Their integrable character in the geometric context – usually associated with the presence of a Lax pair and a spectral deformation – is nothing but the flatness condition of a naturally occurring connection. Examples of interest to geometers generally come from the integrability equations of special surfaces in various ambient spaces: among the best known examples is the sinh-Gordon equation describing constant mean curvature tori in 3-space. In fact, most of the classically studied surfaces, such as isothermic surfaces, surfaces of constant curvature and Willmore surfaces, give rise to such completely integrable PDE. A common thread in many of these examples is the appearance of a harmonic map into some symmetric space, which is well-known to admit a Lax representation with spectral parameter. Once the geometric problem is formulated this way algebro-geometric integration techniques give explicit parameterizations of the surfaces in question in terms of theta functions.

One contribution of the geometric view of these integrable PDE is a much better understanding of the meaning of the hierarchy of flows associated to these equations. In mathematical physics these hierarchies are obtained by deformations of the Lax operators preserving their eigenvalues. This is rather unsatisfactory from the geometric viewpoint, where one wants to see these flows as geometric deformations on the surfaces. A good example of the geometric derivation of the KdV flows can be found in [**17**] (see also [**21**]) and the flows related to the sinh-Gordon equation are explicitly derived from geometric deformations in [**18**].

Recently two more integrable hierarchies, well known to mathematical physicists, have made their appearance in surface geometry: these are the Novikov–Veselov and Davey–Stewartson equations [**13, 23, 11, 12**]. They are derived as

2000 *Mathematics Subject Classification.* Primary 53A30.
Authors supported by SFB288 at TU-Berlin. First author partially supported by The European Contract Human Potential Programme, Research Training Network HPRN-CT-2000-0010. Second author additionally supported by US National Science Foundation grants DMS-90011083 and DMS-9705479.

kernel dimension preserving deformations of the Dirac operator. Since surfaces in 3− and 4-space can be presented by "spinors", these flows in principle give an integrable hierarchy on the space of conformally parametrized tori in 3− and 4-space. The relation to surface geometry leaves a lot to ask for: what is the geometric meaning of these flows? Can one derive them from some geometric principles? What kind of geometry is involved? Even though existing literature [**11, 12, 23**] uses Euclidean surface geometry, it turns out that these flows are in fact Möbius invariant. Thus there should be some Möbius invariant setting from which these flows can be derived. However, the essential new insight is that there is a completely integrable hierarchy on *all* conformally immersed 2-tori, and not just on special classes given by certain geometric conditions, such as constant mean curvature tori etc. One clearly expects that these new flows reduce in some way to the known ones when restricted to the corresponding special surface classes.

The picture emerging here is that the manifold of conformally parametrized 2-tori in 3-space is stratified by finite dimensional submanifolds, namely those tori stationary with respect to some higher flow in the hierarchy. These "finite type" tori can then be parametrized by theta functions on some finite genus Riemann surface. As said, this is just a picture and we feel there still is a lot more explaining to be done before it becomes a useful mathematical theory.

This present note aims to give a self contained, low technology approach to the above mentioned topics. The natural setting of our discussion is Möbius invariant surface geometry. Since we do not claim to have a complete theory, the exposition will at times be sketchy and leave the reader, hopefully, with some urge to dwell further on the issues.

The most popular integrable hierarchy is the KdV hierarchy. Therefore we begin this paper by deriving this hierarchy as natural flows on holomorphic maps into the Riemann sphere. The basic invariant of a holomorphic map f, its Schwarzian derivative $S_z(f)$ with respect to a coordinate z, turns out to be the function satisfying the equations of the KdV hierarchy. The Schwarzian occurs naturally in a Hill equation

$$\psi_{zz} + \tfrac{1}{2}S_z(f)\psi = 0$$

for some appropriate homogeneous lift $[\psi] = f$. Notice that under a change of coordinates the Schwarzian transforms as

$$S_w(f)dw^2 = (S_z(f) - S_z(w))dz^2\,,$$

which makes it conceptually harder to understand what object the Schwarzian really is (see, however [**6**]). This approach to KdV is well known [**21**], but our derivations are closer to the Möbius geometry of conformally immersed Riemann surfaces into the n-sphere.

The latter theme gets introduced in the second section, where we derive the invariants, structure and integrability equations of a conformal surface $f : M \to S^n$ under the Möbius group. In this case the basic equation defining the invariants is an inhomogeneous Hill equation

$$\psi_{zz} + \tfrac{1}{2}S_z(f)\psi = \kappa\,,$$

for an appropriate lift ψ of f into the forward light cone—we view the conformal n-sphere as the projectivized light cone in Lorentz space. The invariant κ is essentially the trace free second fundamental form, or *Hopf differential*, of f, a well known Möbius invariant, and $S_z(f)$ is the *Schwarzian derivative*[1] of the surface f. It transforms under coordinate changes just like the classical Schwarzian above. Of course, if $\kappa = 0$ we reduce to the classical case of holomorphic maps into S^2. The Hopf differential and Schwarzian form a complete set of invariants for surfaces in 3-space with respect to the Möbius group. Their integrability equations, the conformal Gauss and Codazzi equations, are

$$\tfrac{1}{2}S_z(f)_{\bar{z}} = 3\bar{\kappa}\kappa_z + \bar{\kappa}\kappa_z$$
$$\mathrm{Im}\,(\kappa_{\bar{z}\bar{z}} + \tfrac{1}{2}\overline{S_z(f)}\kappa) = 0\,,$$

which gives the fundamental theorem of Möbius invariant surface theory (Theorem 3.1).

To demonstrate the effectiveness of our setup, we discuss a number of natural problems from conformally invariant surface theory: cyclides of Dupin are characterized by holomorphic Schwarzians $S_z(f)$; isothermic surfaces by real Hopf differentials κ; Willmore surfaces are characterized by

$$\kappa_{\bar{z}\bar{z}} + \tfrac{1}{2}\overline{S_z(f)}\kappa = 0\,,$$

a stronger condition than the Codazzi equation. From these conditions we easily see the associated families of isothermic and Willmore surfaces, crucial to their completely integrable character. Moreover, unless a surface is isothermic, the Hopf differential alone determines it uniquely up to Möbius transformations. We also discuss results of Thomsen [**24**] and Richter [**19**] characterising minimal, respectively constant mean curvature, surfaces in 3-dimensional space forms as isothermic Willmore, respectively constrained Willmore, surfaces and indicate a generalisation to isothermic Willmore surfaces in n-space. We include these sometimes very classical results since they amplify the "retro" charm of our viewpoint on Möbius geometry: all calculations are local and done in a fixed coordinate, just as they were 100 years ago, but due to our non-redundant set of invariants turn out to be extraordinarily short and efficient.

Having done that much, we finally get to a geometric derivation of the Novikov–Veselov and Davey–Stewartson equations in the last section. We follow our initial derivation of the KdV hierarchy in the first section to derive the Novikov–Veselov flow for surfaces in n-space. This flow is Möbius invariant by construction and preserves the Willmore energy, at least in 3-space. Moreover, it also preserves isothermic surfaces in any codimension giving a vector version of the modified Novikov–Veselov flow. Again, when the Hopf differential $\kappa = 0$, so that we deal with holomorphic maps into S^2, these flows reduce to the KdV flows. Whereas the Novikov–Veselov flows can be defined in any codimension, the Davey–Stewartson hierarchy needs the additional datum of a complex structure in the normal bundle, which is the case for surfaces in 4-space. This flow is defined in terms of double derivatives of the Hopf differential and thus simpler than the Novikov–Veselov

[1]The appearance of Schwarzian derivatives in conformal surface geometry is not new: they were introduced in a closely related context by Calderbank [**8**] who has also developed, with the first author, an independent and more invariant formulation of the fundamental theorem of Möbius invariant surface theory [**6**].

flow, which is given in terms of triple derivatives of the Hopf differential. Surfaces in 3-space are generally moved into 4-space by the Davey–Stewartson flow, which therefore does not induce flows on surfaces in 3-space. Nevertheless, the flow preserves Willmore surfaces and isothermic surfaces in 4-space.

Even though we mention the word "hierarchies" frequently, we do not really have an explicit scheme to construct all the higher flows for Novikov–Veselov and Davey–Stewartson. This is partially due to the inherently non-explicit character of the higher flows —one needs to solve a $\bar{\partial}$-problem for each flow— and partially due to our insufficient understanding of invariant multilinear differential operators on Riemann surfaces. The latter seems to be an important enough issue to return to. At present there are two attempts in that direction, one [**4**] using deformation theory on quaternionic holomorphic line bundles for surfaces in $S^4 = \mathbb{HP}^1$, and another [**6**] using Cartan connections and Möbius structures for surfaces in S^n.

This work began during the conference "Integrable Systems in Differential Geometry" in Tokyo in the summer of 2000 and has been completed during January 2001, when all authors met again at the SFB288 at TU-Berlin. We take this opportunity to thank the organizers of the Tokyo conference and the SFB288 for their support.

2. The classical Schwarzian and the KdV hierarchy

2.1. The classical Schwarzian derivative. Given a meromorphic function f on a Riemann surface M and a holomorphic coordinate z the *Schwarzian derivative of f with respect to z* is given by the expression

$$S_z(f) = (\frac{f''}{f'})' - \tfrac{1}{2}(\frac{f''}{f'})^2 . \tag{1}$$

One can quickly check that the Schwarzian $S_z(f) = 0$ vanishes if and only if $f(z) = \frac{az+b}{cz+d}$ is fractional linear in z. Moreover, two meromorphic functions f and g are related by a fractional linear transformation

$$g = \frac{af+b}{cf+d}$$

if and only if their Schwarzians $S_z(f) = S_z(g)$ agree. The last property indicates that the Schwarzian is an invariant for holomorphic maps into $\mathbb{CP}^1$. We will develop this viewpoint further since it motivates most of our constructions on conformally immersed surfaces later on.

Let $f : M \to \mathbb{CP}^1$ be a holomorphic immersion, i.e., $df_p \neq 0$ for all $p \in M$, and fix a holomorphic coordinate z on M. A *lift* of f is a holomorphic map $\psi : M \to \mathbb{C}^2 \setminus \{0\}$ so that $f(p)$ is the line spanned by $\psi(p)$. In other words, ψ is a holomorphic section of the pull-back via f of the tautological line bundle over $\mathbb{CP}^1$. Our assumption that f is an immersion is equivalent to ψ and ψ_z being linearly independent at each point $p \in M$. Using a fixed complex volume form det on $\mathbb{C}^2$ we can find, up to sign, a unique *normalized* lift ψ of f by demanding that

$$\det(\psi, \psi_z) = 1 . \tag{2}$$

Indeed, two lifts $\tilde{\psi} = \psi\lambda$ are related by a nowhere vanishing scale $\lambda : M \to \mathbb{C}$ and

$$\det(\tilde{\psi}, \tilde{\psi}_z) = \lambda^2 \det(\psi, \psi_z)\,,$$

which proves our assertion. Taking the derivative of (2) we obtain $\det(\psi, \psi_{zz}) = 0$, so that ψ satisfies Hill's equation

$$\psi_{zz} + \tfrac{c}{2}\psi = 0 \tag{3}$$

for some holomorphic function $c : M \to \mathbb{C}$. From our construction it is clear that c depends only on f and the coordinate z. If the holomorphic map f into $\mathbb{CP}^1$ is written in homogeneous coordinates $[f : 1]$ then one easily checks that

$$c = S_z(f) \tag{4}$$

is the Schwarzian derivative. The above mentioned properties of the Schwarzian now follow immediately from (3). Moreover, given c we solve Hill's equation for

$$\psi = \begin{pmatrix} \psi_1 \\ \psi_2 \end{pmatrix},$$

and then $f = [\psi_1 : \psi_2]$ gives a map into $\mathbb{CP}^1$, unique up to the action of $\mathbf{Sl}(2, \mathbb{C})$, with Schwarzian $S_z(f) = c$. Thus the Schwarzian is an invariant, with respect to the Möbius group $\mathbf{PSl}(2, \mathbb{C})$, for holomorphic maps into $\mathbb{CP}^1$.

To obtain a better understanding of what type of object the Schwarzian is, we differentiate with respect to another holomorphic coordinate w on M. An elementary calculation using (3) reveals the transformation formula

$$S_w(f)dw^2 = (S_z(f) - S_z(w))dz^2\,. \tag{5}$$

Recall that a *projective structure* on M is given by a holomorphic atlas $\mathcal{A}$ whose coordinate transition functions are fractional linear. Given two projective structures $\mathcal{A}$ and $\tilde{\mathcal{A}}$ the holomorphic quadratic differential $Q := S_z(\tilde{z})dz^2$, where z and $\tilde{z}$ are charts in $\mathcal{A}$ and $\tilde{\mathcal{A}}$, is well-defined by (5). Therefore the space of all projective structures on a Riemann surface is an affine space modelled on the vector space $H^0(K^2)$ of holomorphic quadratic differentials. If we have fixed a projective structure on M, the transformation rule (5) implies that the Schwarzian of a function may be regarded as a holomorphic quadratic differential.

2.2. The KdV hierarchy. Consider a deformation $f(t) : M \to \mathbb{CP}^1$ of $f = f(0)$ with corresponding lift $\psi(t)$, which we may assume to be normalized as in (2). Then ψ and ψ_z are a unimodular basis for $\mathbb{C}^2$ for any z and t, and the infinitesimal variation of ψ can be expressed as

$$\psi_t = a\psi + b\psi_z\,. \tag{6}$$

Here a and b are functions of t and z which have to satisfy

$$b_z = -2a \tag{7}$$

due to (2). For the infinitesimal variation of the Schwarzian, we use Hill's equation (3) and obtain

$$\psi_{tzz} + \tfrac{c_t}{2}\psi + \tfrac{c}{2}\psi_t = 0\,.$$

Calculating ψ_{tzz} from (6) and (7) the ψ-part of this relation unravels to

$$c_t = b_{zzz} + 2b_z c + b c_z\,. \tag{8}$$

So far we have been describing general deformations of f. To obtain geometrically meaningful deformations the "radial variation" a, and therefore the "tangential variation" b, in (6) have to be chosen geometrically. The most naive choice of variation is *no* radial variation at all, i.e., $a_1 = 0$. Then, up to a constant, $b_1 = 1$ by (7) and from (8) we get $c_{t_1} = c_z$, which is translational flow along z. To obtain a more interesting flow we use the *old* variation of c as the *new* radial variation, i.e., we put $a_3 = -\frac{1}{2}c_{t_1}$. Then $b_3 = c$ by (7) and from (8) we obtain

$$c_{t_3} = c_{zzz} + 3cc_z\,,$$

which we recognize as the KdV equation. Continuing our recursion scheme $a_{2k+1} = -\frac{1}{2}c_{t_{2k-1}}$, we arrive at an infinite hierarchy of odd order (in z) flows, the KdV hierarchy. For example, the 5th-order flow computes to

$$c_{t_5} = c_{zzzzz} + 5c_{zzz}c + 10c_{zz}c_z + \tfrac{15}{2}c_z c^2\,.$$

Notice that at each step we have constants of integration, which amounts to considering a given order flow modulo the lower order flows.

A map $f : M \to \mathbb{CP}^1$ has *finite type* n if the $(2n-1)^{st}$–order flow is stationary modulo the lower order flows:

$$c_{t_{2n-1}} + \sum_{k=1}^{n-1} \lambda_k c_{t_{2k-1}} = 0\,. \tag{9}$$

For example, type 1 maps have $c_z = 0$ or $c = c_0$ constant, and thus are the loxodromes $f = \exp(i\sqrt{2c_0}z)$. In general, finite type n maps are given by θ–functions on a genus $n-1$ Riemann surface, and the flows in the hierarchy become linear flows on the Jacobian of this Riemann surface. The above geometric construction of an integrable hierarchy and its algebraic-geometric integration is folklore by now. Our exposition follows the spirit of [**18**], where a similar construction was made for the sinh–Gordon hierarchy.

We conclude this section with a remark on the closely related mKdV hierarchy. Whereas the KdV flows commute with the action of the Möbius group $\mathbf{Psl}(2,\mathbb{C})$, the mKdV flows arise by "symmetry breaking": fixing a point at infinity on $\mathbb{CP}^1$ we consider $f : M \to \mathbb{C} \subset \mathbb{CP}^1$ where $\mathbb{C}$ now carries the flat Euclidean structure. In other words, we have reduced the symmetry group of our target space from the Möbius group to the Euclidean group. The fundamental Euclidean invariant of f is now given by the induced metric

$$|df|^2 = f_z\overline{f_z}|dz|^2\,, \qquad f_z\overline{f_z} = e^{2v}$$

where $v : M \to \mathbb{R}$ is a real valued function. Using the formula (1) for $S_z(f)$, it is easy to check that

$$S_z(f) = 2v_{zz} - 2(v_z)^2\,. \tag{10}$$

In particular, this combination of derivatives of the real valued function v is holomorphic. Putting $p := 2v_z$ the above relation reads

$$S_z(f) = p_z - \tfrac{1}{2}p^2\,,$$

which we recognize as the *Miura transform* [**15**]. Therefore the Miura transform is just the expression of the Möbius invariant Schwarzian of f in terms of the Euclidean invariant induced metric of f. It is a well known fact [**15**] that the Miura transform interpolates between the mKdV and KdV flows so that p satisfies the mKdV hierarchy.

3. Conformally immersed surfaces and Schwarzians

The constructions of the previous section carry over almost verbatim to conformally immersed surfaces in n-space. As a result we obtain an elegant description of Möbius invariant surface geometry. Later we will indicate how to construct Möbius invariant flows on conformally immersed surfaces.

3.1. The light-cone model. Following Darboux [**9**], we linearize the action of the Möbius group on the sphere $S^n \subset \mathbb{R}^{n+1,1}$ via the diffeomorphism

$$S^n \cong \mathbb{P}(\mathcal{L}) : x \leftrightarrow [1 : x] \tag{11}$$

between the sphere and the projectivized light cone, where $\mathcal{L} \subset \mathbb{R}^{n+1,1}$ is the null cone in $(n+2)$–dimensional Minkowski space for the quadratic form $\langle v, v\rangle = -v_0^2 + \sum_{k=1}^{n+1} v_k^2$. The projective action of the Lorentz group on $\mathbb{P}(\mathcal{L})$ is by conformal diffeomorphisms, giving rise to a double covering of the Möbius group by the Lorentz group.

The isomorphism (11) is a special case of a more general construction which realizes all space forms as conic sections of $\mathcal{L}$: for non-zero $v_0 \in \mathbb{R}^{n+1,1}$, set

$$S_{v_0} = \{v \in \mathcal{L} : \langle v, v_0\rangle = -1\}.$$

S_{v_0} inherits a positive definite metric of constant curvature $-\langle v_0, v_0\rangle$ from $\mathbb{R}^{n+1,1}$ and is a copy of a sphere, Euclidean space or ball according to the sign of $-\langle v_0, v_0\rangle$. For (11), $v_0 = e_0$.

In this picture, it is easy to describe the m-spheres (totally umbilic submanifolds) in S^n [**20**]: they are all given by a decomposition $\mathbb{R}^{n+1,1} = V \oplus V^\perp$ with $V^\perp$ space-like of dimension $(n-m)$. The m-dimensional sphere is then $\mathbb{P}(\mathcal{L} \cap V)$. Viewed as a submanifold $S_{v_0} \cap V$ of the conic section S_{v_0}, one readily shows that this m-sphere has mean curvature vector

$$H_v = -v_0^\perp - \langle v_0^\perp, v_0^\perp\rangle v \tag{12}$$

at v, where $v_0 = v_0^T + v_0^\perp$ according to the decomposition $V \oplus V^\perp$.

3.2. The Schwarzian derivative of a conformally immersed surface. Let $f : M \to S^n \subset \mathbb{R}^{n+1}$ be a conformal immersion of a Riemann surface M. A *lift* of f is a map $\psi : M \to \mathcal{L}_+$ into the forward light cone such that the null line spanned by $\psi(p)$ is $f(p)$. Perhaps the most naive choice for a lift is the *Euclidean lift*

$$\phi = (1, f)\,, \tag{13}$$

but any positive scaling of this lift will suffice. Note that the induced metric of a scaled lift $\phi\lambda$, with $\lambda : M \to \mathbb{R}_+$, is given by

$$|d(\phi\lambda)|^2 = \lambda^2 |df|^2\,,$$

where the induced metric $|df|^2$ is computed with respect to the standard round metric on S^n which we also denote by $\langle\,,\,\rangle$. Therefore the various lifts of f give rise to the various metrics in the conformal class of $|df|^2$. From this we see that there is a unique *normalized* lift ψ, with respect to a given holomorphic coordinate z on M, for which the induced metric

$$|d\psi|^2 = |dz|^2 \tag{14}$$

is the standard flat metric of the coordinate z. Since (14) is invariant under Lorentz transformations the normalized lift ψ is Möbius invariant.

If we pass to another holomorphic coordinate w the normalized lift with respect to this new coordinate

$$\tilde{\psi} = \psi|w_z| \tag{15}$$

is obtained by scaling ψ by the length of the Jacobian of the coordinate transformation. This follows immediately from the fact that ψ is null and therefore

$$|d\tilde{\psi}|^2 = |\, d\psi|w_z| + \psi d|w_z|\,|^2 = |d\psi|^2|w_z|^2 = |dz|^2|w_z|^2 = |dw|^2 .$$

A fundamental construction of Möbius invariant surface geometry is the *mean curvature* or *central sphere congruence* [**24**] which assigns to each point $f(p)$ on the surface the unique 2-sphere $S(p)$ tangent to f at that point and with the same mean curvature vector $H_{S(p)} = H_p$ at $f(p)$ as f. While this is phrased in terms of the Euclidean lift $\phi = (1, f) : M \to S_{e_0}$, the construction is conformally invariant: $S(p)$ corresponds to a decomposition $\mathbb{R}^{n+1,1} = V(p) \oplus V(p)^\perp$ with $V(p) = \operatorname{span}\{\phi(p), d\phi_p, e_0^T\}$ since ϕ is tangent to $V(p) \cap S_{e_0}$ at p. In view of (12), this reads

$$\operatorname{span}\{(1, f(p)), (0, df_p), (-1, -H_p)\} = \operatorname{span}\{(1, f(p)), (0, df_p0), (0, H_p - f(p))\}.$$

However, one readily checks that $(0, H - f)$ is parallel to $\phi_{z\bar{z}}$ for any holomorphic coordinate z so that V is a rank 4 subbundle of $M \times \mathbb{R}^{n+1,1}$ of signature $(3, 1)$ given by

$$V = \operatorname{span}\{\phi, d\phi, \phi_{z\bar{z}}\}. \tag{16}$$

It is clear that V so defined is independent of all choices (of lift or holomorphic coordinate) and so is a conformal invariant.

To facilitate calculations we choose the unique section $\hat{\psi} \in \Gamma(V)$ with

$$\langle \hat{\psi}, \hat{\psi} \rangle = 0\,, \qquad \langle \psi, \hat{\psi} \rangle = -1\,, \qquad \text{and} \qquad \langle \hat{\psi}, d\psi \rangle = 0\,. \tag{17}$$

Then, given a holomorphic coordinate z, we have the Möbius invariant framing

$$\psi\,,\ \psi_z\,,\ \psi_{\bar{z}}\,,\ \hat{\psi}\ \in\ \Gamma(V) \tag{18}$$

of the bundle V, or rather $V \otimes \mathbb{C}$. The orthogonality relations of this frame are given by (14) and (17), namely

$$\langle \psi, \psi \rangle = \langle \hat{\psi}, \hat{\psi} \rangle = 0\,, \qquad \langle \psi, \hat{\psi} \rangle = -1 \tag{19a}$$

$$\langle \psi, d\psi \rangle = \langle \hat{\psi}, d\psi \rangle = \langle d\hat{\psi}, \hat{\psi} \rangle = 0\,, \tag{19b}$$

$$\langle \psi_z, \psi_z \rangle = \langle \psi_{\bar{z}}, \psi_{\bar{z}} \rangle = 0\,, \tag{19c}$$

$$\langle \psi_z, \psi_{\bar{z}} \rangle = \tfrac{1}{2}\,. \tag{19d}$$

We immediately obtain the fundamental equation of Möbius invariant surface geometry, an inhomogeneous Hill's equation,

$$\psi_{zz} + \tfrac{c}{2}\psi = \kappa \tag{20}$$

defining the complex valued function c and the section κ of $V^\perp \otimes \mathbb{C}$. In light of (3) and (4) we call c the *Schwarzian derivative* of the immersion f in the coordinate z, and denote it by

$$S_z(f) := c\,. \tag{21}$$

We shall see below that the section κ can be identified with the normal bundle valued *Hopf differential* of the immersion f, suitably scaled. Note that, by construction, both c and κ are Möbius invariants of the immersion f for a given coordinate z. For surfaces in 3-space they form a complete set of invariants.

If the Hopf differential κ is identically zero, the surface f is totally umbilic and we can view f as a conformal map taking values in $S^2 = \mathbb{CP}^1$. In this case we recover the classical Schwarzian discussed in section 2.1.

Just as in the classical case, it is helpful to understand the behaviour of c and κ under coordinate changes. Let w be another holomorphic coordinate on M. Then the new normalized lift is given by (15). Computing $\tilde{\psi}_{ww}$ and inserting into (20) gives

$$2(\tilde{\kappa} - \kappa)z_w^2|w_z| + (\tilde{c}|w_z| - cz_w^2|w_z| + 2z_w^2|w_z|_{zz} + 2z_{ww}|w_z|_z)\psi = 0\,. \tag{22}$$

Taking the $V^\perp$ part we see that

$$\tilde{\kappa}\frac{dw^2}{|dw|} = \kappa\frac{dz^2}{|dz|}\,. \tag{23}$$

Thus $\kappa\frac{dz^2}{|dz|}$ is a globally defined quadratic differential with values in $LV^\perp \otimes \mathbb{C}$ where L is the real line bundle $(\bar{K}K)^{-1/2}$ of densities of conformal weight 1 [**8**].

The ψ–part of (22) gives $\tilde{c}dw^2 = (c - 2v_{zz} + 2v_z^2)dz^2$, where we have put $|w_z| = e^v$. From (10) we therefore deduce the same transformation rule

$$S_w(f)dw^2 = (S_z(f) - S_z(w))dz^2 \tag{24}$$

for the Schwarzian of a conformal immersion as for the Schwarzian of a holomorphic function.

We conclude this section by expressing the Möbius invariants $S_z(f)$ and κ in terms of Euclidean quantities associated to the immersion $f : M \to S^n$. If $|df|^2 = e^{2u}|dz|^2$ is the induced metric of f in the holomorphic coordinate z on M, then $\psi = (1, f)e^{-u}$ is the normalized lift (14) of f. Let $N_f M \subset f^\perp$ be the Euclidean normal bundle of M. We have an isometric isomorphism $N_f S^n \cong V^\perp$:

$$\xi \mapsto \langle H, \xi\rangle(1, f) + (0, \xi) \tag{25}$$

and so a section $\hat{\kappa}$ of $N_f S^n \otimes \mathbb{C}$ with

$$\kappa = \langle H, \hat{\kappa}\rangle(1, f) + (0, \hat{\kappa})\,.$$

Inserting into (20) we obtain

$$\hat{\kappa}\frac{dz^2}{|dz|} = \frac{II^{(2,0)}}{|df|} \tag{26}$$

and

$$\tfrac{c}{2}\,dz^2 = \langle H, II^{(2,0)}\rangle + (u_{zz} - (u_z)^2)dz^2\,. \tag{27}$$

Here $II^{(2,0)}$ denote the $(2,0)$-part of the normal bundle valued second fundamental form of f. This shows that κ, up to the isomorphism (25), is the trace free part of the second fundamental form, i.e., the normal bundle valued Hopf differential, scaled by the square root of the induced metric. In particular, the *Willmore energy* of the conformal immersion f is given by

$$W(f) = \int |\kappa|^2\,. \tag{28}$$

3.3. The integrability equations. Unlike the case of holomorphic maps, where any Schwarzian can be integrated, the invariants for a conformal immersion $f : M \to S^n$ will have to satisfy integrability conditions. Recall that f gives rise to the Möbius invariant splitting

$$M \times \mathbb{R}^{n+1,1} = V \oplus V^\perp\,,$$

where $\mathbb{P}(\mathcal{L} \cap V)$ is the mean curvature sphere congruence. A choice of holomorphic coordinate z on M yields a unique normalized lift (14) which, by Hill's equation (20), defines the Schwarzian derivative $S_z(f)$ and the Hopf differential κ.

To compute the integrability conditions on the Schwarzian and the Hopf differential we work with the orthogonal frame

$$F = (\psi, \psi_z, \psi_{\bar z}, \hat\psi, \xi) \tag{29}$$

of $V \oplus V^\perp$ where ξ denotes a section of $V^\perp$. Clearly, the integrability equations for this frame are

$$F_{z\bar z} = F_{\bar z z}$$

or, by using reality conditions,

$$\psi_{zz\bar z} = \psi_{z\bar z z}\,, \tag{30a}$$

$$\operatorname{Im}\hat\psi_{z\bar z} = 0\,, \tag{30b}$$

$$\operatorname{Im}\xi_{z\bar z} = 0\,. \tag{30c}$$

To evaluate these equations, we will make frequent use of the formula

$$\phi = -\langle\phi, \hat\psi\rangle\psi - \langle\phi, \psi\rangle\hat\psi + 2\langle\phi, \psi_{\bar z}\rangle\psi_z + 2\langle\phi, \psi_z\rangle\psi_{\bar z}\,,$$

which expresses a section ϕ of V in the basis (18). Together with (20) we therefore get

$$\psi_{zz} = -\tfrac{c}{2}\psi + \kappa\,, \qquad \tfrac{c}{2} = \langle\psi_{zz}, \hat\psi\rangle = -\langle\psi_z, \hat\psi_z\rangle\,.$$

Moreover,

$$\psi_{z\bar z} = q\psi + \tfrac{1}{2}\hat\psi\,, \qquad q = -\langle\psi_{z\bar z}, \hat\psi\rangle = \langle\psi_z, \hat\psi_{\bar z}\rangle\,,$$

for some real valued function q. Therefore $\hat\psi_z = -c\psi_{\bar z} + 2q\psi_z + \chi$ where χ is a section of $V^\perp \otimes \mathbb{C}$. We now evaluate the first integrability equation (30a) $(\kappa - \tfrac{c}{2}\psi)_{\bar z} = (q\psi + \tfrac{1}{2}\hat\psi)_z$ which, using the last relation, unravels to

$$\kappa_{\bar z} = 2q\psi_z + (\tfrac{1}{2}c_{\bar z} + q_z)\psi + \tfrac{1}{2}\chi\,.$$

To compare the V and $V^\perp$-parts of this equation, we note that the z-derivative of a section ξ in $V^\perp \otimes \mathbb{C}$ decomposes as

$$\xi_z = D_z\xi + \langle \xi, \chi\rangle \psi - 2\langle \xi, \kappa\rangle \psi_{\bar z}\,, \tag{31}$$

where D denotes the connection in the bundle $V^\perp$. Therefore,

$$D_{\bar z}\kappa = \tfrac{1}{2}\chi\,, \qquad q = -\langle \bar\kappa, \kappa\rangle\,, \qquad \tfrac{1}{2}c_{\bar z} + q_z = \langle \kappa, \bar\chi\rangle$$

and (30a) is equivalent to the *conformal Gauss equation*

$$\tfrac{1}{2}c_{\bar z} = 3\langle D_z\bar\kappa, \kappa\rangle + \langle \bar\kappa, D_z\kappa\rangle\,.$$

The second integrability condition (30b) becomes

$$0 = \mathrm{Im}\hat\psi_{z\bar z} = \mathrm{Im}(-c_{\bar z}\psi_{\bar z} - c\psi_{\bar z\bar z} + 2q_{\bar z}\psi_z + 2q\psi_{z\bar z} + \chi_{\bar z})\,.$$

By inserting the appropriate terms and keeping track of reality conditions, we therefore conclude that (30b) is equivalent to the *conformal Codazzi equation*

$$\mathrm{Im}(D_{\bar z}D_{\bar z}\kappa + \tfrac{1}{2}\bar c\kappa) = 0\,.$$

It is easy to see that for the remaining integrability equation (30c) only the $V^\perp$-part gives new conditions. Using (31) we immediately obtain the *conformal Ricci equation*

$$D_{\bar z}D_z\xi - D_zD_{\bar z}\xi - 2\langle \xi, \kappa\rangle\bar\kappa + 2\langle \xi, \bar\kappa\rangle\kappa = 0$$

for a section ξ of $V^\perp$. Note that

$$R^D_{\bar z z}\xi = D_{\bar z}D_z\xi - D_zD_{\bar z}\xi$$

is the curvature of the normal bundle $V^\perp$.

For further reference we collect the fundamental equations for a conformal immersion $f : M \to S^n$ in a given holomorphic coordinate z on M: keeping reality conditions in mind the equations of the frame (29) are given by

$$\begin{aligned} \psi_{z\bar z} &= -\langle \kappa, \bar\kappa\rangle\psi + \tfrac{1}{2}\hat\psi\,, \\ \psi_{zz} &= -\tfrac{c}{2}\psi + \kappa\,, \\ \hat\psi_z &= -2\langle \kappa, \bar\kappa\rangle\psi_z - c\psi_{\bar z} + 2D_{\bar z}\kappa\,, \\ \xi_z &= D_z\xi + 2\langle \xi, D_{\bar z}\kappa\rangle\psi - 2\langle \xi, \kappa\rangle\psi_{\bar z}\,. \end{aligned} \tag{32}$$

The resulting integrability conditions, the conformal Gauss, Codazzi and Ricci equations, are

$$\tfrac{1}{2}c_{\bar z} = 3\langle D_z\bar\kappa, \kappa\rangle + \langle \bar\kappa, D_z\kappa\rangle\,, \tag{33a}$$

$$\mathrm{Im}(D_{\bar z}D_{\bar z}\kappa + \tfrac{1}{2}\bar c\kappa) = 0\,, \tag{33b}$$

$$R^D_{\bar z z}\xi = D_{\bar z}D_z\xi - D_zD_{\bar z}\xi = 2\langle \xi, \kappa\rangle\bar\kappa - 2\langle \xi, \bar\kappa\rangle\kappa\,. \tag{33c}$$

These equations can be given an invariant co-ordinate free meaning using the technology of Cartan connections [**6**]. However this is not our aim here, where we intend mostly to focus on classical surface theory in 3-space. In that case the conformal Ricci equation is vacuous and we get the following fundamental theorem of conformal surface theory:

THEOREM 3.1. *Let M be a Riemann surface and let $L = (K\bar{K})^{-1/2}$ denote the 1-density bundle, where K is the canonical bundle of M.*

If $f_1, f_2 : M \to S^3$ are two conformal immersions inducing the same Hopf differentials and Schwarzians, then there is a Möbius transformation $T : S^3 \to S^3$ with $Tf_1 = f_2$.

Conversely, let $\kappa\frac{dz^2}{|dz|} \in \Gamma(LK^2)$ be a quadratic differential with values in L and c a Schwarzian derivative, i.e., c transforms according to (10). If κ and c satisfy the conformal Gauss and Codazzi equations then there exists a conformal immersion $f : M \to S^3$ with Möbius holonomy, Hopf differential κ, and Schwarzian derivative c.

3.4. Applications to old and new results in conformal surface theory. At this stage it is worthwhile to pause and apply our conformally invariant setup to discuss some results of conformal surface theory. We content ourselves by demonstrating the basic ideas and will work locally, leaving the global discussion as an exercise to the interested reader.

The simplest question one can ask in any geometry is to characterize the homogeneous surfaces, i.e., surfaces which are 2-parameter orbits of the symmetry group under consideration. In Möbius geometry such surfaces are called the *cyclides of Dupin.* Clearly, such a surface must have Hopf differential and Schwarzian constant in the coordinates given by the 2-parameter group. Due to the transformation rule (21) the constancy of the Schwarzian has the invariant meaning that the Schwarzian is holomorphic (this amounts to the flatness of a certain normal Cartan connection on M [**6**]).

THEOREM 3.2. *Let $f : M \to S^3$ be a conformal immersion, not contained in a 2-sphere, whose Schwarzian is holomorphic. Then f is a cyclide of Dupin.*

PROOF. It suffices to show that κ and $S_w(f)$ are constant in an appropriate holomorphic coordinate w. Since c is holomorphic the conformal Gauss equation (33a) reads $(\kappa\bar{\kappa}^3)_z = 0$, expressing the fact that $\bar{\kappa}\kappa^3$ is holomorphic. By our assumption f is not contained in a 2-sphere so that κ is non-zero. Therefore $dw = (\bar{\kappa}\kappa^3)^{1/4}dz$ defines a new holomorphic coordinate w on M. Due to the transformation rule (23) for sections of LK^2 we see that $\kappa = (w_z)^{3/2}\bar{w}_z{}^{-1/2}\tilde{\kappa}$. Combining these two equations gives $\tilde{\kappa} = 1$ and the conformal Codazzi equation (33b) forces $\tilde{c}$ to be imaginary, and thus constant. □

A further basic question to ask is how many of the invariants of an immersed surface are actually needed to determine the surface up to symmetry. In the Euclidean setup this is the Bonnet problem: generically the induced metric and the mean curvature determine a surface up to rigid motions. In Möbius geometry one expects generically that the Hopf differential should determine a surface up to Möbius transformations.

However, let us recall the notion of *isothermic* surfaces [**1, 7, 10**]. Such surfaces are conformally parametrized by their curvature lines away from umbilic points. Put differently, away from umbilics there are holomorphic coordinates in which the Hopf differential κ is *real valued.* In this case the conformal Gauss and Codazzi equations simplify to

$$c_{\bar{z}} = 4(\kappa^2)_z\,, \quad \mathrm{Im}\,(\kappa_{\bar{z}\bar{z}} + \tfrac{1}{2}\bar{c}\kappa) = 0\,.$$

Assuming κ non-zero the two equations combine to give Calapso's equation [**7**]

$$\triangle \tfrac{\kappa_{xy}}{\kappa} + 8(\kappa^2)_{xy} = 0$$

where $z = x + iy$. The conformal Gauss and Codazzi equations are invariant under deformations of the Schwarzian by $c_r = c + r$ where $r \in \mathbb{R}$ is a real parameter, the spectral parameter of algebro-geometric integrable system theory. By Theorem 3.1 we thus see that an isothermic surface comes in an associated family parametrized by $\mathbb{R}$: these are the T-transforms of Calapso and Bianchi [**2, 7**]. Note that since all c_r are distinct the surfaces in the associated family are non-congruent with respect to the Möbius group. Therefore the Hopf differential alone does not determine isothermic surfaces. But these surfaces are the only exceptions:

THEOREM 3.3. *Let $f_1, f_2 : M \to S^3$ be two non-congruent conformal immersions inducing the same Hopf differentials. Then f_1 and f_2 are isothermic surfaces in the same associated family.*

PROOF. Since $\kappa_1 = \kappa_2$ the conformal Codazzi equation implies that the difference of the Schwarzians $cdz^2 := c_1 dz^2 - c_2 dz^2 \in H^0(K^2)$ is a holomorphic quadratic differential. Since the f_k are non-congruent, c is not identically zero, and so we can choose holomorphic coordinates such that $c = 1$. The conformal Codazzi equation now implies $\mathrm{Im}\,(\kappa_k) = 0$, so that both surfaces are isothermic and in the same associated family. □

We conclude this section with a brief discussion of (constrained) Willmore surfaces: it can be checked that a conformal immersion $f : M \to S^n$ is Willmore if and only if $\mathrm{Re}\,(D_{\bar{z}} D_{\bar{z}} \kappa + \frac{1}{2}\bar{c}\kappa) = 0$. Therefore the integrability conditions for a Willmore surface are

$$\begin{aligned} D_{\bar{z}} D_{\bar{z}} \kappa + \tfrac{1}{2}\bar{c}\kappa &= 0\,, \\ \tfrac{1}{2} c_{\bar{z}} &= 3\langle D_z \bar{\kappa}, \kappa\rangle + \langle \bar{\kappa}, D_z \kappa\rangle\,, \\ R^D_{\bar{z}z}\xi &= 2\langle \xi, \kappa\rangle \bar{\kappa} - 2\langle \xi, \bar{\kappa}\rangle \kappa\,. \end{aligned}$$

More generally, as we shall see below, a surface is *constrained Willmore*—that is, it extremizes the Willmore functional with respect to variations through conformal immersions—if

$$D_{\bar{z}} D_{\bar{z}} \kappa + \tfrac{1}{2}\bar{c}\kappa = \mathrm{Re}(\bar{q}\kappa) \tag{34}$$

for some holomorphic quadratic differential $qdz^2 \in H^0(K^2)$. These systems have the symmetry

$$\kappa_\lambda = \lambda\kappa, \qquad c_\lambda = c + (\lambda^2 - 1)q \qquad q_\lambda = \lambda^2 q \tag{35}$$

for unitary $\lambda \in S^1$, which describes the associated family of (constrained) Willmore surfaces in our setup. This spectral deformation is particularly simple for Willmore surfaces (q=0): here the Schwarzian is fixed and the scaled Hopf differential $\kappa \frac{dz^2}{|dz|} \in \Gamma(LK^2)$ is rotated by a phase. All this is reminiscent of the theory of constant mean curvature surfaces in 3-space where the associated family is obtained by rotating the Euclidean Hopf differential by a phase and fixing the induced metric. As we shall see, this similarity is not a coincidence.

A classical result of Thomsen [**24**] characterizes isothermic Willmore surfaces in 3-space as minimal surfaces in some space-form (conic section). More generally, Richter [**19**] proves that constant mean curvature surfaces in space forms are isothermic and constrained Willmore. These results are easy to see in our setup: from (12) we see that a surface has mean curvature H in the conic section S_{v_0} exactly when $H = -\langle v_0, N\rangle$ for N a unit length section of $V^\perp$ (in particular, a surface is minimal if V contains a constant section v_0). Moreover, the metric induced from S_{v_0} is given by

$$e^{2u}|dz|^2 = \frac{|dz|^2}{\langle \psi, v_0\rangle^2}. \tag{36}$$

Suppose then that f has constant mean curvature H in S_{v_0} with induced metric $e^{2u}|dz|^2$. We write

$$v_0 = a\psi + b\psi_z + \bar{b}\psi_{\bar{z}} + \hat{a}\hat{\psi} - HN \tag{37}$$

for real valued functions a, $\hat{a}$ and a complex valued function b. Writing out $v_{0,z}$ in terms of our frame and using (32), we see that the constancy of v_0 amounts to

$$b\kappa + 2\hat{a}\kappa_{\bar{z}} = 0 \tag{38a}$$

$$\hat{a}_z + \tfrac{1}{2}\bar{b} = 0 \tag{38b}$$

$$\bar{b}_z - c\hat{a} + 2H\kappa = 0 \tag{38c}$$

$$a + b_z - 2\hat{a}|\kappa|^2 = 0 \tag{38d}$$

$$a_z - \tfrac{c}{2}b - \bar{b}|\kappa|^2 - 2H\kappa_{\bar{z}} = 0. \tag{38e}$$

From (36) we have $\hat{a} = e^{-u}$ and then (38a) and (38b) yield $e^{-2u}(e^u\kappa)_{\bar{z}} = 0$ so that the Hopf differential $e^u\kappa dz^2$ is holomorphic. Thus, away from umbilics, we can choose the holomorphic coordinate so that $e^u\kappa = 1$ or, equivalently, $\hat{a} = \kappa$. In particular, κ is real and the surface is isothermic. Now (38b) gives $b = -2\kappa_{\bar{z}}$ while (38d) gives $a = 2(\kappa^3 + k_{z\bar{z}})$. This leaves (38c) which reads $\kappa_{\bar{z}\bar{z}} + \frac{\bar{c}}{2}\kappa = H\kappa$ so that ψ is Willmore if $H = 0$ and constrained Willmore otherwise, while, in view of the Gauss equation, (38e) is an identity.

Conversely, given a surface with κ real and

$$\kappa_{\bar{z}\bar{z}} + \tfrac{\bar{c}}{2}\kappa = H\kappa \tag{39}$$

for some constant H we define v_0 by (37) with the same choice of coefficients as above and conclude from (38) that v_0 is constant whence ψ gives a surface of constant mean curvature H in the space form S_{v_0} of constant curvature K given by

$$K = -\langle v_0, v_0\rangle = -H^2 + 4\kappa^4 + 4\kappa\kappa_{z\bar{z}} - 4|\kappa_z|^2 \tag{40}$$

(see [**16**] for a recent analysis of the soliton theory of this equation).

The isothermic spectral deformation $c_r = c + r$, $r \in \mathbb{R}$, preserves the class of constant mean curvature surfaces: in view of (39), we have $\kappa_{\bar{z}\bar{z}} + \frac{\bar{c_r}}{2}\kappa = (H + \frac{r}{2})\kappa$ so that the corresponding surface ψ_r has constant mean curvature $H_r = H + \frac{r}{2}$ in a space form of curvature K_r where, in view of (40), $K_r + H_r^2$ is independent of r. Moreover, all these surfaces are isometric with metric $\kappa^{-2}|dz|^2$. Thus our spectral deformation recovers the family of constant mean curvature surfaces related by the Lawson correspondence [**14**].

Moreover, the spectral symmetry (35) of constrained Willmore surfaces preserves the isothermic condition: if κ is real for a coordinate z then, by (23), κ_λ is real in the rotated coordinate $w = \sqrt{\lambda} z$.

For constant mean curvature surfaces, these two deformations can be combined to give an action of $\mathbb{C} \setminus \{0\}$ on such surfaces. Indeed, for a surface of constant mean curvature H in a space form of curvature K , we have seen that there is a coordinate z for which κ is real and $q = H$. For $\lambda \in \mathbb{C} \setminus \{0\}$, set

$$\kappa_\lambda = \tfrac{\lambda}{|\lambda|}\kappa \qquad c_\lambda = c + 2(\lambda - 1)q \qquad q_\lambda = \lambda q.$$

The fact that $\bar{q}\kappa$ is real ensures that each $\kappa_\lambda, c_\lambda, q_\lambda$ solves (33) and (34). Further, in the rotated coordinate $w = \sqrt{\tfrac{\lambda}{|\lambda|}} z$, κ_λ is real (in fact $\kappa_\lambda dz^2 = \kappa dw^2$) while $q_\lambda dz^2 = |\lambda| H dw^2$ so that, with respect to w, we have a solution of (39) with $H_\lambda = |\lambda| H$. From (40), we conclude that the surface ψ_λ has constant mean curvature $|\lambda| H$ and metric $\kappa^{-2}|dz|^2$ in a space form of curvature $K_\lambda = K + (1 - |\lambda|^2)H$. In particular, for $\lambda \in S^1$, we have a family of isometric surfaces in a fixed space form with fixed H and rotated Hopf differential—this is the classical associated family. For minimal surfaces, the action of $\mathbb{R} \setminus \{0\}$ is trivial but when $H \neq 0$ we get (part of) the Lawson deformation so that these two well-known symmetries are unified in a single $\mathbb{C} \setminus \{0\}$ action.

For M a torus, any holomorphic quadratic differential is of the form $q dz^2$ with q constant so we conclude with Richter [**19**] that *all* constrained Willmore isothermic tori have constant mean curvature in some space form. However, without this global hypothesis, there remains the possibility of isothermic constrained Willmore surfaces that are not spectral deformations of isothermic Willmore surfaces.

One might ask if a version of Thomsen's result holds in n-space. We restrict attention to isothermic Willmore surfaces of which obvious examples, in addition to the above mentioned minimal surfaces, are cylinders in $\mathbb{R}^4$ over elastic curves in $\mathbb{R}^3$, cones over elastic curves in S^3 and rotational surfaces about the plane at infinity of elastic curves in hyperbolic 3-space H^3. These surfaces are clearly isothermic and, by the principle of symmetric criticality, Willmore.

THEOREM 3.4. *Let $f : M \to S^n$ be an isothermic Willmore surface with $n \geq 3$. Then either f is minimal in 3-space, or an isothermic Willmore surfaces in 4-space described by an ODE: the equation describing elastica or a Painleve-type equation for the Hopf differential with radial symmetry.*

We begin by collecting the integrability equations for isothermic Willmore surfaces: since $\kappa = \bar{\kappa}$ is real, Ricci's equation immediately implies that the normal bundle is flat. We can therefore assume $V^\perp = M \times \mathbb{R}^{n-2}$ and D is just directional derivative. The two remaining integrability conditions, the Gauss and Codazzi equations, become

$$\kappa_{zz} + \tfrac{c}{2}\kappa = 0\,, \quad c_{\bar{z}} = 4\langle \kappa, \kappa \rangle_z\,,$$

for the $\mathbb{C}$-valued Schwarzian c and the $\mathbb{R}^{n-2}$-valued Hopf differential κ. Differentiating the Codazzi equation twice with respect to $\bar{z}$ and keeping reality conditions and the Gauss equation in mind, we obtain

$$\operatorname{Im} \langle \kappa, \kappa \rangle_z \kappa_{\bar{z}} = 0 . \tag{41}$$

First note that this is automatically satisfied if $n = 3$, in which case we already know that f is minimal. From now on we assume that $n \geq 4$. If κ is a nonzero constant, then Gauss' equation implies that $c = 0$ and by using the frame equations (32), one can check that f is the Clifford torus. The case $\kappa = 0$ results, as we already have discussed, in a holomorphic map f into the 2-sphere. Therefore we may assume that κ is non-constant. If κ has constant length $\langle \kappa, \kappa \rangle \neq 0$ then $\langle \kappa_z, \kappa \rangle = 0$, and differentiating yields

$$0 = \langle \kappa_{zz}, \kappa \rangle + \langle \kappa_z, \kappa_z \rangle = -\tfrac{c}{2} \langle \kappa, \kappa \rangle + \langle \kappa_z, \kappa_z \rangle .$$

From the Gauss and Codazzi equations we see that c is holomorphic and that $\langle \kappa_z, \kappa_z \rangle$ is anti-holomorphic. Therefore both are constant. After a rotation of the coordinate we may assume that c is a real constant. Then κ is contained in a rotated $\mathbb{R}^{n-2}$ in $\mathbb{C}^{n-2}$ and has constant length. Decomposing the complex Codazzi equation according to the splitting $\mathbb{C}^{n-2} = \mathbb{R}^{n-2} \oplus i\mathbb{R}^{n-2}$, we get

$$\kappa_{xx} - \kappa_{yy} + 2c\kappa = 0 , \qquad \text{and} \qquad \kappa_{xy} = 0 .$$

The latter implies $\kappa(x, y) = k_1(x) + k_2(y)$ which, inserted into the former, gives

$$f''(x) + 2cf(x) = g''(y) - 2cg(y) = \text{constant} .$$

So k_1 and k_2 solve linear ordinary differential equations which can be explicitly integrated. If $c = 0$ only constant κ's are solutions of constant length. If $c > 0$ then any constant length solution is of the form

$$\kappa(x, y) = \cos(\sqrt{2c}\, x)\kappa_1 + \sin(\sqrt{2c}\, x)\kappa_2$$

for κ_1, $\kappa_2 \in \mathbb{R}^{n-2}$ two orthogonal vectors of the same length. The corresponding surfaces are the rotational surfaces, cylinders and cones over free elastic helixes in the respective 4-dimensional space forms.

We may assume now that the length of κ is non-constant. In this case (41) is equivalent to

$$\kappa = k(g) \qquad \text{with} \qquad g = \langle \kappa, \kappa \rangle ,$$

for some $\mathbb{R}^{n-2}$-valued function k of one variable. Inserting into the Codazzi equation yields the second order differential equation

$$g_z^2 k'' + g_{zz} k' + \tfrac{c}{2} k = 0$$

for the function k. This shows that the image of κ is contained in an at most 2-dimensional parallel subbundle of $E \subset V^{\perp}$. Using the fourth frame equation in (32), we deduce that the complementary subbundle $E^{\perp} \subset V^{\perp}$ is a constant subspace in the ambient space. We therefore conclude that f takes values in S^4.

Assuming from now on that our isothermic Willmore surface f is fully contained in 4-space, its trivial normal bundle has a well-defined 90-degree rotation J. Since $\langle \kappa, J\kappa \rangle = 0$, the Codazzi equation gives $\langle \kappa_z, J\kappa \rangle_z = 0$. From this we deduce that

$$\langle \kappa_z, J\kappa \rangle = \bar{h}^{-1} , \qquad h_{\bar{z}} = 0$$

for a holomorphic function h. Note that h is nonzero, otherwise κ would have values in a 1-dimensional parallel subbundle of the normal space and thus f would not be full in 4-space. Since $\kappa = k(g)$ we get

$$\tfrac{h}{|h|^2} = \langle \kappa_z, J\kappa \rangle = \langle k'(g), Jk(g) \rangle g_z$$

implying that h is a real multiple of $g_z = \langle \kappa, \kappa \rangle_z$. Multiplying (41) by this real multiple gives $\operatorname{Im} h\kappa_{\bar z} = 0$. But this is equivalent to κ depending only on $\operatorname{Re} w$, where $w_z = h$ is a new holomorphic coordinate. We may thus assume that $\kappa = k(w + \bar w)$ for some $\mathbb{R}^2$-valued function of one variable, which we again call k. Inserting back into the Codazzi equation yields

$$h^2 k'' + h_z k' + \tfrac{c}{2} k = 0 \qquad \text{or} \qquad k'' + \tfrac{h_z}{h^2} k' + \tfrac{c}{2h^2} k = 0\,.$$

Now f is assumed to be full in 4-space so that k and k' are linearly independent. Since k is real, we conclude that

$$h_z = \alpha h^2\,, \qquad \alpha \in \mathbb{R}$$

and then that

$$\frac{c}{2h^2} = \rho(w + \bar w)\,. \tag{42}$$

The first relation tells us that $h = \frac{1}{az+b}$ is a special fractional linear coordinate change. Taking the $\bar z$ derivative of the second relation and applying the Gauss equation gives

$$\rho' = \frac{2\langle k, k\rangle'}{|h|^2}\,.$$

We have two cases to discuss: first, if h is constant the last relation can be integrated to

$$\rho = 2\frac{\langle k, k\rangle + \gamma}{|h|^2}\,, \qquad \gamma \in \mathbb{R}\,,\ h \in \mathbb{C}\,.$$

This yields the formula

$$c = 2\frac{h}{\bar h}(\langle k, k\rangle + \gamma) \tag{43}$$

for the Schwarzian of our surface f. Inserting back into the Codazzi equation gives the second order ODE for spatial elastica

$$k'' + \frac{h}{\bar h}(\langle k, k\rangle + \gamma)k = 0 \tag{44}$$

which the Hopf differential $\kappa = k(hz + \bar h \bar z)$ has to satisfy. In the second case the fractional linear coordinate change is, after translation and real scaling, of the form $h(z) = \frac{1}{z}$. Then $w + \bar w = 2 \ln |z|$ and

$$\rho' = 2\langle k, k\rangle |z|^2\,, \qquad k'' - k + \rho k = 0\,,$$

is an ODE for the Hopf differential k depending on the radius only. This case is similar to the intrinsic rotational constant mean curvature surfaces [**3, 22, 25**].

Finally, solving the respective ODE's determines the Hopf differential, and the Schwarzian is computed from (42). Inserting the invariants into the frame equations (32) allows us to determine the surface f. We leave the detailed calculations to the interested reader.

4. The Novikov–Veselov and Davey–Stewartson flows

Over the last 15 years a number of papers have been written on *the soliton theory of surfaces*. In all cases only special surface classes, like constant (mean) curvature, isothermic or Willmore surfaces, were considered. When the underlying surface is a 2-torus it has been shown that those surfaces allow an infinite hierarchy of commuting flows. Surfaces stationary under a finite number of flows could be explicitly constructed by noticing that those flows are given by the linear flows on the Jacobian of some auxiliary Riemann surface, the so-called *spectral curve*.

More recently, Konopelchenko and his collaborators [**11, 12, 13, 23**] have shown how to construct hierarchies of commuting flows on the space of *all* conformally immersed tori in 3 and 4-dimensional spheres based on the modified Novikov–Veselov and Davey–Stewartson hierarchies.

Following our construction of the KdV-hierarchy, we are now going to give a manifestly conformally invariant construction of such flows and show how the special surface classes mentioned above are invariant under them.

4.1. Deformations of conformal immersions. Let $M = T^2$ be a 2-torus and $X \in H^0(TM)$ a holomorphic vector field so that $\frac{\partial}{\partial z} = \frac{1}{2}(X - iJX)$ for a holomorphic coordinate z on the universal cover $\mathbb{R}^2$. Given a deformation $f(t): M \to S^n$ of conformal immersions with $f = f(0)$, we get the unique normalized lifts $\psi(t): M \to \mathcal{L}_+$ from (14) satisfying $|d\psi(t)|^2 = |dz|^2$. Therefore we have the canonical frame (18) of $V(t)$ for all times t. The deformation of this frame is determined by the variation

$$\psi_t = a\psi + b\psi_z + \bar{b}\psi_{\bar{z}} + \sigma \tag{45}$$

of the normalized lift ψ. Here $\sigma(t) \in \Gamma(V(t)^\perp)$ is the normal variation, the complex valued function $b(t): M \to \mathbb{C}$ determines the tangential and $a(t): M \to \mathbb{R}$ the radial variation. Since ψ is null, there is no $\hat{\psi}$ component. For the deformation $f(t)$ to remain conformal, i.e.

$$|d\psi|_t^2 = 0\,, \tag{46}$$

the radial and tangential deformations a and b have to be determined by the normal deformation σ. To compute those relations, we need

$$\begin{aligned}(\psi_z)_t &= (\psi_t)_z \\ &= (a_z - \tfrac{c}{2}b - \bar{b}|\kappa|^2 + 2\langle\sigma, D_{\bar{z}}\kappa\rangle)\psi + (a + b_z)\psi_z \\ &\quad + (\bar{b}_z - 2\langle\sigma,\kappa\rangle)\psi_{\bar{z}} + \tfrac{\bar{b}}{2}\hat{\psi} + (b\kappa + D_z\sigma)\end{aligned} \tag{47}$$

where we have used the frame equations (32). Inserting into (46) gives

$$\bar{b}_z = 2\langle\sigma,\kappa\rangle \qquad \text{and} \qquad a = -\mathrm{Re}\, b_z\,, \tag{48}$$

of which the first is a $\bar{\partial}$-problem solvable over a torus if and only if the right hand side has vanishing integral $\int_M \langle\sigma,\kappa\rangle = 0$.

Using (48) we can eliminate a in (47) and rewrite

$$(\psi_z)_t = (-\tfrac{1}{2}b_{zz} - \tfrac{c}{2}b - \bar{b}|\kappa|^2 + \langle\sigma, D_{\bar{z}}\kappa\rangle - \langle D_{\bar{z}}\sigma,\kappa\rangle)\psi + (\mathrm{Im}\, b_z)\psi_z + \tfrac{\bar{b}}{2}\hat{\psi} + (b\kappa + D_z\sigma)\,. \tag{49}$$

(Here $\mathrm{Im} b_z = \frac{1}{2}(b_z - \bar{b_z})$ and so is pure imaginary.)

We now can calculate the remaining deformations. First note that

$$\hat{\psi}_t = (\mathrm{Re}\, b_z)\hat{\psi} - 4\mathrm{Re}\,(\langle\hat{\psi}, \psi_{\bar{z}t}\rangle\psi_z) + \tau \tag{50}$$

where, using (49), we have

$$2\langle\hat{\psi}, \psi_{zt}\rangle = b_{zz} + cb + 2\bar{b}|\kappa|^2 - 2\langle\sigma, D_{\bar{z}}\kappa\rangle + 2\langle D_{\bar{z}}\sigma, \kappa\rangle\,.$$

Moreover, the normal variation τ is obtained from (32),

$$\tau = (\hat{\psi}_t)^\perp = 2((\psi_{zt\bar{z}})^\perp + |\kappa|^2\sigma)\,,$$

which unravels to

$$\tfrac{\tau}{2} = D_{\bar{z}}D_z\sigma + 2\langle\sigma, \bar{\kappa}\rangle\kappa + |\kappa|^2\sigma + 2\mathrm{Re}\,(bD_{\bar{z}}\kappa)\,.$$

Note that the latter is indeed real due to the Ricci equation (33c). Finally, if ξ is normal then its tangential deformation calculates to

$$(\xi_t)^T = \langle\xi, \tau\rangle\psi + \langle\xi, \sigma\rangle\hat{\psi} - 4\mathrm{Re}\,(\langle\xi, b\kappa + D_z\sigma\rangle\psi_{\bar{z}})\,. \tag{51}$$

To calculate the deformations of the Hopf differential κ and the Schwarzian c, we use Hill's equation (20) and obtain

$$\kappa_t = \psi_{tzz} + \tfrac{c_t}{2}\psi + \tfrac{c}{2}\psi_t\,.$$

Taking the $V^\perp$ part, together with (49) and (32), gives

$$D_t\kappa = D_zD_z\sigma + \tfrac{c}{2}\sigma + (\mathrm{Im} b_z + b_z)\kappa + bD_z\kappa + \bar{b}D_{\bar{z}}\kappa\,. \tag{52}$$

Similarly, using (50) and taking an inner product with $\hat{\psi}$ we get

$$\begin{aligned}\tfrac{1}{2}c_t =& \tfrac{1}{2}b_{zzz} + cb_z + \tfrac{1}{2}(bc_z + \bar{b}c_{\bar{z}}) + 8\langle\sigma, \bar{\kappa}\rangle\langle\kappa, \kappa\rangle + \\ & 3\langle D_{\bar{z}}D_z\sigma, \kappa\rangle - \langle\sigma, D_{\bar{z}}D_z\kappa\rangle - 3\langle D_z\sigma, D_{\bar{z}}\kappa\rangle + \langle D_{\bar{z}}\sigma, D_z\kappa\rangle\,.\end{aligned} \tag{53}$$

For later use, we need to record the deformation of the Willmore energy (28). Up to exact forms

$$\tfrac{1}{2}|\kappa|^2_t = \mathrm{Re}\langle D_t\kappa, \bar{\kappa}\rangle \equiv \mathrm{Re}\langle D_zD_z\sigma + \tfrac{c}{2}\sigma, \bar{\kappa}\rangle \equiv \langle\sigma, D_zD_z\bar{\kappa} + \tfrac{c}{2}\bar{\kappa}\rangle\,,$$

where we have used (52) together with the product rule. Note that due to Codazzi's equation (33b) the last expression is real. Therefore the Willmore energy deforms by

$$W_t = 2\int\langle\sigma, D_zD_z\bar{\kappa} + \tfrac{c}{2}\bar{\kappa}\rangle\,. \tag{54}$$

Note that $W_t = 0$ for compactly supported variations if and only if

$$D_{\bar{z}}D_{\bar{z}}\kappa + \tfrac{1}{2}\bar{c}\kappa = \mathrm{Re}(\bar{q}\kappa)$$

for some holomorphic quadratic differential qdz^2. This follows from the constraint $\int\langle\sigma, \kappa\rangle = 0$ on the normal variations (48), expressing the fact that we deform via *conformal* immersions $f(t) : M \to S^n$ rather then *all* immersions. The resulting critical points are the *constrained Willmore* immersions. Every Willmore immersion, $q = 0$, clearly is constrained Willmore.

4.2. The KdV flows revisited. As a warm up, we make contact with our discussion of the KdV flows earlier on. We already have indicated that holomorphic maps into $\mathbb{CP}^1$ are characterized by vanishing Hopf differential $\kappa \equiv 0$. In this case the $\bar\partial$-problem (48) has the holomorphic solution $b = c$. Inserting into the variation of the Schwarzian (53), we immediately obtain the KdV equation

$$c_t = c_{zzz} + 3c_z c\,.$$

4.3. The Novikov–Veselov flows. Adapting our recursive scheme in the construction of the KdV hierarchy in section 2.2 to deformations of conformal immersions, we outline the first steps in the construction of a commuting hierarchy of odd-order flows on conformally immersed tori. For these flows to restrict to the known flows on, say, Willmore surfaces, they should at least preserve the Willmore energy (28). In contrast to the existing literature [**23**], where a hierarchy of flows for surfaces in 3-space is described in terms of Euclidean data, our flows are Möbius invariant from the onset.

Any conformal deformation $f(t) : M \to S^n$ is determined, up to the solution of a $\bar\partial$-problem (48), by the normal variation (45). The most naive flow, which we will label as the first flow, has no normal variation, i.e., $\sigma = 0$, and thus $b = 1$. Then (52), (53) give

$$D_{t_1}\kappa = (D_z + D_{\bar z})\kappa \qquad \text{and} \qquad c_{t_1} = \tfrac{1}{2}(c_z + c_{\bar z})\,,$$

which is just translational flow in X direction.

Our previous scheme—the new normal deformation is given by the old variation of the invariants—suggests $\sigma = D_{t_1}\kappa$ for the construction of the next flow. This is certainly reasonable, since both κ and σ are normal bundle valued. Since σ has to be *real* and the resulting flow should preserve the Willmore energy (28),

$$\sigma = \operatorname{Re} D_z\kappa \tag{55}$$

turns out to be the right choice for the third flow. Let us first check that we can solve the $\bar\partial$-problem $\bar b_z = 2\langle\sigma, \kappa\rangle$: modulo exact forms, we obtain

$$2\langle \operatorname{Re} D_z\kappa, \kappa\rangle \equiv \langle D_{\bar z}\bar\kappa, \kappa\rangle \equiv -\langle \bar\kappa, D_{\bar z}\kappa\rangle \equiv 0\,,$$

where the last identity uses the Gauss equation (33a). Thus we can solve for b, and hence for a, in (48) to get the variation (45) of ψ and f. The solution b of the $\bar\partial$-problem on the torus M is determined only up to a constant for each time t. To obtain a well defined flow in t, we normalize b so that $\int b = 0$.

To see whether the Willmore energy is preserved, we calculate (54) the expression

$$\langle \operatorname{Re} D_z\kappa, D_{\bar z}D_{\bar z}\kappa + \tfrac{1}{2}\bar c\kappa\rangle = \operatorname{Re}\langle D_z\kappa, D_{\bar z}D_{\bar z}\kappa + \tfrac{1}{2}\bar c\kappa\rangle\,, \tag{56}$$

where we note that $D_{\bar z}D_{\bar z}\kappa + \frac{1}{2}\bar c\kappa$ is real by Codazzi's equation (33b). Up to exact forms

$$\begin{aligned}
-\langle D_z\kappa, D_{\bar z}D_{\bar z}\kappa + \tfrac{1}{2}\bar c\kappa\rangle &\equiv \langle D_{\bar z}D_z\kappa, D_{\bar z}\kappa\rangle + \tfrac{1}{4}\bar c_z\langle\kappa,\kappa\rangle \\
&\equiv \langle R^D_{\bar z z}\kappa, D_{\bar z}\kappa\rangle + \big(\tfrac{3}{2}\langle D_{\bar z}\kappa, \bar\kappa\rangle + \tfrac{1}{2}\langle\kappa, D_{\bar z}\bar\kappa\rangle\big)\langle\kappa,\kappa\rangle \\
&\equiv \tfrac{3}{2}\langle R^D_{\bar z z}\kappa, D_{\bar z}\kappa\rangle\,.
\end{aligned}$$

Here we have used Gauss' equation (33a), the Ricci equation (33c) and the formula

$$\langle\kappa,\kappa\rangle\langle\kappa,\bar\kappa\rangle_{\bar z} = -\langle R^D_{\bar z z}\kappa, D_{\bar z}\kappa\rangle + (3\langle D_{\bar z}\kappa,\bar\kappa\rangle + \langle\kappa, D_{\bar z}\bar\kappa\rangle)\langle\kappa,\kappa\rangle\,.$$

Therefore we conclude that our third order flow, which we will call the *Novikov–Veselov flow*, preserves the Willmore energy if flatness of the normal bundle is preserved. This is certainly the case for surfaces in 3-space and also, as we will see later on, for isothermic surfaces in any codimension.

At present we are unable to say more about the general codimension case for the Novikov–Veselov flow. This is mainly due to our low technology setup, which makes any discussion of the normal bundle flow a notational debauch.

There is still much to do here: first we must see that our flow coincides with that defined in Euclidean terms in the existing literature [**13, 23**]. Further, a recursive scheme for constructing higher flows is required. For $n = 3$, the first question is answered in the affirmative by Richter [**19**] who proposes also $\sigma_{2n+1} = \mathrm{Re}(D_z)^n\kappa$ for the higher flows. We shall return to these matters elsewhere.

4.4. The Davey–Stewartson flows. In the previous section we indicated how to construct a hierarchy of Möbius invariant odd order flows on immersed tori in n-space. To obtain even order flows one needs more structure in the normal bundle, which is available for surfaces in 4-space. There, the normal bundle has a complex structure J which we can use to define a zero order flow, namely rotation in the normal bundle by J. The corresponding deformation of the Hopf differential is given by $D_{t_0}\kappa = J\kappa$ and we take $\sigma = \mathrm{Re}\, J\kappa$ as our normal deformation for the construction of the second order flow. To obtain the conformal deformation $f(t) : M \to S^4$, we have to solve for the tangential variation b in (45). As before, this amounts to solving the $\bar\partial$-problem $\bar b_z = 2\langle\sigma,\kappa\rangle$. Now

$$2\langle\sigma,\kappa\rangle = \langle J\kappa,\kappa\rangle + \langle J\bar\kappa,\kappa\rangle = \langle J\bar\kappa,\kappa\rangle$$

since J is skew. But Ricci's equation (33c) implies

$$\int\langle J\bar\kappa,\kappa\rangle = \pi\deg N_f M\,,$$

so that the $\bar\partial$-problem (48) is solvable if and only if the normal bundle of f has degree zero. This condition is preserved under any continuous deformation, and we obtain a unique second order flow by normalizing $\int b = 0$.

This flow also preserves the Willmore energy (54): up to exact forms

$$\langle\mathrm{Re}\, J\kappa, D_{\bar z}D_{\bar z}\kappa + \tfrac{1}{2}\bar c\kappa\rangle = \mathrm{Re}\langle J\kappa, D_{\bar z}D_{\bar z}\kappa\rangle \equiv \mathrm{Re}\langle D_{\bar z}\kappa, JD_{\bar z}\kappa\rangle = 0\,,$$

where we have used the fact that $DJ = 0$ and J is skew.

Again there is more to be done, both in making contact with the existing literature [**11, 12**] and constructing higher flows.

4.5. Flows on isothermic surfaces. As an explicit example of the above discussions, we apply the Novikov–Veselov and Davey–Stewartson flow to isothermic surfaces in S^n [**5**]. Recall that $f : M \to S^n$ is isothermic if the Hopf differential is real, i.e., if $\kappa = \bar\kappa$. Note that in this case Ricci's equation (33c) immediately implies the flatness of the normal bundle. Thus we can define a "vector" version of the Novikov–Veselov flow and the Davey–Stewartson flow in 4-space on isothermic surfaces, both preserving the Willmore energy, once we have shown that the isothermic condition is preserved under the flows.

First we note that in the isothermic case the $\bar{\partial}$-problem can be explicitly solved for both flows. Since κ is real, the integrability equations for an isothermic surface become

$$c_{\bar{z}} = 4\langle \kappa, \kappa \rangle_z\,, \quad \mathrm{Im}\,(D_{\bar{z}} D_{\bar{z}} \kappa + \tfrac{1}{2}\bar{c}\kappa) = 0\,, \quad R^D = 0\,.$$

As discussed in section 4.3 the Novikov–Veselov flow is (up to the factor of 2) given by the normal variation $\sigma = 2\mathrm{Re}\, D_z \kappa$. Therefore

$$\bar{b}_z = 2\langle \sigma, \kappa \rangle = \langle \kappa, \kappa \rangle_z + \langle \kappa, \kappa \rangle_{\bar{z}} = (\langle \kappa, \kappa \rangle + \tfrac{\bar{c}}{4})_z$$

and we obtain the explicit solution $b = \langle \kappa, \kappa \rangle + \frac{c}{4}$ of the $\bar{\partial}$-problem (48). To see that the Novikov–Veselov flow preserves isothermic surfaces, we need to check whether $\kappa = \bar{\kappa}$ is preserved under the flow. Recalling (52) and the fact that κ is real, we immediately calculate

$$\begin{aligned}
\mathrm{Im}(D_{t_3}\kappa) &= \mathrm{Im}(D_z D_z D_z \kappa + D_z D_z D_{\bar{z}} \kappa + \tfrac{c}{2}(D_z + D_{\bar{z}})\kappa + 2b_z \kappa) \\
&= (D_z + D_{\bar{z}})\mathrm{Im}(D_z D_z \kappa + \tfrac{c}{2}\kappa) - \tfrac{1}{2}\mathrm{Im}(c_{\bar{z}} - 4\langle \kappa, \kappa \rangle_z)\kappa \\
&= 0\,,
\end{aligned}$$

where we have used the integrability equations. Therefore isothermic surfaces are preserved under the Novikov–Veselov flow and

$$D_{t_3}\kappa = \mathrm{Re}(2D_z D_z D_z \kappa + \tfrac{3}{2} c D_z \kappa + \tfrac{3}{4} c_z \kappa + 2(\langle \kappa, \kappa \rangle D_z \kappa - \langle \kappa, D_z \kappa \rangle \kappa))\,,$$

which is a modified vector version of the usual Novikov–Veselov equation found in the literature [**23**]. Notice that the last term $\langle \kappa, \kappa \rangle D_z \kappa - \langle \kappa, D_z \kappa \rangle \kappa$ vanishes for surfaces in 3-space.

A similar, but simpler, computation shows that the Davey–Stewartson flow preserves isothermic surfaces in 4-space. In this case the normal deformation is $\sigma = \mathrm{Re}\, J\kappa = J\kappa$ since κ is real. The $\bar{\partial}$-problem $\bar{b}_z = 2\langle \sigma, \kappa \rangle = 2\langle J\kappa, \kappa \rangle = 0$ is solved by $b = 0$. Using (52) again we calculate $\mathrm{Im} D_{t_2}\kappa = \mathrm{Im}(D_z D_z J\kappa + \frac{c}{2} J\kappa) = 0$. This gives $D_{t_2}\kappa = J(D_z D_z \kappa + \frac{c}{2}\kappa)$ for the Davey–Stewartson flow on κ for isothermic surfaces in 4-space.

References

[1] L. Bianchi, *Ricerche sulle superficie isoterme e sulla deformazione delle quadriche*, Ann. di Mat. **11** (1905), 93–157.

[2] ———, *Complementi alle ricerche sulle superficie isoterme*, Ann. di Mat. **12** (1905), 19–54.

[3] A. Bobenko and A. Its, *The Painlevé III equation and the Iwasawa decomposition*, Manuscripta Math. **87** (1995), 369–377.

[4] C. Bohle, *Quaternionic projective structures and the Davey–Stewartson flow on surfaces*, in preparation.

[5] F.E. Burstall, *Isothermic surfaces: conformal geometry, Clifford algebras and integrable systems*, preprint math.DG/0003096, 2000.

[6] F.E. Burstall and D. Calderbank, *Conformal submanifold geometry*, in preparation.

[7] P. Calapso, *Sulle superficie a linee di curvatura isoterme*, Rendiconti Circolo Matematico di Palermo **17** (1903), 275–286.

[8] D.M.J. Calderbank, *Möbius structures and two dimensional Einstein–Weyl geometry*, J. reine angew. Math. **504** (1998), 37–53.

[9] G. Darboux, *Leçons sur la théorie générale des surfaces et les applications géomtriques du calcul infinitésimal. Parts 1 and 2*, Gauthier-Villars, Paris, 1887.
[10] ———, *Sur les surfaces isothermiques*, C.R. Acad. Sci. Paris **128** (1899), 1299–1305.
[11] B. G. Konopelchenko, *Weierstrass representations for surfaces in 4D spaces and their integrable deformations via DS hierarchy*, Ann. Glob. Anal. Geom. **18** (2000), 61–74.
[12] B.G. Konopelchenko and G. Landolfi, *Induced surfaces and their integrable dynamics. II. Generalized Weierstrass representations in 4D spaces and deformations via DS hierarchy*, Stud. Appl. Math. **104** (2000), 129–169.
[13] B.G. Konopelchenko and I.A. Taimanov, *Generalized Weierstrass formulae, soliton equations and Willmore surfaces. I. Tori of revolution and the mKdV equation*, Preprint dg-ga/9506011, 1995.
[14] H.B. Lawson, *Complete minimal surfaces in S^3*, Ann. of Math. **92** (1970), 335–374.
[15] D. Mumford, *Tata lectures on theta. III*, Birkhäuser Boston Inc., Boston, MA, 1991.
[16] E. Musso and L. Nicolodi, *Special isothermic surfaces and solitons*, Preprint math.DG/0108170, 2001.
[17] U. Pinkall, *Hamiltonian flows on the space of star-shaped curves*, Results Math. **27** (1995), 328–332.
[18] U. Pinkall and I. Sterling, *On the classification of constant mean curvature tori*, Ann. of Math. **130** (1989), 407–451.
[19] J. Richter, *Conformal maps of a Riemannian surface onto the space of quaternions*, Berlin thesis, 1997.
[20] B.A. Rozenfel'd, *Conformal differential geometry of families of C_m in C_n*, Mat. Sbornik N.S. **23(65)** (1948), 297–313.
[21] G. Segal, *The geometry of the KdV equation*, Int. J. Modern Phys. A **6** (1991), 2859–2869.
[22] B. Smyth, *A generalization of a theorem of Delaunay on constant mean curvature surfaces*, Statistical thermodynamics and differential geometry of microstructured materials (Minneapolis, MN, 1991), Springer, New York, 1993, pp. 123–130.
[23] I.A. Taimanov, *Modified Novikov-Veselov equation and differential geometry of surfaces*, Amer. Math. Soc. Transl. (2) **179** (1997), 133–151.
[24] G. Thomsen, *Ueber konforme Geometrie I: Grundlagen der konformen Flaechentheorie*, Hamb. Math. Abh. **3** (1923), 31–56.
[25] M. Timmreck, U. Pinkall, and D. Ferus, *Constant mean curvature planes with inner rotational symmetry in Euclidean 3-space*, Math. Z. **215** (1994), 561–568.

Department of Mathematical Sciences, University of Bath, Bath BA2 7AY, UK

E-mail address: `feb@math.bath.ac.uk`

Department of Mathematics, University of Massachusetts, Amherst, MA 01003, USA

E-mail address: `franz@gang.umass.edu`

Fachbereich Mathematik, MA 8-5, Technische Universität Berlin, Strasse des 17. Juni 135, 10633 Berlin, GERMANY

E-mail address: `pinkall@sfb288.math.tu-berlin.de`

Contemporary Mathematics
Volume **308**, 2002

Anti-Self-Dual Metrics on Lie Groups

Vivian De Smedt and Simon Salamon

Introduction

Let G be a Lie group of dimension n equipped with a left-invariant Riemannian metric g. The curvature tensor $R = R(g)$ is completely determined by its value at the identity element of G. In this way, R is an element of the vector space

$$\mathcal{R} = S^2(\Lambda^2\mathfrak{g}^*) \ominus \Lambda^4\mathfrak{g}^*,$$

where $\mathfrak{g}$ is the Lie algebra of G and $A \ominus B$ denotes the orthogonal complement of B in A. Moreover, there is an $O(n)$-invariant decomposition

$$\mathcal{R} = \mathcal{V} \oplus \mathcal{W},$$

where $\mathcal{V} \cong S^2\mathfrak{g}^*$. The component Ric of R in $\mathcal{V}$ is the Ricci tensor, and the component W in $\mathcal{W}$ is the Weyl tensor. If $W = 0$, then M is conformally flat in the sense that there exist local coordinates x^i for which g is a scalar function times $\sum_{i=1}^n dx^i \otimes dx^i$. For example, the compact Lie group $U(2)$ (double covered by $S^1 \times S^3$) acquires such coordinates from its description as a discrete quotient of $\mathbb{R}^4 \setminus \{0\}$.

If G is 4-dimensional and oriented, the Hodge involution $*\colon \Lambda^2\mathfrak{g}^* \to \Lambda^2\mathfrak{g}^*$ gives rise to a decompostion

$$W = W^+ + W^-,$$

where $W^\pm \in S_0^2(\Lambda^\pm\mathfrak{g}^*)$ and $\Lambda^\pm\mathfrak{g}^*$ is the $\pm$-eigenspace of $*$. If $W^+ = 0$ the metric g is said to be *anti-self-dual*, or 'ASD' for short. In this paper, we classify left-invariant metrics on 4-dimensional Lie groups with $W^+ = 0$ and $W^- \neq 0$. Such a group is necessarily solvable, and in §3 we describe the general form of its Lie algebra. We prove that the hypotheses eliminate all but a 1-parameter family of Lie algebras, and that there are essentially only two distinct left-invariant metrics that are anti-self-dual.

The proof of the main theorem (**1.6** below) is accomplished by first eliminating Lie algebras that admit an orientation-reversing automorphism. A key feature of the method is that the Gram-Schmidt process is compatible with the triangular nature

2000 *Mathematics Subject Classification.* Primary 53C26; Secondary 53C55, 22E25.

of the nilpotent Lie algebra structure constants. This enables one to compute the structure constants for an orthonormal basis relative to an arbitrary inner product. Once freedom in the choice of basis has been removed, the equation $W^+ = 0$ is solved by means of a specially adapted Maple program reproduced in the Appendix. No doubt further effort would eliminate the need for a computer analysis, but this aspect of the work may have independent interest.

Acknowledgments. This paper was begun during a visit by the first author in Oxford in 1997, in parallel with the work [**1**] that establishes similar results independently. The second author is grateful for the encouragement received at the Tokyo conference to finally present the material in written form.

1. Summary of results

Let V denote a real 4-dimensional vector space, with basis $(\mathbf{f}_1, \mathbf{f}_2, \mathbf{f}_3, \mathbf{f}_4)$. For $\lambda \in \mathbb{R}$, let $\mathfrak{g}_\lambda$ denote the Lie algebra defined on V by the relations

$$\begin{aligned} [\mathbf{f}_1, \mathbf{f}_2] &= \mathbf{f}_2 - \lambda \mathbf{f}_3, \\ [\mathbf{f}_1, \mathbf{f}_3] &= \lambda \mathbf{f}_2 + \mathbf{f}_3, \\ [\mathbf{f}_1, \mathbf{f}_4] &= 2\mathbf{f}_4, \\ [\mathbf{f}_2, \mathbf{f}_3] &= -\mathbf{f}_4. \end{aligned} \tag{1.1}$$

The Jacobi identity is easily verified. The Lie algebra $\mathfrak{g}_\lambda$ is solvable and $\mathfrak{g}'_\lambda = [\mathfrak{g}_\lambda, \mathfrak{g}_\lambda]$ is isomorphic to the 3-dimensional Heisenberg algebra. Changing the signs of $\mathbf{f}_3, \mathbf{f}_4$ is equivalent to replacing λ by $-\lambda$, though we shall see that $\mathfrak{g}_\lambda$, $\mathfrak{g}_{\lambda'}$ are not isomorphic if $|\lambda| \neq |\lambda'|$.

Later on, it will be convenient to express the relations (1.1) in terms of differential forms. To this end, let $(\mathbf{f}^i)$ denote the dual basis of $\mathfrak{g}_\lambda^*$. With the convention that $d\mathbf{f}^i(u,v) = -\mathbf{f}^i[u,v]$, (1.1) asserts that

$$\begin{aligned} d\mathbf{f}^1 &= 0, \\ d\mathbf{f}^2 &= -\mathbf{f}^{12} - \lambda \mathbf{f}^{13}, \\ d\mathbf{f}^3 &= \lambda \mathbf{f}^{12} - \mathbf{f}^{13}, \\ d\mathbf{f}^4 &= -2\mathbf{f}^{14} + \mathbf{f}^{23}. \end{aligned} \tag{1.2}$$

Let G_λ denote the simply-connected Lie group corresponding to $\mathfrak{g}_\lambda$. We shall not identify the group structure, but (1.2) can be integrated to yield coordinates t, x, y, u on G_λ relative to which

$$\begin{aligned} \mathbf{f}^1 &= dt, \\ \mathbf{f}^2 &= e^{-t}(dx + \lambda y dt), \\ \mathbf{f}^3 &= e^{-t}(dy - \lambda x dt), \\ \mathbf{f}^4 &= e^{-2t}(du + \tfrac{1}{2}(x dy - y dx) - \tfrac{1}{2}\lambda(x^2 + y^2) dt). \end{aligned} \tag{1.3}$$

For $k > 0$, let $\langle\, .\, , .\, \rangle_k$ denote the inner product on V defined by

$$\langle \mathbf{f}_i, \mathbf{f}_j \rangle = \begin{cases} k^2, & i = j = 1, \\ 1, & 2 \leqslant i = j \leqslant 4, \\ 0, & i \neq j. \end{cases} \tag{1.4}$$

This inner product induces a left-invariant metric

$$g_k = k^2\mathbf{f}^1 \otimes \mathbf{f}^1 + \mathbf{f}^2 \otimes \mathbf{f}^2 + \mathbf{f}^3 \otimes \mathbf{f}^3 + \mathbf{f}^4 \otimes \mathbf{f}^4$$

on G_λ. We shall show that the corresponding metrics are all isometric for fixed k. The associated curvature tensor does not therefore depend upon λ. Using the program in the Appendix, it is easy to verify

1.5 Proposition. Relative to the orthonormal basis $(\frac{1}{k}\mathbf{f}_1, \mathbf{f}_2, \mathbf{f}_3, \mathbf{f}_4)$ of $\mathfrak{g}_\lambda$, the Ricci and Weyl tensors of g_k have the diagonal forms

$$\mathrm{Ric} = \begin{pmatrix} \frac{6}{k^2} & 0 & 0 & 0 \\ 0 & \frac{4}{k^2} + \frac{1}{2} & 0 & 0 \\ 0 & 0 & \frac{4}{k^2} + \frac{1}{2} & 0 \\ 0 & 0 & 0 & \frac{8}{k^2} - \frac{1}{2} \end{pmatrix},$$

$$W^\pm = \frac{k^2 \mp 3k + 2}{3k^2} \begin{pmatrix} -1 & 0 & 0 \\ 0 & -1 & 0 \\ 0 & 0 & 2 \end{pmatrix}.$$

Observe that Ric is degenerate if $k = 4$, and $W^+ = 0$ if $k \in \{1, 2\}$. We shall prove that, conformally speaking, there are no other non-flat ASD examples:

1.6 Theorem. Let G be an oriented four-dimensional Lie group admitting a left-invariant Riemannian metric g such that $W^+ = 0$ and $W^- \neq 0$. Then the Lie algebra of G is isomorphic to $\mathfrak{g}_\lambda$ for some $\lambda \geqslant 0$, and g is locally homothetic to either g_1 or g_2.

The distinction between W^+, W^- depends upon a choice of orientation, and in this sense inclusion of the prefix 'anti' in the title is purely a matter of taste. However, an orientation *is* distinguished by the presence of a complex structure. The set of positively-oriented almost complex structures on G compatible with a metric is isomorphic to $SO(4)/U(2) \cong S^2$, and the equation $W^+ = 0$ is precisely the integrability condition for a tautological almost complex structure on $G \times S^2$ [**3**]. A special situation in which $W^+ = 0$ occurs when each element in S^2 represents an integrable complex structure on G, which is then called *hypercomplex.*

Four-dimensional Lie groups with a left-invariant hypercomplex structure were classified by Barberis [**4**], and the above theorem can therefore be viewed as a generalization of this work. Indeed, [**4**] asserts that G_λ admits a left-invariant hypercomplex structure if and only if $\lambda = 0$, and that (G_0, g_1) is hyperhermitian. On the other hand, (G_0, g_2) is isometric to the complex hyperbolic plane $\mathbb{CH}^2 = SU(2,1)/S(U(2) \times U(1))$ with its symmetric metric, which is well known to satisfy $W^- = 0$ (with respect to the orientation relative to the natural complex structure). This metric appeared in Jensen's classification of Einstein metrics on Lie groups [**8**], which provides a starting point for an alternative approach to classifying self-dual metrics and other structures on 4-dimensional Lie groups [**1**].

The fact that there are just two values of the parameter k that solve our problem is reminiscent of the existence of two Einstein metrics on certain sphere bundles.

For example, it is well known that $\mathbb{CP}^3$ (as a bundle over S^4) has its standard Kähler-Einstein metric as well as a nearly-Kähler metric with weak holonomy $U(3)$.

2. Building Lie algebras

Let $\mathfrak{a}_k$ denote the abelian Lie algebra whose underlying vector space is $\mathbb{R}^k$. Let $\mathfrak{h}_3$ denote the Lie algebra of the Heisenberg group; thus $\mathfrak{h}_3$ has a basis $(\mathbf{f}_1, \mathbf{f}_2, \mathbf{f}_3)$ satisfying $\mathbf{f}_1 = [\mathbf{f}_2, \mathbf{f}_3] = -[\mathbf{f}_3, \mathbf{f}_2]$ and all other brackets zero.

Given a Lie algebra $\mathfrak{h}$, let

$$\operatorname{der}\mathfrak{h} = \{D\colon \mathfrak{h} \to \mathfrak{h} : D[x,y] = [Dx,y] + [x,Dy]\}$$

denote the set of derivations of $\mathfrak{h}$. Suppose that $\rho\colon \mathfrak{b} \to \operatorname{der}\mathfrak{h}$ is a homomorphism of Lie algebras. The *extension of* $\mathfrak{h}$ *by* ρ is the Lie algebra $\mathfrak{g}$ with underlying vector space $\mathfrak{b}\oplus\mathfrak{h}$, in which $\mathfrak{b}$ and $\mathfrak{h}$ are both subalgebras and $[x,h] = \rho(x)(h)$ for all $x \in \mathfrak{b}$ and $h \in \mathfrak{h}$. We write

$$\mathfrak{g} = \mathfrak{b} \oplus_\rho \mathfrak{h},$$

and if $\mathfrak{b}$ is abelian, we say that $\mathfrak{g}$ is an *abelian extension* of $\mathfrak{h}$.

2.1 Proposition. If $\mathfrak{g}$ is a 4-dimensional Lie algebra with zero centre then $\mathfrak{g}$ is isomorphic to an abelian extension of $\mathfrak{a}_2$, $\mathfrak{a}_3$ or $\mathfrak{h}_3$.

Proof. If $\mathfrak{g}$ is not solvable, its radical $\mathfrak{r}$ is a proper ideal. The Levi decomposition gives $\mathfrak{g} = \mathfrak{s} \oplus_\rho \mathfrak{r}$ where $\mathfrak{s}$ is a semi-simple algebra and $\rho\colon \mathfrak{s} \to \operatorname{der}\mathfrak{r}$. It follows that $\dim\mathfrak{s} = 3$ and $\dim\mathfrak{r} = 1$. But 1-dimensional representations of a semi-simple Lie algebra are trivial, so the centre $\mathfrak{z}$ of $\mathfrak{g}$ equals $\mathfrak{r}$. We may therefore assume that $\mathfrak{g}$ is solvable.

A Lie algebra $\mathfrak{g}$ is solvable if and only if the ideal $\mathfrak{g}' = [\mathfrak{g},\mathfrak{g}]$ is nilpotent. The only non-abelian nilpotent Lie algebra of dimension less than 4 is $\mathfrak{h}_3$, so $\mathfrak{g}'$ is one of 0, $\mathfrak{a}_1$, $\mathfrak{a}_2$, $\mathfrak{a}_3$, $\mathfrak{h}_3$.

If $\mathfrak{g}' = 0$ then $\mathfrak{z} = \mathfrak{g} \neq 0$.

If $\mathfrak{g}' \cong \mathfrak{a}_1$ then $\mathfrak{g}$ has a basis $(\mathbf{f}_1, \mathbf{f}_2, \mathbf{f}_3, \mathbf{f}_4)$ with $\mathfrak{g}' = \langle \mathbf{f}_1 \rangle$. Unless $\mathbf{f}_1 \in \mathfrak{z}$, we may suppose that $\mathbf{f}_2, \mathbf{f}_3, \mathbf{f}_4$ are chosen such that $[\mathbf{f}_1, \mathbf{f}_2] = \mathbf{f}_1$ and $[\mathbf{f}_1, \mathbf{f}_i] = 0$ for $i = 3, 4$. We may further modify $\mathbf{f}_3$ in order that $[\mathbf{f}_2, \mathbf{f}_3] = 0$. But then the Jacobi identity

$$[[\mathbf{f}_2, \mathbf{f}_3], \mathbf{f}_4] + [[\mathbf{f}_3, \mathbf{f}_4], \mathbf{f}_2] + [[\mathbf{f}_4, \mathbf{f}_2], \mathbf{f}_3] = 0$$

implies that $[\mathbf{f}_3, \mathbf{f}_4] = 0$ and $\mathbf{f}_3 \in \mathfrak{z}$.

If $\mathfrak{g}' \cong \mathfrak{a}_2$ then there is an isomorphism $\mathfrak{g} \cong \mathbb{R}^2 \oplus \mathfrak{g}'$ of vector spaces, and the bracket determines a linear mapping $\rho\colon \mathbb{R}^2 \to \operatorname{der}(\mathfrak{g}')$. If $\ker\rho$ is non-zero then it contains a 1-dimensional subspace that when added to $\mathfrak{g}'$ yields an abelian algebra $\mathfrak{a}_3$ of which $\mathfrak{g}$ is an abelian extension. If ρ is injective then $\mathfrak{g}'_{\mathbb{C}}$ has a basis of eigenvectors for the commuting elements in $\operatorname{Im}\rho$. It follows that there exists a real basis $(\mathbf{f}_1, \mathbf{f}_2, \mathbf{f}_3, \mathbf{f}_4)$ of $\mathfrak{g}$ such that $\mathfrak{g}' = \langle \mathbf{f}_3, \mathbf{f}_4 \rangle$ and $\operatorname{ad}\mathbf{f}_1$ restricts to the identity on $\mathfrak{g}'$. Then $\mathbf{f}_2$ may be modified so that $[\mathbf{f}_1, \mathbf{f}_2] = 0$, and $\mathfrak{g} = \mathbb{R}^2 \oplus_\rho \mathfrak{g}'$ is an abelian extension.

If $\mathfrak{g}' \cong \mathfrak{a}_3$ or $\mathfrak{g}' \cong \mathfrak{h}_3$ then $\mathfrak{g}$ is immediately an abelian extension of $\mathfrak{g}'$. □

Four-dimensional solvable Lie algebras can be broadly divided into 7 classes according to the triple (d', d'', d''') of dimensions of $\mathfrak{g}'$, $\mathfrak{g}''=[\mathfrak{g}',\mathfrak{g}']$, $\mathfrak{g}'''=[\mathfrak{g}'',\mathfrak{g}'']$. In [**1**], curvature computations are carried out for each of these classes in turn. For our purposes, the following observation helps to restrict the range of algebras that need to be considered.

2.2 Lemma. Let G be a simply-connected Lie group with a left-invariant metric g. There exists an orientation-reversing isometry of (G, g) in either of the two cases: (i) the Lie algebra $\mathfrak{g}$ has non-zero centre, or (ii) $\mathfrak{g}$ is an abelian extension of $\mathfrak{a}_3$.

Proof. We shall exhibit an orthogonal transformation of $(\mathfrak{g}, \langle\,.\,,.\,\rangle)$ which reverses the orientation of $\mathfrak{g}$ in each case. This automorphism will induce the desired isometry of (G, g).

(i) If $\mathfrak{z}\neq 0$, choose an orthonormal basis $(\mathbf{f}_1,\mathbf{f}_2,\mathbf{f}_3,\mathbf{f}_4)$ of $\mathfrak{g}$ with $\mathbf{f}_1\in\mathfrak{z}$. It is then immediate that the linear mapping $\phi\colon \mathfrak{g}\to\mathfrak{g}$ defined by

$$\phi(\mathbf{f}_i)=\begin{cases}-\mathbf{f}_1 & i=1,\\ \mathbf{f}_i & i\neq 1,\end{cases}$$

is a Lie algebra automorphism.

(ii) If $\mathfrak{g}=\mathbb{R}\oplus_\rho\mathfrak{a}_3$, choose an orthonormal basis $(\mathbf{f}_1,\dots,\mathbf{f}_4)$ with $\mathbf{f}_2,\mathbf{f}_3,\mathbf{f}_4\in\mathfrak{a}_3$. This time,

$$\psi(\mathbf{f}_i)=\begin{cases}\mathbf{f}_1 & i=4,\\ -\mathbf{f}_i & i\geqslant 2\end{cases}$$

is the required automorphism. □

Combined with Proposition 2.1, this yields

2.3 Corollary. Let G be an oriented Lie group with a left-invariant Riemannian metric that satisfies $W^+=0$ and $W^-\neq 0$. Then $\mathfrak{g}$ is an abelian extension of $\mathfrak{a}_2$ or $\mathfrak{h}_3$.

2.4 Proposition. The Lie algebras $\mathfrak{g}_\lambda$, $\mathfrak{g}_{\lambda'}$ are isomorphic if and only if $\lambda=\pm\lambda'$.

Proof. The derived algebra $\mathfrak{g}'_\lambda=[\mathfrak{g}_\lambda,\mathfrak{g}_\lambda]$ in (1.1) has a basis $(\mathbf{f}_2,\mathbf{f}_3,\mathbf{f}_4)$. With respect to this basis, $\operatorname{ad}\mathbf{f}_1$ acts on $\mathfrak{g}'_\lambda$ as

$$\begin{pmatrix}1 & \lambda & 0\\ -\lambda & 1 & 0\\ 0 & 0 & 2\end{pmatrix}.$$

It follows that if f is any element of $\mathfrak{g}_\lambda\setminus\mathfrak{g}'_\lambda$ then

$$\frac{\det(\operatorname{ad}(f)|_{\mathfrak{g}'_\lambda})}{\operatorname{tr}(\operatorname{ad}(f)|_{\mathfrak{g}'_\lambda})}=\tfrac12(1+\lambda^2),$$

so $\mathfrak{g}_\lambda,\mathfrak{g}_{\lambda'}$ cannot be isomorphic unless $\lambda=\pm\lambda'$. We observed in §1 that they *are* isomorphic if $\lambda=-\lambda'$. □

2.5 Proposition. Let $\lambda \in \mathbb{R}$ and $k, k' > 0$. The Riemannian manifolds (G_λ, g_k) and $(G_{\lambda'}, g_{k'})$ are isometric if and only if $k = k'$.

Proof. The fact that (G_λ, g_k) is not isometric to $(G_{\lambda'}, g_{k'})$ if $k \neq k'$ follows immediately from **1.5**, the computations of which can be carried out by starting from (3.7) below. It therefore suffices to show that (G_λ, g_k) is isometric to (G_0, g_k) for fixed $k > 0$.

Define $\mathfrak{u}$ as the abelian extension

$$\mathfrak{u} = \langle \mathbf{f}_0 \rangle \oplus_\rho \mathfrak{g}_0,$$

where ρ is given by

$$\begin{array}{ll} [\mathbf{f}_0, \mathbf{f}_2] = -\mathbf{f}_3, & [\mathbf{f}_0, \mathbf{f}_1] = 0, \\ [\mathbf{f}_0, \mathbf{f}_3] = \mathbf{f}_2, & [\mathbf{f}_0, \mathbf{f}_4] = 0. \end{array} \tag{2.6}$$

Let U denote the simply connected Lie group with Lie algebra $\mathfrak{u}$.

Comparing (2.6) with (1.1), we see that the mapping $\mathfrak{g}_\lambda \to \mathfrak{u}$ defined by

$$\begin{array}{rcl} \mathbf{f}_1 & \mapsto & \lambda \mathbf{f}_0 + \mathbf{f}_1, \\ \mathbf{f}_i & \mapsto & \mathbf{f}_i, \quad 2 \leqslant i \leqslant 4, \end{array}$$

is a Lie algebra homomorphism. We denote the induced immersion $G_\lambda \to U$ by i_λ.

The 1-parameter subgroup $\mathbb{R}$ of U generated by $\mathbf{f}_0$ acts on the coordinates x, y by

$$\begin{array}{rcl} x & \mapsto & x \cos s + y \sin s, \\ y & \mapsto & -x \sin s + y \cos s. \end{array} \tag{2.7}$$

This gives rise to a submersion $\pi : U \to U/\mathbb{R}$ onto a homogeneous manifold for which $\pi \circ i_\lambda$ is a diffeomorphism for each λ. By construction,

$$(\pi \circ i_\lambda)_* \mathbf{f}_i = (\pi \circ i_0)_* \mathbf{f}_i, \quad 1 \leqslant i \leqslant 4,$$

and $\phi = (\pi \circ i_\lambda)^{-1} \circ (\pi \circ i_0)$ is the required isometry $(G_0, g_k) \to (G_\lambda, g_k)$.

More explicitly, we may identify U with $\mathbb{R} \times G_0$ by means of coordinates s, t, x, y, u relative to which $\mathbf{f}^0 = ds$ and (1.3) holds with $\lambda = 0$. Then (2.7) and the fact that $i_\lambda^*(ds) = \lambda dt$ imply that

$$\begin{aligned} \phi^*(dx^2 + dy^2) &= (dx + \lambda y dt)^2 + (dy - \lambda x dt)^2, \\ \phi^*(du + \tfrac{1}{2}(x dy - y dx)) &= du + \tfrac{1}{2}(x dy - y dx) - \tfrac{1}{2}\lambda(x^2 + y^2) dt, \end{aligned}$$

whence $\phi^* g_k = g_k$. □

3. Proof of the theorem

Let G be a 4-dimensional Lie group with a chosen orientation, and let g be a left-invariant Riemannian metric. The latter corresponds to an inner product $\langle\, .\, , .\, \rangle$ on the Lie algebra $\mathfrak{g}$ of G. The curvature tensor of g and its components $W^\pm$ are completely determined by the structure constants of the Lie algebra $\mathfrak{g}$ and its inner product, so the problem is purely algebraic.

Because we are only interested in metrics that are not conformally flat, the Corollary tells us that $\mathfrak{g}$ must be an abelian extension of $\mathfrak{a}_2$ or $\mathfrak{h}_3$. This information alone is sufficient to proceed with the calculations, though these are simplified by appealing to a classification of 4-dimensional solvable Lie algebras that appears for example in [**12**]. The algebras we need to consider are listed in the following table, with the structure constants encapsulated in the exterior derivatives of a basis $(\mathbf{f}^1, \mathbf{f}^2, \mathbf{f}^3, \mathbf{f}^4)$ of $\mathfrak{g}^*$. In each case, $d\mathbf{f}^1 = 0$. The real parameters α, β satisfy $0 \leqslant \alpha \leqslant 2$ and $\beta \geqslant 0$.

type	$(d\mathbf{f}^2, d\mathbf{f}^3, d\mathbf{f}^4)$	extension of
$\mathfrak{g}_2 \oplus \mathfrak{g}_2$	$(0, \mathbf{f}^{13}, \mathbf{f}^{24})$	$\mathfrak{a}_2$
$\mathfrak{g}_{4,1}$	$(0, \mathbf{f}^{13}, \mathbf{f}^{14} + \mathbf{f}^{23})$	$\mathfrak{a}_2$, $\mathfrak{h}_3$
$\mathfrak{g}_{4,2}$	$(0, \mathbf{f}^{13} - \mathbf{f}^{42}, \mathbf{f}^{14} - \mathbf{f}^{23})$	$\mathfrak{a}_2$
$\mathfrak{g}_{4,9}(\alpha)$	$((1-\alpha)\mathbf{f}^{12}, -\mathbf{f}^{13}, -\alpha\mathbf{f}^{14} - \mathbf{f}^{23})$	$\mathfrak{h}_3$
$\mathfrak{g}_{4,10}$	$(\mathbf{f}^{12}, \mathbf{f}^{12} + \mathbf{f}^{13}, \mathbf{f}^{23} + 2\mathbf{f}^{14})$	$\mathfrak{h}_3$
$\mathfrak{g}_{4,11}(\beta)$	$(\beta\mathbf{f}^{12} + \mathbf{f}^{13}, -\mathbf{f}^{12} + \beta\mathbf{f}^{13}, -\mathbf{f}^{23} + 2\beta\mathbf{f}^{14})$	$\mathfrak{h}_3$

Let $\mathfrak{g}$ be one of $\mathfrak{g}_{4,1}$, $\mathfrak{g}_{4,9}(\alpha)$, $\mathfrak{g}_{4,10}$, $\mathfrak{g}_{4,11}(\beta)$. In each case, the above basis satisfies

$$(3.1)\qquad \begin{aligned} d\mathbf{f}^1 &= 0,\\ d\mathbf{f}^2 &\in \langle \mathbf{f}^{12}, \mathbf{f}^{13}\rangle,\\ d\mathbf{f}^3 &\in \langle \mathbf{f}^{12}, \mathbf{f}^{13}\rangle,\\ d\mathbf{f}^4 &\in \langle \mathbf{f}^{23}, \mathbf{f}^{14}\rangle. \end{aligned}$$

Apply the Gram-Schimdt process to obtain an orthonomal basis $(\mathbf{e}^1, \mathbf{e}^2, \mathbf{e}^3, \mathbf{e}^4)$ satisfying

$$(3.2)\qquad \mathbf{e}^j = \sum_{i=1}^{j} a_i^j \mathbf{f}^i, \quad a_j^j \neq 0.$$

The relations (3.1) then become

$$(3.3)\qquad \begin{aligned} d\mathbf{e}^1 &= 0,\\ d\mathbf{e}^2 &= c_{12}^2\mathbf{e}^{12} + c_{13}^2\mathbf{e}^{13},\\ d\mathbf{e}^3 &= c_{12}^3\mathbf{e}^{12} + c_{13}^3\mathbf{e}^{13},\\ d\mathbf{e}^4 &= c_{12}^4\mathbf{e}^{12} + c_{13}^4\mathbf{e}^{13} + c_{14}^4\mathbf{e}^{14} + c_{23}^4\mathbf{e}^{23}, \end{aligned}$$

for certain real constants c_{ij}^k. In this new scheme, each line on the right-hand side must include all terms from the previous line. Even if the isomorphism class of $\mathfrak{g}$ is fixed, the structure constants c_{ij}^k are allowed to vary to reflect all possible choices of inner product. However, we must first ensure that (3.3) still describes a Lie algebra.

3.4 Lemma. In addition to (3.3), we may suppose that (i) $c_{23}^4 = 1$, (ii) $c_{14}^4 = c_{12}^2 + c_{13}^3$, and (iii) $c_{12}^4 = 0$.

Proof. The Jacobi identity is equivalent to the equations $d(d\mathbf{e}^i) = 0$ for $1 \leqslant i \leqslant 4$. In our case, only the equation $d(d\mathbf{e}^4) = 0$ is in doubt, and this yields

$$0 = c_{23}^4(c_{12}^2 + c_{13}^3 - c_{14}^4).$$

For the Lie algebras under consideration, $\mathbf{f}^{23}$ appears with a non-zero coefficient in $d\mathbf{f}^4$. It follows from (3.2) that $c_{23}^4 \neq 0$, whence (ii).

Given the similar nature of $d\mathbf{e}^2, d\mathbf{e}^3$, the form of (3.3) is preserved by a rotation of the form

$$\begin{array}{rcl} \mathbf{e}^2 & \mapsto & (\cos\theta)\mathbf{e}^2 + (\sin\theta)\mathbf{e}^3, \\ \mathbf{e}^3 & \mapsto & -(\sin\theta)\mathbf{e}^2 + (\cos\theta)\mathbf{e}^3. \end{array} \tag{3.5}$$

Applying this substitution,

$$d\mathbf{e}^4 = (c_{12}^4 \cos\theta - c_{13}^4 \sin\theta)\mathbf{e}^{12} + \cdots$$

and we can choose θ so that (iii) is satisfied relative to the new orthonormal basis.

Finally, performing the overall scaling $\mathbf{e}^i \mapsto c_{23}^4 \mathbf{e}^i$ gives (i). □

The Weyl tensor can now be computed in terms of the c_{ij}^k. The orientation of $(\mathbf{e}^i)$ can be reversed by changing the sign of $\mathbf{e}^1$. This will alter the signs of the c_{1j}^k, but will not interfere with the above assumptions. It therefore suffices to compute W^+, whose vanishing is not affected by the rescaling that accompanied condition (i).

We now proceed to solve $W^+ = 0$, that represents $\dim S_0^2(\Lambda^+\mathfrak{g}^*) = 5$ quadratic equations in the coefficients of (3.3). There remain 5 unknowns, namely

$$c_{12}^2,\ c_{13}^2,\ c_{12}^3,\ c_{13}^3,\ c_{13}^4.$$

3.6 Proposition. Given the above assumptions, there are only two real solutions of $W^+ = 0$, namely

$$\begin{array}{ll} \text{(i)} & a_{13}^3 = -1,\ a_{13}^4 = 0,\ a_{12}^3 = -a_{13}^2,\ a_{12}^2 = -1; \\ \text{(ii)} & a_{13}^3 = -\frac{1}{2},\ a_{13}^4 = 0,\ a_{12}^3 = -a_{13}^2,\ a_{12}^2 = -\frac{1}{2}. \end{array}$$

Proof. This involves an application of the Appendix. □

Setting $a_{13}^2 = \lambda/k$ with $k = 1$ or 2 gives

$$\begin{array}{rcl} d\mathbf{e}^1 & = & 0, \\ k\, d\mathbf{e}^2 & = & -\mathbf{e}^{12} - \lambda\mathbf{e}^{13}, \\ k\, d\mathbf{e}^3 & = & \lambda\mathbf{e}^{12} - \mathbf{e}^{13}, \\ k\, d\mathbf{e}^4 & = & -2\mathbf{e}^{14} + k\mathbf{e}^{23}. \end{array} \tag{3.7}$$

Let $(\mathbf{e}_i)$ denote the dual basis. Then $(\mathbf{f}_1, \mathbf{f}_2, \mathbf{f}_3, \mathbf{f}_4) = (k\mathbf{e}_1, \mathbf{e}_2, \mathbf{e}_3, \mathbf{e}_4)$ satisfies (1.2) and (1.4) (the new f's are unrelated to those in (3.2)). In terms of the table, it is easy to verify that

$$\mathfrak{g}_\lambda \cong \begin{cases} \mathfrak{g}_{4,9}(2), & \text{if } \lambda = 0, \\ \mathfrak{g}_{4,11}(\frac{1}{\lambda}), & \text{if } \lambda > 0. \end{cases}$$

To complete the proof of the theorem, similar arguments are required for $\mathfrak{g}_2 \oplus \mathfrak{g}_2$ and $\mathfrak{g}_{4,2}$. In the first case, the analogue of (3.3) is

$$\begin{aligned} d\mathbf{e}^1 &= 0, \\ d\mathbf{e}^2 &= 0, \\ d\mathbf{e}^3 &= c^3_{12}\mathbf{e}^{12} + c^3_{13}\mathbf{e}^{13}, \\ d\mathbf{e}^4 &= c^4_{12}\mathbf{e}^{12} + c^4_{13}\mathbf{e}^{13} + c^4_{14}\mathbf{e}^{14} + c^4_{23}\mathbf{e}^{23} + c^4_{24}\mathbf{e}^{24}. \end{aligned}$$

The Jacobi identity implies that

$$a^4_{23}(a^3_{13} - a^4_{14}) + a^4_{13}a^4_{24} = 0.$$

An overall scaling can be used to set $a^4_{24} = 1$ and this leaves 5 unknowns. There are no solutions to $W^+ = 0$.

The equations for $\mathfrak{g}_{4,2}$ are initially more complicated. Gram-Schmidt yields

$$\begin{aligned} d\mathbf{e}^1 &= 0, \\ d\mathbf{e}^2 &= 0, \\ d\mathbf{e}^3 &= c^3_{12}\mathbf{e}^{12} + c^3_{13}\mathbf{e}^{13} + c^3_{14}\mathbf{e}^{14} + c^3_{23}\mathbf{e}^{23} + c^3_{24}\mathbf{e}^{24}, \\ d\mathbf{e}^4 &= c^4_{12}\mathbf{e}^{12} + c^4_{13}\mathbf{e}^{13} + c^4_{14}\mathbf{e}^{14} + c^4_{23}\mathbf{e}^{23} + c^4_{24}\mathbf{e}^{24}. \end{aligned}$$

A rotation in the plane generated by $\mathbf{e}^1, \mathbf{e}^2$, in analogy to (3.5), enables us to arrange that $c^3_{14} = 0$ and $c^3_{24} \neq 0$. After an overall scaling of the basis we may assume that $c^3_{24} = 1$. The Jacobi identity then implies that

$$c^4_{13} = 0, \quad c^4_{13} = c^3_{13}.$$

This leaves 6 unknowns, sufficiently few for the program to reveal that once again there are no solutions with $W^+ = 0$.

4. Further properties

Invariant complex structures. To check whether the manifold (G_λ, g_k) is hyperhermitian, we take a standard basis

$$\begin{aligned} \omega_1 &= \mathbf{e}^{12} + \mathbf{e}^{34}, \\ \omega_2 &= \mathbf{e}^{13} + \mathbf{e}^{42}, \\ \omega_3 &= \mathbf{e}^{14} + \mathbf{e}^{23} \end{aligned}$$

of self-dual 2-forms. Let I_1, I_2, I_3 denote the corresponding left-invariant almost complex structures, and $\theta_1, \theta_2, \theta_3$ the Lee forms defined by $d\omega_i = \omega_i \wedge \theta_i$. An easy consequence of (3.7) is that

$$\begin{aligned} \theta_1 &= -\frac{3}{k}\mathbf{e}^1 - \lambda\mathbf{e}^4 = \theta_2, \\ \theta_3 &= -(1 + \frac{2}{k})\mathbf{e}^1. \end{aligned}$$

The equality between θ_1 and θ_2 implies (via some standard theory [**11**]) that I_3 is always integrable. On the other hand, θ_1 and θ_3 are only equal if $\lambda = 0$ and $k = 1$.

4.1 Corollary. The Lie group G_λ admits an invariant complex structure I_3 for any λ, but an invariant hypercomplex structure (I_1, I_2, I_3) only for $\lambda = 0$. This hypercomplex structure is compatible with g_1.

Properties of g_1. Suppose that $\lambda = 0$ and $k = 1$. Using the coordinates t, x, y, u as in (1.3) and the substitution $v = \mathbf{e}^t$, take

$$\mathbf{e}^1 = dt, \quad \mathbf{e}^2 = \frac{1}{v}dx, \quad \mathbf{e}^3 = \frac{1}{v}dy, \quad \mathbf{e}^4 = \frac{1}{v^2}(du + \tfrac{1}{2}(xdy - ydx)).$$

If we further set

$$\begin{aligned} \tilde{\omega}_1^{\pm} &= vdv \wedge dx \pm (dy \wedge du + \tfrac{1}{2}ydy \wedge dx) \\ \tilde{\omega}_2^{\pm} &= vdv \wedge dy \pm (du \wedge dx + \tfrac{1}{2}xdx \wedge dy) \\ \tilde{\omega}_3^{\pm} &= dv \wedge du - \tfrac{1}{2}(xdy \wedge dv - ydx \wedge dv) \pm vdxdy, \end{aligned}$$

then $\tilde{\omega}_i^+ = v^3\omega_i$. The following is immediate.

4.2 Lemma. Five of the six 2-forms $\tilde{\omega}_i^{\pm}$ are closed, all except $\tilde{\omega}_3^-$.

The fact that $d\tilde{\omega}_i^+ = 0$ for all i implies that the almost-complex structures I_1, I_2, I_3 are all integrable, and $v^3 g_1$ is a hyperkähler metric. This observation is contained in [**5**]; in fact

$$v^3 g_1 = v(dx^2 + dy^2 + dv^2) + \frac{1}{v}(du + \tfrac{1}{2}(xdy - ydx))^2 \tag{4.3}$$

is in the form of the Gibbons-Hawking ansatz for a hyperkähler metric with S^1 action. On the other hand, if $\sigma = (\cos\theta)\omega_1^- + (\sin\theta)\omega_2^-$ then (for any θ) $(v^3 g_1, \sigma)$ is an almost-Kähler Einstein structure which is not Kähler. There is a family of similar local counterexamples to the Goldberg conjecture [**2**].

Links with other geometries. Let $z = x + iv$, and consider the upper half plane $H = \{z \in \mathbb{C} : v > 0\}$. Then $v^3 g_1$ coincides with the natural metric on the cotangent bundle T^*H induced from a 'special Kähler' metric on H. This aspect of the above example is explained in [**11**]. Here, we merely remark that special Kähler metrics are determined locally by a 'holomorphic prepotential' $\mathcal{F}$ that here we take to be $\frac{1}{6}z^3$ [**6**]. This allows one to introduce a dual or 'conjugate' holomorphic coordinate

$$w = \tfrac{1}{2}z^2 = \tfrac{1}{2}(x^2 - v^2) - xv\, i,$$

whose real part we denote by $\tilde{x}$. The (incomplete) special Kähler metric $v(dx^2 + dv^2)$ on H has associated 2-form

$$\omega = dx \wedge d\tilde{x} = \frac{v}{2i}dz \wedge d\bar{z}.$$

Decreeing the 1-forms $dx, d\tilde{x}$ to be parallel defines a flat symplectic connection on H which can be used to construct the hyperkähler metric (4.3) in a functorial manner.

The Lie algebra $\mathfrak{g}_0$ also features in [**7**], as one example in a study of Einstein metrics constructed from left-invariant metrics on nilpotent groups. Higher dimensional examples give rise to Einstein metrics on solvable extensions, and related metrics with exceptional holonomy.

Quadratic components of W. In four dimensions, the space $\mathcal{W}$ of Weyl tensors has symmetric square

$$S^2\mathcal{W} \cong S^2\mathcal{W}^+ \oplus (\mathcal{W}^+ \otimes \mathcal{W}^-) \oplus S^2\mathcal{W}^-, \tag{4.4}$$

and there exist $SO(4)$-irreducible 9-dimensional representations $\mathcal{V}^\pm$ for which

$$S^2\mathcal{W}^\pm \cong \mathcal{V}^\pm \oplus \mathcal{W}^\pm \oplus \mathbb{R}$$

(see for example [**10**]). Thus, the vanishing of $\mathcal{W}^+$ reduces the number of non-zero components of $W \otimes W$ in (4.4) from 7 to 3. Moreover, if W^+ has repeated eigenvalues as in **1.5** then the $\mathcal{W}^+$-component of $W^+ \otimes W^+$ is proportional to W^+ (this also implies that certain components of $W^+ \otimes W^+ \otimes W^+$ vanish in $S^3\mathcal{W}^+$).

On a higher dimensional manifold M, it is known that the Weyl tensor W is irreducible. Appropriate analogues of anti-self-duality can in theory be defined in terms of components of $S^2\mathcal{W}$. Of special interest are likely to be conditions imposed on W by the existence of one or more orthogonal complex structures on M.

5. Appendix

The essential part of the Maple program used to solve the equation $W^+ = 0$ is reproduced below. It is set up to mimic the assumptions (3.3) and **3.4**, though the remaining cases of $\mathfrak{g}_2 \oplus \mathfrak{g}_2$ and $\mathfrak{g}_{4,2}$ are similar.

Once the solutions are found it is an easy matter to substitute back using for example `assign(so[1])` and determine W^- by setting `z:=-1`.

```
# Define the structure constants
   for i to 4 do defform(e[i]=1) od:
   for i to 5 do defform(a[i]=const) od:
   defform(d(e[1])=0):
   defform(d(e[2])=a[1]*e[1]&^e[2]+a[2]*e[1]&^e[3]):
   defform(d(e[3])=a[3]*e[1]&^e[2]+a[4]*e[1]&^e[3]):
   defform(d(e[4])=a[5]*e[1]&^e[3]+(a[1]+a[4])*e[1]&^e[4]+e[2]&^e[3]):

# Define the set of connection coefficients
   gset:={}:
   for k to 4 do g[k]:=array(antisymmetric,1..4,1..4) od:
      for k to 4 do for j to 4 do for i to j-1 do
         defform(g[k][i,j]=const); gset:=gset union {g[k][i,j]}
      od od od:

# Define and solve the Cartan equations
   fset:={}:
   for k to 4 do
      de[k]:=sum(sum(g[m][k,n]*e[m]&^e[n],m=1..4),n=1..4)
   od:
   c:=(k,i,j)->coeff(de[k]-d(e[k]),e[i]&^e[j]):
   for k to 4 do for j to 4 do for i to j-1 do
      fset:=fset union {c(k,i,j)=c(k,j,i)}
   od od od:
```

```
    sol:=solve(fset,gset);
    assign(sol):

# Define the Riemann curvature tensor
    r1:=(i,j)->sum(g[p][i,j]*d(e[p]),p=1..4):
    r2:=(i,j)->sum(sum(sum(g[u][i,r]*g[v][r,j]*e[u]&^e[v],u=1..4),
                                                  v=1..4),r=1..4):
    r:=(i,j,x,y)->coeff(simpform(r1(i,j)-r2(i,j)),e[x]&^e[y]):
    for i to 4 do for j to 4 do
    R[i,j]:= array(antisymmetric,1..4,1..4):
       for y to 4 do
          for x to y-1 do R[i,j][x,y]:=r(i,j,x,y)-r(i,j,y,x) od
    od:
    od od:

# Define the Ricci, scalar and Weyl+ curvature
    ric:=array(symmetric,1..4,1..4):
    for j to 4 do for i to j do
       ric[i,j]:= sum(R[i,p][j,p],p=1..4)
    od od:
    sc:=sum(ric[q,q],q=1..4):
    W:=array(symmetric,1..3,1..3): z:=1:
       W[1,1]:= R[1,2][1,2]+R[3,4][3,4]+2*z*R[1,2][3,4]-sc/6:
       W[2,2]:= R[1,3][1,3]+R[4,2][4,2]+2*z*R[1,3][4,2]-sc/6:
       W[3,3]:= R[1,4][1,4]+R[2,3][2,3]+2*z*R[1,4][2,3]-sc/6:
       W[1,2]:= R[1,2][1,3]+z*R[1,2][4,2]+z*R[3,4][1,3]+R[3,4][4,2]:
       W[1,3]:= R[1,2][1,4]+z*R[1,2][2,3]+z*R[3,4][1,4]+R[3,4][2,3]:
       W[2,3]:= R[1,3][1,4]+z*R[1,3][2,3]+z*R[4,2][1,4]+R[4,2][2,3]:

# Solve Weyl+=0
      hset:={W[1,1]=0,W[1,2]=0,W[1,3]=0,W[2,2]=0,W[2,3]=0}:
      so:=solve(hset):
```

References

[1] E. Abbena: Classification of 4-dimensional self-dual Lie groups, unpublished.

[2] J. Armstrong: An ansatz for almost-Kähler Einstein manifolds, J. Reine Angew. Math. **542** (2002).

[3] M.F. Atiyah, N.J. Hitchin and I.M. Singer: Self-duality in four-dimensional Riemannian geometry, Proc. Roy. Soc. Lond. **A362** (1978), 425–461.

[4] M.L. Barberis: Hypercomplex structures on four dimensional Lie groups, Proc. Amer. Math. Soc. **125** (1997) 1043–1054.

[5] M.L. Barberis: Homogeneous hyper-Hermitian metrics which are conformally hyper-Kähler, `math.DG/0009035`.

[6] D. Freed: Special Kähler manifolds, Commun. Math. Phys. **203** (1999), 31–52.

[7] G.W. Gibbons, H. Lü, C.N. Pope and K.S. Stelle: Supersymmetric domain walls from metrics of special holonomy, `hep-th/0108191`.

[8] G.R. Jensen: Homogeneous Einstein spaces of dimension four, J. Diff. Geom. **3** (1969) 309–349.

[9] P. Nurowski and M. Przanowski: A four-dimensional example of Ricci-flat metric admitting almost-Kähler non-Kähler structure, Class. Quantum Grav. **16** (1999), L9–L13.

[10] S. Salamon: Harmonic 4-spaces, Math. Ann. **269** (1984), 169–178.

[11] S.M. Salamon: Hermitian geometry, in 'Invitations to Geometry and Topology', eds. M.R. Bridson and S.M. Salamon, Oxford University Press, 2002.
[12] M. Vergne et al.: Représentations des groupes de Lie résolubles, Monographie Soc. Math. France, Vol. 4, Dunod, Paris, 1972.

RUE EMILE BOUILLIOT 25, 1050 BRUXELLES, BELGIUM

MATHEMATICAL INSTITUTE, 24-29 ST GILES, OXFORD OX1 3LB, UK

Current address: Dipartimento di Matematica, Politecnico di Torino, 10129 Torino, ITALY

E-mail address: salamon@calvino.polito.it

Contemporary Mathematics
Volume **308**, 2002

Weierstraß-Type Representation of Timelike Surfaces with Constant Mean Curvature

Josef Dorfmeister, Jun-ichi Inoguchi, and Magdalena Toda

The second author would like to dedicate this paper to the memory of his father, Kan

Introduction

It is well known that every solution of the elliptic sinh-Gordon (or sinh-Laplace) equation

$$\omega_{z\bar{z}} + \sinh \omega = 0$$

describes a constant mean curvature (CMC) surface in Euclidean 3-space $\mathbb{E}^3$, where the symmetric quadratic form $I = e^{\omega} dz d\bar{z}$ is the induced metric of the CMC surface. With respect to the conformal structure determined by I, the Gauß map is a harmonic map into S^2.

A loop group theoretic Weierstraß-type representation for harmonic maps of Riemann surfaces into compact Riemannian symmetric spaces was established in [**14**]. This is frequently referred to in the literature as the *DPW-method*, and CMC surfaces were investigated intensively by this method in [**12**]. A simple way for determining a normal form of the Weierestraß data for CMC surfaces was given in [**38**]. In [**13**], a reinterpretation of the classical Weierstraß representation of minimal surfaces was given in terms of the DPW-method. We refer to [**9**] for a survey of results about CMC surfaces via the DPW-method.

On the other hand, every solution to the sine-Gordon equation

$$\phi_{xy} + \sin \phi = 0$$

describes a (weakly regular) pseudospherical surface in Euclidean 3-space $\mathbb{E}^3$, where ϕ is the angle between the two asymptotic directions. The quadratic form $II = \sin \phi \, dxdy$ is the second fundamental form of the surface. With respect to the Lorentzian conformal structure determined by II, the Gauß map is a (Lorentzian) harmonic map into S^2.

2000 *Mathematics Subject Classification.* Primary 53A10, 22E67, 37K25, 53B30.

The second author was partially supported by Grant-in-Aid for the Encouragement of Young Scientists 12740051 of the Japan Society for Promotion of Science. Most of this work was done when the second author visited the University of Kansas.

M. Melko and I. Sterling [**28**], [**29**] presented a modern approach to pseudospherical surfaces via the theory of finite type (Lorentzian) harmonic maps into S^2. Recently the third author established a Weierstraß-type representation for pseudospherical surfaces in $\mathbb{E}^3$ in her thesis [**36**]. In particular, she showed that there exist certain normalized potentials for pseudospherical surfaces. Finally, in [**15**] it was shown that finite type pseudospherical surfaces can be constructed from particularly simple normalized potentials.

Both the sine-Gordon and the elliptic sinh-Gordon equation are real forms of the complexified sine-Gordon equation. But the complexified sine-Gordon equation has other real forms: the hyperbolic sinh-Gordon and the hyperbolic cosh-Gordon equations. Solutions to these equations describe CMC surfaces not in $\mathbb{E}^3$ but in Minkowski 3-space $\mathbb{E}^3_1$. In fact, let M be a timelike CMC surface in $\mathbb{E}^3_1$ parametrized locally by null coordinates (x, y). Denote by D the discriminant of the characteristic equation for the shape operator. Then the Gauß-Codazzi equations of M become

$$\omega_{xy} + H^2 \sinh \omega = 0, \text{ if } D > 0,$$

$$\omega_{xy} + \frac{H^2}{2} e^{\omega} = 0, \text{ if } D = 0,$$

$$\omega_{xy} + H^2 \cosh \omega = 0, \text{ if } D < 0.$$

At this point we would like to point out a certain similarity between timelike CMC surfaces and CMC surfaces in hyperbolic 3-space. Note that both H^3 and $\mathbb{E}^3_1$ are naturally embedded in Minkowski 4-space $\mathbb{E}^4_1$. Let M be a CMC surface in hyperbolic 3-space H^3 parametrized by isothermic coordinates (isothermal curvature-line coordinates) $z = x + \sqrt{-1}y$, then its Gauß-Codazzi equations become

$$\omega_{z\bar{z}} + (H^2 - 1) \sinh \omega = 0, \text{ if } H^2 > 1,$$

$$\omega_{z\bar{z}} - \frac{1}{2} e^{-\omega} = 0, \text{ if } H^2 = 1,$$

$$\omega_{z\bar{z}} + (H^2 - 1) \cosh \omega = 0, \text{ if } H^2 < 1.$$

Therefore, at the level of the Gauß-Codazzi equations, timelike CMC surface geometry can be considered as a "hyperbolic version" of CMC surface geometry in H^3. There is another similarity between timelike surface geometry in $\mathbb{E}^3_1$ and surface geometry in H^3. In fact, timelike HIMC surfaces, i.e., timelike surfaces with harmonic inverse mean curvature in $\mathbb{E}^3_1$ [**16**] and Bonnet surfaces in H^3 (which are not Willmore surfaces) [**7**] are described by the same integrable equations, namely the *Painlevé equations of type* V and VI.

The hyperbolic sinh-Gordon equation and the Liouville equation have been studied extensively by the soliton theoretic approach. For instance, (the infinitesimal version of) dressing transformations for the hyperbolic sinh-Gordon equation were studied in [**1**], and Bäcklund transformations for the hyperbolic sinh-Gordon and the Liouville equation via the classical theory of line-congruences were studied in [**27**]. A Darboux form for Bäcklund transformations of the hyperbolic sinh-Gordon equation was given in [**22**], [**24**]. Here, Bäcklund transformations were reformulated as *transformations on extended framings*.

However as far as the authors know, few studies of the cosh-Gordon equation are available. Finite-gap solutions to the elliptic cosh-Gordon equation were obtained in [**2**]. Minimal surfaces in H^3 in terms of finite-gap solutions of the elliptic cosh-Gordon equation were studied in [**3**], and radial-symmetric solutions to the elliptic cosh-Gordon equation in [**32**]. The radial symmetry reduces the elliptic cosh-Gordon equation to the third Painlevé equation. Moreover, the minimal surfaces in H^3 corresponding to these solutions of the elliptic cosh-Gordon equation were examined in [**31**].

In this paper we establish a Weierstraß-type representation for timelike CMC surfaces in Minkowski 3-space. The Weierstraß-type representation gives a unified construction of solutions to the sinh-Gordon, the Liouville and the cosh-Gordon equations. Moreover our Weierstraß-type representation can be regarded as a *non-linear d'Alembert formula* for these three nonlinear wave equations.

The paper is organized as follows: In Section 1, we fix notation and recall for the convenience of the reader basic facts on the geometry of surfaces in Minkowski 3-space. We devote Section 2 to results from loop group theory. We introduce certain loop groups (different from the ones usually used) and prove Birkhoff and Iwasawa splitting theorems (Theorem 2.5 and Theorem 2.6). In Section 3, we rewrite the relevant set of (harmonic map) equations as *one* differential equation with free *real* (loop) parameter. Based on this set-up, we derive a correspondence between harmonic maps into the pseudosphere S^2_1 with appropriate regularity and certain connection one-forms.

The Weierstraß-type representation for (Lorentzian) harmonic maps into S^2_1 is introduced in Section 4. In particular, we associate to each timelike constant mean curvature surface a "normalized potential" which is smooth around a base point, and, starting from any smooth potential, we construct a timelike constant mean curvature surface. Our results are only local at this point. It would be interesting to develop a global study of the results obtained in Section 4.

We apply the Weierstraß-type representation for constructing timelike CMC surfaces in Section 5. Moreover we exhibit a formula which allows us to find explicitly the normalized potentials for timelike constant mean curvature surfaces. The usefulness of the results in Section 5 is demonstrated in the final section. We discuss fundamental examples of timelike CMC surfaces via the Weierstraß-type representation.

1. Timelike surfaces

1.1 We start with preliminaries on the geometry of timelike surfaces in Minkowski 3-space. Let $\mathbb{E}^3_1$ be *Minkowski 3-space* with Lorentzian metric $\langle \cdot, \cdot \rangle = -du_1^2 + du_2^2 + du_3^2$ in terms of the natural coordinate system (u_1, u_2, u_3) of the Cartesian 3-space $\mathbb{R}^3$. Let M be a connected orientable 2-manifold and $\varphi : M \to \mathbb{E}^3_1$ an immersion. The immersion φ is said to be *timelike* if the induced metric I of M is Lorentzian. The induced Lorentzian metric I determines a Lorentz conformal structure $\mathcal{C}_I$ on M. We treat $(M, \mathcal{C}_I)$ as a Lorentz surface and φ as a conformal immersion. For the general theory of Lorentz surfaces, we refer to [**37**]. Hereafter we will assume that M is an orientable timelike surface in $\mathbb{E}^3_1$ (immersed by φ). It is worthwhile remarking that there exists no compact timelike surface in $\mathbb{E}^3_1$. (See [**33**], p.125.)

Let (x, y) be a *null coordinate system* with respect to the conformal structure $\mathcal{C}_I$. Then the first fundamental form I can be written in terms of (x, y) as follows:

$$I = e^{\omega}\, dxdy.$$

Now let N be a unit normal vector field of M, i.e. such that $\langle N, N\rangle = 1$, $\langle \varphi_x, N\rangle = \langle \varphi_y, N\rangle = 0$. The *second fundamental form* $I\!I$ of M derived from N is defined by

$$I\!I = -\langle \mathrm{d}\varphi,\ \mathrm{d}N\rangle.$$

The *shape operator* S of M derived from N is defined by $S := -\mathrm{d}N$. The shape operator S is related to $I\!I$ by $I\!I(X, Y) = \langle SX, Y\rangle$ for all vector fields X, Y on M. The mean curvature H of M is defined by $H := \mathrm{tr}\, S/2$. Note that the mean curvature H and Gaußian curvature K of M are given by $H = \mathrm{tr}(I\!I \cdot I^{-1})/2$, $K = \det S = \det(I\!I \cdot I^{-1})$. (See [**33**], page 107.) The characteristic values of S, i.e., the (complex) solutions to

$$\det(t\,\mathrm{I} - S) = 0,\ \mathrm{I} = \text{identity transformation of } TM.$$

are called the *principal curvatures.* Since the metric I is indefinite, both principal curvatures may be *non-real* complex numbers.

A point p of M is said to be an *umbilic point* if $I\!I$ is proportional to I at p. Equivalently, p is an umbilic point if and only if the two principal curvatures at p are the same and the corresponding eigenspace is 2-dimensional. A timelike surface is said to be a *totally umbilic surface* if all the points are umbilical. It is known that every totally umbilic timelike surface in $\mathbb{E}^3_1$ is congruent to an open portion of a pseudosphere $S^2_1(r) := \{\mathbf{u} \in \mathbb{E}^3_1 \mid \langle \mathbf{u}, \mathbf{u}\rangle = r^2\}$ of radius $r > 0$ or a timelike plane. Let $\varphi : M \to \mathbb{E}^3_1$ be a timelike surface with unit normal vector field N as before. Then, on a simply connected null coordinate region $\mathbb{D}$, we can define an orthonormal frame field $\mathcal{F}$ defined by

$$\mathcal{F} = (e^{-\omega/2}(-\varphi_x + \varphi_y), e^{-\omega/2}(\varphi_x + \varphi_y), N) : \mathbb{D} \to \mathrm{O}^{++}_1(3),$$

where $\mathrm{O}^{++}_1(3)$ denotes the identity component of the Lorentz group:

$$\mathrm{O}_1(3) = \left\{A \in \mathrm{GL}(3;\mathbb{R}) \mid \langle A\mathbf{u}, A\mathbf{v}\rangle = \langle \mathbf{u}, \mathbf{v}\rangle,\ \mathbf{u}, \mathbf{v} \in \mathbb{E}^3_1\ \right\}.$$

1.2 Throughout this paper, we identify $\mathbb{E}^3_1$ with the Lie algebra $\mathfrak{g} = \mathfrak{sl}(2;\mathbb{R})$. We take the following basis $\{\,\mathbf{i},\ \mathbf{j}',\ \mathbf{k}'\}$ of $\mathfrak{sl}(2;\mathbb{R})$:

$$\mathbf{i} = \begin{pmatrix} 0 & -1 \\ 1 & 0 \end{pmatrix},\quad \mathbf{j}' = \begin{pmatrix} 0 & 1 \\ 1 & 0 \end{pmatrix},\quad \mathbf{k}' = \begin{pmatrix} -1 & 0 \\ 0 & 1 \end{pmatrix}.$$

The basis $\{\,\mathbf{i},\ \mathbf{j}',\ \mathbf{k}'\}$ satisfies the relations $\mathbf{i}^2 = -\mathbf{1}$, $\mathbf{j}'^2 = \mathbf{k}'^2 = \mathbf{1}$, and $\mathbf{i}\mathbf{j}' = -\mathbf{j}'\mathbf{i} = \mathbf{k}'$, $\mathbf{j}'\mathbf{k}' = -\mathbf{k}'\mathbf{j}' = -\mathbf{i}$, $\mathbf{k}'\mathbf{i} = -\mathbf{i}\mathbf{k}' = \mathbf{j}'$, where $\mathbf{1}$ denotes the identity matrix. Hereafter we identify $\mathbb{E}^3_1$ with $\mathfrak{g}$ via

$$(u_1, u_2, u_3) \longleftrightarrow u_1\mathbf{i} + u_2\mathbf{j}' + u_3\mathbf{k}'. \tag{1.1}$$

The real algebra $\mathbb{H}'$ generated by $\{\,\mathbf{1},\ \mathbf{i},\ \mathbf{j}',\ \mathbf{k}'\}$ is called the algebra of *split quaternions.* The algebra $\mathbb{H}'$ is isomorphic to the algebra $\mathrm{M}(2;\mathbb{R})$ of all 2 by 2 real matrices. The commutation relations of $\mathfrak{g}$ are

$$[\,\mathbf{i},\ \mathbf{j}'\,] = 2\,\mathbf{k}',\ [\,\mathbf{j}',\ \mathbf{k}'\,] = -2\,\mathbf{i},\ [\,\mathbf{k}',\ \mathbf{i}\,] = 2\,\mathbf{j}'.$$

By the linear isomorphism (1.1) the Lorentz metric $\langle\cdot,\cdot\rangle$ corresponds to the scalar product:

$$\langle X, Y\rangle = \frac{1}{2}\mathrm{tr}(XY), \quad X, Y \in \mathfrak{g}. \tag{1.2}$$

This scalar product induces a biinvariant Lorentz metric of constant curvature -1 on the special linear group $G = \mathrm{SL}(2;\mathbb{R})$. Hence G is identified with the anti-de Sitter 3-space H^3_1. (See [**33**].)

The special linear group G acts isometrically on $\mathfrak{g}$ via the Ad-action:

$$\mathrm{Ad} : G \times \mathfrak{g} \to \mathfrak{g}; \quad \mathrm{Ad}(g)X = gXg^{-1}.$$

The Ad-action induces a double covering $G \to \mathrm{O}^{++}_1(3)$ of the Lorentz group $\mathrm{O}^{++}_1(3)$. By using this double covering we can find a lift $\hat{\Phi}$ (called a *coordinate frame*) of $\mathcal{F}$ to $\mathrm{SL}(2;\mathbb{R})$:

$$\mathrm{Ad}(\hat{\Phi})(\mathbf{i}, \mathbf{j}', \mathbf{k}') = \mathcal{F}. \tag{1.3}$$

The coordinate frame $\hat{\Phi}$ satisfies the following Frenet (or Gauß-Weingarten) equations [**23**]:

$$\frac{\partial}{\partial x}\hat{\Phi} = \hat{\Phi}\hat{U}, \quad \frac{\partial}{\partial y}\hat{\Phi} = \hat{\Phi}\hat{V}, \tag{1.4}$$

where

$$\hat{U} = \begin{pmatrix} -\frac{1}{4}\omega_x & -Qe^{-\frac{\omega}{2}} \\ \frac{H}{2}e^{\frac{\omega}{2}} & \frac{1}{4}\omega_x \end{pmatrix}, \quad \hat{V} = \begin{pmatrix} \frac{1}{4}\omega_y & -\frac{H}{2}e^{\frac{\omega}{2}} \\ Re^{-\frac{\omega}{2}} & -\frac{1}{4}\omega_y \end{pmatrix}, \tag{1.5}$$

and $Q := \langle \varphi_{xx}, N\rangle$, $R := \langle \varphi_{yy}, N\rangle$, $H = 2e^{-\omega}\langle \varphi_{xy}, N\rangle$. It is easy to see that $Q^\# := Q\mathrm{d}x^2$ and $R^\# := R\mathrm{d}y^2$ are globally defined on M. The quadratic differentials $Q^\#$ and $R^\#$ are called the *Hopf differentials of* M.

The second fundamental form $I\!I$ of M is related to Q and R by $I\!I = Q^\# + R^\# + HI$. This formula implies that the common zeros of Q and R coincide with the umbilic points of M. The Gauß equation is $H^2 - K = 4e^{-2\omega}QR$. Note that the condition $QR = 0$ does not imply the umbilicity of M. (See [**30**].)

The Gauß-Codazzi equation, i.e., the integrability condition of the Frenet equations $\hat{V}_x - \hat{U}_y + [\,\hat{U}, \hat{V}\,] = 0$ has the following form:

$$\omega_{xy} + \frac{1}{2}H^2e^{\omega} - 2QRe^{-\omega} = 0, \tag{G}$$

$$H_x = 2e^{-\omega}Q_y, \quad H_y = 2e^{-\omega}R_x. \tag{C}$$

The Codazzi equations (C) show that the constancy of the mean curvature H is equivalent to the condition $Q_y = R_x = 0$, i.e., $Q = Q(x)$, $R = R(y)$.

REMARK 1.1. *Let $(M, \mathcal{C})$ be a Lorentz surface and (x, y) a null coordinate system. Then the following differential operators are well defined:*

$$\mathrm{d}' := \frac{\partial}{\partial x}\mathrm{d}x, \quad \mathrm{d}'' := \frac{\partial}{\partial y}\mathrm{d}y. \tag{1.6}$$

A function $f : M \to \mathbb{R}$ is said to be a Lorentz holomorphic function [respectively, Lorentz anti-holomorphic function] if $\mathrm{d}''f = 0$ [respectively, $\mathrm{d}'f = 0$].

Next a 1-form $A = A_x\mathrm{d}x + A_y\mathrm{d}y$ is said to be a Lorentz holomorphic 1-form if $A = A_x\mathrm{d}x$ and A_x is a Lorentz holomorphic function. Similarly A is said to be a Lorentz anti-holomorphic 1-form if $A = A_y\mathrm{d}y$ and A_y is a Lorentz anti-holomorphic function.

According to this terminology, the constancy of mean curvature is characterized as follows: Let $\varphi : M \to \mathbb{E}^3_1$ *be a timelike surface. Then* (M, φ) *is of constant mean curvature if and only if* Q *is a Lorentz holomorphic function and* R *is a Lorentz anti-holomorphic function.*

REMARK 1.2. *Let* $(M, \mathcal{C})$ *be a Lorentz surface and* (x, y), $(\tilde{x}, \tilde{y})$ *null coordinate systems. Then these two coordinate systems are related by* $\partial\tilde{x}/\partial y = 0$, $\partial\tilde{y}/\partial x = 0$. *Namely* $\tilde{x}$ *and* $\tilde{y}$ *depend only on* x *and* y *respectively.*

On timelike surfaces of constant mean curvature $H \geq 0$, a special (local) coordinate system is available ([**16**], [**21**], [**30**], [**37**]).

PROPOSITION 1.3. *Let* $\varphi : M \to \mathbb{E}^3_1$ *be a timelike surface of constant mean curvature* $H \neq 0$. *Assume that* (M, φ) *has real distinct principal curvatures. Then there exists a local coordinate system* (x, y) *such that*

$$I = e^{\omega}\,\mathrm{d}x\mathrm{d}y, \quad II = \frac{H}{2}\left(\mathrm{d}x^2 + 2e^{\omega}\mathrm{d}x\mathrm{d}y + \mathrm{d}y^2\right). \tag{1.7}$$

With respect to this local coordinate system, the Gauß-Codazzi equation becomes

$$\omega_{xy} + H^2 \sinh\omega = 0. \tag{shG}$$

The partial differential equation (shG) is called the *hyperbolic sinh-Gordon equation* or *affine Toda field equation* of type $A_1^{(1)}$.

REMARK 1.4. *Let* (x, y) *be the local coordinate system in the preceding Proposition. Introduce a local coordinate system* (u, v) *by* $x = u + v$, $y = -u + v$. *Then* I *and* II *can be represented as*

$$I = e^{\omega}(-\mathrm{d}u^2 + \mathrm{d}v^2), \quad II = 2He^{\frac{\omega}{2}}\left(-\sinh\frac{\omega}{2}\mathrm{d}u^2 + \cosh\frac{\omega}{2}\mathrm{d}v^2\right).$$

The local coordinate system (u, v) *is (Lorentz) isothermal and a curvature-line coordinate system. Such a coordinate system* (u, v) *is called an isothermic coordinate system.*

Timelike CMC surfaces with real distinct principal curvatures are called *isothermic timelike CMC surfaces.* Note that an isothermic coordinate system can be characterized as a local null coordinate system (x, y) such that $Q = R \neq 0$. See [**16**]. A very different situation is discussed in the following proposition.

PROPOSITION 1.5. *Let* $\varphi : M \to \mathbb{E}^3_1$ *be a timelike surface of constant mean curvature* $H \neq 0$. *Assume that* (M, φ) *has non-real complex principal curvatures. Then there exists a local coordinate system* (x, y) *such that*

$$I = e^{\omega}\mathrm{d}x\mathrm{d}y, \quad II = \frac{H}{2}\left(\mathrm{d}x^2 + 2e^{\omega}\mathrm{d}x\mathrm{d}y - \mathrm{d}y^2\right). \tag{1.8}$$

With respect to this local coordinate system, the Gauß-Codazzi equations become

$$\omega_{xy} + H^2 \cosh\omega = 0. \tag{chG}$$

The partial differential equation (chG) is called the *hyperbolic cosh-Gordon equation.* The local coordinate system (x, y) is called an *anti-isothermic coordinate system*; this may be characterized as a local null coordinate system such that $Q = -R \neq 0$. Timelike CMC surfaces with imaginary principal curvatures are called *anti-isothermic timelike CMC surfaces.* The notion of "anti-isothermic coordinate" has been introduced in [**16**] (Definition 4.13).

REMARK 1.6. *Let* $\varphi : M \to \mathbb{E}^3_1$ *be a timelike surface of constant mean curvature* H. *Assume that* (M, φ) *has two equal and real principal curvatures. Then the Gauß equation of* (M, φ) *becomes the Liouville equation:*

$$\omega_{xy} + \frac{H^2}{2} e^{\omega} = 0. \tag{L}$$

1.3 Next, we shall define the Gauß map of a timelike surface. Let M be a timelike surface and N a unit normal vector field to M. The *Gauß map* ψ of M is a smooth map of M into S^2_1, which assigns to each $p \in M$ the point $\psi(p) \in \mathbb{E}^3_1$ obtained by parallel translation of the unit normal vector N_p of M at p to the origin of $\mathbb{E}^3_1$. The constancy of the mean curvature is characterized by the harmonicity of the Gauß map (cf. [**30**]).

PROPOSITION 1.7. *The Gauß map of a timelike surface is harmonic if and only if the mean curvature is constant.*

REMARK 1.8. *Let* (M_1, g_1) *and* (M_1, g_2) *be (semi-) Riemannian manifolds and* $\psi : M_1 \to M_2$ *a smooth map. Then* ψ *is said to be a harmonic map if its tension field*

$$\tau(\psi) := \mathrm{tr}(\nabla \mathrm{d}\psi)$$

vanishes. In the case $\dim M_1 = 2$, *the harmonicity of a smooth map* ψ *is invariant under the conformal transformation of* (M_1, g_1). *In particular, when* (M_1, g_1) *is a Lorentzian 2-manifold, a harmonic map* ψ *is often called a (Lorentzian) harmonic map (or wave map). Note that the constancy of the Gaußian curvature is also characterized by the following proposition.*

PROPOSITION 1.9. *Let* M *be a timelike surface. Assume that the Gaußian curvature has constant sign on* M. *Then the second fundamental form* II *gives* M *another (semi-) Riemannian metric. With respect to the conformal structure determined by* II, *the Gauß map of* M *is harmonic if and only if* K *is constant.*

The Ad-action of G on S^2_1 is transitive and isometric. The isotropy subgroup K of G at $\mathbf{k}'$ is $K = \{u_0 \mathbf{1} + u_3 \mathbf{k}' \mid u_0^2 - u_3^2 = -1\}$. The isotropy subgroup K is isomorphic to the multiplicative group $\mathbb{R}^*$. The natural projection $\pi : G = H^3_1 \to S^2_1$, given by $\pi(g) = \mathrm{Ad}(g)\mathbf{k}'$, $g \in G$, defines a principal $\mathbb{R}^*$-bundle H^3_1 over S^2_1. The fibering $\pi : H^3_1 \to S^2_1$ is called the *Hopf-fibering of* S^2_1. The Lie algebra $\mathfrak{k}$ of K is given by $\mathfrak{k} = \mathbb{R}\,\mathbf{k}'$. The tangent space of S^2_1 at the origin $\mathbf{k}'$ is $\mathfrak{m} = \mathbb{R}\mathbf{i} \oplus \mathbb{R}\mathbf{j}'$. Let σ be the involution of $\mathfrak{g}$ defined by $\sigma = \mathrm{Ad}(\mathbf{k}') = \Pi_{\mathfrak{k}} - \Pi_{\mathfrak{m}}$, where $\Pi_{\mathfrak{k}}$ and $\Pi_{\mathfrak{m}}$ are the projections from $\mathfrak{g}$ onto $\mathfrak{k}$ and $\mathfrak{m}$ respectively. Then the pair $(\mathfrak{g}, \sigma)$ is a symmetric Lie algebra data associated with the Lorentzian symmetric space $S^2_1 = G/K$.

2. Loop groups

2.1 To study timelike CMC surfaces in the spirit of [**14**], we need to introduce some notation involving loop groups. Let us denote the polynomial loop algebra of $\mathfrak{g} = \mathfrak{sl}(2; \mathbb{R})$ by

$$\Lambda_{\mathrm{pol}}\, \mathfrak{g} = \{\xi(\lambda) = \sum_{\mathrm{finite}} \xi_j \lambda^j : S^1 \to \mathfrak{g}\},$$

where S^1 denotes the unit circle in $\mathbb{C}$. Let σ be the involution of $\mathfrak{g}$ corresponding to the Lorentzian symmetric space $S^2_1 = G/K$ defined above. Then the *polynomial twisted loop algebra* of $\mathfrak{g}$ is defined by

$$\Lambda_{\mathrm{pol}}\,\mathfrak{g}_\sigma = \{\xi(\lambda) \in \Lambda_{\mathrm{pol}}\mathfrak{g} \mid \sigma(\xi(\lambda)) = \xi(-\lambda)\}\,.$$

For the purposes of this paper we need a certain Banach-completion of $\Lambda_{\mathrm{pol}}\,\mathfrak{g}_\sigma$. To this end we introduce the norm $|\cdot|$ for $\mathfrak{g}$:

$$|A| := \max_j \{\sum_{i=1}^{2} |a_{ij}|\ \}, \quad A = (a_{ij}) \in \mathfrak{g}.$$

We extend this norm to the polynomial loop algebra in the following way:

$$||\xi|| = \sum_j |\xi_j|, \quad \xi = \sum_j \xi_j \lambda^j \in \Lambda_{\mathrm{pol}}\,\mathfrak{g}.$$

Denote the completions of $\Lambda_{\mathrm{pol}}\,\mathfrak{g}$ and $\Lambda_{\mathrm{pol}}\,\mathfrak{g}_\sigma$ with respect to the norm $||\cdot||$ by $\Lambda\mathfrak{g}$ and $\Lambda\mathfrak{g}_\sigma$ respectively. Then the Lie algebras $\Lambda\mathfrak{g}$ and $\Lambda\mathfrak{g}_\sigma$ are Banach Lie algebras (cf. Proposition 4.2.1 in [**36**]). Actually, these are Banach algebras of continuous functions on S^1. Moreover since the involution σ is inner, these Banach Lie algebras are isomorphic.

Next we introduce the following Lie subalgebras of $\Lambda\mathfrak{g}$:

$$\Lambda^+\mathfrak{g} = \{\xi(\lambda) = \sum_{j\geq 0} \xi_j\lambda^j \in \Lambda\mathfrak{g}\}, \quad \Lambda^-\mathfrak{g} = \{\xi(\lambda) = \sum_{j\leq 0} \xi_j\lambda^j \in \Lambda\mathfrak{g}\},$$

$$\Lambda^+_*\mathfrak{g} = \{\xi(\lambda) = \sum_{j>0} \xi_j\lambda^j \in \Lambda\mathfrak{g}\}, \quad \Lambda^-_*\mathfrak{g} = \{\xi(\lambda) = \sum_{j<0} \xi_j\lambda^j \in \Lambda\mathfrak{g}\}.$$

Then we have the following decompositions as direct sums of linear spaces: $\Lambda\mathfrak{g} = \Lambda^+_*\mathfrak{g} \oplus \Lambda^-\mathfrak{g} = \Lambda^-_*\mathfrak{g} \oplus \Lambda^+\mathfrak{g}$. Similarly, we introduce the Lie subalgebras $\Lambda^+\mathfrak{g}_\sigma, \Lambda^-\mathfrak{g}_\sigma$, $\Lambda^+_*\mathfrak{g}_\sigma, \Lambda^-_*\mathfrak{g}_\sigma$ of the twisted loop algebra, and these satisfy $\Lambda\mathfrak{g}_\sigma = \Lambda^+_*\mathfrak{g}_\sigma \oplus \Lambda^-\mathfrak{g}_\sigma = \Lambda^-_*\mathfrak{g}_\sigma \oplus \Lambda^+\mathfrak{g}_\sigma$. One can define analogous connected Banach Lie groups: ΛG, $\Lambda^\pm G$ and $\Lambda^\pm_* G$, whose Lie algebras are $\Lambda\mathfrak{g}$, $\Lambda^\pm\mathfrak{g}$ and $\Lambda^\pm_*\mathfrak{g}$ respectively, and similarly for the twisted case.

2.2 In this subsection we recall the classically known Birkhoff decomposition theorem for loop groups of the *complex* special linear group $\mathrm{SL}(2;\mathbb{C})$. We use this result in order to prove that a similar factorization holds for the loop groups $\tilde{\Lambda}G$, and $\tilde{\Lambda}G_\sigma$ defined below.

THEOREM 2.1. (*Birkhoff decomposition of* $\Lambda G^{\mathbb{C}}$) *We have:*

$$\Lambda G^{\mathbb{C}} = \bigsqcup_{w\in\mathfrak{T}} \Lambda^- G^{\mathbb{C}} \cdot w \cdot \Lambda^+ G^{\mathbb{C}}.$$

Here $\mathfrak{T}$ denotes the group of homomorphisms from S^1 into the subgroup of diagonal matrices of $\mathrm{SL}(2;\mathbb{C})$, *that is,*

$$\mathfrak{T} = \left\{ \left(\begin{array}{cc} \lambda^a & 0 \\ 0 & \lambda^{-a} \end{array} \right) \middle| \, a > 0 \right\}.$$

Moreover, the multiplication maps

$$\Lambda^-_* G^{\mathbb{C}} \times \Lambda^+ G^{\mathbb{C}} \to \Lambda G^{\mathbb{C}}, \quad \Lambda^+_* G^{\mathbb{C}} \times \Lambda^- G^{\mathbb{C}} \to \Lambda G^{\mathbb{C}}$$

are diffeomorphisms onto the open dense subsets $\mathcal{B}^\circ_\Lambda(-,+)$ *and* $\mathcal{B}^\circ_\Lambda(+,-)$ *of* $\Lambda G^{\mathbb{C}}$, *called the big cells of* $\Lambda G^{\mathbb{C}}$. *In particular if* γ *is an element of* $\mathcal{B}_\Lambda = \mathcal{B}^\circ_\Lambda(-,+) \cap \mathcal{B}^\circ_\Lambda(+,-)$, *then* γ *has unique decompositions*:

$$\gamma = \gamma_- \cdot \ell_+ = \gamma_+ \cdot \ell_-, \quad \gamma_\pm \in \Lambda_*^\pm G^{\mathbb{C}}, \ \ell_\pm \in \Lambda^\pm G^{\mathbb{C}}.$$

Here

$$\Lambda_*^+ G^{\mathbb{C}} = \{\gamma \in \Lambda^+ G \mid \gamma(\lambda) = \mathbf{1} + \sum_{k\geq 1} \gamma_k \lambda^k \ \},$$

and

$$\Lambda_*^- G^{\mathbb{C}} = \{\gamma \in \Lambda^- G \mid \gamma(\lambda) = \mathbf{1} + \sum_{k\leq -1} \gamma_k \lambda^k \ \}.$$

Next let $\tilde{\Lambda} G$ be the subset of ΛG whose elements, as maps defined on S^1, admit analytic continuations to $\mathbb{C}^*$. Similarly we define $\tilde{\Lambda} G_\sigma$. It is easy to check that $\tilde{\Lambda} G$ is a subgroup of ΛG and that $\tilde{\Lambda} G_\sigma$ is a subgroup of ΛG_σ. For $\tilde{\Lambda} G$ we will use the topology induced from ΛG. Then we obtain the following theorem.

THEOREM 2.2. (*Birkhoff decomposition of* $\tilde{\Lambda} G$) *We have:*

$$\tilde{\Lambda} G = \bigsqcup_{w \in \mathcal{T}} \tilde{\Lambda}^- G \cdot w \cdot \tilde{\Lambda}^+ G.$$

Here $\mathcal{T}$ *denotes the group of homomorphisms from* S^1 *into the subgroup of diagonal matrices of* $\mathrm{SL}(2;\mathbb{C})$. *Moreover, the multiplication maps*

$$\tilde{\Lambda}_*^- G \times \tilde{\Lambda}^+ G \to \tilde{\Lambda} G, \quad \tilde{\Lambda}_*^+ G \times \tilde{\Lambda}^- G \to \tilde{\Lambda} G.$$

are diffeomorphisms onto the open dense subsets $\mathcal{B}^\circ(-,+)$ *and* $\mathcal{B}^\circ(+,-)$ *of* $\tilde{\Lambda} G$, *called the big cells of* $\tilde{\Lambda} G$. *In particular if* γ *is an element of* $\mathcal{B}^\circ = \mathcal{B}^\circ(-,+) \cap \mathcal{B}^\circ(+,-)$, *then* γ *has unique decompositions*:

$$\gamma = \gamma_- \cdot \ell_+ = \gamma_+ \cdot \ell_-, \quad \gamma_\pm \in \tilde{\Lambda}_*^\pm G, \ \ell_\pm \in \tilde{\Lambda}^\pm G.$$

PROOF. The idea of this proof is similar to the one presented in the Appendix of [**36**]. □

Finally, we consider the twisted loop groups defined earlier. We also define the twisted analytic loop groups derived from $\tilde{\Lambda} G_\sigma$ in the obvious way. Then we have

THEOREM 2.3. (*Birkhoff decomposition of* $\tilde{\Lambda} G_\sigma$) *Theorem* 2.2 *also holds for the twisted groups.*

PROOF. We note that the twisting involution σ is given by an inner automorphism of G. Therefore, the twisted loop group and the untwisted loop group are isomorphic. Actually, the isomorphism from the untwisted loop group to the twisted loop group is easily given: powers λ^k on the diagonal are doubled, powers λ^k in the $(1,2)$-position are replaced by λ^{2k+1} and in the $(2,1)$-position by λ^{2k-1}. The claim now follows. □

REMARK 2.4. *The proof actually gives the Birkhoff decomposition theorem for* ΛG_σ *as well.*

2.3 We have the following fundamental decomposition theorem:

THEOREM 2.5. (*Iwasawa decomposition of* $\Lambda G_\sigma \times \Lambda G_\sigma$) *Let* $\Delta(\Lambda G_\sigma \times \Lambda G_\sigma)$ *denote the diagonal subgroup of* $\Lambda G_\sigma \times \Lambda G_\sigma$. *Then we have*

$$\Lambda G_\sigma \times \Lambda G_\sigma = \bigsqcup \Delta(\Lambda G_\sigma \times \Lambda G_\sigma) \cdot (\mathbf{1}, w) \cdot (\Lambda^- G_\sigma \times \Lambda^+ G_\sigma),$$

where $w \in \mathfrak{T}$ *is as in Theorem 2.2. Moreover, the multiplication maps*

$$\Delta(\Lambda G_\sigma \times \Lambda G_\sigma) \times (\Lambda_*^- G_\sigma \times \Lambda^+ G_\sigma) \to \Lambda G_\sigma \times \Lambda G_\sigma,$$
$$\Delta(\Lambda G_\sigma \times \Lambda G_\sigma) \times (\Lambda_*^+ G_\sigma \times \Lambda^- G_\sigma) \to \Lambda G_\sigma \times \Lambda G_\sigma$$

are diffeomorphisms onto the open dense subsets $\mathfrak{I}_\Lambda(+,-)$ *and* $\mathfrak{I}_\Lambda(-,+)$ *of* $\Lambda G_\sigma \times \Lambda G_\sigma$*– called the big cells of* $\Lambda G_\sigma \times \Lambda G_\sigma$.

PROOF. Take $(g,h) \in \Lambda G_\sigma \times \Lambda G_\sigma$. Decompose $g^{-1} \cdot h$ according to the Birkhoff decomposition of ΛG_σ (Theorem 2.2): $g^{-1} \cdot h = u_- \, w \, u_+$, $u_\pm \in \Lambda G_\sigma$, $w \in \mathfrak{T}$. It is easy to verify that the splitting $(g,h) = (gu_-, gu_-)(\mathbf{1}, w)(\,(u_-)^{-1}, u_+)$ gives the Iwasawa decomposition of (g,h) in the *un*twisted loop group $\Lambda G \times \Lambda G$. (cf. Theorem 4.1 in [**6**].) Since the factors gu_- and $u_\pm$ are σ-twisted, this splitting is the required (Iwasawa) splitting in the *twisted* loop group $\Lambda G_\sigma \times \Lambda G_\sigma$.

For the second claim we note that our definitions imply

$$\Delta(\Lambda G_\sigma \times \Lambda G_\sigma) \bigcap (\Lambda_*^- G_\sigma \times \Lambda^+ G_\sigma) = \{\mathbf{1}\}.$$

Thus the splitting is unique, whence the map is a bijection onto its image. The proof that it is a diffeomorphism is almost the same as for the Birkhoff decomposition, as in [**11**]. It remains to show that the big cell is dense. But if (g,h) is given, then $g^{-1} \cdot h$ is in $\Lambda G_\sigma \cong \Lambda G$ and from the Birkhoff decomposition theorem 2.3, we know that in every neighbourhood of this element there is an element in the big cell. Therefore, in every neighbourhood of g and h there exists some g' and h' such that $(g')^{-1} \cdot h'$ is in the big cell of ΛG_σ. But the proof above shows that then (g', h') is in the big cell of $\Lambda G_\sigma \times \Lambda G_\sigma$. □

2.4 For twisted loop groups of elements with analytic extensions, we have the following decomposition theorem:

THEOREM 2.6. (*Iwasawa decomposition of* $\tilde{\Lambda} G_\sigma \times \tilde{\Lambda} G_\sigma$) *Let* $\Delta(\tilde{\Lambda} G_\sigma \times \tilde{\Lambda} G_\sigma)$ *denote the diagonal subgroup of* $\tilde{\Lambda} G_\sigma \times \tilde{\Lambda} G_\sigma$. *Then we have*

$$\tilde{\Lambda} G_\sigma \times \tilde{\Lambda} G_\sigma = \bigsqcup \Delta(\tilde{\Lambda} G_\sigma \times \tilde{\Lambda} G_\sigma) \cdot (\mathbf{1}, w) \cdot (\tilde{\Lambda}^- G_\sigma \times \tilde{\Lambda}^+ G_\sigma),$$

where $w \in \mathfrak{T}$ *is as in Theorem 2.2. Moreover, the multiplication maps*

$$\Delta(\tilde{\Lambda} G_\sigma \times \tilde{\Lambda} G_\sigma) \times (\tilde{\Lambda}_*^- G_\sigma \times \tilde{\Lambda}^+ G_\sigma) \to \tilde{\Lambda} G_\sigma \times \tilde{\Lambda} G_\sigma,$$
$$\Delta(\tilde{\Lambda} G_\sigma \times \tilde{\Lambda} G_\sigma) \times (\Lambda_*^+ G_\sigma \times \tilde{\Lambda}^- G_\sigma) \to \tilde{\Lambda} G_\sigma \times \tilde{\Lambda} G_\sigma$$

are diffeomorphisms onto the open dense subsets $\mathfrak{I}(+,-)$ *and* $\mathfrak{I}(-,+)$ *of* $\tilde{\Lambda} G_\sigma \times \tilde{\Lambda} G_\sigma$*– called the big cells of* $\tilde{\Lambda} G_\sigma \times \tilde{\Lambda} G_\sigma$.

PROOF. Take $(g,h) \in \tilde{\Lambda} G_\sigma \times \tilde{\Lambda} G_\sigma$. Decompose $g^{-1} \cdot h$ according to the Birkhoff decomposition of $\tilde{\Lambda} G_\sigma$ (Theorem 2.2):

$$g^{-1} \cdot h = u_- \, w \, u_+, \quad u_\pm \in \tilde{\Lambda} G_\sigma, \quad w \in \mathfrak{T}.$$

As in the proof of Theorem 2.5, the splitting $(g,h) = (gu_-, gu_-)(\mathbf{1}, w)((u_-)^{-1}, u_+)$ gives the Iwasawa decomposition of (g,h) in the *twisted* loop group $\Lambda G_\sigma \times \Lambda G_\sigma$.

We need to check that u_- and u_+ are in the "$\sim$"-group. By definition, u_- has a holomorphic extension to the exterior of the unit disk and is finite at ∞. Thus $u_+ = (gu_-w)^{-1}h$ has a holomorphic extension to the exterior of the unit disk. Thus u_+ has holomorphic extension to $\mathbb{C}^*$. Similarly u_- has also a holomorphic extension to $\mathbb{C}^*$. This proves the claim. The remaining assertions follow from the previous theorem, since we use the induced topology. □

REMARK 2.7. *The classical Iwasawa decomposition for untwisted loop groups of compact simple Lie groups was generalized in* [**25**] *to the case of loop groups of general simple Lie groups, and in* [**4**], *of general Lie groups.*

REMARK 2.8. *I. T. Gohberg and his collaborators investigated splittings for matrix valued functions over (separate) contours and more general Banach algebras of functions. For instance, the Birkhoff splitting for matrix valued functions with coefficients in the Wiener algebra:*

$$\mathcal{A} = \{f(\lambda) = \sum f_j\,\lambda^j : S^1 \to \mathbb{C} \;\Big|\; \sum |f_j| < \infty\}$$

was established by Gohberg in [**17**]. *Splittings for matrix valued functions with coefficients in the Wiener algebra on the real line were obtained* [**8**]. *For more information, we refer to* [**17**] *and* [**8**] *and the references therein.*

For our geometric purposes—a Weierstraß-type representation for timelike surfaces in Minkowski space— we need real Banach algebras of functions and in fact we also need a real loop parameter λ. Fortunately all the loop group elements occurring in our geometric context (extended framings etc.) have analytic extensions to $\mathbb{C}^*$, i.e. these geometric loop group elements are all contained in $\tilde{\Lambda}G$. For this reason we have presented in this section the Birkhoff and Iwasawa decomposition theorems for the twisted analytic loop group $\tilde{\Lambda}G_\sigma$.

3. Harmonic maps into S^2_1

3.1 In this section we shall derive a correspondence between harmonic maps from a simply connected Lorentz surface $\mathbb{D}$ to S^2_1 and certain flat connections (the so-called *zero curvature representation*). We note that since the harmonic map equation is a local condition, it suffices to consider harmonic maps from simply connected Lorentz surfaces into S^2_1. For the rest of this paper $\mathbb{D}$ will always denote a simply connected region of the Minkowski plane $(\mathbb{R}^2(x,y), dxdy)$ containing the origin.

The following result is the starting point of our approach.

PROPOSITION 3.1. *A smooth map $\psi : \mathbb{D} \to S^2_1 \subset \mathbb{E}^3_1$ is harmonic if and only if $\psi_{xy} = \rho\,\psi$ for some function ρ on $\mathbb{D}$.*

3.2 Let $\psi : \mathbb{D} \to S^2_1$ be a smooth map and $\pi : G \to S^2_1$ the Hopf fibration as before. Since $\mathbb{D}$ is simply connected, ψ has a smooth lift $\Psi : \mathbb{D} \to G$ unique up to the right K-action. Such a lift Φ is called a *framing* of ψ. Note that the framing Ψ is related to ψ by $\psi = \mathrm{Ad}(\Psi)\mathbf{k}'$. Since $\mathbb{D}$ is simply connected, the pull-back bundle $\psi^*\,G$ is necessarily a trivial bundle $\mathbb{D} \times K$. So the group $\mathcal{G}$ of gauge transformations of $\psi^*\,G$ can be identified with $\mathrm{C}^\infty(\mathbb{D}, K)$.

We wish to describe the harmonicity of ψ in terms of a framing. Let μ_G be the Maurer-Cartan form of G. It is well-known that μ_G satisfies the *Maurer-Cartan equation*:

$$d\mu_G + \frac{1}{2}[\mu_G \wedge \mu_G] = 0.$$

The 1-form $\alpha = \Psi^*\mu_G = \Psi^{-1}d\Psi$ of μ_G by Ψ then satisfies

$$d\alpha + \frac{1}{2}[\alpha \wedge \alpha] = 0. \tag{3.1}$$

The identity (3.1) is equivalent to the integrability condition for the existence of a smooth map $\Psi : \mathbb{D} \to G$ such that $\alpha = \Psi^*\mu_G$ (the Frobenius theorem). By definition, α is a $\mathfrak{g}$-valued 1-form on $\mathbb{D}$. The $\mathfrak{g}$-valued 1-form α has a type decomposition $\alpha = \alpha_0 + \alpha_1$ with respect to $\mathfrak{g} = \mathfrak{k} \oplus \mathfrak{m}$. Write

$$\alpha_0 = \alpha'_{\mathfrak{k}}\, dx + \alpha''_{\mathfrak{k}}\, dy, \quad \alpha_1 = \alpha'_{\mathfrak{m}}\, dx + \alpha''_{\mathfrak{m}}\, dy.$$

Then α_0 and α_1 can be decomposed with respect to the conformal structure of $\mathbb{D}$ as follows:

$$\alpha_0 = \alpha'_0 + \alpha''_0, \; \alpha_1 = \alpha'_1 + \alpha''_1,$$
$$\alpha'_0 = \alpha'_{\mathfrak{k}}\, dx, \quad \alpha''_0 = \alpha''_{\mathfrak{k}}\, dy, \quad \alpha'_1 = \alpha'_{\mathfrak{m}}\, dx, \quad \alpha''_1 = \alpha''_{\mathfrak{m}}\, dy.$$

Define α' and α'' by $\alpha' := \alpha'_0 + \alpha'_1$, $\alpha'' := \alpha''_0 + \alpha''_1$. The 1-forms α' and α'' are called the (1,0)-*part* and (0,1)-*part* of α respectively. With respect to the conformal structure of $\mathbb{D}$, we decompose the exterior differential operator d as $d = d' + d''$ according to (1.6). Then we have $\alpha' = \Psi^{-1}\, d'\Psi$, $\alpha'' = \Psi^{-1}\, d''\Psi$. By the usual computations we obtain the following (cf. [**28**])

PROPOSITION 3.2. *Let $\psi : \mathbb{D} \to S^2_1$ be a smooth map with framing Ψ. Then ψ is harmonic if and only if $d(*\alpha_1) + [\alpha_0 \wedge *\alpha_1] = 0$ for $\alpha = \Psi^{-1}d\Psi$. Here $*$ denotes the Hodge star operator acting on ($\mathfrak{g}$-valued) one forms on $\mathbb{D}$ defined by $*\, dx = dx$, $*\, dy = -dy$.*

3.3 Let $A^1(\mathbb{D}; \mathfrak{g})$ be the space of all $\mathfrak{g}$-valued one-forms and $\mathcal{A}$ the affine space of all connection 1-forms on the product bundle $\mathbb{D} \times G$. The space $\mathcal{A}$ is an affine space associated to the linear space $A^1(\mathbb{D}; \mathfrak{g})$. We shall choose the trivial flat connection as the origin of $\mathcal{A}$, then the space $\mathcal{A}$ is identified with $A^1(\mathbb{D}; \mathfrak{g})$. Hereafter we shall identify $\mathcal{A}$ with $A^1(\mathbb{D}; \mathfrak{g})$ in this way.

DEFINITION 3.3. *A connection $\alpha \in \mathcal{A} = A^1(\mathbb{D}; \mathfrak{g})$ is admissible provided that*

$$d\alpha + \frac{1}{2}[\alpha \wedge \alpha] = 0, \quad d(*\alpha_1) + [\alpha_0 \wedge *\alpha_1] = 0.$$

The space of all admissible connections on $\mathbb{D}$ is denoted by $\mathcal{A}^*$. The Frobenius theorem implies the following:

LEMMA 3.4. *Let α be an admissible connection. Then there exists a harmonic map $\psi : \mathbb{D} \to S^2_1$ such that for any framing Ψ of ψ we have $\Psi^{-1}d\Psi = \alpha$.*

Recall that framings are uniquely determined (as a solution to $\Psi^{-1}d\Psi = \alpha$) up to multiplication on the left by matrices independent of x and y. We will remove this freedom in the loop group formalism discussed below.

3.4 For an admissible connection α, we define a scalar field $\mathcal{S}(\alpha)$ by

$$\mathcal{S}(\alpha)(x, y) := \langle \alpha'_{\mathfrak{m}}(x, y),\ \alpha'_{\mathfrak{m}}(x, y)\rangle \cdot \langle \alpha''_{\mathfrak{m}}(x, y),\ \alpha''_{\mathfrak{m}}(x, y)\rangle.$$

Let us define the linear subspaces $\mathcal{A}_+$, $\mathcal{A}_0$ and $\mathcal{A}_-$ of $\mathcal{A}^*$ by

$$\mathcal{A}_+ := \{\alpha \in \mathcal{A}^* \mid \mathcal{S}(\alpha) > 0\}, \quad \mathcal{A}_- := \{\alpha \in \mathcal{A}^* \mid \mathcal{S}(\alpha) < 0\},$$
$$\mathcal{A}_0 := \{\alpha \in \mathcal{A}^* \mid \mathcal{S}(\alpha) = 0\}.$$

It is easily checked that $\mathcal{A}^*$, $\mathcal{A}_\pm$ and $\mathcal{A}_0$ are invariant under the action of the gauge group as well as under conformal changes of $\mathbb{D}$, where the gauge group $\mathcal{G}$ acts on $\mathcal{A}$ as follows:

$$g^*\alpha = g^{-1}\,\mathrm{d}g + \mathrm{Ad}(g^{-1})\alpha, \; g \in \mathcal{G}, \; \alpha \in \mathcal{A}.$$

PROPOSITION 3.5. *Let $\mathcal{H}^*$ be the space of all harmonic maps from $\mathbb{D}$ to S^2_1. Define the linear subspaces $\mathcal{H}_+$, $\mathcal{H}_-$ and $\mathcal{H}_0$ of $\mathcal{H}^*$ by*

$$\begin{aligned}
\mathcal{H}_+ &= \{\psi \in \mathcal{H}^* \mid \langle \psi_x, \psi_x\rangle \langle \psi_y, \psi_y\rangle > 0 \text{ on } \mathbb{D}\},\\
\mathcal{H}_- &= \{\psi \in \mathcal{H}^* \mid \langle \psi_x, \psi_x\rangle \langle \psi_y, \psi_y\rangle < 0 \text{ on } \mathbb{D}\},\\
\mathcal{H}_0 &= \{\psi \in \mathcal{H}^* \mid \langle \psi_x, \psi_x\rangle \langle \psi_y, \psi_y\rangle = 0 \text{ on } \mathbb{D}\}.
\end{aligned}$$

Then there are the following bijective correspondences:

$$\mathcal{H}^* \longleftrightarrow \mathcal{A}^*/\mathcal{G}, \quad \mathcal{H}_\pm \longleftrightarrow \mathcal{A}_\pm/\mathcal{G}, \quad \mathcal{H}_0 \longleftrightarrow \mathcal{A}_0/\mathcal{G}.$$

Here $\mathcal{G}$ denotes the gauge group of the bundle $\psi^*G = \mathbb{D}\times K$. These correspondences are given by $\alpha = \Psi^{-1}\mathrm{d}\Psi$ via a framing Ψ of ψ.

REMARK 3.6. *Here is a differential geometric interpretation of the subspaces $\mathcal{H}_\pm$ and $\mathcal{H}_0$. We know that to any harmonic map $\psi \in \mathcal{H}_\pm$, there corresponds a timelike immersion φ. (See Section 4.) The condition $\psi \in \mathcal{H}_+$ [respectively, $\psi \in \mathcal{H}_-$] is equivalent to the positivity [respectively, negativity] of the discriminant for the characteristic equation of the shape operator of φ. Of course the condition $\psi \in \mathcal{H}_0$ corresponds to the property "φ has real repeated principal curvatures".*

3.5 To close this section, we introduce the so-called *spectral parameter.*

DEFINITION 3.7. *Let $\alpha \in \mathcal{A}$ be a connection. A loop α_λ of connections through α is defined by $\alpha_\lambda = \alpha_0 + \lambda\alpha_1' + \lambda^{-1}\alpha_1''$, $\lambda \in \mathbb{R}^+$.*

Note that the spectral deformation preserves $\mathcal{S}$, i.e., $\mathcal{S}(\alpha_\lambda) \equiv \mathcal{S}(\alpha)$. The following observation is fundamental for our approach.

PROPOSITION 3.8. *A connection $\alpha \in \mathcal{A}$ is admissible if and only if the loop α_λ through α satisfies*

$$\mathrm{d}\alpha_\lambda + \frac{1}{2}[\alpha_\lambda \wedge \alpha_\lambda] = 0 \tag{3.2}$$

for every λ.

It is clear that each α_λ is admissible whenever α is, and α_λ generates the same loop. For every α_λ satisfying (3.2), there exists a one-parameter family of smooth maps $\Psi_\lambda : \mathbb{D} \to G$ depending smoothly on λ such that $\Psi_\lambda^{-1}\mathrm{d}\Psi_\lambda = \alpha_\lambda$. From now on we shall normalize Ψ_λ by $\Psi_\lambda(0,0) \equiv \mathbf{1}$. Such a normalized one-parameter family $\{\Psi_\lambda\}_{\lambda\in\mathbb{R}^+}$ of maps is called an *extended framing* of the harmonic map ψ. To every harmonic map $\psi : \mathbb{D} \to S^2_1$, there is a naturally associated one-parameter family of harmonic maps $\{\psi_\lambda\}$ such that $\psi_1 = \psi$, parametrized by $\lambda \in \mathbb{R}^+$. We shall therefore refer to ψ_λ as a *loop of harmonic maps* (*through* ψ). The extended framing Ψ_λ of ψ has its values in $\tilde{\Lambda}G_\sigma$ and can therefore be regarded as a mapping $\Psi = \Psi_\lambda : \mathbb{D} \to \tilde{\Lambda}G_\sigma$ into the twisted loop group.

REMARK 3.9. *The Cauchy problem for Lorentzian harmonic maps from* $\mathbb{E}^2_1$ *into* S^2_1 *was investiaged in* [**19**].

4. Weierstraß-type representation for harmonic maps

4.1 In this section, we establish a Weierstraß-type representation for Lorentzian harmonic maps into $S^2_1 = G/K$. We start by introducing the following linear spaces:

$$\Lambda_{-\infty,1} := \{\xi' \in \Lambda\mathfrak{g}_\sigma \mid \xi' = \sum_{k=-\infty}^{1} \xi'_k \lambda^k \},$$

$$\Lambda_{-1,\infty} := \{\xi'' \in \Lambda\mathfrak{g}_\sigma \mid \xi'' = \sum_{k=-1}^{\infty} \xi''_k \lambda^k \}.$$

DEFINITION 4.1. *Let us denote the space of all* $\Lambda_{-\infty,1}$*-valued Lorentz holomorphic* 1*-forms on* $\mathbb{D}$ *by* $\mathcal{P}'$. *Similarly we denote by* $\mathcal{P}''$ *the space of* $\Lambda_{-1,\infty}$*-valued Lorentz anti-holomorphic* 1*-forms on* $\mathbb{D}$.

By definition the elements $\xi' \in \mathcal{P}'$ and $\xi'' \in \mathcal{P}''$ have the following forms:

$$\xi' = \left(\sum_{k=-\infty}^{1} \xi'_k(x)\, \lambda^k \right) \mathrm{d}x, \quad \xi'' = \left(\sum_{k=-1}^{\infty} \xi''_k(y)\, \lambda^k \right) \mathrm{d}y. \tag{4.1}$$

Here the coefficients $\xi'_k(x)$ [respectively, $\xi''_k(y)$] are smooth functions of x [respectively, y]. We call elements of $\mathcal{P}'$ and $\mathcal{P}''$ $(1,0)$*-potentials* and $(0,1)$*-potentials* for harmonic maps.

4.2 We will show below that the space $\mathcal{P}' \times \mathcal{P}''$ can serve as the space of *Weierstraß data* for the construction of harmonic maps from $\mathbb{D}$ into S^2_1.

REMARK 4.2. *In the case of CMC surfaces in Euclidean* 3*-space* $\mathbb{E}^3$ *the corresponding potentials are the holomorphic potentials for harmonic Gauß maps from Riemann surfaces into* S^2. *See* [**12**].

THEOREM 4.3. (Weierstraß-type representation of harmonic maps) *Let* $\{\xi', \xi''\} \in \mathcal{P}' \times \mathcal{P}''$ *be a potential. Solve the two independent initial value problems* :

$$\mathrm{d}'\Psi' = \Psi'\, \xi', \quad \mathrm{d}''\Psi'' = \Psi''\, \xi'',$$
$$\Psi'(x=0) = \Psi''(y=0) = \mathbf{1}.$$

Then the Iwasawa decomposition

$$(\Psi', \Psi'') = (\Psi, \Psi)(L_-^{-1}, L_+^{-1}) \in \Delta(\tilde{\Lambda}G_\sigma \times \tilde{\Lambda}G_\sigma) \cdot \tilde{\Lambda}^-_* G_\sigma \times \tilde{\Lambda}^+ G_\sigma$$

gives an extended framing Ψ *defined on a region containing the origin. Hence* $\psi := \mathrm{Ad}(\Psi)\mathbf{k}'$ *gives a loop of harmonic maps into* S^2_1.

PROOF. From the Iwasawa splitting (Theorem 2.6) of (Ψ', Ψ'') we know that, locally around $(0,0)$, $\Delta(\tilde{\Lambda}G_\sigma \times \tilde{\Lambda}G_\sigma) \ni (\Psi, \Psi) = (\Psi', \Psi'')(L_-, L_+)$. The Maurer-Cartan form $(\Psi, \Psi)^{-1}\mathrm{d}(\Psi, \Psi)$ may be computed as

$$(\Psi, \Psi)^{-1}\, \mathrm{d}(\Psi, \Psi) = (L_-^{-1}\xi' L_- + L_-^{-1}\mathrm{d}L_-, L_+^{-1}\xi'' L_+ + L_+^{-1}\mathrm{d}L_+), \tag{4.2}$$

where we have used $(\Psi')^{-1}\mathrm{d}\Psi' = \xi'$, $(\Psi'')^{-1}\mathrm{d}\Psi'' = \xi''$. By construction, the two components of (4.2) are equal. We call them α. Evidently the $\mathrm{d}x$-part of the first component of the right hand side of (4.2) only involves positive powers of λ with exponents equal to or smaller than 1, while the dx-part of the second component involves only exponents equal to or greater than 0. Thus in α only the exponents 0 and 1 occur in the dx-part. A similar argument for the dy-part shows that it only involves the terms with exponents -1 and 0. Thus we obtain

$$\alpha = \Psi^{-1}\mathrm{d}\Psi = \lambda^{-1}(\xi''_{-1}\mathrm{d}y) + (\xi'_0\mathrm{d}x + \xi''_0\mathrm{d}y) + \lambda(\xi'_1\mathrm{d}x).$$

Hence Ψ is an extended framing, since $\xi'_1,\ \xi''_{-1} \in \mathfrak{m},\ \ \xi'_0,\ \xi''_0 \in \mathfrak{k}$. □

4.3 This section is in some sense the *converse* of the previous one. However, we shall compute potentials only in some normal form, which, in general, will involve singularities. A relation between smooth surfaces and singularity-free potentials of the form (4.1) has not been established yet.

Let $\psi : \mathbb{D} \to G/K$ be a harmonic map with an extended framing Ψ. Then we may perform both type of Birkhoff decompositions (Theorem 2.3) for Ψ as long as its image lies in both big cells. First we consider the Birkhoff decomposition of the following type: $\tilde\Lambda^-_*G_\sigma \times \tilde\Lambda^+G_\sigma \subset \tilde\Lambda G_\sigma$. Since the big cell $\mathcal{B}(-,+)$ of $\tilde\Lambda G_\sigma$ is open and Ψ is continuous, the set

$$D_1 := \{(x,y) \in \mathbb{D} \mid \Psi(x,y) \text{ belongs to the big cell } \mathcal{B}(-,+)\ \}$$

is an open subset of $\mathbb{D}$. Note that D_1 contains $(0,0)$. Set $\mathcal{S}_1 := \mathbb{D} \setminus D_1$. Similarly, we have D_2 and $\mathcal{S}_2$ for the splitting $\tilde\Lambda^+_*G_\sigma \times \tilde\Lambda^-G_\sigma \subset \tilde\Lambda G_\sigma$. We can perform the two Birkhoff splittings on the extended framing Ψ over $\mathbb{D} \setminus \mathcal{S},\ \ \mathcal{S} := \mathcal{S}_1 \cup \mathcal{S}_2$:

$$\Psi = \Psi_-\,L_+ = \Psi_+\,L_-,\ \ \Psi_\pm \in \tilde\Lambda^\pm_*G_\sigma,\ L_\pm \in \tilde\Lambda^\pm G_\sigma. \tag{4.3}$$

From the splitting (4.3), we obtain

$$\Psi_-^{-1}\mathrm{d}\Psi_- = L_+(\Psi^{-1}\mathrm{d}\Psi)L_+^{-1} - \mathrm{d}L_+\,L_+^{-1}. \tag{4.4}$$

Since $\alpha_\lambda = \alpha_0 + \lambda\alpha'_1 + \lambda^{-1}\alpha''_1$, we have $L_+(\Psi^{-1}\mathrm{d}\Psi)L_+^{-1} = L_+(\alpha_0 + \lambda\alpha'_1 + \lambda^{-1}\alpha''_1)L_+^{-1}$. Since $L_+ \in \tilde\Lambda^+G_\sigma$, we can write $L_+ = \sum_{k\geq 0} L_k^+(x,y)\lambda^k$. The left hand side of (4.4) contains negative powers of λ. On the other hand, the right hand side contains nonnegative powers of λ only. Thus we have $\Psi_-^{-1}\frac{\partial\Psi_-}{\partial x} = 0$. Therefore Ψ_- depends only on y and λ. Next by comparing the left and right hand sides of the $(0,1)$-part of (4.4), we obtain: $\Psi_-^{-1}\,\mathrm{d}''\Psi_- = \lambda^{-1}\{\,\mathrm{Ad}\,(\,L_0^+\,)\ \alpha''_1\}\,\mathrm{d}y$. The left hand side of (4.4) depends only on y (and λ). Hence $\eta'' := \mathrm{Ad}\,(\,L_0^+\,)\ \alpha''_1\ \mathrm{d}y$ is an $\mathfrak{m}$-valued anti-holomorphic 1-form on $\mathbb{D} \setminus \mathcal{S}$. By similar arguments for $\Psi_+ = \Psi\ L_-^{-1}$, where $L_- = \sum_{k\leq 0} L_k^-\lambda^k$, we have $\Psi_+^{-1}\frac{\partial\Psi_+}{\partial y} = 0$. Thus Ψ_+ depends only on x (and λ). Moreover the 1-form $\eta' := \mathrm{Ad}\,(\,L_0^-\,)\ \alpha'_1\ \mathrm{d}x$ is an $\mathfrak{m}$-valued holomorphic 1-form on $\mathbb{D}\setminus\mathcal{S}$. Obviously $\xi' := \lambda\eta' \in \mathcal{P}',\ \ \xi'' := \lambda^{-1}\eta'' \in \mathcal{P}''$. One can check that the pair of potentials $\{\xi',\ \xi''\}$ reproduces the harmonic map ψ via Weierstraß representation. (cf. Lemma 4.5 and Theorem 4.10 in [**14**] and Theorem 2.1 in [**38**].)

THEOREM 4.4. (The normalized potentials) *Let $\psi : \mathbb{D} \to G/K$ be a harmonic map with $\psi(0,0) = K$ and Ψ an extended framing of ψ. Then there exists an open subset $(0,0) \in \mathbb{D} \setminus \mathcal{S}$ on which Ψ splits into*

$$\Psi = \Psi_-L_+ = \Psi_+L_-,\quad \Psi_\pm \in \tilde\Lambda^\pm_*G_\sigma,\quad L_\pm \in \tilde\Lambda^\pm G_\sigma.$$

The 1-forms η', η'' defined by

$$\eta'(x) = \Psi_+^{-1}\mathrm{d}\,\Psi_+\,\lambda^{-1},\quad \eta''(y) = \Psi_-^{-1}\mathrm{d}\,\Psi_-\,\lambda$$

are, respectively, $\mathfrak{m}$*-valued holomorphic and anti-holomorphic* 1*-forms locally defined around* $(0,0)$.

Conversely, any harmonic map $\psi : \mathbb{D} \to G/K$ *with* $\psi(0,0) = K$ *can be constructed in a sufficiently small neighborhood of* $(0,0)$ *from a pair of* $\mathfrak{m}$*-valued* 1*-forms* η', η'' *which are holomorphic and anti-holomorphic respectively. The harmonic map* ψ *is constructed via the Weierstraß representation with potentials* $\xi' := \lambda\,\eta'$, $\xi'' := \lambda^{-1}\,\eta''$.

REMARK 4.5. *(1) The potentials* η' *and* η'' *used in the preceding theorem are smooth only close to* $(0,0)$. *In Euclidean CMC surface geometry, the corresponding "holomorphic potentials" can be extended meromorphically to the whole Riemann surface* $\mathbb{D}$. *The poles of the potentials are in the singular set* $\mathcal{S}$. *However, in the case under consideration, the type of singularities that may occur on all of* $\mathbb{D}$ *still need to be investigated. Therefore we formulate our claims only locally.*

(2) It is easy to see that the pair $\{\eta', \eta''\}$ *is, locally around* $(0,0)$, *uniquely determined by the harmonic map* ψ *locally around* $(0,0)$. *(cf.* [**14**]*). Following,* [**9**], [**36**] *and* [**38**], *we call the pair* $\{\eta',\ \eta''\}$ *(or the pair* $\{\xi',\ \xi''\}$ *defined by* $\xi' = \lambda\,\eta'$, $\xi'' := \lambda^{-1}\,\eta''$*) the normalized potential for* ψ *with the origin as the reference point.*

Up to now we have only very little information on the singular set. Thus in this paper, we restrict our attention to holomorphic and anti-holomorphic potentials $\{\xi', \xi''\}$.

REMARK 4.6. *A Weierstraß-type representation of harmonic maps from Riemann surfaces into noncompact Riemannian symmetric spaces was studied in* [**5**].

5. Normalized potentials for timelike CMC surfaces

5.1 In this section, we shall apply the Weierstraß representation for harmonic maps into S^2_1 to timelike CMC surfaces.

Let $\varphi : \mathbb{D} \to \mathbb{E}^3_1$ be a timelike CMC surface with Gauß map ψ and coordinate frame $\hat{\Phi}$ (See (1.3)). The Gauß-Codazzi equations (G) and (C) of φ are invariant under the deformation

$$Q \longmapsto Q_\lambda := \lambda Q, \quad R \longmapsto R_\lambda := \lambda^{-1} R, \ \lambda \in \mathbb{R}^+. \tag{5.1}$$

Integrating the Frenet equations (1.4) with Q_λ and R_λ, one obtains a one-parameter family of timelike surfaces $\{\hat{\varphi}_\lambda\}$. This deformation does not affect the induced metric and the mean curvature. Hence all the surfaces $\{\hat{\varphi}_\lambda\}$ are isometric and have the same constant mean curvature. The family $\{\hat{\varphi}_\lambda\}$ is called the *associated family* of φ.

This one-parameter deformation of $\hat{\varphi}$ satisfies the following Lax equations:

$$\frac{\partial}{\partial x}\hat{\Phi}_\lambda = \hat{\Phi}_\lambda \hat{U}(\lambda), \quad \frac{\partial}{\partial y}\hat{\Phi}_\lambda = \hat{\Phi}_\lambda \hat{V}(\lambda), \tag{5.2}$$

$$\hat{U}(\lambda) = \begin{pmatrix} -\frac{1}{4}\omega_x & -\lambda Q e^{-\frac{\omega}{2}} \\ \frac{H}{2}e^{\frac{\omega}{2}} & \frac{1}{4}\omega_x \end{pmatrix}, \quad \hat{V}(\lambda) = \begin{pmatrix} \frac{1}{4}\omega_y & -\frac{H}{2}e^{\frac{\omega}{2}} \\ \lambda^{-1} R e^{-\frac{\omega}{2}} & -\frac{1}{4}\omega_y \end{pmatrix}.$$

A solution $\hat{\Phi}_\lambda$ to (5.2) is not an extended framing for the harmonic Gauß map ψ in the sense of the definition given in Section 3.5, since the λ-distribution does not fit. To relate these two $\mathbb{R}^+$-families to each other we perform the transformation

$$\Phi := g(\lambda)^{-1}\,\hat{\Phi}_{\lambda^2}\,g(\lambda),\quad g(\lambda) = \begin{pmatrix} 0 & -\sqrt{\lambda} \\ 1/\sqrt{\lambda} & 0 \end{pmatrix},\quad \lambda \in \mathbb{R}^+. \tag{5.3}$$

Then the coefficient matrices of the Lax pair $\{\hat{U}, \hat{V}\}$ change to

$$U = g(\lambda)^{-1}\hat{U}(\lambda^2)g(\lambda) = \begin{pmatrix} \frac{1}{4}\omega_u & -\lambda\frac{H}{2}e^{\frac{\omega}{2}} \\ \lambda Q e^{-\frac{\omega}{2}} & -\frac{1}{4}\omega_u \end{pmatrix}, \tag{5.4}$$

$$V = g(\lambda)^{-1}\hat{V}(\lambda^2)g(\lambda) = \begin{pmatrix} -\frac{1}{4}\omega_v & -\lambda^{-1}Re^{-\frac{\omega}{2}} \\ \lambda^{-1}\frac{H}{2}e^{-\frac{\omega}{2}} & \frac{1}{4}\omega_v \end{pmatrix}.$$

Comparing this with [**23**] we conclude:

PROPOSITION 5.1. (*Sym formula*) *Let* Φ *be a solution to the Lax equations* (5.3) and (5.4). *Then*

$$\varphi_\lambda = -\frac{1}{H}\left\{\frac{\partial}{\partial t}\Phi\cdot\Phi^{-1} + \frac{1}{2}\mathrm{Ad}(\Phi)\mathbf{k}'\right\},\quad \lambda = e^t \in \mathbb{R}^+ \tag{S}$$

describes a real loop of timelike surfaces of constant mean curvature H. *The first fundamental form* I *and the Gauß map* N_λ *of* φ_λ *are given by* $I = e^\omega dxdy$ *and* $N_\lambda = \mathrm{Ad}(\Phi)\mathbf{k}'$. *The logarithmic derivative part* $\varphi_\lambda^K = -\frac{1}{H}\frac{\partial}{\partial t}\Phi\cdot\Phi^{-1}$ *of* φ_λ *describes a real loop of timelike surfaces with constant Gaußian curvature* $4H^2$.

5.2 Next we calculate the normalized potential $\{\xi',\ \xi''\}$ in terms of the fundamental quantites of the timelike CMC surface φ. More precisely we shall clarify the role of the normalized potentials in the construction of timelike CMC surfaces. First decompose $\alpha := \Phi^{-1}d\Phi$ as $\alpha = \alpha_0 + \lambda\alpha_1' + \lambda^{-1}\alpha_1''$. Next perform the Birkhoff decompositions of Φ with regard to $\tilde{\Lambda}_*^+G_\sigma \times \tilde{\Lambda}^-G_\sigma \subset \tilde{\Lambda}G_\sigma$, $\tilde{\Lambda}_*^-G_\sigma \times \tilde{\Lambda}^+G_\sigma \subset \tilde{\Lambda}G_\sigma$, over the big cells $\mathcal{B}(-,+)$ and $\mathcal{B}(+,-)$ of $\tilde{\Lambda}G_\sigma$:

$$\Phi = \Phi_+\,L_- = \Phi_-\,L_+,\quad L_+ = \sum_{k\geq 0} L_k^+\lambda^k,\quad L_- = \sum_{k\leq 0} L_k^-\lambda^k, \tag{5.5}$$

The potentials $\xi' := \Phi_+^{-1}\,d'\Phi_+$ and $\xi'' := \Phi_-^{-1}\,d''\Phi_-$ have the form

$$\xi' = \lambda\ \left\{L_0^+(x,y)\,\alpha_1'\,L_0^+(x,y)^{-1}\right\},$$

$$\xi'' = \lambda^{-1}\ \left\{L_0^-(x,y)\,\alpha_1''\,L_0^-(x,y)^{-1}\right\}.$$

More explicitly we have

$$\Phi_+^{-1}\frac{\partial\Phi_+}{\partial x} = \lambda\left\{L_0^-\begin{pmatrix} 0 & -\frac{H}{2}e^{\omega/2} \\ Q(x)e^{-\omega/2} & 0 \end{pmatrix}(L_0^-)^{-1}\right\},$$

$$\Phi_-^{-1}\frac{\partial\Phi_-}{\partial y} = \lambda^{-1}\left\{L_0^+\begin{pmatrix} 0 & -R(y)e^{-\omega/2} \\ \frac{H}{2}e^{\omega/2} & 0 \end{pmatrix}(L_0^+)^{-1}\right\}. \tag{5.6}$$

Comparing the two sides of (5.6) we see that the left hand side depends on y and λ only, while the right hand side depends on x, y and λ. Recall from the Codazzi equation (C) that, since the mean curvature H is constant, R depends only on y. Hence

$$\Phi_-^{-1}\frac{\partial\Phi_-}{\partial y} = \lambda^{-1}\left\{L_0^+(0,y)\begin{pmatrix} 0 & -R(y)e^{-\omega(0,y)/2} \\ \frac{H}{2}e^{\omega(0,y)/2} & 0 \end{pmatrix}L_0^+(0,y)^{-1}\right\}.$$

We abbreviate $L_0^+(0,y)$ by $W_0(y)$ and compute $W_0(y)$ more explicitly. From (5.5) we obtain the Maurer-Cartan equation:

$$L_+ \, \alpha \, L_+^{-1} - \mathrm{d}L_+ \, L_+^{-1} = \Phi_-^{-1}\mathrm{d}\Phi_-. \tag{5.7}$$

Comparing the λ^0-terms of both sides in (5.7) we obtain $0 = W_0 \, \alpha_0'' \, W_0^{-1} - \mathrm{d}W_0 \, W_0^{-1}$. Hence $\mathrm{d}W_0 = W_0 \, \alpha_0''$. Namely we have

$$\frac{\mathrm{d}W_0}{\mathrm{d}y} = W_0 \left\{ \frac{\omega_y(0,y)}{4} \right\} \mathbf{k}'.$$

Thus $W_0(y)$ is given explicitly by

$$W_0(y) = \begin{pmatrix} e^{-\omega(0,y)/4+c_1} & 0 \\ 0 & e^{\omega(0,y)/4+c_2} \end{pmatrix}, \quad c_1, \ c_2 \in \mathbf{R}.$$

Since we require the initial condition $W_0(0) = \mathbf{1}$, we see that the $(0,1)$-potential ξ'' is given by

$$\xi'' = \lambda^{-1} \begin{pmatrix} 0 & -R(y)e^{\{-\omega(0,y)+\omega(0,0)/2\}} \\ \frac{H}{2}e^{\{\omega(0,y)-\omega(0,0)/2\}} & 0 \end{pmatrix} \mathrm{d}y.$$

By similar arguments for $\Phi_+ = \Phi \, L_-^{-1}$, we obtain

$$\Phi_+^{-1}\frac{\partial \Phi_+}{\partial x} = \lambda \left\{ L_0^-(x,0) \begin{pmatrix} 0 & -\frac{H}{2}e^{\omega(x,0)/2} \\ Q(x)e^{-\omega(x,0)/2} & 0 \end{pmatrix} L_0^-(x,0)^{-1} \right\}.$$

The function $\Gamma_0(x) := L_0^-(x,0)$ is a solution to

$$\frac{\mathrm{d}\Gamma_0}{\mathrm{d}x} = \Gamma_0 \left\{ \frac{-\omega_x(x,0)}{4} \right\} \mathbf{k}'.$$

Thus we have

$$\Gamma_0(x) = \begin{pmatrix} e^{\omega(x,0)/4+c_3} & 0 \\ 0 & e^{-\omega(x,0)/4+c_4} \end{pmatrix}, \quad c_3, \ c_4 \in \mathbb{R}.$$

Since we require the initial condition $\Gamma_0(0,0) = \mathbf{1}$, the $(1,0)$-potential ξ' is given by

$$\xi' = \lambda \begin{pmatrix} 0 & -\frac{H}{2}e^{\{\omega(x,0)-\omega(0,0)/2\}} \\ Q(x)e^{\{-\omega(x,0)+\omega(0,0)/2\}} & 0 \end{pmatrix} \mathrm{d}x.$$

THEOREM 5.2. *The normalized potentials with the origin as a reference point are given by*

$$\xi' = \lambda \begin{pmatrix} 0 & -\frac{H}{2}f(x) \\ Q(x)/f(x) & 0 \end{pmatrix} \mathrm{d}x,$$

$$\xi'' = \lambda^{-1} \begin{pmatrix} 0 & -R(y)/g(y) \\ \frac{H}{2}g(y) & 0 \end{pmatrix} \mathrm{d}y.$$

Here the functions $f(x)$ *and* $g(y)$ *are given by* $f(x) = \exp\{\omega(x,0) - \omega(0,0)/2\}$, $g(y) = \exp\{\omega(0,y) - \omega(0,0)/2\}$.

REMARK 5.3. *From this result one sees that the normalized potentials are essentially the Hopf differentials together with the values of the conformal factor of the metric restricted to one pair of null coordinate lines.*

REMARK 5.4. (Nonlinear d'Alembert formula) *It is known that every solution of the linear wave equation: $\omega_{xy} = 0$ can be written as the sum of a Lorentz holomorphic function and a Lorentz anti-holomorphic function by the so called d'Alembert formula:*

$$\omega(x, y) = f(x) + g(y).$$

The Weierstraß-type representation (Theorem 4.3) together with Theorem 5.2 provide us with nonlinear analogues of the d'Alembert formula for the hyperbolic sinh-Gordon equation, the Liouville equation and the hyperbolic cosh-Gordon equation (cf. [**26**]*). More precisely for any initial data $f(x)$ and $g(y)$, the normalized potentials with $Q = H/2$, $R = \epsilon H/2$, $\epsilon = 0, \pm 1$ produce solutions ω to the hyperbolic sinh-Gordon equation ($\epsilon = 1$), the Liouville equation ($\epsilon = 0$) and the hyperbolic cosh-Gordon equation ($\epsilon = -1$) via the Weierstraß-type representation (Theorem 4.3). (See Proposition 1.3, Proposition 1.5 and Remark 1.6.)*

6. Examples

Even though the results of the previous sections give a one-to-one correspondence between special potentials and timelike CMC surfaces (cf. Remark 4.5), making this correspondence explicit is a different matter. To give examples we start with the easiest case from the potential point of view.

6.1 (Cylinders and pseudospheres)

EXAMPLE 6.1. (*Hyperbolic cylinders*) *Let us take the following potentials:*

$$\xi' = \frac{\lambda}{4}\begin{pmatrix} 0 & -1 \\ -1 & 0 \end{pmatrix} \mathrm{d}x, \quad \xi'' = \frac{\lambda^{-1}}{4}\begin{pmatrix} 0 & 1 \\ 1 & 0 \end{pmatrix} \mathrm{d}y.$$

Then the normalized potentials $\{\xi', \xi''\}$ produce the associated family of the following timelike CMC surface $\varphi : \mathbb{R}^2 \to \mathbb{E}^3_1$ of mean curvature $1/2$:

$$\varphi(x, y) = (\sinh \frac{x-y}{2}, \frac{x+y}{2}, \cosh \frac{x-y}{2}).$$

The image of φ is a timelike hyperbolic cylinder. The induced metric of $\mathbb{R}^2$ is $I = \mathrm{d}x\mathrm{d}y$. Thus φ is an isometric embedding of the Minkowski plane $(\mathbb{R}^2, \mathrm{d}x\mathrm{d}y)$ into $\mathbb{E}^3_1$.

EXAMPLE 6.2. (*Circular cylinders*) *The normalized potentials $\{\xi', \xi''\}$ defined by*

$$\xi' = \frac{\lambda}{4}\begin{pmatrix} 0 & -1 \\ 1 & 0 \end{pmatrix} \mathrm{d}x, \quad \xi'' = \frac{\lambda^{-1}}{4}\begin{pmatrix} 0 & -1 \\ 1 & 0 \end{pmatrix} \mathrm{d}y$$

produce the associated family of timelike CMC surface $\varphi : \mathbb{R}^2 \to \mathbb{E}^3_1$ of mean curvature $1/2$:

$$\varphi(x, y) = (\frac{x-y}{2}, \sin \frac{x+y}{2}, -\cos \frac{x+y}{2}).$$

The image of φ is a timelike circular cylinder. The induced metric of $\mathbb{R}^2$ is $I = \mathrm{d}x\mathrm{d}y$. Thus φ is an isometric immersion of the Minkowski plane $(\mathbb{R}^2, \mathrm{d}x\mathrm{d}y)$ into $\mathbb{E}^3_1$. Note that both a timelike hyperbolic cylinder and a timelike circular cylinder correspond to the vacuum solution of the hyperbolic sinh-Gordon equation.

EXAMPLE 6.3. *(Totally umbilical pseudosphere) The normalized potentials $\{\xi', \xi''\}$ with $Q = R = 0$ produce a totally umbilical pseudosphere in $\mathbb{E}^3_1$.*

6.2 Next we shall give examples of timelike CMC surfaces with repeated real principal curvatures, namely, timelike CMC surfaces with $QR = 0$. Such surfaces have no Euclidean counterparts. As in the case of indefinite affine spheres with $AB = 0$ (See Section 9.2 in [**10**]), timelike CMC surfaces with $QR = 0$ have specific shapes. To investigate such surfaces, we recall the notion of a B-scroll introduced in [**18**].

DEFINITION 6.4. *Let $\gamma = \gamma(s)$ be a smooth curve in $\mathbb{E}^3_1$ defined on an interval $I \subset \mathbb{R}$. Then γ is said to be a null Frenet curve if*

(1) $\langle \gamma', \gamma' \rangle = 0$,

(2) there exist vector fields A, B, C along γ and two functions κ and τ such that $A = \gamma'$, $\langle A, B\rangle = 1$, $\langle A, A\rangle = \langle B, B\rangle = 0, \langle C, C\rangle = 1, \langle A, C\rangle = \langle B, C\rangle = 0$ and

$$\frac{\mathrm{d}}{\mathrm{d}s}(A, B, C) = (A, B, C)\begin{pmatrix} 0 & 0 & -\tau \\ 0 & 0 & -\kappa \\ \kappa & \tau & 0 \end{pmatrix}.$$

The frame field $L = (A, B, C)$ along γ is called the null Frenet frame field of γ. The two functions κ and τ are called the curvature and torsion of γ respectively.

DEFINITION 6.5. *Let γ be a null Frenet curve in $\mathbb{E}^3_1$. The ruled surface*

$$\varphi(s, t) = \gamma(s) + tB(s) : I \times \mathbb{R}^* \to \mathbb{E}^3_1$$

is called the B-scroll of γ.

Since $I = (t\tau)^2\mathrm{d}s^2 + 2\mathrm{d}s\,\mathrm{d}t$, every B-scroll is timelike. (In fact $\det I = -1$.) The mean curvature of φ is $H(s,t) = \tau(s)$. Thus φ is of constant mean curvature if and only if τ is constant.

A specific example of a null Frenet curve with constant torsion 1 is

$$\gamma(s) = \left(\frac{\sinh(2s)}{2}, \frac{\cosh(2s)}{2}, s\right)$$

with null Frenet frame field:

$$\begin{aligned} A(s) &= (\cosh 2s, \sinh 2s, 1)\,, \\ B(s) &= \frac{1}{2}\left(-\cosh(2s), -\sinh(2s), 1\right), \\ C(s) &= (-\cosh 2s, -\sinh 2s, 0)\,. \end{aligned}$$

The B-scroll φ of γ,

$$\varphi(s,t) = \left(\frac{\sinh(2s) - t\cosh(2s)}{2}, \frac{\cosh(2s) - t\sinh(2s)}{2}, s + \frac{t}{2}\right),$$

is a timelike surface with constant mean curvature 1 ([**27**], p.33).

Now let φ be a constant mean curvature B-scroll of a null Frenet curve γ. The local coordinate system (x, y) defined by

$$x := s - \frac{2}{\tau^2 t} - \frac{1}{2}, \quad y := s + \frac{1}{2}$$

is a null coordinate system of φ. Then

$$I = e^{\omega(x,y)}\,\mathrm{d}x\mathrm{d}y, \quad e^{\omega(x,y)} = \frac{4}{H^2\{(x-y)+1\}^2},$$

$$Q(x) = 0,\ R(y) = H(y - \frac{1}{2}).$$

Thus the normalized potentials are given by

$$\begin{aligned} \xi' &= \lambda \begin{pmatrix} 0 & (x+1)^{-2} \\ 0 & 0 \end{pmatrix} \mathrm{d}x, \\ \xi'' &= \lambda^{-1} \begin{pmatrix} 0 & -R(y)H(1-y)^2/2 \\ (1-y)^{-2} & 0 \end{pmatrix} \mathrm{d}y. \end{aligned}$$

Conversely, we shall prove that timelike CMC surfaces derived from normalized potentials of the form:

$$\xi' = \lambda \begin{pmatrix} 0 & -\frac{H}{2} f(x) \\ 0 & 0 \end{pmatrix} \mathrm{d}x, \quad \xi'' = \lambda^{-1} \begin{pmatrix} 0 & -R(y)/g(y) \\ \frac{H}{2} g(y) & 0 \end{pmatrix} \mathrm{d}y. \tag{6.1}$$

are B-scrolls of constant mean curvature H. To this end we consider the following initial value problems:

$$\mathrm{d}'\Phi_+ = \Phi_+\, \xi', \quad \mathrm{d}''\Phi_- = \Phi_-\, \xi'', \quad \Phi_+(x=0) = \Phi_-(y=0) = \mathbf{1}$$

with potential (6.1). The solution Φ_+ is easily obtained as

$$\Phi_+ = \begin{pmatrix} 1 & -\lambda F(x) \\ 0 & 1 \end{pmatrix}, \quad F(x) = \frac{H}{2} \int_0^x f(x)\mathrm{d}x.$$

We shall obtain the extended framing Φ derived from $\{\xi',\ \xi''\}$ by the technique used in [**12**]. (See section 3.6 of [**12**].) Express Φ_- as

$$\Phi_- = \begin{pmatrix} a & b \\ c & d \end{pmatrix} = \mathbf{1} + \sum_{j=1}^{\infty} \begin{pmatrix} a_i & b_i \\ c_i & d_i \end{pmatrix} \lambda^{-j}.$$

Take a $\tilde{\Lambda} G_\sigma^+$-valued map $\mathcal{G}$

$$\mathcal{G}(x,y) = \begin{pmatrix} 1 & -\lambda F/(1+c_1 F) \\ 0 & 1 \end{pmatrix}.$$

Then $\tilde{\Phi} := \Phi_+^{-1}\, \Phi_-\, \mathcal{G}$ can be computed as

$$\tilde{\Phi} = \tilde{\Phi}_0 + \sum_{j \geq 1} \tilde{\Phi}_j \lambda^{-j} = \begin{pmatrix} 1 + c_1 F & 0 \\ 0 & 1/(1+c_1 F) \end{pmatrix} + \text{negative powers}.$$

Put $\tilde{\Phi}_+ = \tilde{\Phi}_0\, \mathcal{G}^{-1}$ and $\tilde{\Phi}_- := \tilde{\Phi}\, \tilde{\Phi}_0^{-1}$. Then $\tilde{\Phi}_- \in \tilde{\Lambda}^- G_\sigma$, $\tilde{\Phi}_+ \in \tilde{\Lambda}_*^+ G_\sigma$ and ${\Phi_+}^{-1}\Phi_- = \tilde{\Phi}_- \tilde{\Phi}_+$. Thus we obtain $\tilde{\Phi}_- = L_-$. The extended framing Φ may be computed as $\Phi = \Phi_+ L_- = \Phi_+ \tilde{\Phi}_- = \Phi_- \tilde{\Phi}_+^{-1}$. By the Sym formula (Proposition 5.1), we obtain the constant mean curvature immersion

$$\varphi(x,y) = \gamma(y) + q(x,y)B(y),$$

$$\gamma(y) := -\frac{1}{H}\left\{ \frac{\partial \Phi_-}{\partial t} \Phi_-^{-1} + \frac{1}{2}\mathrm{Ad}(\Phi_-)\mathbf{k}' \right\},$$

$$B(y) := \frac{\lambda}{H^2}\mathrm{Ad}(\Phi_-)(\mathbf{j}' - \mathbf{i}), \quad q(x,y) = \frac{F(x)}{H\{1 + c_1(y)F(x)\}}.$$

A direct computation shows that $\gamma(y)$ is a null Frenet curve and φ is the B-scroll of γ.

References

[1] O. Babelon and D. Bernard, Affine solitons: a relation between tau functions, dressing and Bäcklund transformations, Inter. J. Modern Phys. A **8** (1993), 507-543.

[2] M.V. Babich, Real finite-gap solutions of the equation $\Delta u = \text{ch}u$, Mat. Zametki **50** (1991), 3–9, 158, English translation: Math. Notes **50** (1991), 663–667.

[3] M. Babich and A.I. Bobenko, Willmore tori with umbilic lines and minimal surfaces in hyperbolic space, Duke Math. J. **72** (1993), 151–185.

[4] V. Balan and J. Dorfmeister, Birkhoff decomposition and Iwasawa decomposition for loop groups, Tôhoku Math. J., **53** (2001), 593–615.

[5] V. Balan and J. Dorfmeister, A Weierstrass-type representation for harmonic maps from Riemann surface into semisimple symmetric spaces, in preparation.

[6] M.J. Bergvelt and M.A. Guest, Actions of loop groups on harmonic maps, Trans. Amer. Math. Soc. **326** (1991), 861–886.

[7] A.I. Bobenko and U. Eitner, *Painlevé Equations in Differential Geometry of Surfaces*, Lecture Notes in Math., **1753**, Springer Verlag, 2000.

[8] K.F. Clancey and I. Gohberg, Factorization of matrix functions and singular integral operators, *Operator Theory: Advances and Applications*, **3**, Birkhäuser Verlag, Basel-Boston, Mass., 1981.

[9] J. Dorfmeister, Generalized Weierstraß representations of surfaces, these proceedings.

[10] J. Dorfmeister and U. Eitner, Weierstraß-type representation of affine spheres, Abh. Math. Sem. Hamburg **71** (2001), 225–250.

[11] J. Dorfmeister, H. Gradl and J. Szmigielski, Systems of PDEs obtained from factorization in loop groups, Acta Appl. Math. **53** (1998), 1–58.

[12] J. Dorfmeister and G. Haak, Meromorphic potentials and smooth CMC surfaces, Math. Z., **224** (1997), 603–640.

[13] J. Dorfmeister, F. Pedit and M. Toda, Minimal surfaces via loop groups, Balkan J. Geom. Appl. **2** (1997), 25–40, (http://www.emis.de/journals/BJGA/2.1/4.html).

[14] J. Dorfmeister, F. Pedit and H. Wu, Weierstrass type representations of harmonic maps into symmetric spaces, Comm. Analysis and Geom. **6** (1998), 633-668.

[15] J. Dorfmeister and I. Sterling, Finite type Lorentz harmonic maps and the method of Symes, Differential Geom. Appl., to appear.

[16] A. Fujioka and J. Inoguchi, Timelike surfaces with harmonic inverse mean curvature, these proceedings.

[17] I.T. Gohberg, A factorization problem in normed rings, functions of isometric and symmetric operators and singular integral equations, Russian Math. Surveys **19** (1964), 63–114.

[18] L. Graves, Codimension one isometric immersions between Lorentz spaces, Trans. Amer. Math. Soc. **252** (1979), 367-392.

[19] C.-H. Gu, On the harmonic maps from $\mathbb{R}^{1,1}$ to $S^{1,1}$, J. reine Angew. Math. **346** (1984), 101-109.

[20] M.A. Guest, *Harmonic Maps, Loop Groups and Integrable Systems*, London Math. Soc. Student Texts **38**, Cambridge Univ. Press, Cambridge, 1997.

[21] H. Hu, Sine-Laplace equations, Sinh-Laplace equations and harmonic maps, Manuscripta Math. **40** (1982), 205-216.

[22] H. Hu, On the geometry of sinh-Gordon equation, *Qualitative Aspects and Applications of Nonlinear Evolution Equations* (Trieste 1993) World Sci. Publishing, River Edge, NJ., 1994, pp.35–47.

[23] J. Inoguchi, Timelike surfaces of constant mean curvature in Minkowski 3-space, Tokyo J. Math. **21** (1998), 141-152.

[24] J. Inoguchi, Darboux transformations on timelike constant mean curvature surfaces, J. Geom. Phys. **32** (1999), 57–78.

[25] P. Kellersch, *Eine Verallgemeinerung der Iwasawa Zerlegung in Loop Gruppen*, Ph.D. Thesis, Technische Universität München, 1999.

[26] I. M. Kričever, An analogue of the d'Alembert formula for the equations of a principal chiral field and the sine-Gordon equation (Russian), Dokl. Akad. Nauk SSSR **253** (1980), 288–292 English translation: Soviet Math. Dokl. **22** (1980), 79–84 (1981).

[27] L. McNertney, *One-parameter families of surfaces with constant curvature in Lorentz 3-space*, Ph.D. Thesis, Brown Univ., 1980.

[28] S. Melko and I. Sterling, Applications of soliton theory to the construction of pseudospherical surfaces in $\mathbf{R}^3$, Ann. of Global Anal. Geom. **11** (1993), 65–107.

[29] S. Melko and I. Sterling, Integrable systems, harmonic maps and the classical theory of surfaces, *Harmonic Maps and Integrable Systems* (A. P. Fordy and J. C. Wood eds.), Aspects of Math. E **23**, Viewig, Braunschweig, (1994), 129–144.

[30] T.K. Milnor, Harmonic maps and classical surface theory in Minkowski 3-space, Trans. Amer. Math. Soc. **280** (1983), 161–185.

[31] V.Y. Novokshenov, Minimal surfaces in the hyperbolic space and radial-symmetric solution of the cosh-Laplace equation, *Algebraic and Geometric Methods in Mathematical Physics* (Kaciveli, 1993), Math. Phys. Stud. **19**, Kluwer Acad. Publ., Dordrecht, 1996, pp. 357–370.

[32] V.Y. Novokshenov, Radial-symmetric solution of the cosh-Laplace equation and the distribution of its singularities, Russian J. Math. Phys. **5** (1997), 211–226.

[33] B. O'Neill, *Semi-Riemannian Geometry with Applications to Relativity*, Pure and Applied Math., vol. 130, Academic Press, Orlando, 1983.

[34] D. Petersen and V. Kac, Infinite flag varieties and conjugacy theorems, Proc. Nat. Acad. Sci. U.S.A., **80** (1983), 1778–1782.

[35] A. Pressley and G. Segal, *Loop Groups*, Oxford Math. Monographs, Oxford University Press, 1986.

[36] M. Toda, *Pseudo spherical surfaces via moving frames and loop groups*, Ph.D. Thesis, University of Kansas (2000).

[37] T. Weinstein, *An Introduction to Lorentz Surfaces*, de Gruyter Expositions in Math. vol. 22, Walter de Gruyter, Berlin, 1996.

[38] H. Wu, A simple way for determining the normalized potentials for harmonic maps, Ann. Global Anal. Geom. **17** (1999), 189–199.

Department of Mathematics, University of Kansas, Lawrence, Kansas 66045-7567, USA

Current address: Fakultät für Mathematik, Technische Universität München, Arcisstrasse 21, 80290 München, GERMANY

E-mail address: `dorfm@ma.tum.de`

Department of Applied Mathematics, Fukuoka University, Fukuoka 814-0180, JAPAN

Current address: Department of Mathematics Education, Faculty of Education, Utsunomiya University, Mine-machi 350, Utsunomiya 321–8505, JAPAN

E-mail address: `inoguchi@cc.utsunomiya-u.ac.jp`

Department of Mathematics, University of Kansas, Lawrence, Kansas 66045-7567, USA

Current address: Department of Mathematics and Statistics, Texas Tech University, Lubbock, TX 79407, USA

E-mail address: `mtoda@math.ttu.edu`

Contemporary Mathematics
Volume **308**, 2002

A Differential-Geometric Schottky Problem, and Minimal Surfaces in Tori

Norio Ejiri

§1 Introduction.

Schoen and Yau [S-Y], and Sacks and Uhlenbeck [S-U1] defined an energy function E_ϕ on the Teichmüller space $T(R)$ for a fixed Riemann surface R of genus γ associated with harmonic maps and showed the existence of an incompressible minimal surface. One of their important ideas is that a minimum point of the energy function corresponds to a compact orientable branched minimal surface. If the ambient spaces are negatively curved manifolds or Riemannian flat tori, then the energy function is smooth and every critical point corresponds to a compact orientable branched minimal surface. The index ($index_E$) and the nullity ($nullity_E$) for the hessian of the energy function are closely related to the index ($index_A$) and the nullity ($nullity_A$) of the Jacobi operator of the corresponding surface. For example:

THEOREM 3.4. *$index_E = index_A$. In particular, $index_A \leq 6\gamma - 6$.*

On the other hand, there are some differences between the two situations, for instance:

COROLLARY 3.20. *(i) If N is negatively curved, E_ϕ is nondegenerate if and only if the corresponding compact orientable branched minimal surface has no branch points and $nullity_A = 0$. (ii) If N is a Riemannian flat torus, E_ϕ is nondegenerate if and only if the corresponding compact orientable branched minimal surface has no branch points and only trivial Jacobi fields.*

In this paper, we shall investigate the energy function E_ϕ when the ambient space is a Riemannian flat torus. Taking the Riemann matrix for a point of the Teichmüller space $T(R)$, we can construct a holomorphic mapping κ of $T(R)$ into the Siegel upper half space H_γ. Furthermore, defining the energy function on H_γ

2000 *Mathematics Subject Classification.* Primary 49Q05; Secondary 58E12.

by $E_{\widetilde{L}}(\tau) = \frac{1}{2}\mathrm{tr}(P(\tau)\widetilde{L})$, where $\widetilde{L}$ is determined by the periods of ϕ and

$$P(\tau) = \begin{pmatrix} (\mathrm{Im}\tau) + (\mathrm{Re}\tau)(\mathrm{Im}\tau)^{-1}(\mathrm{Re}\tau) & -(\mathrm{Re}\tau)(\mathrm{Im}\tau)^{-1} \\ -(\mathrm{Im}\tau)^{-1}(\mathrm{Re}\tau) & (\mathrm{Im}\tau)^{-1} \end{pmatrix},$$

we discover $E_\phi = E_{\widetilde{L}} \circ \kappa$. Thus the investigation of the energy function E_ϕ is reduced to the study of the energy function $E_{\widetilde{L}}$ on H_γ and differential-geometric properties of $\kappa(T(R))$, which is the set RM of Riemann matrices in H_γ. Furthermore, using the embedding P of H_γ into the space S^2_R of real symmetric matrices of size 2γ, we obtain two submanifolds of $P(RM)$, namely, the image $P(RM_{non-hyper})$ of Riemann matrices of non-hyperelliptic Riemann surfaces and the image $P(RM_{hyper})$ of Riemann matrices of hyperelliptic Riemann surfaces. The energy function becomes the height function on $P(RM)$ in the direction of a positive semi-definite symmetric matrix. So its critical points correspond to compact orientable branched minimal surfaces. Even for height functions in the direction of other non-positive semi-definite symmetric matrices we shall find a relation between the critical points and multivalued weakly conformal harmonic maps of Riemann surfaces into $\mathbf{R}^{p+q}$ with inner product of type (p, q), a space which will be denoted by $L_{p,q}$. We call the exterior differentials of these multivalued maps $L_{p,q}$-valued weakly conformal harmonic forms.

THEOREM 8.1. *Let τ be an element of $RM_{non-hyper}$. Then there exist a unique Riemann surface M and two canonical homology bases $\{A_i, B_i\}$, $\{-A_i, -B_i\}$ up to automorphisms of M such that $dS = \pm LT_\tau^{-1}\Psi$, where $L \in F_{p,q}$, is weakly conformal, if and only if the height function corresponding to $\widetilde{L}$ on $P(RM_{non-hyper})$ is critical at $P(\tau)$. Similarly, let τ be an element of RM_{hyper}. Then there exist a unique Riemann surface M and a unique canonical homology basis $\{A_i, B_i\}$ up to automorphisms of M such that $dS = LT_\tau^{-1}\Psi$, where $L \in F_{p,q}$, is weakly conformal if and only if the height function corresponding to $\widetilde{L}$ on $P(RM_{hyper})$ is critical at $P(\tau)$.*

Thus we should treat minimal surfaces (weakly conformal harmonic maps of Riemann surfaces) in indefinite flat tori. Furthermore, our point of view makes explicit the catastrophe manifold **Cat** as critical points of energy functions on the Teichmüller space with control parameters S^2_R.

COROLLARY 8.2. **Cat** *is the sum of the unit normal bundles of $P(RM_{non-hyper})$ and $P(RM_{hyper})$ in $\mathbf{R}^{\gamma(2\gamma+1)}$.*

We find a new criterion to determine whether a compact orientable branched minimal surface in an indefinite flat torus is a holomorphic curve:

PROPOSITION 9.3 (CRITERION). *$P(\tau)$ is a critical point of a height function on $P(H_\gamma)$ for $\widetilde{L}$ if and only if there exists a multivalued holomorphic map Ω of M into $\mathbf{C}^{p+q}$ with Hermitian inner product of type (p, q) such that $\widetilde{L}$ corresponds to the period of Ω. In particular, $L \in F_{2p,2q}$ are lattice vectors if and only if Ω is a holomorphic map into a complex torus with Hermitian inner product of type (p, q).*

For a non-hyperelliptic minimal surface in a Riemannian flat torus, $index_A = index_E = index_{\widetilde{L}}$ and $nullity_E = nullity_{\widetilde{L}}$. On the other hand, the situation is different for a hyperelliptic minimal surface in a Riemannian flat torus. Let $[(M,f)]$ express an element of $T(R)$ and $HT(R)$ the submanifold of $T(R)$ consisting of those $[(M,f)]$ such that M is a hyperelliptic Riemann surface. Then the degeneracy of κ induces a symmetric bilinear form B and the hessian of the energy function is the direct sum of the hessian of the height function on $P(RM_{hyper})$ and $B(\mu,\nu)E_{\tilde{L}}$, so $index_A = index_E \geq index_{\tilde{L}}$, $nullity_E \geq nullity_{\tilde{L}}$. This inequality gives rise to two classes of hyperelliptic minimal surfaces. A hyperelliptic minimal surface with $index_{\tilde{L}} = 0$ in a Riemannian flat torus is said to be **hyperelliptically stable**. It has minimum area among nearby hyperelliptic surfaces if $nullity_{\tilde{L}} = 0$. Moreover, if it is a minimum of the height function on $P(RM_{hyper})$, it is said to be **hyperelliptically area-minimizing**. Ross [Ro] proved that the Schwarz P-surface and D-surface, and Schoen's gyroid satisfy $index_A = 1$ with only trivial Jacobi fields. We shall see that these surfaces are hyperelliptically stable ($index_{\tilde{L}} = 0$) and satisfy $nullity_{\tilde{L}} = 0$. Furthermore, the Schwarz P-surface is not hyperelliptically area-minimizing. On the other hand, Micallef [Mi] proved that all hyperelliptic, stable minimal surfaces in Riemannian flat tori are holomorphic.

We shall study deformations of a compact orientable branched minimal surface in the torus (Corollary 8.4). If $nullity_{\tilde{L}} = 0$, then critical points smoothly change (see, for example, [P-S]). Interesting examples with $nullity_{\tilde{L}} = 0$ and non-trivial Jacobi fields are Albanese maps of hyperelliptic Riemann surfaces.

THEOREM 10.4 (A BIFURCATION FOR THE ALBANESE MAP OF A HYPERELLIPTIC RIEMANN SURFACE). *There exists a $(\gamma-1)(\gamma-2)$-parameter deformation for the Albanese map of a hyperelliptic Riemann surface, where the corresponding hyperelliptic minimal surfaces are non-holomorphic. If τ_0, a minimum point of the height function on $P(H_\gamma)$, remains in $RM_{non-hyper}$, then we have the Albanese map of the non-hyperelliptic Riemann surface corresponding to τ_0 except for the above hyperelliptic example. Furthermore, if $\gamma = 3$, then the only minimal surfaces appearing in the deformation are these examples.*

Generally, we are not concerned with $index_A$ and $nullity_A$ of a minimal surface in an indefinite flat torus (except for Riemannian flat tori), while $nullity_{\tilde{L}}$ of a compact orientable branched minimal surface in an indefinite flat torus plays an important role as mentioned above. For example, we can construct a family of hyperelliptic minimal surfaces in a 4-dimensional Minkowski flat torus with $nullity_{\tilde{L}} = 0$.

COROLLARY 11.3. *The Schwarz P-surface gives a one parameter family of full isometric embedded minimal surfaces in a 4-dimensional Minkowski flat torus with $nullity_{\tilde{L}} = 0$. There exists a 9-dimensional perturbation space of each minimal surface, which contains the one parameter family.*

Thus the differential-geometric properties of $P(RM)$ in $\mathbf{R}^{\gamma(2\gamma+1)}$ are very important in the study of compact orientable branched minimal surfaces in indefinite flat tori and hence we pose the following **differential-geometric Schottky problem**:

MAIN PROBLEM. *Characterize $P(RM)$, $P(RM_{non-hyper})$ and $P(RM_{hyper})$ differential-geometrically.*

Buser and Sarnak [B-S] studied height functions in the direction of $\widetilde{L} = n^t n$, $n \in Z^{2\gamma}$, whose rank is 1, and proved that the moduli space $\mathbf{M}_\gamma$ of Riemann surfaces lies in a very small neighborhood of the boundary of the moduli space $A_\gamma = H_\gamma / Sp(2\gamma, \mathbf{Z})$ of principally polarized abelian varieties for large γ. Their ideas are related to the fact that $E_\phi = E_{\widetilde{L}} \circ \kappa$. On the other hand, the height function in the direction of a positive definite symmetric matrix $\widetilde{L}$ on H_γ is an unbounded positive proper function, $E_{\widetilde{L}}(\tau) = \frac{1}{2}\mathrm{tr}\{(\mathrm{Im}\tau)^{-1}(\tau - \tau_0)\overline{\Phi(\tau - \tau_0)}\} + \mathrm{tr}\{\mathrm{Im}\tau_0 \mathrm{Re}\Phi\}$, where $\tau_0 \in H_\gamma$ and Φ is the positive definite Hermitian matrix determined by $\widetilde{L}$. So $E_{\widetilde{L}}$ on the closure of $P(RM)$ and $P(RM_{hyper})$ must have a minimum. Let $G(\widetilde{L})$ be the subgroup of $Sp(2\gamma, \mathbf{Z})$ defined by $\{h \in Sp(2\gamma, \mathbf{Z}) : {}^t h \widetilde{L} h = \widetilde{L}\}$ and $\widehat{G}(\widetilde{L})$ the subgroup

$$\left\{ \begin{pmatrix} -a & b \\ c & -d \end{pmatrix} : \begin{pmatrix} a & b \\ c & d \end{pmatrix} \in G(\widetilde{L}) \right\}.$$

Then $E_{\widetilde{L}}$ is $\widehat{G}(\widetilde{L})$-equivariant and hence $\widehat{G}(\widetilde{L})$ preserves the critical points with the same value. We have congruent minimal surfaces whenever an element of $\widehat{G}(\widetilde{L})$ interchanges two minimum points.

PROBLEM 6. *Let $\widetilde{L}$ be a positive definite symmetric matrix. Then, is the minimum point for the height function corresponding to $\widetilde{L}$ on the closure of $P(RM)$ in $\mathbf{R}^{\gamma(2\gamma+1)}$ unique up to $\widehat{G}(\widetilde{L})$?*

Note that, if a minimum point is a point of $P(RM)$ such that $\widetilde{L}$ is a normal vector of $P(H_\gamma)$ at the minimum point, then Problem 6 is easy, because the minimum point is $P(\tau_0)$ (see Lemma 10.1) and the corresponding minimal surface is a holomorphic curve in $T^{2\gamma}$ by the Criterion of Proposition 9.3. In fact, Problem 6 without "up to $\widehat{G}(\widetilde{L})$" is a translation of Arezzo's conjecture. He has attempted to establish the Torelli theorem for homotopically area-minimizing minimal surfaces in Riemannian flat tori as follows:

AREZZO'S CONJECTURE. *Is a generalized homotopically area-minimizing map given in Theorem [Ar] unique?*

We shall study homotopically area-minimizing minimal surfaces of genus 4 in the torus whose lattice is associated with E_8 and prove the following:

THEOREM 12.8. *The answer to Problem 6 is negative for $\gamma = 4$. In particular, Arezzo's conjecture fails.*

To investigate the behaviour of $RM_{non-hyper}$ near RM_{hyper} in H_γ, the range of B is an important geometric object:

COROLLARY 14.2. *All stable branched minimal immersions of a hyperelliptic Riemann surface (M, f) into Riemannian flat tori are holomorphic if and only if the range of B at $[(M, f)]$ is full.*

A result of Micallef asserts that the range of B for any hyperelliptic Riemann surface is full. On the other hand, Oort and Steenbrink [O-S] proved that the

Torelli mapping of the moduli space of Riemann surfaces into the moduli space of principally polarized abelian varieties is a locally closed immersion. This solves the local Torelli problem for algebraic curves over **C**. We note that the fullness of the range of B for any hyperelliptic Riemann surface is equivalent to Oort and Steenbrink's result, and we obtain a differential-geometric solution of the local Torelli problem.

COROLLARY 14.3. *The following statements are equivalent.*

1. *All stable hyperelliptic minimal surfaces in Riemannian flat tori are holomorphic.*
2. *The Torelli mapping of the moduli space of Riemann surfces into the moduli space of principally polarized abelian varieties is a locally closed immersion.*

The author is grateful to Professor Y. Namikawa and Professor K. Yosikawa for their comments on the local Torelli problem for algebraic curves over **C**.

§2 Energy functions on Teichmüller spaces.

Let N be an n-dimensional compact Riemannian manifold, M a compact Riemann surface of genus γ, and ϕ a continuous map of M into N. Then ϕ induces a map $\phi_{\#}$ of $\pi_1(M,*)$ into $\pi_1(N,\phi(*))$, where $*$ is a fixed point of M. Let $L_1^2(M,N)$ denote the space of maps of M into N having square integrable first derivatives in the distribution sense. Schoen and Yau [S-Y] proved that there exists an energy-minimizing harmonic map among $\{f \in L_1^2(M,N) : f_{\#} = \tau^{-1}\phi_{\#}\tau\}$, where τ is some curve from $f(*)$ to $\phi(*)$. We call it a harmonic map for ϕ. Let R be a fixed compact Riemann surface of genus γ. Let (M,f) be a compact Riemann surface M of genus γ and a diffeomorphism f of R onto M. We define an equivalence relation for such pairs as follows: (M_2,f_2) is equivalent to (M_1,f_1) if and only if there exists a biholomorphic map h of M_1 onto M_2 such that h is homotopic to $f_2f_1^{-1}$. The space of equivalence classes is called the Teichmüller space $T(R)$ of R.

Let $[(M,f)]$ denote a point of $T(R)$ and ϕ a smooth map of R into an n-dimensional Riemannian manifold N. Let $\widetilde{f_1}$ be the energy-minimizing harmonic map of M_1 into N for ϕf_1^{-1} and (M_2,f_2) equivalent to (M_1,f_1) by a biholomorphic map h. If $\widetilde{f_2}$ is an energy-minimizing harmonic map of M_2 into N for ϕf_2^{-1}, then $\widetilde{f_2}h$ becomes an energy-minimizing harmonic map of M_1 into N for ϕf_1^{-1}. Thus the energies of $\widetilde{f_1}$ and $\widetilde{f_2}$ are the same, which defines the energy function E_ϕ on $T(R)$ by giving the energy of a corresponding energy-minimizing harmonic map at $[(M,f)]$. Let ζ be a parameter of a neighborhood of a point $[(M,f)]$. Then there is a Riemannian metric g_ζ on R whose scalar curvature is -1, which gives the complex structure corresponding to ζ (see, for example, Theorem 2.4.2 of Tromba [T1]). Let $[g_\zeta]$ denote the corresponding point of $T(R)$. Furthermore, for the almost complex structure J_ζ corresponding to g_ζ, $[J_\zeta]$ also denotes the same point of $T(R)$. We denote by (R,g_ζ) the Riemann surface compatible with g_ζ. Eells-Lemaire [E-L] and Koiso [K] proved that there exists a hamonic map $S(g_\zeta)$ of (R,g_ζ) into N with smooth dependence on ζ for a harmonic map of (R,g_{ζ_0}) into N whose second variation is nondegenerate. If N is negatively curved, then it is well-known that the second variation of a non-constant harmonic map is positive. Thus energy functions on $T(R)$ are smooth (see Theorem 5.1 of Sacks and Uhlenbeck [S-U2]).

On the other hand, I do not know whether this holds for N with non-positive sectional curvatures. However, we can show that energy functions are smooth if N is a Riemannian flat torus as follows: First of all, Eells and Lemaire [E-L] proved that there exists a hamonic map of (R, g_ζ) into S^1 with smooth dependence on ζ for a harmonic map of (R, g_{ζ_0}) onto S^1. This implies that the induced harmonic 1-forms on (R, g_ζ) are smooth on ζ and so are holomorphic 1−forms on (R, g_ζ). The rest of the proof will be given in Section 7.

§3 First and second variational formulae of the energy function.

Let g_t be a one parameter family in g_ζ and $S(g_t) = S_t$ the harmonic map for ϕ with respect to g_t (we denote g_0 and S_0 by g and S, respectively). Then the energy function along t is defined by the energy $E(g_t, S_t)$ of S_t with respect to g_t:

$$E_\phi([g_t]) = E(g_t, S_t) = \frac{1}{2}\int_R g_t^{ij} < \frac{\partial S_t}{\partial x^i}, \frac{\partial S_t}{\partial x^j} > \sqrt{g_t}dx^1dx^2, \tag{3.1}$$

where (x^1, x^2) is a local coordinate system of R, $\frac{\partial S_t}{\partial x^i} = S_{t_*}(\frac{\partial}{\partial x^i})$ and $\sqrt{g_t}$ means $\sqrt{\det(g_{tij})}$. The harmonic map equation is

$$\frac{1}{\sqrt{g_t}}\frac{\partial}{\partial x^j}(g_t^{ij}\sqrt{g_t}\frac{\partial S_t^\alpha}{\partial x^i}) + \Gamma^\alpha_{\gamma\beta}(S_t)\frac{\partial S_t^\gamma}{\partial x^i}\frac{\partial S_t^\beta}{\partial x^j}g_t^{ij} = 0, \tag{3.2}$$

where $(y^1, \ldots, y^n)$ is a local coordinate system of N and the $\Gamma^\alpha_{\beta\gamma}$ are the Christoffel symbols with respect to this local coordinate system.

Let h_{tij} be the first variation of g_t. Then h_{tij} is trace free for g_{tij}, because g_t has constant scalar curvature −1 (see, for example, Theorem 2.4.1 of Tromba [T1]). We may assume $h_{ij}(= h_{0ij})$ is divergence free for g_{ij} (Theorem 2.4.2 of Tromba [T1]), which implies that $h_{zz}dz^2(h_{zz} = \frac{1}{4}(h_{11} - h_{22} - 2ih_{12}))$ is a holomorphic quadratic differential for (R, g). The first and second differentials of g_t^{ij} are

$$(g_t^{ij})' = -g_t^{il}g_t^{jm}(g_{tlm})',\ (g_t^{ij})''_{t=0} = h^{il}g^{jm}h_{lm} + g^{il}h^{jm}h_{lm} - g^{il}g^{jm}(g_{tlm})''_{t=0}.$$

Since $(\sqrt{g_t})' = \frac{1}{2}g_t^{ij}, (g_{tij})'\sqrt{g_t}(\sqrt{g_t})''_{t=0} =$

$$-\frac{1}{2}h^{ij}h_{ij}\sqrt{g} + \frac{1}{2}g^{ij}(g_{tij})''_{t=0}\sqrt{g} + \frac{1}{2}g^{ij}(g_{tij})'_{t=0}(\sqrt{g_t})'_{t=0},$$

and h_{tij} is trace free for g_{tij}, we see that

$$-h^{ij}h_{ij} + g^{ij}(g_{tij})''_{t=0} = 0. \tag{3.3}$$

Let $V(h)$ be the variation vector field $(S_t)'_{t=0}$. Then the first variation of $E_\phi([g_t])$ is given by

$$DE_\phi([g_t])h_t = \frac{1}{2}\int_R (g_t^{ij})' < \frac{\partial S_t}{\partial x^i}, \frac{\partial S_t}{\partial x^j} > \sqrt{g_t}dx^1dx^2 + \int_R < g_t^{ij}\sqrt{g_t}\frac{\partial S_t}{\partial x^i}, (\frac{\partial S_t}{\partial x^j})' > dx^1dx^2.$$

Since S_t is a harmonic map with respect to g_t, we obtain

$$DE_\phi([g_t])h_t = \frac{1}{2}\int_R (g_t^{ij})' < \frac{\partial S_t}{\partial x^i}, \frac{\partial S_t}{\partial x^j} > \sqrt{g_t}dx^1dx^2.$$

Thus the second variation $D^2E_\phi([g])(h,h)$ of $E_\phi([g_t])$ at $t=0$ becomes

$$\frac{1}{2}\int_R [(g_t^{ij})''_{t=0} < \frac{\partial S}{\partial x^i}, \frac{\partial S}{\partial x^j} > \sqrt{g} + 2(g_t^{ij})'_{t=0} < (\frac{\partial S_t}{\partial x^i})'_{t=0}, \frac{\partial S}{\partial x^j} > \sqrt{g}]dx^1dx^2,$$

which implies

$$\begin{aligned} D^2&E_\phi([g])(h,h) \\ &= \frac{1}{2}\int_R [(h^{il}g^{jm}h_{lm} + g^{il}h^{jm}h_{lm} - g^{il}g^{jm}(g_{tlm})''_{t=0}) \\ &\qquad < \frac{\partial S}{\partial x^i}, \frac{\partial S}{\partial x^j} > \sqrt{g} - 2h^{ij} < \frac{\partial V(h)}{\partial x^i}, \frac{\partial S}{\partial x^j} > \sqrt{g}]dx^1dx^2. \end{aligned}$$

Hence:

PROPOSITION 3.1.

$$DE_\phi([g])h = -\frac{1}{2}\int_R h^{ij} < \frac{\partial S}{\partial x^i}, \frac{\partial S}{\partial x^j} > d\mu_g,$$

$$\begin{aligned} D^2E_\phi([g])(h,h) = \frac{1}{2}\int_R \Big[&(h^{il}g^{jm}h_{lm} + g^{il}h^{jm}h_{lm} - g^{il}g^{jm}(g_{tlm})''_{t=0}) \\ &< \frac{\partial S}{\partial x^i}, \frac{\partial S}{\partial x^j} > \sqrt{g} - 2h^{ij} < \frac{\partial V(h)}{\partial x^i}, \frac{\partial S}{\partial x^j} > \sqrt{g}\Big]dx^1dx^2. \end{aligned}$$

THEOREM 3.2 (SACKS-UHLENBECK [S-U1], SCHOEN-YAU [S-Y]). *Assume that N has negative sectional curvatures or N is a Riemannian flat torus. If ζ_0 is a critical point of E_ϕ, then the corresponding harmonic map of (R, g_{ζ_0}) into N is weakly conformal, namely, a branched minimal immersion.*

PROOF. Let $z = x^1 + ix^2$ be a complex local coordinate system of (R, g_{ζ_0}) and $\lambda \mid dz \mid^2$ the metric g_{ζ_0}. We set

$$\xi(z) =< \frac{\partial S}{\partial x^1}, \frac{\partial S}{\partial x^1} > - < \frac{\partial S}{\partial x^2}, \frac{\partial S}{\partial x^2} > -2i < \frac{\partial S}{\partial x^1}, \frac{\partial S}{\partial x^2} >= 4 < \frac{\partial S}{\partial z}, \frac{\partial S}{\partial z} > .$$

Then $\xi(z)dz^2$ is a holomorphic quadratic differential because of the harmonicity of S. Note that $\xi(z)dz^2$ is 0 if and only if S is weakly conformal. Since

$$DE_\phi([g])h = -\mathrm{Re}\int_R \frac{1}{\lambda} h_{zz}\overline{\xi(z)}dx^1dx^2 = - < h_{zz}dz^2, \xi(z)dz^2 >_{WP},$$

where $< \ , \ >_{WP}$ is the Weil-Peterson metric on $T(R)$, $[g_{\zeta_0}]$ is a critical point of E_ϕ if and only if $< h_{zz}dz^2, \xi(z)dz^2 >_{WP} = 0$ holds for all h, which is equivalent to $\xi(z)dz^2 = 0$. □

Let ∇ be the induced covariant differentiation on the tangent bundle of s and $\nabla^\perp$ the induced covariant differentiation of the normal bundle. Proposition 3.3 follows from Proposition 3.1 and Theorem 3.2.

PROPOSITION 3.3. *Let $[g]$ be a critical point of E_ϕ. Then*

$$D^2E_\phi([g])(h,h) = \frac{1}{2}\int_R [\mu h^{ij}h_{ij} - 2h^{ij} < \frac{\partial V(h)}{\partial x^i}, \frac{\partial S}{\partial x^j} > \sqrt{g}]dx^1dx^2,$$

where $\mu((dx^1)^2 + (dx^2)^2)$ is the metric induced on the coordinate neighborhood of a complex local coordinate system $z = x^1 + ix^2$.

Let $[g_t]$ be a curve in $T(R)$. Let us suppose that $[g]$ is a critical point of E_ϕ whose second variation along $[g_t]$ is negative. As $A(S(g_t)) \leq E_\phi([g_t]) = E(g_t, S(g_t)) \leq E_\phi([g]) = E(g, S(g)) = A(S(g))$, where $A(S(g))$ means the area of $S(g)$, we obtain

$$D^2A(W,W) \leq D^2E_\phi([g])(h,h), \tag{3.4}$$

where W is the normal component of the variation vector field along $S(g)$. Thus the second variation of areas is negative. Next, let s be a section of the normal bundle and S_t a variation in the direction s in N. Then induced tensor field $\widetilde{g_{tij}}$ is given by $< \frac{\partial S_t}{\partial x^i}, \frac{\partial S_t}{\partial x^j} >$, whose first variation $\widetilde{h_{ij}}$ is $-2\sigma^s_{ij}$, where σ^s_{ij} is the second fundamental form in the direction of s. Note that $\widetilde{g_{tij}}$ may be degenerate on branch points. Thus we do not know whether $\widetilde{g_{tij}}$ defines an almost complex structure on R with smooth dependence on t. Assume that s is zero on a neighborhood of the branch points. Since $S_t = S(g)$ here, we have the almost complex structures J_t associated with $\widetilde{g_{tij}}$ which is smoothly dependent on t. $[J_t]$ is the smooth curve above in $T(R)$ such that $[J_0] = [g]$. We have to consider the tangent vectors of $[J_t]$. Let $[g_t]$ be a point corresponding to $[J_t]$ in $T(R)$. Then there exists a non-negative function ρ_t such that $\widetilde{g_t} = \rho_t g_t$, where ρ_0 is only zero for branch points. Its differential at $t = 0$ is $\widetilde{h} = (\rho_t)'_{t=0}g + \rho_0 h$. By minimality, $\mathrm{tr}_{\widetilde{g}}\widetilde{h} = 0$ holds except at branch points and hence $\mathrm{tr}_g\widetilde{h} = 0$, and furthermore, $\mathrm{tr}_g h = 0$. These give $\widetilde{h} = \rho_0 h$.

The holomorphic quadratic differential h as the tangent vector of $[g_t]$ at $t = 0$ is given by the orthogonal projection of

$$\frac{-2\sigma^s_{zz}dz^2}{\rho_0} \tag{3.5}$$

on to the space Q of holomorphic quadratic differentials on (R, g). Note that (3.5) is well-defined, because s is zero on the neighborhood of the branch points. Since $E_\phi([g_t]) \leq E(g_t, S_t) = E(\widetilde{g_t}, S_t) = A(S_t)$,

$$D^2E_\phi([g])(h,h) \leq D^2A(s,s). \tag{3.6}$$

These results yield:

THEOREM 3.4. *$index_E = index_A$. In particular, $index_A \leq 6\gamma - 6$.*

PROOF. $index_E \leq index_A$: Let T be the maximal subspace of the tangent space where D^2E_ϕ is negative definite. Then $\dim T = index_E$. When we transfer an element of T to the normal component of the variation vector field along $S(g)$, we obtain a linear map. If there exists a non-zero element X of the kernel, then, by (3.4), $0 = D^2A(0,0) \leq D^2E_\phi([g])(X,X) < 0$, which is a contradiction.

$index_E \geq index_A$: Let V be the maximal subspace of the space of sections of the normal bundle where D^2A is negative definite. Then $\dim V = index_A$ is finite.

So there exists a cut-off function φ such that φ is zero on some neighborhood of branch points and D^2A is negative definite on the subspace φV. Let s be an element of φV. Then we can transfer s to h given in (3.5). This map is linear and injective, because, by (3.6), $D^2E_\phi([g])(h,h) \leq D^2A(s,s) < 0$. □

Remark: We should also use a cut-off function in the proof of $index_E \leq index_A$, because we do not use the smoothness of the area functional A. However the proof is the same as the proof of $index_E \geq index_A$.

Proposition 3.3 establishes the following second variational formula:

THEOREM 3.5.

$$\begin{aligned}
&D^2E_\phi(\tau,h)\\
&= \frac{1}{2}\int_R [\frac{\mu}{\lambda}\tau^{ij}h_{ij} + 2 < \widetilde{\Delta}V(\tau), V(h) > + \frac{2}{\lambda} < \widetilde{R}_{\frac{\partial S}{\partial x^i}V(\tau)}\frac{\partial S}{\partial x^i}, V(h) >]d\mu_g\\
&\qquad = \mathrm{Re}[\int_R \frac{4}{\lambda}\tau_{\bar{z}\bar{z}}(\frac{\mu}{\lambda}h_{zz} - 2 < \widetilde{\nabla}_{\frac{\partial}{\partial z}}V(h), \frac{\partial S}{\partial z} >)dx^1dx^2].
\end{aligned}$$

Here $[\frac{\mu}{\lambda}h_{zz} - 2 < \widetilde{\nabla}_{\frac{\partial}{\partial z}}V(h), \frac{\partial S}{\partial z} >]dz^2$ *is holomorphic.*

PROOF. Differentiating the harmonic map equation (3.2) at $t=0$, we obtain

$$\begin{aligned}
(3.7)\qquad & -\sum_{i,j}\frac{\partial}{\partial x^i}(\frac{1}{\lambda})h_{ij}\frac{\partial S^\alpha}{\partial x^j} - \frac{1}{\lambda}\sum_{i,j}h_{ij}\frac{\partial^2 S^\alpha}{\partial x^i\partial x^j} + \sum_i \frac{\partial^2 V(h)^\alpha}{\partial x^i\partial x^i}\\
&+V(h)^\delta(\frac{\partial}{\partial y^\delta}\Gamma^\alpha_{\gamma\beta})\sum_i\frac{\partial S^\gamma}{\partial x^i}\frac{\partial S^\beta}{\partial x^i} + 2\Gamma^\alpha_{\gamma\beta}\sum_i\frac{\partial V(h)^\gamma}{\partial x^i}\frac{\partial S^\beta}{\partial x^i}\\
&-\frac{1}{\lambda}\sum_{i,j}h_{ij}\Gamma^\alpha_{\gamma\beta}\frac{\partial S^\gamma}{\partial x^i}\frac{\partial S^\beta}{\partial x^j} = 0.
\end{aligned}$$

On the other hand:

$$(3.8)\qquad \widetilde{\nabla}_{\frac{\partial}{\partial x^i}}\frac{\partial S}{\partial x^j} = \frac{\partial^2 S^\alpha}{\partial x^i\partial x^j}\frac{\partial}{\partial y^\alpha} + \Gamma^\alpha_{\gamma\beta}\frac{\partial S^\gamma}{\partial x^i}\frac{\partial S^\beta}{\partial x^j}\frac{\partial}{\partial y^\alpha},$$

$$\begin{aligned}
\widetilde{\nabla}_{\frac{\partial}{\partial x^i}}\widetilde{\nabla}_{\frac{\partial}{\partial x^i}}V(h) &= \frac{\partial^2 V^\alpha(h)}{\partial x^i\partial x^i}\frac{\partial}{\partial y^\alpha} + 2\frac{\partial V^\alpha(h)}{\partial x^i}\frac{\partial S^\beta}{\partial x^i}\Gamma^\delta_{\alpha\beta}\frac{\partial}{\partial y^\delta} + V^\beta(h)\frac{\partial^2 S^\gamma}{\partial x^i\partial x^i}\Gamma^\alpha_{\gamma\beta}\frac{\partial}{\partial y^\alpha}\\
&+V^\beta(h)\frac{\partial S^\gamma}{\partial x^i}\frac{\partial S^\varepsilon}{\partial x^i}(\frac{\partial}{\partial y^\varepsilon}\Gamma^\alpha_{\gamma\beta})\frac{\partial}{\partial y^\alpha} + V^\beta(h)\frac{\partial S^\gamma}{\partial x^i}\Gamma^\alpha_{\gamma\beta}\frac{\partial S^\delta}{\partial x^i}\Gamma^\varepsilon_{\alpha\delta}\frac{\partial}{\partial y^\varepsilon}.
\end{aligned}$$

Since the harmonic map equation is $\sum_i(\frac{\partial^2 S^\alpha}{\partial x^i\partial x^i} + \frac{\partial S^\gamma}{\partial x^i}\frac{\partial S^\beta}{\partial x^i}\Gamma^\alpha_{\gamma\beta}) = 0$,

$$(3.9)\qquad \widetilde{\nabla}_{\frac{\partial}{\partial x^i}}\widetilde{\nabla}_{\frac{\partial}{\partial x^i}}V(h)$$

$$\begin{aligned}
&= \frac{\partial^2 V^\alpha(h)}{\partial x^i\partial x^i}\frac{\partial}{\partial y^\alpha} + 2\frac{\partial S^\beta}{\partial x^i}\frac{\partial V^\alpha(h)}{\partial x^i}\Gamma^\delta_{\alpha\beta}\frac{\partial}{\partial y^\delta} + V^\beta(h)\Gamma^\alpha_{\gamma\beta}(-\frac{\partial S^\delta}{\partial x^i}\frac{\partial S^\varepsilon}{\partial x^i}\Gamma^\gamma_{\delta\varepsilon})\frac{\partial}{\partial y^\alpha}\\
&+V^\beta(h)\frac{\partial S^\gamma}{\partial x^i}\frac{\partial S^\varepsilon}{\partial x^i}(\frac{\partial}{\partial y^\varepsilon}\Gamma^\alpha_{\gamma\beta})\frac{\partial}{\partial y^\alpha} + V^\beta(h)\frac{\partial S^\gamma}{\partial x^i}\Gamma^\alpha_{\gamma\beta}\frac{\partial S^\delta}{\partial x^i}\Gamma^\varepsilon_{\alpha\delta}\frac{\partial}{\partial y^\varepsilon}.
\end{aligned}$$

Substituting these equations into (3.7) and using the curvature tensor $\widetilde{R}$ of N, which is defined by $\widetilde{R}_{XY}Z = \widetilde{\nabla}_{[X,Y]}Z - \widetilde{\nabla}_X\widetilde{\nabla}_Y Z + \widetilde{\nabla}_Y\widetilde{\nabla}_X Z$, we obtain

$$(3.10)\qquad -\sum_{i,j}\frac{\partial}{\partial x^i}(\frac{1}{\lambda})h_{ij}\frac{\partial S^\alpha}{\partial x^j}\frac{\partial}{\partial y^\alpha} - \frac{1}{\lambda}\sum_{i,j}h_{ij}\widetilde{\nabla}_{\frac{\partial}{\partial x^i}}\frac{\partial S}{\partial x^j} +$$

$$\sum_i \widetilde{\nabla}_{\frac{\partial}{\partial x^i}}\widetilde{\nabla}_{\frac{\partial}{\partial x^i}}V(h) + \sum_i \widetilde{R}_{\frac{\partial S}{\partial x^i}V(h)}\frac{\partial S}{\partial x^i} = 0.$$

For a complex local coordinate system z, (3.10) can be expressed as

$$(3.11)\qquad -h_{\bar{z}\bar{z}}(\frac{\partial}{\partial z}\frac{1}{\lambda})\frac{\partial S}{\partial z} - h_{zz}(\frac{\partial}{\partial \bar{z}}\frac{1}{\lambda})\frac{\partial S}{\partial \bar{z}}$$

$$-(\frac{1}{\lambda})h_{\bar{z}\bar{z}}\widetilde{\nabla}_{\frac{\partial}{\partial z}}\frac{\partial S}{\partial z} - (\frac{1}{\lambda})h_{zz}\widetilde{\nabla}_{\frac{\partial}{\partial \bar{z}}}\frac{\partial S}{\partial \bar{z}} + \widetilde{\nabla}_{\frac{\partial}{\partial \bar{z}}}\widetilde{\nabla}_{\frac{\partial}{\partial z}}V(h) + \widetilde{R}_{\frac{\partial S}{\partial \bar{z}}V(h)}\frac{\partial S}{\partial z} = 0.$$

Furthermore, taking the complex bilinear inner-product of $\frac{\partial S}{\partial z}$ and the left hand side (3.11), we obtain

$$(3.12)\qquad -h_{zz}(\frac{\partial}{\partial \bar{z}}\frac{1}{\lambda})\frac{\mu}{2} - (\frac{1}{\lambda})h_{zz}(\frac{1}{2}\frac{\partial}{\partial \bar{z}}\mu) + <\widetilde{\nabla}_{\frac{\partial}{\partial \bar{z}}}\widetilde{\nabla}_{\frac{\partial}{\partial z}}V(h), \frac{\partial S}{\partial z}> = 0,$$

which implies

$$(3.13)\qquad \frac{\partial}{\partial \bar{z}}[<\widetilde{\nabla}_{\frac{\partial}{\partial z}}V(h), \frac{\partial S}{\partial z}> - \frac{\mu}{2\lambda}h_{zz}] = 0.$$

Since

$$\int_R h^{ij} <\widetilde{\nabla}_{\frac{\partial}{\partial x^i}}V(\tau), \frac{\partial S}{\partial x^j}> \lambda dx^1 dx^2$$
$$= \int_R -\frac{\partial}{\partial x^i}(h^{ij}\lambda) < V(\tau), \frac{\partial S}{\partial x^j}> - \lambda h^{ij} < V(\tau), \widetilde{\nabla}_{\frac{\partial}{\partial x^i}}\frac{\partial S}{\partial x^j}> dx^1 dx^2$$
$$= \int_R < V(\tau), -\widetilde{\Delta}V(h) - (\frac{1}{\lambda})\widetilde{R}_{\frac{\partial S}{\partial x^i}V(h)}\frac{\partial S}{\partial x^i}> d\mu_g,$$

Proposition 3.3 gives $4D^2E_\phi([g])(\tau, h) =$

$$\int_R \{h^{ij}(\mu\tau_{ij} - 2 < \frac{\partial V(\tau)}{\partial x^i}, \frac{\partial S}{\partial x^j}> \lambda) + \tau^{ij}(\mu h_{ij} - 2 < \frac{\partial V(h)}{\partial x^i}, \frac{\partial S}{\partial x^j}> \lambda)\}dx^1 dx^2.$$

Since two terms of the right hand side are equivalent,

$$D^2E_\phi([g])(\tau, h) = \frac{1}{2}\int_R \tau^{ij}(\mu h_{ij} - 2 < \frac{\partial V(h)}{\partial x^i}, \frac{\partial S}{\partial x^j}> \lambda)dx^1 dx^2$$

holds. □

Now we shall investigate the null vectors of D^2E_ϕ.

COROLLARY 3.6. *$h_{zz}dz^2$ is a null vector of D^2E_ϕ at a critical point if and only if*

$$\frac{\mu}{\lambda}h_{zz} - 2 < \widetilde{\nabla}_{\frac{\partial}{\partial z}} V(h), \frac{\partial S}{\partial z} >= 0.$$

Let $V(h)^\top, V(h)^\perp$ be the tangent component and the normal component of $V(h)$, respectively. Assume that h is a null vector of D^2E_ϕ. It follows from Corollary 3.6 that $\frac{\mu}{\lambda}h_{zz} - 2 < \widetilde{\nabla}_{\frac{\partial}{\partial z}} V(h), \frac{\partial S}{\partial z} >= 0$, which is equivalent to

$$\nabla_{\frac{\partial}{\partial z}}(V(h)^{\top(01)}) = [\frac{1}{\lambda}h_{zz} + \frac{2}{\mu}\sigma_{zz}^{V(h)^\perp}]\frac{\partial S}{\partial \bar{z}}. \tag{3.14}$$

Note that the sum of the order of the zeros for $\sigma_{zz}^{V(h)^\perp}$ and $\frac{\partial S}{\partial \bar{z}}$ is the order of zeros for μ hence (3.14) is well-defined. The normal component of (3.11) gives

$$-\frac{1}{\lambda}h_{\bar{z}\bar{z}}\sigma_{zz} - \frac{1}{\lambda}h_{zz}\sigma_{\bar{z}\bar{z}} + (\widetilde{\nabla}_{\frac{\partial}{\partial \bar{z}}}\widetilde{\nabla}_{\frac{\partial}{\partial z}}V(h))^\perp + (\widetilde{R}_{\frac{\partial S}{\partial \bar{z}}V(h)}\frac{\partial S}{\partial z})^\perp = 0,$$

which, together with

$$(\widetilde{\nabla}_{\frac{\partial}{\partial \bar{z}}}\widetilde{\nabla}_{\frac{\partial}{\partial z}}V(h))^\perp = [\frac{1}{\lambda}h_{\bar{z}\bar{z}} + \frac{2}{\mu}\sigma_{\bar{z}\bar{z}}^{V(h)^\perp}]\sigma_{zz} + [\frac{1}{\lambda}h_{zz} + \frac{2}{\mu}\sigma_{zz}^{V(h)^\perp}]\sigma_{\bar{z}\bar{z}}$$

$$+(\nabla'_{\frac{\partial}{\partial \bar{z}}}\sigma)(\frac{\partial}{\partial z}, V(h)^\top) + \nabla^\perp_{\frac{\partial}{\partial \bar{z}}}\nabla^\perp_{\frac{\partial}{\partial z}}V(h)^\perp - \frac{2}{\mu} < V(h)^\perp, \sigma_{zz} > \sigma_{\bar{z}\bar{z}},$$

implies

$$\nabla^\perp_{\frac{\partial}{\partial z}}\nabla^\perp_{\frac{\partial}{\partial \bar{z}}}V(h)^\perp + \frac{2}{\mu}\sigma_{zz}^{V(h)^\perp}\sigma_{\bar{z}\bar{z}} + (\nabla'_{\frac{\partial}{\partial z}}\sigma)(\frac{\partial S}{\partial \bar{z}}, V(h)^\top)$$

$$+(\widetilde{R}_{\frac{\partial S}{\partial z}V(h)^\top}\frac{\partial S}{\partial \bar{z}})^\perp + (\widetilde{R}_{\frac{\partial S}{\partial z}V(h)^\perp}\frac{\partial S}{\partial \bar{z}})^\perp = 0,$$

where $(\nabla'_X\sigma)(Y,Z) = \nabla^\perp_X\sigma(Y,Z) - \sigma(\nabla_XY,Z) - \sigma(Y,\nabla_XZ)$.

On the other hand, let J be the Jacobi operator acting on the space of sections of the normal bundle of the compact orientable branched minimal surface. Then:

LEMMA 3.7. *For a normal section s,*

$$Js = -\frac{4}{\mu}\nabla^\perp_{\frac{\partial}{\partial \bar{z}}}\nabla^\perp_{\frac{\partial}{\partial z}}s - \frac{8}{\mu^2} < s, \sigma_{\bar{z}\bar{z}} > \sigma_{zz} - \frac{4}{\mu}(\widetilde{R}_{\frac{\partial S}{\partial \bar{z}}s}\frac{\partial S}{\partial z})^\perp.$$

It follows from the Codazzi equation and minimality that

$$(\nabla'_{\frac{\partial}{\partial z}}\sigma)(\frac{\partial S}{\partial \bar{z}}, V(h)^\top) + (\widetilde{R}_{\frac{\partial S}{\partial z}V(h)^\top}\frac{\partial S}{\partial \bar{z}})^\perp = 0,$$

which, together with Lemma 3.7, yields the following:

LEMMA 3.8. *The normal component of* (3.11) *is* $-\frac{\mu}{4}JV(h)^\perp$.

We shall study the relation between the null spaces of $D^2E_\phi([g])$ and D^2A.

PROPOSITION 3.9. *For a given null vector $h_{zz}dz^2$, $V(h)^{\perp}$ is a Jacobi field of the compact orientable branched minimal surface and we have $\nabla_{\frac{\partial}{\partial z}}(V(h)^{\top(01)}) = [\frac{1}{\lambda}h_{zz} + \frac{2}{\mu}\sigma_{zz}^{V(h)^{\perp}}]\frac{\partial S}{\partial \bar{z}}$.*

Let D be the Fredholm operator $\xi \mapsto \nabla_{\frac{\partial}{\partial z}}\xi dz$ taking sections of $S_*T^{(01)}R$ to sections of $S_*T^{(01)}R \otimes T^{*(10)}R$, and D^* the adjoint operator. Then the kernel $K(D^*)$ of D^* is the space of holomorphic sections of $S_*T^{(01)}R \otimes T^{*(10)}R$. We can define a map of $K(D^*)$ into Q by

$$\xi_z dz \mapsto < \xi_z, \frac{\partial S}{\partial z} > dz^2. \tag{3.15}$$

LEMMA 3.10. *This map (3.15) is an injective linear map. S is an immersion if and only if it is surjective.*

PROOF. Let b be the number of branch points. Then $c_1(S_*T^{(0,1)}R \otimes T^{*(1,0)}R) = -b+4(\gamma-1)$. The Clifford theorem (see, for example, [G-H]) shows that $\dim_{\mathbf{C}} K(D^*)$ is $3\gamma - 3 - b$ if $b \leq 2\gamma - 4$, $2\gamma - 1 - [\frac{b}{2}]$ if $2\gamma - 3 \leq b \leq 4\gamma - 4$ and 1 if $b \geq 4\gamma - 3$. Note that $\dim Q = 3\gamma - 3 \geq \dim K(D^*)$. □

Let $Z = \{q_{zz}dz^2 \in Q : \frac{1}{\lambda}q_{zz}\frac{\partial S}{\partial \bar{z}}dz$ is orthogonal to $K(D^*)\}$.

LEMMA 3.11. *Let s be a section of S^*TN. Then there exists $h_{zz}dz^2 \in Q$ such that $(\frac{1}{\lambda}h_{zz} + \frac{2}{\mu}\sigma_{zz}^{s^{\perp}})\frac{\partial S}{\partial \bar{z}}dz$ is orthogonal to $K(D^*)$. In particular, $h_{zz}dz^2$ is determined up to elements of Z.*

LEMMA 3.12. *Let s be a Jacobi field and $h_{zz}dz^2$ as in Lemma 3.11. Then we have a tangent vector field W such that $D(W^{(01)}) = [\frac{1}{\lambda}h_{zz} + \frac{2}{\mu}\sigma_{zz}^{s}]\frac{\partial S}{\partial \bar{z}}dz$. In particular, W is determined up to the tangent vector fields η such that, for $q_{zz}dz^2 \in Z$, $D(\eta^{\top(01)}) = \frac{1}{\lambda}q_{zz}\frac{\partial S}{\partial \bar{z}}dz$.*

PROOF. The Fredholm Alternative Theorem gives W. □

LEMMA 3.13. *Let s be a Jacobi field and W the element obtained in Lemma 3.12. Then $s + W$ satisfies (3.11). Let h be associated with s as in Lemma 3.11. Then $s + W - V(h)$ is parallel.*

PROOF. Since $V(h)$ also satisfies (3.11), for $\widetilde{V} = s + W - V(h)$, we have $\widetilde{\nabla}_{\frac{\partial}{\partial \bar{z}}}\widetilde{\nabla}_{\frac{\partial}{\partial z}}\widetilde{V} + \widetilde{R}_{\frac{\partial S}{\partial \bar{z}}\widetilde{V}}\frac{\partial S}{\partial z} = 0$. Since N is non-positively curved, $\widetilde{V}$ is parallel. In particular, if N is negatively curved, then $s + W = V(h)$. □

LEMMA 3.14. *Assume that N is a Riemannian flat torus. Let Y be a parallel vector field on N. Then $\nabla_{\frac{\partial}{\partial z}}(Y^{\top(01)}) = \frac{2}{\mu}\sigma_{zz}^{Y^{\perp}}\frac{\partial S}{\partial \bar{z}}$.*

Let J_a be the set of Jacobi fields and P_a the set of normal components of parallel fields. Then $P_a \subset J_a$ and $\dim P_a = n$. We denote by *nullE* the set of null vectors of D^2E_ϕ. Let s be in J_a and $h_{zz}dz^2$ be associated with s as in Lemma 3.11. Then we can prove $h_{zz}dz^2$ is in $nullE$ as follows. Suppose that $h_{zz}dz^2$ is not zero. Then by Lemma 3.12, we obtain W such that $\nabla_{\frac{\partial}{\partial z}} W^{\top(01)} = [\frac{1}{\lambda}h_{zz} + \frac{2}{\mu}\sigma^s_{zz}]\frac{\partial S}{\partial \bar{z}}$. By Lemma 3.13, we have a parallel field Y such that $W + s - V(h) = Y$ and hence $V(h)^\top = W - Y^\top, V(h)^\perp = s - Y^\perp$. Since, by Lemma 3.14,

$$\nabla_{\frac{\partial}{\partial z}} V(h)^{\top(01)} = [\frac{1}{\lambda}h_{zz} + \frac{2}{\mu}\sigma^s_{zz}]\frac{\partial S}{\partial \bar{z}} - \frac{2}{\mu}\sigma^{Y^\perp}_{zz}\frac{\partial S}{\partial \bar{z}} = [\frac{1}{\lambda}h_{zz} + \frac{2}{\mu}\sigma^{V(h)^\perp}_{zz}]\frac{\partial S}{\partial \bar{z}},$$

$h_{zz}dz^2$ is a null vector by Corollary 3.6 and (3.14). We can define a linear map $\widetilde{J_E}$ of J_a into $nullE/Z$. For $Y^\perp \in P_a$, consider a corresponding $h_{zz}dz^2$. Since, by Lemma 3.14, $(\frac{2}{\mu}\sigma^{Y^\perp}_{zz})\frac{\partial S}{\partial \bar{z}}dz$ is orthogonal to $K(D^*)$, so is $(\frac{1}{\lambda}h_{zz})\frac{\partial S}{\partial \bar{z}}dz$. We find $h_{zz}dz^2 \in Z$, namely, $\widetilde{J_E}(Y^\perp) = 0$. We can define a linear map J_E of J_a/P_a into $nullE/Z$.

PROPOSITION 3.15. *J_E is an isomorphism.*

PROOF. Injection: Let s be an element of J_a and $h_{zz}dz^2$ an element ($\in Q$) obtained for s as in Lemma 3.11. If $h_{zz}dz^2$ is in Z, then $(\frac{1}{\lambda}h_{zz})\frac{\partial S}{\partial \bar{z}}dz$ is orthogonal to $K(D^*)$ and so is $(\frac{2}{\mu}\sigma^s_{zz})\frac{\partial S}{\partial \bar{z}}dz$. Then we obtain W such that $\nabla_{\frac{\partial}{\partial z}} W^{(01)} = (\frac{2}{\mu}\sigma^s_{zz})\frac{\partial S}{\partial \bar{z}}$. By Lemma 3.13, $W + s$ is a parallel field. Surjection: Let $h_{zz}dz^2$ be in $nullE$ and not in Z. If $V(h)^\perp = 0$, then, $\nabla_{\frac{\partial}{\partial z}}(V(h)^{\top(01)}) = \frac{1}{\lambda}h_{zz}\frac{\partial S}{\partial \bar{z}}$ holds by (3.14). Hence $h_{zz}dz^2 \in Z$, which is a contradiction. So $V(h)^\perp$ is not zero and $\widetilde{J_E}V(h)^\perp$ belongs to the equivalence class of $h_{zz}dz^2$. □

If N is negatively curved, we have:

THEOREM 3.16. *$nullity_E = nullity_A + 6\gamma - 6 - 2\dim_{\mathbf{C}} K(D^*)$ if N is negatively curved. In particular, if the map is an immersion, then $nullity_E = nullity_A$.*

Since $nullity_E + index_E \leq 6\gamma - 6$, we obtain:

COROLLARY 3.17. *Assume that N is negatively curved. Then $index_A + nullity_A \leq 2\dim_{\mathbf{C}} K(D^*)$.*

Furthermore, when N is a Riemannian flat torus, we have:

THEOREM 3.18. *$nullity_E = nullity_A + 6\gamma - 6 - 2\dim_{\mathbf{C}} K(D^*) - n$ for $n \geq 3$ and $nullity_E = 6\gamma - 6 - 2\dim_{\mathbf{C}} K(D^*)$ for $n = 2$, if N is a Riemannian flat torus of dimension n. If the map is an immersion, then $nullity_E = nullity_A - n$ for $n \geq 3$ and $nullity_E = 0$ for $n = 2$.*

COROLLARY 3.19. *$index_A + nullity_A \leq 2\dim_{\mathbf{C}} K(D^*) + n$ if N is a Riemannian flat torus.*

Beeson [Be] proved that Dim Ker$D^2E(u) = 3 + M +$ Dim Ker$D^2A(u)$, where E is the Dirichlet integral, A is the area and M is the number of forced Jacobi fields, namely $M =$ sum of the orders of the boundary branch points plus twice the orders of the interior branch points. A forced Jacobi field means that the infinitesimal variation vector field with respect to a one parameter family of harmonic maps given by the null vector of the hessian is tangent to the surface. Let $q(\neq 0)$ be an element of Z. If N is negatively curved, then $V(q)$ has no normal component. If N is a Riemannian flat torus, then choosing $S(g_\varsigma)$, we have $V(q)^\perp = 0$. Thus elements of Z give forced Jacobi fields.

The meaning of non-degeneracy of the hessian at a critical point is as follows (cf. the last statement in [T2]).

COROLLARY 3.20. *(i) If N is negatively curved, E_ϕ is nondegenerate if and only if the corresponding compact orientable branched minimal surface has no branch points and nullity$_A = 0$. (ii) If N is a Riemannian flat torus, E_ϕ is nondegenerate if and only if the corresponding compact orientable branched minimal surface has no branch points and only trivial Jacobi fields.*

We do not know whether the energy function is smooth for non-positively curved Riemannian manifolds. However we note that energy functions are smooth for Riemannian products of negatively curved manifolds and Riemannian flat tori and critical points also corresponds to branched minimal immersions. As an application, we obtain:

COROLLARY 3.21. *Let N_1, N_2 be negatively curved Riemannian manifolds or Riemannian flat tori and $\widetilde{\phi} = (\phi_1, \phi_2)$ a continuous map of a Riemann surface R into a Riemannian product of $N_1 \times N_2$. Then, if there exists a compact orientable immersed minimal surface in N_1 which is homotopic to ϕ_1 and has no Jacobi fields if N_1 is negatively curved, only trivial Jacobi fields if N_1 is a Riemannian flat torus, then for c close to 0, there exists a compact orientable immersed minimal surface in $N_1 \times N_2(c)$ which is homotopic to $\widetilde{\phi}$, where $N_2(c)$ is homothetic to N_2 with homothetic constant c.*

PROOF. When the ambient space is $N_1 \times N_2(c)$, the energy function $E_{\widetilde{\phi}}$ is $E_{\phi_1} + cE_{\phi_2}$. Since E_{ϕ_1} is nondegenerate at the critical point corresponding to the compact orientable immersed minimal surface in N_1, critical points of $E_{\widetilde{\phi}}$ vary smoothly for c close to 0. □

Note that c in Corollary 3.21 may be negative, because Theorem 3.2 also holds for $c < 0$. When $M = N_1$ and ϕ is the identity, we obtain a minimal map of M into $N_2(c)$ homotopic to ϕ_2 (see [Sc]). As an application, we can construct minimal graphs on the Poincaré disc in the Lorentzian product of the Poincaré disc and **R**.

§4 Riemann matrices.

Let $\{A_i, B_i\}$ be a fixed canonical system of generators of $\pi_1(R, *)$ on R and $[(M, f)]$ a point of $T(R)$. Then $\{f(A_i), f(B_i)\}$ in M gives rise to a canonical homology basis, also denoted by $\{A_i, B_i\}$. Let $\psi_1, \dots, \psi_\gamma$ be a canonical basis of the holomorphic differentials of first kind for the canonical homology basis: $\int_{A_j} \psi_i = \delta_{ij}$. Then the matrix $\tau = (\tau_{ij})$ of size γ defined by $\tau_{ij} = \int_{B_j} \psi_i$ is called the Riemann matrix, which provides an element of the Siegel upper half space H_γ of symmetric matrices τ of size γ such that $\mathrm{Im}\tau$ is positive (see, for example, [F-K]). Let κ be a map of $T(R)$ into H_γ defined by $[(M,f)] \mapsto \tau$. Then Rauch (see, for example, [I-T]) proved that $d\kappa(\mu)_{jk} = \int_M \psi_j \psi_k \mu$, where μ is a harmonic Beltrami differential. Let $HT(R)$ denote $\{[(M,f)] \in T(R) : M$ is a hyperelliptic Riemann surface $\}$. Then Ahlfors proved:

THEOREM 4.1([A]). *$HT(R)$ is a complex submanifold of* $\dim_{\mathbf{C}} HT(R) = 2\gamma - 1$ *in $T(R)$, κ is holomorphic, κ has full rank $3\gamma - 3$ for a point $\in T(R) \setminus HT(R)$, $\kappa \mid_{HT(R)}$ has full rank $2\gamma - 1$, and κ is degenerate in the direction of the complement of $HT(R)$.*

Furthermore, there exists a natural holomorphic covering space projection π of $T(R)$ onto $U(\gamma)$, which is called the Torelli space, and a holomorphic map $^{\#}\kappa$ of $U(\gamma)$ into H_γ such that $\kappa = {}^{\#}\kappa\pi$. Let $\widetilde{H}$ be $\pi(HT(R))$. Then $^{\#}\kappa$ is $2:1$ on $U(\gamma)\backslash\widetilde{H}$ and $1:1$ on $\widetilde{H}$. In particular, we have an involutory modular transformation ϑ of $U(\gamma)$ (see, for example, [Ra]). Then $U(\gamma)/\vartheta$ may be identified with the set RM of Riemann matrices (as a subset) in H_γ, the famous Torelli Theorem. Let $\{p_i\}$ be a divergent sequence in $T(R)$ such that $\lim \kappa(p_i) = \tau \in H_\gamma$. Using the irreducibility of Theta divisors of Riemann surfaces, we know that τ is not a Riemann matrix [I-T].

§5 The embedding P of H_γ into S^2_R and a differential-geometric Schottky problem.

The special linear group. $SL(2\gamma, \mathbf{R})$ acts on itself as follows: for $g \in SL(2\gamma, \mathbf{R})$, $f \in SL(2\gamma, \mathbf{R}) \mapsto gf^tg \in SL(2\gamma, \mathbf{R})$. Let $P_{2\gamma}$ be the set of positive definite symmetric matrices of size 2γ and determinant one. Then $P_{2\gamma}$ is a subset of $SL(2\gamma, \mathbf{R})$ which is the orbit through the identity by the above action. Thus we may conclude that $P_{2\gamma}$ is $SL(2\gamma, \mathbf{R})/SO(2\gamma, \mathbf{R})$. Let $Sp(2\gamma, \mathbf{R})$ be the symplectic group defined by

$$\left\{ g \in SL(2\gamma, \mathbf{R}) : {}^t gJg = J = \begin{pmatrix} 0 & I_\gamma \\ -I_\gamma & 0 \end{pmatrix} \right\}$$

and let $\natural_{2\gamma} = \{g \in P_{2\gamma} : gJg = J\}$. Then $\natural_{2\gamma}$ is the orbit of $Sp(2\gamma, \mathbf{R})$ through the identity of $SL(2\gamma, \mathbf{R})$ and hence $\natural_{2\gamma}$ is $Sp(2\gamma, \mathbf{R})/U(\gamma)$. It is easy to see that $g \in \natural_{2\gamma}$ has the expression:

$$\begin{pmatrix} Y + XY^{-1}X & -XY^{-1} \\ -Y^{-1}X & Y^{-1} \end{pmatrix},$$

where X, Y are symmetric matrices of size γ and Y is positive definite, that is, $X + iY \in H_\gamma$. Thus we can define a bijection P of H_γ onto $\natural_{2\gamma}$ and a map P of H_γ into $P_{2\gamma}$. Since $P_{2\gamma}$ is a submanifold in S^2_R, we have a map P of H_γ into S^2_R. Let $g \in Sp(2\gamma, \mathbf{R})$ be

$$g = \begin{pmatrix} a & b \\ c & d \end{pmatrix}, \ {}^t ca = {}^t ac, \ {}^t db = {}^t bd, \ -{}^t cb + {}^t ad = I_\gamma.$$

Then $Sp(2\gamma, \mathbf{R})$ acts on H_γ as follows: $g(z) = (az+b)(cz+d)^{-1}$. Observe that (i) g acts identically on H_γ if and only if $g = \pm I_{2\gamma}$ and (ii) $Sp(2\gamma, \mathbf{R})$ acts transitively on H_γ. We define an embedding of H_γ onto $\natural_{2\gamma}$ such that $Q(z) = P(-X+iY)$ for $z = X + iY \in H_\gamma$. Then Q satisfies

$$Q(g(z)) = gQ(z)^t g \tag{5.1}$$

and hence Q is an $Sp(2\gamma, \mathbf{R})$-equivariant map of H_γ onto $\natural_{2\gamma}$. In particular, P is an embedding of H_γ into S^2_R.

Remark: For example, $P(H_1)$ is one sheet of a hyperboloid of revolution of two sheets in $\mathbf{R}^3$ and hence positively curved. Since the second fundamental form σ of $\natural_{2\gamma}$ satisfies $\sigma(JX, JY) = \sigma(X, Y)$, the sectional curvature of the plane spanned by X, JX is non-negative, by the Gauss equation.

Let M_γ be the moduli space of Riemann surfaces of genus γ, i.e. the quotient space of $T(R)$ by the Teichmüller modular group (the mapping class group). A complex torus with an ample real Riemann form is called an abelian variety, which is equivalent to it being algebraic. A pair $(T^\gamma, A(\xi,\eta))$ of a γ-dimensional complex torus T^γ and an ample real Riemann form $A(\xi,\eta)$ is called a polarized abelian variety. The form $A(\xi,\eta)$ is called an ample real Riemann form if $A(\xi,\eta)$ is alternating, $A(\xi, \sqrt{-1}\eta)$ is positive definite, and $A(\xi,\eta) \in \mathbf{Z}$ for all lattice vectors ξ, η.

Choosing a suitable basis of the covering vector space and a suitable basis of the lattice group, we can consider $T^\gamma = \mathbf{C}^\gamma/\langle G\rangle, G = (z, e)$, where $z \in H_\gamma$, and $A(x,y) = {}^t x A y$, where A is the skew-symmetric integer matrix

$$A = \begin{pmatrix} 0 & -e \\ e & 0 \end{pmatrix}, \quad e = \begin{pmatrix} \epsilon_1 & 0 & \dots & 0 \\ 0 & \epsilon_2 & \dots & 0 \\ \vdots & \vdots & \ddots & \vdots \\ 0 & 0 & \dots & \epsilon_\gamma \end{pmatrix}, \quad \epsilon_i > 0, \ \epsilon_i | \epsilon_{i+1}.$$

In particular, if e is the identity, then $(T^\gamma, A(\xi,\eta))$ is called a principally polarized abelian variety. It is well-known that $A(\xi,\eta)$ is the imaginary part of the positive definite Hermitian form $H(\xi,\eta)$ (ample complex Riemann form) defined by $H(\xi,\eta) = {}^t\bar{\xi}(\mathrm{Im} z)^{-1}\eta$ and $H(zm+n, zm+n)$ is

$$({}^t m, {}^t n)\begin{pmatrix} Y + XY^{-1}X & XY^{-1} \\ Y^{-1}X & Y^{-1} \end{pmatrix}\begin{pmatrix} m \\ n \end{pmatrix} = ({}^t m, {}^t n)Q(z)\begin{pmatrix} m \\ n \end{pmatrix},$$

where $m, n \in \mathbf{Z}^\gamma$. $H_\gamma/Sp(2\gamma, \mathbf{Z})$ is known as the moduli space A_γ of principally polarized abelian varieties. The map κ induces a map of M_γ into A_γ which is called the Torelli mapping t. Thus, we associate a Riemann surface to the Jacobian variety . Note that t is injective (Torelli Theorem).

The Schottky problem is to describe $t(M_\gamma) \subset A_\gamma$**.** A very interesting characterization of $t(M_\gamma)$ in terms of properties of theta divisors and in terms of the KP equation is known ([Gu], [Sh]). Let X be a principally polarized abelian variety given by $\mathbf{C}^\gamma/\langle G\rangle$, $G = (z, e)$, $z \in H_\gamma$ and ϑ Riemann's theta function for X: $\vartheta(\zeta) = \sum_{m \in \mathbf{Z}^\gamma} \exp(2\pi i {}^t m\zeta + \pi i {}^t m z m)$. Shiota [Sh] proved:

THEOREM (SHIOTA). *The following two conditions for X are equivalent:*

(A) i) There exist vectors $a_1 (\neq 0), a_2, a_3 \in \mathbf{C}^\gamma$ and a quadratic form $Q(t) = \sum_{i,j}^3 Q_{ij} t_i t_j$, where $Q_{ij} \in \mathbf{C}$, such that for any $\zeta \in \mathbf{C}^\gamma, \tau(t) = \tau(t,\zeta) = \exp(Q(t))$

$\vartheta(t_1a_1 + t_2a_2 + t_3a_3 + \zeta)$ gives a solution to the Kadomtsev-Petviashivili (KP) equation in Hirota's form: $(D_1^4 + 3D_2^2 - 4D_1D_3)\tau \cdot \tau = 0$, and ii) the theta divisor Θ of X is irreducible.

(B) X is isomorphic to the Jacobian variety of a complete smooth curve of genus γ over $\mathbf{C}$.

Let $RM_{non-hyper}$ and RM_{hyper} be the spaces of Riemann matrices of non-hyperelliptic surfaces and hyperelliptic surfaces. Then $\dim_{\mathbf{C}} RM_{non-hyper} = 3\gamma - 3$ and $\dim_{\mathbf{C}} RM_{hyper} = 2\gamma - 1$ and $RM = RM_{non-hyper} \bigcup RM_{hyper}$. We view $P(RM)$ as the set of Riemann matrices in S^2_R. It is natural to study $P(RM)$ as the union of $P(RM_{non-hyper})$ and $P(RM_{hyper})$. Now we can present **a differential-geometric Schottky problem**:

MAIN PROBLEM. *Characterize $P(RM)$, $P(RM_{non-hyper})$ and $P(RM_{hyper})$ differential-geometrically.*

Remark: Let $\{A_i, B_i\}$ be a canonical homology basis. Then $g = \begin{pmatrix} a & b \\ c & d \end{pmatrix} \in Sp(2\gamma, \mathbf{Z})$ acts on $\{A_i, B_i\}$ as $\{A_i, B_i\} \longmapsto \{A_i, B_i\}g$ and hence the Riemann matrix τ corresponding to $\{A_i, B_i\}$ is transferred to $(a + \tau c)^{-1}(b + \tau d)$ corresponding to $\{A_i, B_i\}g$. Thus the Riemann matrix corresponding to $\{A_i, B_i\}(-g^{-1})$ is $\begin{pmatrix} -a & b \\ c & -d \end{pmatrix}(\tau)$.

§6 Symmetric matrices of type (p, q, r) as control parameters.

Let $L_{p,q}$ be the space R^{p+q} with inner product

$$< x, y >_{p,q} = (x^1 \cdots x^{p+q}) I_{p,q} \begin{pmatrix} y^1 \\ \vdots \\ y^{p+q} \end{pmatrix}, \quad I_{p,q} = \begin{pmatrix} I_p & 0 \\ 0 & -I_q \end{pmatrix}.$$

Let $(\ell_1, \dots, \ell_{2\gamma})$ be a sequence of vectors which span $L_{p,q}$. Then the matrix of size 2γ defined by $(< \ell_i, \ell_j >_{p,q})$ is of type (p, q, r), where p is the number of positive eigenvalues, q is the number of negative eigenvalues and $r = 2\gamma - p - q$ is the number of 0-eigenvalues. Conversely, for any symmetric matrix (p_{ij}) of type (p, q, r), we can find a sequence $(\ell_1, \dots, \ell_{2\gamma})$ such that $\ell_1, \dots, \ell_{2\gamma}$ span $L_{p,q}$ and $(p_{ij}) = (< \ell_i, \ell_j >_{p,q})$ holds up to the isometry group $O_{p,q}$ of $L_{p,q}$. Let $F_{p,q}$ be the set of sequences $(\ell_1, \dots, \ell_{2\gamma})$ such that $\ell_1, \dots, \ell_{2\gamma}$ span $L_{p,q}$. Then the above argument implies:

LEMMA 6.1. *$O_{p,q} \backslash F_{p,q}$ parametrizes the subset consisting of matrices of type (p, q, r).*

For $(\ell_1, \dots, \ell_{2\gamma}) \in F_{p,q}$, in general, $\ell_1, \dots, \ell_{2\gamma}$ are not lattice vectors which provide an indefinite flat torus of dimension $p + q$. Let $\ell'_1, \dots, \ell'_{2\gamma}$ be lattice vectors. Then there exist lattice vectors $\ell_1, \dots, \ell_{p+q}$ such that $(\ell'_1, \dots, \ell'_{2\gamma}) = (\ell_1, \dots, \ell_{p+q})g$, $(\ell'_1, \dots, \ell'_{2\gamma})g' = (\ell_1, \dots, \ell_{p+q})$, where g is a $(p + q, 2\gamma)$-matrix and g' is a $(2\gamma, p+q)$-matrix whose components are integers. Then the symmetric matrix corresponding to $(\ell'_1, \dots, \ell'_{2\gamma})$ is ${}^t gSg$, where $S = (< \ell_i, \ell_j >_{p,q})$. Let $S_g = \{{}^t gSg : S \text{ is a regular symmetric matrix of type } (p,q), \|{}^t gSg\| = 1\}$, where g is a $(p+q, 2\gamma)$-matrix with integer components such that there exists a $(2\gamma, p+q)$-matrix g' with integer components and $gg' = 1_{p+q}$. Then the dimension of S_g is $\frac{1}{2}(p+q)(p+q+1) - 1$ and $S_{p,q} = \bigcup_g S_g$ parametrizes the set of matrices of type (p,q,r) which gives the set of lattice vectors.

LEMMA 6.2. *If $S_g \bigcap S_h \neq \emptyset$, then $S_g = S_h$.*

§7 Branched minimal immersions into an indefinite flat torus.

Let ϕ be a full continuous map of R into a torus $T^n_{p,q} = \mathbf{R}^n/G$, where G denotes a lattice. Then we set $\phi(A_j) = \ell_j \in G, \dots, \phi(B_j) = \ell_{\gamma+j} \in G$ in the homology sense and obtain an $(n, 2\gamma)$-matrix$(\ell_1, \dots, \ell_{2\gamma})$ which is denoted by L. Choosing a suitable covering torus $\widetilde{T^n_{p,q}}$ of $T^n_{p,q}$, ϕ may be lifted to a map $\widetilde{\phi}$ into $\widetilde{T^n_{p,q}}$ such that $\ell_1, \dots, \ell_{2\gamma}$ for the lifted map $\widetilde{\phi}$ generate the lattice. Thus, $\pi_1(R, *)$ is mapped onto $\pi_1(\widetilde{T^n_{p,q}}, \widetilde{\phi}(*))$. Since we can also lift the harmonic map for ϕ, hereafter it is enough to study harmonic maps for ϕ where $\pi_1(R, *)$ is mapped surjectively. Maps like this are called primitive [N-S1]. In this paper, we only treat primitive harmonic maps. Naturally, primitive maps are full. Then, by the lattice vectors $\ell'_1, \dots, \ell'_n \in L_{p,q}$, $L = (L_1, L_2) = (\ell'_1, \dots, \ell'_n)g$ holds, where L_1, L_2 are (n, γ) matrices and g is a $(p+q, 2\gamma)$-matrix with integer components and there exists a $(2\gamma, p+q)$-matrix g' with integers components and $gg' = I_{p+q}$.

Let S be a harmonic map of (M, f) into $T^n_{p,q}$ for ϕ. Then, as the differential dS is a $\mathbf{R}^n$-valued harmonic 1-form, there exists a $(n, 2\gamma)$-matrix Q such that

$$dS = Q\Psi, \tag{7.1}$$

where Ψ is the $\mathbf{R}^{2\gamma}$-valued 1-form given by ${}^t(\mathrm{Re}\psi_1, ..., \mathrm{Re}\psi_\gamma, \mathrm{Im}\psi_1, ..., \mathrm{Im}\psi_\gamma)$. Since S is a harmonic map for ϕ and $\int_{A_j} dS = \ell_j, \int_{B_j} dS = \ell_{\gamma+j}$, $j = 1, \dots, \gamma$ holds, (7.1) yields $QT_\tau = L$, where T_τ denotes the matrix

$$\begin{pmatrix} I & \mathrm{Re}\tau \\ 0 & \mathrm{Im}\tau \end{pmatrix}$$

and hence $dS = LT_\tau^{-1}\Psi$. If $L = T_\tau$, then S is called the Albanese map of (M, f) into the Jacobian variety whose lattice vectors are generated by column vectors of T_τ. Since $\Psi^{(10)}$ is $\frac{1}{2}{}^t(\psi_1, ..., \psi_\gamma, (-i)\psi_1, ..., (-i)\psi_\gamma)$, $dS^{(10)} = \frac{1}{2}(L_1(I + i(\mathrm{Re}\tau)(\mathrm{Im}\tau)^{-1}) - iL_2(\mathrm{Im}\tau)^{-1}){}^t(\psi_1 \cdots \psi_\gamma)$.

Remark: $dS = LT_\tau^{-1}\Psi$ implies smooth dependence on the parameter ζ for harmonic maps $S(\zeta)$ of (R, g_ζ) into tori, which completes the last statement in Section 2. When we write $dS = {}^t(ds_1 \cdots ds_n)$, it follows from Riemann's period relation (see, for example, [F-K]) that

$$\int ds_\ell \wedge *ds_\ell = \sum_{j=1}^{\gamma} \Big[\int_{A_j} ds_\ell \int_{B_j} *ds_\ell - \int_{B_j} ds_\ell \int_{A_j} *ds_\ell \Big],$$

which leads us to

$$2E_\phi = \sum_{k,\ell,i,j} (I_{p,q})_{k,\ell} L_{ki} L_{\ell j} (\mathrm{Im}\tau)_{ij} + \sum_{k,\ell,i,j} (I_{p,q})_{k,\ell} L_{k\gamma+i} L_{\ell\gamma+j} (\mathrm{Im}\tau)^{-1}_{ij}$$

$$+ \sum_{k,\ell,i,j,a} (I_{p,q})_{k,\ell} L_{ki} L_{\ell j} (\mathrm{Re}\tau)_{ja} ((\mathrm{Re}\tau)(\mathrm{Im}\tau)^{-1})_{ia}$$

$$- \sum_{k,\ell,i,j} (I_{p,q})_{k,\ell} 2 L_{ki} L_{\ell\gamma+j} ((\mathrm{Re}\tau)(\mathrm{Im}\tau)^{-1})_{ij}.$$

THEOREM 7.1. *The energy of S is given by $\frac{1}{2}\mathrm{tr}(P(\tau)\widetilde{L})$, where $L_{11} = {}^tL_1 I_{p,q} L_1$, $L_{12} = {}^tL_1 I_{p,q} L_2$, $L_{21} = {}^tL_2 I_{p,q} L_1$, $L_{22} = {}^tL_2 I_{p,q} L_2$ and*

$$\widetilde{L} = \begin{pmatrix} L_{11} & L_{12} \\ L_{21} & L_{22} \end{pmatrix}.$$

The map S is weakly conformal if and only if ${}^t dS^{(1,0)} I_{p,q} dS^{(1,0)} = 0$, that is, $\sum (I_{p,q})_{i,j} A_{ik} A_{j\ell} \psi_k \psi_\ell = 0$, where $A = \frac{1}{2}(L_1(I + i(\mathrm{Re}\tau)(\mathrm{Im}\tau)^{-1}) - iL_2(\mathrm{Im}\tau)^{-1})$. For $\frac{1}{2}\gamma(\gamma+1)$ holomorphic quadratic differentials $\psi_j\psi_k$, it is well-known that if (M, f) is not hyperelliptic, then there are $\frac{1}{2}(\gamma-2)(\gamma-3)$ linearly independent relations such that

$$\sum A^\ell_{ij} \psi_i \psi_j = 0, \tag{7.2}$$

where A^ℓ, $1 \le \ell \le \frac{1}{2}(\gamma-2)(\gamma-3)$, are symmetric matrices of size γ. Thus S is a branched minimal immersion if and only if there exist complex numbers $\alpha_1, \ldots, \alpha_{\frac{1}{2}(\gamma-2)(\gamma-3)}$ such that ${}^tAI_{p,q}A = \sum \alpha_\ell A^\ell$. If (M, f) is hyperelliptic, then there are $\frac{1}{2}(\gamma-1)(\gamma-2)$ linearly independent relations such that $\sum B^\ell_{ij}\psi_i\psi_j = 0$, where B^ℓ, $1 \le \ell \le \frac{1}{2}(\gamma-1)(\gamma-2)$, are symmetric matrices of size γ. Thus S is a branched minimal immersion if and only if there exist complex numbers $\alpha_1, \ldots, \alpha_{\frac{1}{2}(\gamma-1)(\gamma-2)}$ such that

$${}^tAI_{p,q}A = \sum \alpha_\ell B^\ell. \tag{7.3}$$

We define a real linear map Θ_τ of the space S^2_R of real symmetric matrices of size 2γ with standard Euclidean inner product into the space S^2_C of complex symmetric matrices of size γ with standard Hermitian inner product by $\widetilde{L} \mapsto {}^tAI_{p,q}A$, that is,

$$\begin{aligned}\widetilde{L} \longrightarrow & \frac{1}{4}[(I + i(\mathrm{Im}\tau)^{-1}(\mathrm{Re}\tau))L_{11}(I + i(\mathrm{Re}\tau)(\mathrm{Im}\tau)^{-1}) \\ & - i(I + i(\mathrm{Im}\tau)^{-1}(\mathrm{Re}\tau))L_{12}(\mathrm{Im}\tau)^{-1} \\ & - i(\mathrm{Im}\tau)^{-1}L_{21}(I + i(\mathrm{Re}\tau)(\mathrm{Im}\tau)^{-1}) - (\mathrm{Im}\tau)^{-1}L_{22}(\mathrm{Im}\tau)^{-1}].\end{aligned}$$

LEMMA 7.2. *Θ_τ is defined for all $\tau \in H_\gamma$ and is onto. In particular, the dimension of $\mathrm{Ker}\Theta_\tau$ is γ^2.*

PROOF. If $\widetilde{L}$ satisfies $L_{11} = L_{12} = 0$, then $\Theta_\tau(\widetilde{L}) = -\frac{1}{4}(\mathrm{Im}\tau)^{-1}L_{22}(\mathrm{Im}\tau)^{-1}$. Thus the image of Θ_τ contains all real symmetric matrices. For $\widetilde{L}$ satisfying $L_{11} = L_{22} = 0$, $\Theta_\tau(\widetilde{L}) = \frac{1}{4}((\mathrm{Im}\tau)^{-1}(\mathrm{Re}\tau)L_{12}(\mathrm{Im}\tau)^{-1}+(\mathrm{Im}\tau)^{-1}L_{21}(\mathrm{Re}\tau)(\mathrm{Im}\tau)^{-1} - i(L_{12}(\mathrm{Im}\tau)^{-1} + (\mathrm{Im}\tau)^{-1}L_{21}))$. Since the image of Θ_τ contains all real symmetric matrices of size γ, the image of Θ_τ includes $i(L_{12}(\mathrm{Im}\tau)^{-1} + (\mathrm{Im}\tau)^{-1}L_{21})$, which covers all symmetric matrices of size γ with purely imaginary components. Thus Θ_τ is onto. □

Let τ be a Riemann matrix corresponding to $[(M, f)]$ and W_τ the complex subspace of S^2_C spanned by $A^\ell, 1 \leq \ell \leq \frac{1}{2}(\gamma-2)(\gamma-3)$ if (M, f) is non-hyperelliptic, or spanned by $B^\ell, 1 \leq \ell \leq \frac{1}{2}(\gamma-1)(\gamma-2)$ if (M, f) is hyperelliptic. By Rauch's result, $\mathrm{tr}(A^\ell dk(\mu)) = 0$ and $\mathrm{tr}(B^\ell dk(\mu)) = 0$, so that we can find the following geometric meaning of W_τ:

LEMMA 7.3. *$\overline{W_\tau}$ is the normal space $(\kappa(T(R)))^\perp$ of $\kappa(T(R))$ in H_γ at $[(M, f)]$ if M is not hyperelliptic and the normal space $(\kappa(HT(R)))^\perp$ of $\kappa(HT(R))$ in H_γ at $[(M, f)]$ if M is hyperelliptic.*

From (7.2), (7.3) and Lemma 7.3 we obtain:

LEMMA 7.4. *dS is weakly conformal if and only if $\overline{{}^tAI_{p,q}A}$ is a normal vector at τ of $RM_{non-hyper}$, RM_{hyper} in H_γ.*

We define an energy function $E_{\widetilde{L}}$ on H_γ by $\frac{1}{2}\mathrm{tr}(P(\tau)\widetilde{L})$. Let $G(\widetilde{L})$ be the subgroup of $Sp(2\gamma, \mathbf{Z})$ defined by $\{h \in Sp(2\gamma, \mathbf{Z}) : {}^th\widetilde{L}h = \widetilde{L}\}$ and $\widehat{G}(\widetilde{L})$ the subgroup defined by

$$\left\{\widehat{h} = \begin{pmatrix} -a & b \\ c & -d \end{pmatrix} : h = \begin{pmatrix} a & b \\ c & d \end{pmatrix} \in G(\widetilde{L})\right\}.$$

Then it is easy to obtain the following by (5.1):

LEMMA 7.5. *$E_{\widetilde{L}}$ is $\widehat{G}(\widetilde{L})$-equivariant.*

PROOF. $2E_{\widetilde{L}}(\tau) = \mathrm{tr}(P(\tau)\widetilde{L}) = \mathrm{tr}(P(\tau){}^th\widetilde{L}h) = \mathrm{tr}(hP(\tau){}^th\widetilde{L}) = \mathrm{tr}(hQ(-\overline{\tau}){}^th\widetilde{L}) = \mathrm{tr}(Q(h(-\overline{\tau}))\widetilde{L}) = \mathrm{tr}(Q(-\overline{\widehat{h}(\tau)})\widetilde{L}) = \mathrm{tr}(P(\widehat{h}(\tau))\widetilde{L}) = 2E_{\widetilde{L}}(\widehat{h}(\tau))$. □

We shall calculate the gradient of $E_{\widetilde{L}}(\tau)$. Note that S^2_C is the tangent space of H_γ. It is easy to see the following:

LEMMA 7.6.

$$E_{\widetilde{L}}(\tau) = \frac{1}{2}\mathrm{tr}\left\{\begin{pmatrix} I & \mathrm{Re}\tau \\ 0 & -I \end{pmatrix}\begin{pmatrix} \mathrm{Im}\tau & 0 \\ 0 & (\mathrm{Im}\tau)^{-1} \end{pmatrix}^t \begin{pmatrix} I & \mathrm{Re}\tau \\ 0 & -I \end{pmatrix}\widetilde{L}\right\}.$$

Let θ be a real symmetric matrix of size γ. Then, by a simple calculation, the differentials of $E_{\widetilde{L}}(\tau)$ in the directions of $i\theta$ and θ are given by

$$\frac{1}{2}\mathrm{tr}\left\{\begin{pmatrix} \theta & 0 \\ 0 & -(\mathrm{Im}\tau)^{-1}\theta(\mathrm{Im}\tau)^{-1} \end{pmatrix}^t \begin{pmatrix} I & \mathrm{Re}\tau \\ 0 & -I \end{pmatrix}\widetilde{L}\begin{pmatrix} I & \mathrm{Re}\tau \\ 0 & -I \end{pmatrix}\right\}$$

and

$$\operatorname{tr}\left\{\begin{pmatrix}0 & \theta\\ 0 & 0\end{pmatrix}\begin{pmatrix}\operatorname{Im}\tau & 0\\ 0 & (\operatorname{Im}\tau)^{-1}\end{pmatrix}^t\begin{pmatrix}I & \operatorname{Re}\tau\\ 0 & -I\end{pmatrix}\widetilde{L}\right\},$$

respectively. Using these, we obtain:

LEMMA 7.7. $\operatorname{grad}E_{\widetilde{L}}(\tau) = 2i\overline{\Theta_\tau(\widetilde{L})}$.

We have the following geometric meaning of critical points of $E_{\widetilde{L}}$ restricted to $RM_{non-hyper}$ and RM_{hyper}.

PROPOSITION 7.8. *Let τ be an element of $RM_{non-hyper}$. Then the restriction of $E_{\widetilde{L}}$ to $RM_{non-hyper}$ is critical at τ if and only if there exist a unique Riemann surface M and two canonical homology bases $\{A_i, B_i\}$, $\{-A_i, -B_i\}$ up to automorphisms of M such that $dS = \pm LT_\tau^{-1}\Psi$, where $L \in F_{p,q}$, is weakly conformal. Similarly, let τ be an element of RM_{hyper}. Then the restriction of $E_{\widetilde{L}}$ to RM_{hyper} is critical at τ if and only if there exist a unique Riemann surface M and a unique canonical homology basis $\{A_i, B_i\}$ up to automorphisms of M such that $dS = LT_\tau^{-1}\Psi$, where $L \in F_{p,q}$, is weakly conformal.*

PROOF. For the non-hyperelliptic case: $E_{\widetilde{L}}|_{RM_{non-hyper}}$ has a critical point τ if and only if $\overline{\Theta_\tau(\widetilde{L})}$ is a normal vector of $RM_{non-hyper}$ in H_γ at τ. The existence of the unique Riemann surface and the two canonical homology bases follow from the property of ${}^{\#}\kappa$. The result follows from Lemma 7.4. □

We call such a dS a weakly conformal harmonic form. Let ρ be an automorphism of M and $\{A_i', B_i'\}$ equal to $\{\rho_*(A_i), \rho_*(B_i)\}$. Then the weakly conformal harmonic form corresponding to $\{A_i', B_i'\}$ is $(\pm 1)LT_\tau^{-1}(\rho^{-1})^*\Psi$. Furthermore, it is a critical point of $E_{{}^th^{-1}\widetilde{L}h^{-1}} = E_{\widetilde{L}} \circ \widehat{h}^{-1}$, where $h \in Sp(2\gamma, \mathbf{Z})$. We have the following geometric meaning of $G(\widetilde{L})$. Let $\widehat{h}$ be an element of $\widehat{G}(\widetilde{L})$ and τ a critical point of $E_{\widetilde{L}}$. Then $\widehat{h}(\tau)$ is also a critical point. By Proposition 7.8, we have a Riemann surface M and canonical homology bases $\{A_i, B_i\}, \{\widehat{A}_i, \widehat{B}_i\}$ corresponding to $\tau, \widehat{h}(\tau)$, respectively. It follows from the Remark in Section 5 that we may consider $\{\widehat{A}_i, \widehat{B}_i\} = \{A_i, B_i\}(-h^{-1})$. Assume that there exists an isometry $O \in O(n)$ such that $OL = Lh$. Naturally, if the rank of $\widetilde{L} = 2\gamma$, then O exists. It seems that we have two weakly conformal harmonic forms with the same periods L corresponding to $\widetilde{L}$. However, note that they may correspond to different canonical homology bases $\tau, \widehat{h}(\tau)$. Let dS be the weakly conformal harmonic form corresponding to the critical point τ. Then, since $(-O)^{-1}L = L(-h^{-1})$, $(-O)dS$ is the weakly conformal harmonic form corresponding to $\widehat{h}(\tau)$ with periods L. In particular, if dS gives a minimal surface, then $(-O)dS$ yields a congruent minimal surface. Let H be a subgroup of $G(\widetilde{L})$ such that $\widehat{H}$ fixes τ and there exists $O(h)$ satisfying $O(h)L = Lh$ for each $h \in H$. Then we obtain an $O(H)$-equivariant weakly conformal harmonic form.

Remark: For $\tau \in RM_{non-hyper}$, the two weakly conformal harmonic forms correspond under inversion, so we may identify them.

Let σ be the second fundamental form of $\kappa(T(R))$ at $[(M,f)]$ in H_γ if (M,f) is non-hyperelliptic, and the second fundamental form of $\kappa(HT(R))$ at $[(M,f)]$ in H_γ if (M,f) is hyperelliptic. Then we obtain the following formulae at $[(M,f)]$ by Theorem 7.1 and Lemma 7.7:

THEOREM 7.9. *If (M,f) is non-hyperelliptic, then*

$$D^2E_\phi(\mu,\nu) = D^2E_{\widetilde{L}}(\kappa_*\mu,\kappa_*\nu) + 2\mathrm{Im}(\mathrm{tr}(\sigma(\mu,\nu)\Theta_\tau(\widetilde{L}))),$$

where μ,ν are tangent vectors at $[(M,f)]$.

THEOREM 7.10. *Let S be a branched minimal immersion of a hyperelliptic Riemann surface (M,f) into $T^n_{p,q}$ which corresponds to a critical point of E_ϕ. We can decompose the tangent space at the critical point in $T(R)$ into the tangent space $HT(R)^\top$ of $HT(R)$ and the kernel $\mathrm{Ker}\kappa_*$ of κ_*. Then the hessian of E_ϕ for the decomposition satisfies the following: $D^2E_\phi(\mu,\nu) = D^2E_{\widetilde{L}}(\kappa_*\mu,\kappa_*\nu) + 2\mathrm{Im}(\mathrm{tr}(\sigma(\mu,\nu)\Theta_\tau(\widetilde{L})))$, where $\mu,\nu \in HT(R)^\top$, $D^2E_\phi(HT(R)^\top,\mathrm{Ker}\kappa_*) = 0$ and there exists a basis of $\mathrm{Ker}\kappa_*$ such that the hessian on $\mathrm{Ker}\kappa_*$ is a diagonal matrix with eigenvalues $\lambda_1,-\lambda_1,\ldots,\lambda_{\gamma-2},-\lambda_{\gamma-2}$.*

PROOF. Let X be a cross section of $HT(R)^\top$ and Y a cross section of $\mathrm{Ker}\kappa_*$. Then $D^2E_\phi(X,Y) = X(YE_\phi) = X(d\kappa_*(Y)E_{\widetilde{L}}) = 0$ holds. Let μ,ν be in $\mathrm{Ker}\kappa_*$ and X,Y the tangent vector fields on $T(R)|_{HT(R)}$ such that $X([(M,f)]) = \mu$, $Y([(M,f)]) = \nu$. Then we can define a symmetric bilinear map B of $\mathrm{Ker}\kappa_*$ into the normal space of RM_{hyper} at $\kappa([(M,f)])$ by $B(\mu,\nu) = (X(Y\kappa))^\perp$. Immediately, we see that $B(\mathbf{J}\mu,\mathbf{J}\nu) = -B(\mu,\nu)$ holds, where $\mathbf{J}$ is the almost complex structure of $T(R)$ and H_γ. So $D^2E_\phi(\mu,\nu) = B(\mu,\nu)E_{\widetilde{L}}$ and this completes Theorem 7.10. □

If E_ϕ restricted to $HT(R)$ has positive semi-definite hessian at a critical point, then the corresponding compact orientable branched minimal surface is said to be **hyperelliptically stable**. If E_ϕ restricted to $HT(R)$ has a minimum point, then the corresponding compact orientable branched minimal surface is said to be **hyperelliptically area-minimizing**.

COROLLARY 7.11. *Let S be a minimal immersion of a hyperelliptic Riemann surface into a Riemannian flat torus with only trivial Jacobi fields. Then $5\gamma-4 \geq index_A \geq \gamma-2$. In particular, if $index_A = \gamma-2$, then the minimal surface is hyperelliptically stable.*

Ross's result and Corollary 7.11 use the fact that these minimal surfaces are hyperelliptically stable.

PROBLEM 1. *Are the Schwarz P-surface, the D-surface, and Schoen's gyroid hyperelliptically area-minimizing?*

We can prove the following (another proof is given in Section 13):

PROPOSITION 7.12. *The Schwarz P-surface is not hyperelliptically area-minimizing.*

PROOF. We consider two parallel squares of side $H/\sqrt{2}$ separated by distance $H/2$. Using the square catenoid whose boundary is the boundaries of the above two squares, we construct the Schwarz P-surface in the torus whose lattice vectors are orthogonal vectors with length H. In place of the square catenoid, we consider two squares minus a disc with radius ϵ whose centre is the centre of the squares and attach the boundary of a cylinder of radius ϵ with length $H/2$ to the boundaries of the above discs. After smoothing it, we obtain a compact orientable surface in the torus. Note that the surface is hyperelliptic, because the hyperelliptic involution is given by the inversion in the torus at the centre of one side of the square. If ϵ goes to zero, then the area converges to H^2. Ross (page 192 of [Ro]) has shown that the Schwarz P-surface has area greater than H^2, which completes the proof of Proposition 7.12. □

§8 On the differential-geometric Schottky problem (1).

THEOREM 8.1. *Let τ be an element of $RM_{non-hyper}$. Then there exist a unique Riemann surface M and two canonical homology bases $\{A_i, B_i\}$, $\{-A_i, -B_i\}$ up to automorphisms of M such that $dS = \pm LT_\tau^{-1}\Psi$, where $L \in F_{p,q}$, is weakly conformal, if and only if the height function corresponding to $\widetilde{L}$ on $P(RM_{non-hyper})$ is critical at $P(\tau)$. Similarly, let τ be an element of RM_{hyper}. Then there exist a unique Riemann surface M and a unique canonical homology basis $\{A_i, B_i\}$ up to automorphisms of M such that $dS = LT_\tau^{-1}\Psi$, where $L \in F_{p,q}$, is weakly conformal if and only if the height function corresponding to $\widetilde{L}$ on $P(RM_{hyper})$ is critical at $P(\tau)$.*

Let **Cat** be the set of $(P(\tau), \widetilde{L})$ such that $dS = LT_\tau^{-1}\Psi$ is weakly conformal and $\|\widetilde{L}\| = 1$.

COROLLARY 8.2. ***Cat** is the sum of the unit normal bundles of $P(RM_{non-hyper})$ and $P(RM_{hyper})$ in $\mathbf{R}^{\gamma(2\gamma+1)}$.*

We obtain the Gauss map of **Cat** into the unit sphere of S^2_R. Now we define $index_{\widetilde{L}}$ and $nullity_{\widetilde{L}}$ for $dS = LT_\tau^{-1}\Psi$, which is weakly conformal, by the index and the nullity of the height function corresponding to $\widetilde{L}$ at the critical point $P(\tau)$. If the column vectors of L are lattice vectors and the corresponding minimal surface is hyperelliptic, then $index_{\widetilde{L}} = 0$ means that the hyperelliptic minimal surface is hyperelliptically stable. For example, the Schwartz P-surface, the D-surface and Schoen's gyroid satisfy $index_{\widetilde{L}} = 0$ and $nullity_{\widetilde{L}} = 0$. On the other hand, $nullity_{\widetilde{L}} = 0$ is equivalent to the fact that the Gauss map is nondegenerate at $(P(\tau), \widetilde{L})$, that is, it is a regular point. Thus critical points are smoothly dependent on a small smooth perturbation of the height function.

COROLLARY 8.3. *Let S be a branched minimal immersion of a Riemann surface into an indefinite flat torus $T^n_{p,q}$ such that $dS = LT_\tau^{-1}\Psi$. If S has $nullity_{\widetilde{L}} = 0$, then there exists a $(\frac{1}{2}n(n+1) - 1)$-dimensional perturbation of $T^n_{p,q}$ and the corresponding perturbation of a compact orientable branched minimal surface. In particular, if the given minimal surface is hyperelliptic, then so are the perturbed minimal surfaces.*

PROOF. There exists g such that $\widetilde{L} \in S_g$. □

Even for a collapse or a swell of a torus, we can treat the deformation of minimal surfaces. For example, we obtain the following.

COROLLARY 8.4. *Let T_c be a one-dimensional torus with length c and $\widetilde{\phi}$ a primitive map of a Riemann surface R into $T^n \times T_c$ such that the projection ϕ of R into T^n is also primitive. Then, if there exists a compact orientable branched minimal surface with $nullity_{\widetilde{L}} = 0$ homotopic to ϕ in T^n, then for c close to 0, there exists a compact orientable branched minimal surface homotopic to $\widetilde{\phi}$ in $T^n \times T_c$.*

Note that the metric of T_c may be negative definite (see Corollary 3.21). Generally, minimal surfaces obtained in Corollary 8.4 may have other conformal structures. We shall give examples (see Corollary 11.3) of compact orientable minimal surfaces with the same conformal structure. Albanese maps of hyperelliptic Riemann surfaces satisfy $nullity_{\widetilde{L}} = 0$ and $nullity_A = 2(\gamma - 1)$ (see Section 10). Compact orientable minimal surfaces given in Corollary 8.3 are hyperelliptic and further, $nullity_A = 2(\gamma - 1)$ brings about a bifurcation (see Section 10). See [J-J-P], [W] for deformations of minimal surfaces under deformations of Riemannian metrics in ambient Riemannian manifolds. Meeks [Me] constructed two 5-dimensional families V of embedded hyperelliptic minimal surfaces of genus 3 in 3-dimensional Riemannian flat tori which contain the Schwarz P-surface and the D-surface ($nullity_{\widetilde{L}} = 0$), so that we find that the neighborhood of V at these surfaces are embedded hyperelliptic minimal surfaces constructed by deformation of the ambient tori without bifurcations. Furthermore, the Schwartz P-surface has countably many associate minimal surfaces [Me], [N-S2] and hence $nullity_{\widetilde{L}} = 0$ holds for these associate minimal surfaces. Thus each associate minimal surface gives rise to a 5-dimensional family of hyperelliptic minimal surfaces. Note that V contains a CLP-surface with Gauss map whose branch points lie in a great circle of $S^2(1)$. By [M-R], [N], this CLP-surface has $index_A = 3$ and only trivial Jacobi fields. Meeks [Me] found the interesting fact that the conjugate surface of the CLP-surface is a CLP-surface up to an automorphism and is contained in another family of V. It seems that, by a deformation of the ambient torus of the CLP-surface, we may obtain two 5-dimensional families of hyperelliptic minimal surfaces and a bifurcation in spite of $nullity_{\widetilde{L}} = 0$. However note that the two families are different (see the Remark following Proposition 7.8) subsets in **Cat**. By Sard's Theorem, we obtain:

COROLLARY 8.5. *There exists a measure zero set in the unit sphere of S^2_R such that the compact orientable branched minimal surfaces corresponding to the critical points of the height function for $\widetilde{L}$ except for a set of measure zero admit $nullity_{\widetilde{L}} = 0$.*

It is interesting to study regular values of the Gauss map of **Cat** into the unit sphere in S^2_R.

PROBLEM 2. *Find a regular value $\widetilde{L}$ such that* $\mathrm{rank}(\widetilde{L}) < 2\gamma$ *and such that L corresponding to $\widetilde{L}$ gives a lattice.*

Note that the definitions of $index_{\widetilde{L}}$ and $nullity_{\widetilde{L}}$ are different from $index_A$ and $nullity_A$ of compact orientable branched minimal surfaces. However, we have found the following for a compact orientable branched minimal surface in a Riemannian flat torus: If the surface is non-hyperelliptic, then $index_A = index_{\widetilde{L}}$, and if the map is immersion, then $nullity_A = nullity_{\widetilde{L}} + \mathrm{rank}(\widetilde{L})$. On the other hand, if the surface is hyperelliptic, then $index_A \geq index_{\widetilde{L}}$, and if the map is an immersion, then $nullity_A \geq nullity_{\widetilde{L}} + \mathrm{rank}(\widetilde{L})$. We have studied the critical points of energy functions with control parameters (the unit sphere of S_R^2) on the Teichmüller space (Theorem 3.2). We may meet a catastrophe of compact orientable branched minimal surfaces in tori as Besson and Tromba found a cusp catastrophe for a piece of the Enneper surface [B-T1]. So the study of the singularities of the Gauss map of **Cat** into the unit sphere of S_R^2 is very important, and is part of the differential-geometric Schottky problem. Since the perturbation space of 3-dimensional tori has dimension 5 and Thom classified seven elementary catastrophes for 5-control parameters (see, for example, [P-S]), the following interesting problem arises:

PROBLEM 3. *Find an elementary catastrophe for compact orientable minimal surfaces with $nullity_{\widetilde{L}} \neq 0$ in 3-dimensional Riemannian flat tori. Which of the seven elementary catastrophes occurs?*

Examples of $nullity_{\widetilde{L}} \neq 0$. Recall V: two real 5-dimensional families of hyperelliptic minimal surfaces of genus 3 in 3-tori and V contains a CLP-surface which has $index_A = 3$ and only trivial Jacobi fields, while there exists a compact orientable minimal surface in V with $index_A = 1$ and only trivial Jacobi fields (Schwartz P-surface, D-surface). The continuity of the spectrum of Jacobi operators implies that there exist compact orientable minimal surfaces in V with $index_A = 1$ and a non-trivial Jacobi field ($nullity_{\widetilde{L}} \neq 0$). What happens at these hyperelliptic minimal surfaces? We have two possibilities. One is that the hyperelliptic minimal surface is not isolated in the ambient torus. Another is a bifurcation at the hyperelliptic minimal surface. So we may find an elementary catastrophe. The classification of embedded hyperelliptic minimal surfaces of genus 3 in 3-dimensional Riemannian flat tori is very important. For a fixed hyperelliptic minimal surface with $nullity_{\widetilde{L}} = 0$ in a 3-dimensional Riemannian flat torus, we obtain the deformations of hyperelliptic minimal surfaces under small deformations of the Riemannian flat torus. However we may meet various phenomena: a bifurcation, a topologically broken minimal surface, etc. under large deformations of the Riemannian flat torus. In spite of these phenomena, we believe that the others may be obtained by deformations of Riemannian flat tori.

PROBLEM 4. *Is the space of embedded hyperelliptic minimal surfaces of genus* 3 *in* 3*-dimensional Riemannian flat tori a* 5*-dimensional path-connected submanifold (with singularities) in* **Cat**? *Conversely, is the catastrophe manifold around an example with $nullity_{\widetilde{L}} \neq 0$ in V only an open set of V?*

Let PS_R^2 be the set of positive semi-definite symmetric matrices of size 2γ. Then we have the space of branched minimal immersions of (M, f) into Riemannian flat 2γ-tori as a convex set.

THEOREM 8.6. *Assume that τ is a Riemann matrix of (M, f). Then, at $P(\tau)$, $PS^2_R \cap (P(RM_{non-hyper}))^\perp$ is a convex set of dimension $\gamma(\gamma+1)+\gamma^2 - 6(\gamma-1)$ if (M, f) is non-hyperelliptic, and $PS^2_R \cap (P(RM_{hyper}))^\perp$ at $P(\tau)$ is a convex set of dimension $\gamma(\gamma+1)+\gamma^2-(4\gamma-2)$ if (M, f) is hyperelliptic. The interior point $\widetilde{L}$ corresponds to the minimal embedding of (M, f) into a 2γ-dimensional torus whose lattice vectors are given by column vectors of (L_1, L_2) (Lemma 6.1) corresponding to $\widetilde{L}$. If column vectors corresponding to the boundary point $\widetilde{L}$ span lattice vectors in $\mathbf{R}^n$, then the boundary point corresponds to a branched minimal immersion of (M, f) into an n-dimensional Riemannian flat torus.*

PROOF. When the rank of $\widetilde{L}$ is 2γ, dS is injective. Furthermore, the fact that the Albanese map is an embedding asserts that the compact orientable minimal surfaces corresponding to interior points are embedded under the assumption of primitive maps. □

A compact orientable branched minimal surface corresponding to a point of the boundary is obtained by collapsing the Jacobi variety for the Albanese map. So it is very interesting to study the boundary of $PS^2_R \bigcap (P(RM_{non-hyper}))^\perp$ and $PS^2_R \bigcap (P(RM_{hyper}))^\perp$ for each fixed conformal structure on R. Which element of the boundary gives a torus T^n, $n < 2\gamma$? What is the minimum such n? This is Question 8.1 in [Me]. It is interesting to investigate the subset of elements giving rise to tori in the boundary. Meeks [Me] proved that there exists an element of the boundary which gives a $(2\gamma-1)$-dimensional torus if $\gamma \geq 4$.

§9 Holomorphic maps into complex tori with indefinite flat metrics.

We extend $< \ , \ >_{p,q}$ to a Hermitian inner product on $\mathbf{C}^n$ such that $< X, Y >_{p,q} = {}^tX I_{p,q} \overline{Y}$. Real linearly independent $2n$-vectors in $\mathbf{C}^n$ give a complex torus $T^n_{\mathbf{C}}$ with indefinite Hermitian inner product $< \ , \ >_{p,q}$. Then $T^n_{\mathbf{C}}$ may be a $2n$-dimensional indefinite flat torus T^{2n} with inner product $< \ , \ >_{2p,2q}$ and the orthogonal complex structure J such that $J^2 = -I_{2p+2q}$, ${}^tJ I_{2p,2q} J = I_{2p,2q}$. It is well-known that holomorphic curves in $T^n_{\mathbf{C}}$ are compact orientable branched minimal surfaces in T^{2n}. We shall prove that holomorphic curves in complex tori with indefinite Hermitian inner product correspond to critical points of the height function on $P(H_\gamma)$. By Lemma 7.7, we obtain the following:

LEMMA 9.1. *$P(\tau)$ is a critical point of a height function on $P(H_\gamma)$ for $\widetilde{L}$ if and only if $\widetilde{L} \in \mathrm{Ker}\Theta_\tau$.*

Next we shall study $\mathrm{Ker}\Theta_\tau$. For instance, the $\widetilde{L}$ for Albanese maps are in $\mathrm{Ker}\Theta_\tau$. Now $\widetilde{L}$ is in $\mathrm{Ker}\Theta_\tau$ if and only if

$$\begin{aligned}(9.1)\qquad & L_{11} - (\mathrm{Im}\tau)^{-1}(\mathrm{Re}\tau)L_{11}(\mathrm{Re}\tau)(\mathrm{Im}\tau)^{-1} + (\mathrm{Im}\tau)^{-1}(\mathrm{Re}\tau)L_{12}(\mathrm{Im}\tau)^{-1} \\ & +(\mathrm{Im}\tau)^{-1}L_{21}(\mathrm{Re}\tau)(\mathrm{Im}\tau)^{-1} - (\mathrm{Im}\tau)^{-1}L_{22}(\mathrm{Im}\tau)^{-1} = 0,\end{aligned}$$

$$(9.2)\qquad (\mathrm{Im}\tau)^{-1}(\mathrm{Re}\tau)L_{11} + L_{11}(\mathrm{Re}\tau)(\mathrm{Im}\tau)^{-1} - L_{12}(\mathrm{Im}\tau)^{-1} - (\mathrm{Im}\tau)^{-1}L_{21} = 0,$$

which are equivalent to $(\mathrm{Im}\tau)^{-1}((\mathrm{Re}\tau)L_{11} - L_{21})$ being skew and

$$L_{11} - (\mathrm{Im}\tau)^{-1}[(\mathrm{Re}\tau)L_{11}(\mathrm{Re}\tau) - L_{21}(\mathrm{Re}\tau) - (\mathrm{Re}\tau)L_{12} + L_{22}](\mathrm{Im}\tau)^{-1} = 0.$$

We set $\Xi = (\text{Im}\tau)^{-1}((\text{Re}\tau)L_{11} - L_{21})$. Then $\widetilde{L}$ is in $\text{Ker}\Theta_\tau$ if and only if there exists a skew symmetric matrix Ξ such that $L_{12} = L_{11}(\text{Re}\tau) + \Xi(\text{Im}\tau)$, $L_{22} = (\text{Im}\tau)L_{11}(\text{Im}\tau) + (\text{Re}\tau)L_{11}(\text{Re}\tau) - (\text{Im}\tau)\Xi(\text{Re}\tau) + (\text{Re}\tau)\Xi(\text{Im}\tau)$. Furthermore, we put $\Phi = L_{11} - i\Xi$ (Hermitian matrix). Then we obtain $L_{11} = \text{Re}\Phi$, $L_{22} = \text{Re}(\tau\overline{\Phi}\tau)$, $L_{12} = \text{Re}(\Phi\tau)$.

LEMMA 9.2. *$\widetilde{L} \in Ker\Theta_\tau$ if and only if there exists an Hermitian matrix Φ of size γ such that $L_{11} = \text{Re}\Phi$, $L_{22} = \text{Re}(\tau\overline{\Phi}\tau)$, $L_{12} = \text{Re}(\Phi\tau)$.*

It is easy to see that, for an Hermitian matrix Φ of size γ which has p positive eigenvalues and q negative eigenvalues and $\gamma - p - q$ null eigenvalues, there exists a $(p+q, \gamma)$-matrix a such that $\Phi = a^* I_{p,q} a$. Note that the matrix a is unique up to isometries of $< \ , \ >_{p,q}$. Thus we obtain

$$L_{11} = {}^t(\text{Re}a)I_{p,q}(\text{Re}a) + {}^t(\text{Im}a)I_{p,q}(\text{Im}a),$$

$$\begin{aligned} L_{12} = &({}^t(\text{Re}a)I_{p,q}(\text{Re}a) + {}^t(\text{Im}a)I_{p,q}(\text{Im}a))(\text{Re}\tau) \\ &-({}^t(\text{Re}a)I_{p,q}(\text{Im}a) - {}^t(\text{Im}a)I_{p,q}(\text{Re}a))(\text{Im}\tau), \end{aligned}$$

$$(*) \qquad \begin{aligned} L_{22} = (\text{Re}\tau)\{&({}^t(\text{Re}a)I_{p,q}(\text{Re}a) + {}^t(\text{Im}a)I_{p,q}(\text{Im}a))(\text{Re}\tau) \\ &-({}^t(\text{Re}a)I_{p,q}(\text{Im}a) - {}^t(\text{Im}a)I_{p,q}(\text{Re}a))(\text{Im}\tau)\} \\ +(\text{Im}\tau)\{&({}^t(\text{Re}a)I_{p,q}(\text{Re}a) + {}^t(\text{Im}a)I_{p,q}(\text{Im}a))(\text{Im}\tau) \\ &+({}^t(\text{Re}a)I_{p,q}(\text{Im}a) - {}^t(\text{Im}a)I_{p,q}(\text{Re}a))(\text{Re}\tau)\}. \end{aligned}$$

Using the matrix $a = (a_{ij})$, we can define a multivalued holomorphic map Ω of M into $\mathbf{C}^{p+q}$ with Hermitian inner product of type (p,q) as follows: Let $\omega_i = \sum_k a_{ik}\psi_k$. Then ${}^t(\int \omega_1 \cdots \int \omega_{p+q})$ gives a multivalued holomorphic map Ω whose real period (L_1, L_2) is

$$\begin{pmatrix} (\text{Re}a) & (\text{Re}a)(\text{Re}\tau) - (\text{Im}a)(\text{Im}\tau) \\ (\text{Im}a) & (\text{Re}a)(\text{Im}\tau) + (\text{Im}a)(\text{Re}\tau) \end{pmatrix}.$$

Thus $\widetilde{L}$ corresponding to (L_1, L_2) coincides with the given $\widetilde{L} \in \text{Ker}\Theta_\tau$ in $(*)$.

Conversely, the period L for a multivalued holomorphic map of M into $\mathbf{C}^{p+q}$ with Hermitian inner product $< \ , \ >_{p,q}$ gives $\widetilde{L} = {}^tLI_{2p,2q}L \in \text{Ker}\Theta_\tau$, where

$$I_{2p,2q} = \begin{pmatrix} I_{p,q} & 0 \\ 0 & I_{p,q} \end{pmatrix}.$$

PROPOSITION 9.3 (CRITERION). *$P(\tau)$ is a critical point of a height function on $P(H_\gamma)$ for $\widetilde{L}$ if and only if there exists a multivalued holomorphic map Ω of M into $\mathbf{C}^{p+q}$ with Hermitian inner product of type (p,q) such that $\widetilde{L}$ corresponds to the period of Ω. In particular, $L \in F_{2p,2q}$ are lattice vectors if and only if Ω is a holomorphic map into a complex torus with Hermitian inner product of type (p,q).*

Let τ be an element of H_γ and $\widetilde{L}$ a normal vector of $P(H_\gamma)$ at $P(\tau)$. Then we have the Hermitian matrix Φ as in Lemma 9.2 and hence a, which give a multivalued holomorphic map of the principally polarized abelian variety corresponding to τ into $\mathbf{C}^{p+q}$ whose periods correspond to $\widetilde{L}$. Since RM for genus 2 is open and dense in

H_2, a critical point of a height function on $P(RM)$ is also a critical point of the height function on $P(H_2)$. Corollary 9.4 follows from the Criterion. Meeks [Me] proved Corollary 9.4 for Riemannian flat tori.

COROLLARY 9.4. *Let $n \geq 3$. Let S be a full branched minimal immersion of a Riemann surface of genus 2 in an indefinite flat torus $T^n_{p,q}$. Then there exists an orthogonal complex structure on $T^n_{p,q}$ such that S is holomorphic. In particular, $n = 4, p = q = 2$ or $p = 4, q = 0$ or $p = 0, q = 4$.*

COROLLARY 9.5. *There does not exist a compact orientable branched minimal surface of genus 2 in 3, 4-dimensional Lorentzian flat tori.*

Similarly, since $RM_{non-hyper}$ for genus 3 is open and dense in H_3, a critical point of a height function on $P(RM_{non-hyper})$ is also a critical point of the height function on $P(H_3)$. The Criterion yields Corollary 9.6. Micallef and Meeks [M-M] have proved Corollary 9.6 for Riemannian flat tori.

COROLLARY 9.6. *Let $n \geq 3$. Let S be a full branched minimal immersion of a non-hyperelliptic Riemann surface of genus 3 in an indefinite flat torus $T^n_{p,q}$. Then there exists an orthogonal complex structure on $T^n_{p,q}$ such that S is holomorphic. In particular, $n = 4$ or 6 holds, $p = q = 2$ or $p = 4, q = 0$ or $p = 0, q = 4$ if $n = 4$ and $p = 6, q = 0$ or $p = 4, q = 2$ or $p = 2, q = 4$ or $p = 0, q = 6$ if $n = 6$.*

COROLLARY 9.7. *There does not exist a full non-hyperelliptic, compact orientable branched minimal surface of genus 3 in 3, 4, 5, 6-dimensional Lorentzian flat tori.*

§10 The nullity of Albanese maps.

By a simple calculation, we obtain:

LEMMA 10.1. *Let $\tau_0 \in H_\gamma$ and Φ a positive semi-definite Hermitian matrix of size γ. We set $L_{11} = \mathrm{Re}\Phi$, $L_{12} = \mathrm{Re}(\Phi\tau_0)$, $L_{22} = \mathrm{Re}(\tau_0\overline{\Phi\tau_0})$. Then $E_{\tilde{L}}(\tau) = \frac{1}{2}\mathrm{tr}\{(\mathrm{Im}\tau)^{-1}(\tau - \tau_0)\overline{\Phi(\tau - \tau_0)}\} + \mathrm{tr}\{\mathrm{Im}\tau_0\mathrm{Re}\Phi\}$. In particular, $E_{\tilde{L}}(\tau) \geq \mathrm{tr}\{\mathrm{Im}\tau_0\mathrm{Re}\Phi\}$. Furthermore, τ_0 is a critical point of $E_{\tilde{L}}(\tau)$ and $D^2E_{\tilde{L}}(\tau_0)(s,s) = \mathrm{tr}\{(\mathrm{Im}\tau_0)^{-1}s\overline{\Phi s}\}$, where s is a symmetric matrix. When $\mathrm{rank}(\Phi) = \gamma$, τ_0 is the only critical point of $E_{\tilde{L}}$.*

It is well-known that a section of the normal bundle of a complex submanifold in a Kaehler manifold is a Jacobi field if and only if the section is holomorphic [Si]. Let τ be the Riemann matrix for a non-hyperelliptic Riemann surface (M, f) or $\gamma = 2$. Then, by Lemma 10.1, $D^2E_{\widetilde{T_\tau}}$ is positive definite, which, together with Lemma 9.1 and Theorem 7.9, implies that D^2E_ϕ is positive definite. Theorem 3.18 yields the following:

COROLLARY 10.2. *For the Albanese map of a non-hyperelliptic Riemann surface or $\gamma = 2$, holomorphic sections of the normal bundle are the normal components of parallel vector fields, which implies that the dimension of the space of holomorphic sections is γ.*

Let τ be the Riemann matrix for a hyperelliptic Riemann surface (M, f). Then it follows from Theorem 7.10 and Lemma 10.1 that D^2E_ϕ is positive-definite for directions of $HT(R)^\top$ and 0 (i.e.,$B(\mu,\nu)E_{\widetilde{T_\tau}} = 0$) holds for directions of $\mathrm{Ker}\kappa_*$. Theorem 3.18 establishes the following:

COROLLARY 10.3. *The dimension of the space of holomorphic sections is* $2(\gamma - 1)$ *for the Albanese map of a hyperelliptic Riemann surface* $(\gamma \geq 3)$.

For Albanese maps of hyperelliptic Riemann surfaces we have $nullity_{\widetilde{T_{\tau_0}}} = 0$. Varying τ_0 in H_γ for $E_{\widetilde{T_{\tau_0}}}(\tau) = \frac{1}{2}\mathrm{tr}\{(\mathrm{Im}\tau)^{-1}(\tau-\tau_0)\overline{(\tau-\tau_0)}\} + \mathrm{tr}\{\mathrm{Im}\tau_0\}$, we obtain hyperelliptic minimal surfaces by Corollary 8.3. When we move τ_0 in $H_\gamma \setminus RM_{hyper}$, the resulting hyperelliptic minimal surfaces are not holomorphic, because τ_0 is the only critical point of $E_{\widetilde{L}}$ on H_γ. Furthermore, when we move τ_0 in $RM_{non-hyper}$, we obtain another compact orientable minimal surface (the Albanese map). If $\gamma = 3$, then the resulting compact orientable minimal surfaces are just the above two minimal surfaces. (If another compact orientable minimal surface appears, then it is not hyperelliptic and hence holomorphic by Corollary 9.6, and gives the same Albanese map as one of the two minimal surfaces, which is a contradiction.)

THEOREM 10.4 (A BIFURCATION FOR THE ALBANESE MAP OF A HYPERELLIPTIC RIEMANN SURFACE). *There exists a* $(\gamma-1)(\gamma-2)$*-parameter deformation for the Albanese map of a hyperelliptic Riemann surface, where the corresponding hyperelliptic minimal surfaces are non-holomorphic. If* τ_0*, a minimum point of the height function on* $P(H_\gamma)$*, remains in* $RM_{non-hyper}$*, then we have the Albanese map of the non-hyperelliptic Riemann surface corresponding to* τ_0 *except for the above hyperelliptic example. Furthermore, if* $\gamma = 3$*, then the only minimal surfaces appearing in the deformation are these examples.*

Arezzo [Ar2] has obtained a similar result. Hyperelliptic minimal surfaces appearing under the deformation are hyperelliptically area-minimizing (see Section 12).

PROBLEM 5. *How many compact orientable minimal surfaces are there for* $\gamma \geq 4$*?*

We note that Problem 5 is a differential-geometric counterpart of the local Torelli problem for algebraic curves over **C** (see Section 14).

§11 Branched minimal surfaces in Lorentzian flat tori.

From the Criterion in Section 9, full compact orientable branched minimal surfaces in Lorentzian flat tori are not holomorphic, so it is interesting to study such tori.

THEOREM 11.1. *For a Riemann surface* M *of genus* $\gamma \geq 4$*, there exists a full minimal immersion of* M *into* $T^{2\gamma}_{2\gamma-1,1}$.

PROOF. By [Me] (see the last remark in Section 8), we have a full minimal immersion of M into a $(2\gamma-1)$-dimensional Riemannian flat torus. So this gives a critical point of the height function corresponding to $\widetilde{L}$ of type $(2\gamma-1,0,1)$. Let $\widetilde{L}'$ be the normal vector corresponding to the Albanese map of M. Then $\widetilde{L}'$ is positive definite and we can choose a positive number ϵ such that $\widetilde{L}-\epsilon\widetilde{L}'$ is a normal vector of type $(2\gamma-1,1,0)$, which gives the required minimal surface. □

Using Meeks' idea (Corollary 9.1 in [Me]), we shall construct compact orientable embedded minimal surfaces of genus 3 in 4-dimensional Minkowski flat tori which are isometric to the Schwarz P-surface. Let χ be a branched minimal immersion of M into a Riemannian flat torus T^3 represented by $\mathrm{Re}^t(\int\omega_1,\int\omega_2,\int\omega_3)$. Assume that χ satisfies Property **P** (see [Me]). We define an $\mathbf{R}^5$-valued 1-form on M by $d\chi_c = {}^t(\mathrm{Re}\omega_1, \mathrm{Re}\omega_2, \mathrm{Re}\omega_3, c\mathrm{Re}\omega_3, c\mathrm{Im}\omega_3)$, hence $d\chi_c^{(1,0)} = \frac{1}{2}{}^t(\omega_1,\omega_2,\omega_3,c\omega_3,-ic\omega_3)$. So $d\chi_c$ is weakly conformal with respect to an inner product of type $(3,2)$. By Property **P**, $\int\omega_3$ provides a holomorphic map into T^2 and hence we obtain a branched minimal immersion of M into $T^3\times T^2$ with inner product of type $(3,2)$, which is not full. Indeed, when ${}^t(Z_1,Z_2,Z_3,Z_4,Z_5)$ is the canonical coordinate system of $\mathbf{R}^5$, the lattice vectors are in the hyperplane defined by $cZ_3=Z_4$. Thus the image of χ_c lies in a 4-dimensional torus T^4 and we denote by $\widetilde{S_c}$ the map of M into T^4. We consider a map of $\mathbf{R}^4$ into $\mathbf{R}^5$ by $Z_1=Y_1, Z_2=Y_2, Z_3=Y_3, Z_4=cY_3, Z_5=Y_4$, where $(Y_1,\dots,Y_4)$ is a coordinate system of $\mathbf{R}^4$. Then the induced inner product is given by the diagonal matrix with eigenvalues $1,1,1-c^2,-1$. The induced metric on T^4 is type $(3,1)$ if $|\,c\,|<1$, degenerate if $c=\pm1$, type $(2,2)$ if $|\,c\,|>1$. The map $\widetilde{S_c}$ of M into T^4 satisfies $d\widetilde{S_c} = {}^t(\mathrm{Re}\omega_1,\mathrm{Re}\omega_2,\mathrm{Re}\omega_3,c\mathrm{Im}\omega_3)$.

Now we consider the case $|\,c\,|<1$. The new coordinate system $Y_1=X_1, Y_2=X_2, Y_3=\sqrt{1-c^2}X_3, Y_4=X_4$ of $\mathbf{R}^4$ gives a map S_c of M into a 4-dimensional Minkowski flat torus:

$$dS_c = \begin{pmatrix} \mathrm{Re}\omega_1 \\ \mathrm{Re}\omega_2 \\ \sqrt{1-c^2}\mathrm{Re}\omega_3 \\ c\mathrm{Im}\omega_3 \end{pmatrix}, \quad dS_c^{(1,0)} = \frac{1}{2}\begin{pmatrix} \omega_1 \\ \omega_2 \\ \sqrt{1-c^2}\omega_3 \\ -ic\omega_3 \end{pmatrix}.$$

Note that the induced metric of S_c is the same as the induced metric of χ.

THEOREM 11.2. *Let M be a compact orientable branched minimal surface in a Riemannian flat 3-torus with Property* **P**. *Then M admits a one parameter family of full isometric branched minimal immersions in 4-dimensional Minkowski flat tori, which swell from the Riemannian flat 3-torus.*

We have only used the property that $\int\omega_3$ covers an elliptic curve. It is well-known that several classical examples of periodic surfaces satisfy Property **P**. A typical example is the Schwarz P-surface. Recall that the Schwarz P-surface satisfies $index_{\widetilde{L}}=0$ and $nullity_{\widetilde{L}}=0$. Let $\widetilde{L_c}\in S_R^2$ correspond to S_c. Since $\widetilde{L_c}$ is continuous on c, S_c has $index_{\widetilde{L_c}}=0$ and $nullity_{\widetilde{L_c}}=0$ for small enough c. Since the Schwarz P-surface is embedded, S_c is embedded.

COROLLARY 11.3. *The Schwarz P-surface gives a one parameter family of full isometric embedded minimal surfaces in a 4-dimensional Minkowski flat torus with* *nullity*$_{\widetilde{L}}$ $= 0$. *There exists a 9-dimensional perturbation space of each minimal surface, which contains the one parameter family.*

§12 On the differential-geometric Schottky problem (2).

Let $p = (p_{ij}) \in P_{2\gamma}$. We define a flat Riemannian metric on $R^{2\gamma}/Z^{2\gamma}$ by $\sum p_{ij}dx^i dx^j$. Furthermore $F_{2\gamma} = P_{2\gamma}/SL(2\gamma, \mathbf{Z})$ parametrizes the space of 2γ-dimensional Riemannian flat tori with unit volume. Define, for $p \in P_{2\gamma}$, $m(p) = \min_{n_1 \in \mathbf{Z}^{2\gamma}, n_1 \neq 0} {}^t n_1 p n_1$. It is easy to see that $m(p)$ is the square of the length of the shortest closed geodesic of the above Riemannian flat torus. Define the constants $r_{2\gamma}, \delta_{2\gamma}, \eta_{2\gamma}$ by $r_{2\gamma} = \max_{p \in P_{2\gamma}} m(p)$, $\delta_{2\gamma} = \max_{p \in \natural_{2\gamma}} m(p)$ and $\eta_{2\gamma} = \sup_{p \in Q(RM)} m(p) = \sup_{p \in P(RM)} m(p)$. Then $r_{2\gamma}$ is Hermite's constant which gives the maximum density of spheres with centres at lattice points. We note the following: $m(p) = \min_{0 \neq \lambda:\text{lattice vector}} H(\lambda, \lambda)$ for $p = Q(z)$, $z \in H_\gamma$. Since $m(p) \mapsto 0$ if and only if $p \mapsto \partial(F_{2\gamma})$ (Mahler's compactness theorem), m is the distance from $\partial(F_{2\gamma})$. Since ${}^t npn = \mathrm{tr}(pn^t n)$, $m(p)$ is the minimum of heights in directions of $n^t n$ of $p \in P_{2\gamma}$ in $\mathbf{R}^{\gamma(2\gamma+1)}$. In [B-S], Buser and Sarnak proved the following beautiful result.

THEOREM([B-S]). $(\frac{\sigma_{2\gamma}}{2})^{-\frac{1}{\gamma}} \leq \delta_{2\gamma}\,(\sigma_{2\gamma} = \frac{\pi^\gamma}{\Gamma(\gamma+1)})$, $\eta_{2\gamma} \leq \frac{3}{\pi}\log(4\gamma+3)$.

They stated that if $\gamma \geq 38$, then the above inequalities are effective, that is, $\eta_{2\gamma} \leq \frac{3}{\pi}\log(4\gamma+3) < (\sigma_{2\gamma}/2)^{-1/\gamma} \leq \delta_{2\gamma}$ holds and the following: $\eta_2 = \delta_2 = r_2 = \frac{2}{\sqrt{3}}(A_2), \eta_4 = \delta_4 = r_4 = \sqrt{2}(D_4)$, $1.511.. = \frac{4}{\sqrt{7}} \leq \eta_6 = \delta_6 < r_6 = 1.665..(E_6)$, $\eta_8 < \delta_8 = r_8 = 2(E_8), \eta_{24} < \delta_{24} = r_{24} = 4(\Lambda_{24})$. We shall consider the geometric meaning of $m(P(\tau))$, where $\tau \in RM$. Since ${}^t n_1 P(\tau) n_1$ is $\mathrm{tr}(P(\tau) n_1{}^t n_1)$, $\frac{1}{2}{}^t n_1 P(\tau) n_1$ is the energy of a harmonic map of M into S^1 with length one whose periods are given by ${}^t n_1$. So $\frac{1}{2}m(P(\tau))$ is the minimum energy of a homotopically non-trivial harmonic map of M into S^1. Thus we know that the minimum energy of a homotopically non-trivial harmonic map of M into S^1 is not greater than $\frac{3}{2\pi}\log(4\gamma+3)$. In fact, they proved the result for $\eta_{2\gamma}$ by estimating the energy of some test map of M into S^1. Their result describes the behaviour of height functions in the direction of $n_1{}^t n_1$ and is a partial answer for the differential-geometric Schottky problem. Let $\widetilde{L}$ be a positive definite symmetric matrix of size 2γ. Then it is easy to see that the height function for $\widetilde{L}$ on $P(H_\gamma)$ is a positive function, increasing to ∞. So there exists a minimum point $P(\tau_0) \in P(H_\gamma)$. Since $\widetilde{L}$ is a normal vector at $P(\tau_0)$, we obtain a positive definite Hermitian matrix Φ such that $L_{11} = \mathrm{Re}\Phi$, $L_{12} = \mathrm{Re}(\Phi\tau_0)$, $L_{22} = \mathrm{Re}(\tau_0\Phi\overline{\tau_0})$. Hence the height function on H_γ is given as follows:

LEMMA 12.1. *For a positive definite symmetric matrix* $\widetilde{L}$ *of size* 2γ,

$$E_{\widetilde{L}}(\tau) = \frac{1}{2}\mathrm{tr}\{(\mathrm{Im}\tau)^{-1}(\tau - \tau_0)\overline{\Phi(\tau - \tau_0)}\} + \mathrm{tr}\{\mathrm{Im}\tau_0\mathrm{Re}\Phi\}.$$

By Lemma 9.2, the normal space of $P(H_\gamma)$ at $P(\tau)$ is identified with the set **Her** of all Hermitian matrices of size γ and hence the normal bundle of $P(H_\gamma)$ is $H_\gamma \times \mathbf{Her}$. The normal map is given by $(\tau, \Phi) \in H_\gamma \times \mathbf{Her} \longmapsto \widetilde{L} \in S^2_R$, where $L_{11} = \mathrm{Re}\Phi$, $L_{22} = \mathrm{Re}(\tau\overline{\Phi\tau})$, $L_{12} = \mathrm{Re}(\Phi\tau)$. Furthermore, for a positive definite symmetric matrix $\widetilde{L}$, there exists only one element $(\tau, \Phi) \in H_\gamma \times \mathbf{Her}$ whose image is $\widetilde{L}$.

COROLLARY 12.2. *The normal Gauss map restricted to positive definite Hermitian matrices is a diffeomorphism to the space of positive definite symmetric matrices.*

COROLLARY 12.3. *$G(\widetilde{L})$ and $\widehat{G}(\widetilde{L})$ are finite subgroups of $Sp(2\gamma, \mathbf{Z})$ for a positive definite symmetric matrix $\widetilde{L}$.*

PROOF. Let τ be an element of H_γ and $I(\tau)$ the isotropy subgroup at τ of $Sp(2\gamma, \mathbf{Z})$. It is well-known that $I(\tau)$ is a finite group (see, for example, [Mu]). On the other hand, by Lemma 10.1 and Lemma 7.5, $\widehat{G}(\widetilde{L})$ must fix τ_0. Thus $\widehat{G}(\widetilde{L})$ is a finite group and so is $G(\widetilde{L})$. □

We define $\ell_1(\widetilde{L})$ and $\ell_2(\widetilde{L})$ to be the minima of the height function on $P(H_\gamma)$ on the closure of $P(RM)$; $\ell_1(\widetilde{L}) = \mathrm{tr}\{\mathrm{Im}\tau_0 \mathrm{Re}\Phi\}$. It is clear that $\ell_1(\widetilde{L}) = \ell_2(\widetilde{L})$ for $\gamma \leq 3$, because RM is open and dense in H_γ for $\gamma \leq 3$. The difference $\ell_2(\widetilde{L}) - \ell_1(\widetilde{L})$ is very interesting. Namely, if the height of $P(\tau)$ in the direction of $\widetilde{L}$ is less than $\ell_2(\widetilde{L})$, then τ is not a Riemann matrix. For $\ell_2(\widetilde{L})$ there are two possibilities. If the height function on the closure of $P(RM)$ has a minimum at a point of $P(RM)$, then by Theorem 8.1, we obtain a homotopically area-minimizing minimal surface. Otherwise, the minimum point is not a point of $P(RM)$. What is happening here? Arezzo [Ar1] has answered this as follows:

THEOREM ([AR1]). *Let Σ be a closed topological surface of genus $r \geq 1$, M a Riemannian manifold of dimension at least 3, and let $u : \Sigma \longrightarrow M$ be a continuous map such that $u_* : H_1(\Sigma, \mathbf{Z}) \longrightarrow H_1(M, \mathbf{Z})$ is injective. Then there exists $\widetilde{\Sigma}$ a Riemann surface with nodes obtained by collapsing a set of homologically trivial curves γ_k, whose parts are $\Sigma_{r_i} \setminus \{p_1^i, \cdots, p_s^i\}$, with the following properties:*

1. For all i there exists some $f_i : \Sigma_{r_i} \longrightarrow M$ such that f_i is a branched conformal minimal immersion and $\mathrm{Area}(f_i) \leq \mathrm{Area}(g)$ for every C^∞-map $g : \Sigma_{r_i} \longrightarrow M$ with the same action on $H_1(\Sigma_{r_i}, \mathbf{Z})$ as the one induced by the restriction of u to Σ_{r_i}.

2. If $f : \bigcup_i \Sigma_{r_i} \longrightarrow M$ is the map given by the maps f_i on each Σ_{r_i}, we have $f_* = u_*$ on $H_1(\Sigma, \mathbf{Z})$ in the sense above. Furthermore, after having defined $\mathrm{Area}(f) = \sum_i \mathrm{Area}(f_i)$, we obtain $\mathrm{Area}(f) \leq \mathrm{Area}(s)$ for every C^∞-map $s : \Sigma \longrightarrow M$ with the same action on $H_1(\Sigma, \mathbf{Z})$ as u.

It follows from [Ar1] that we obtain area-minimizing minimal surfaces in the second case. In this paper we call these generalized area-minimizing minimal surfaces. We can reprove the result of [Ar1] when the ambient space is a Riemannian flat torus as follows:

PROOF. Let τ be an element of H_γ such that τ is not a Riemann matrix and is the minimum point of $E_{\widetilde{L}}$ on the closure of $P(RM)$. Then τ is an element of the $Sp(2\gamma, \mathbf{Z})$-orbit of the direct sum of some Riemann matrices [I-T]. To simplify the proof, we assume $\tau = \begin{pmatrix} \tau_1 & 0 \\ 0 & \tau_2 \end{pmatrix}$, where τ_1, τ_2 are Riemann matrices of Riemann surfaces M_1, M_2 of genus γ_1, γ_2, respectively. Then there exists a permutation matrix Q such that

$$\operatorname{tr}(P(\tau)\widetilde{L}) = \operatorname{tr}\Big\{\begin{pmatrix} P(\tau_1) & 0 \\ 0 & P(\tau_2) \end{pmatrix} {}^tQ\widetilde{L}Q\Big\}.$$

We set $\begin{pmatrix} Q_{11} & Q_{12} \\ Q_{21} & Q_{22} \end{pmatrix} = {}^tQ\widetilde{L}Q$, where Q_{11}, Q_{22} are positive definite symmetric matrices of size $2\gamma_1, 2\gamma_2$, respectively. Then $\operatorname{tr}(P(\tau)\widetilde{L}) = \operatorname{tr}(P(\tau_1)Q_{11}) + \operatorname{tr}(P(\tau_2)Q_{22})$. Since

$$\operatorname{tr}(P(\tau)\widetilde{L}) \leq \operatorname{tr}(P(\tau')\widetilde{L}),\ \tau' = \begin{pmatrix} \tau_1' & 0 \\ 0 & \tau_2' \end{pmatrix},$$

we find that $P(\tau_i), i = 1, 2$ is a minimum point of the height function in the direction of Q_{ii} for the closure of the space of Riemann matrices of size γ_i in $\mathbf{R}^{\gamma_i(2\gamma_i+1)}$. Thus, by Theorem 8.1, we obtain two homotopically area-minimizing minimal surfaces in $T^{2\gamma_i}, i = 1, 2$, which satisfy conditions 1, 2 in the above theorem. □

Now we see that $\ell_2(\widetilde{L})$ is closely related to the areas of homotopically area-minimizing minimal surfaces of genus $\gamma'(\gamma' \leq \gamma)$ in $T^{2\gamma'}$. Note that when $\ell_1(\widetilde{L}) = \ell_2(\widetilde{L})$ holds, if the minimum point with value $\ell_2(\widetilde{L})$ is a Riemann matrix, then the corresponding minimal surface is holomorphic by the Criterion (Proposition 9.3). Conversely, if the minimal surface corresponding to a critical point for the height function $\widetilde{L}$ is holomorphic, then $\ell_1(\widetilde{L}) = \ell_2(\widetilde{L})$ and we obtain a unique minimum point for the height function corresponding to $\widetilde{L}$ on $P(H_\gamma)$. When does a minimum point occur at a Riemann matrix? We can answer this question for $\gamma = 4$ as follows:

PROPOSITION 12.4. *Let τ be an element of the closure of RM such that τ is not a Riemann matrix. If τ is a minimum point of a height function in the direction of a positive definite symmetric matrix $\widetilde{L}$ of size 2γ on the closure of $P(RM)$ and the size of the Riemann matrices of the decomposition of τ is not greater than 3, then $\widetilde{L}$ is a normal vector of $P(H_\gamma)$ at $P(\tau)$ and hence τ is the unique minimum point.*

PROOF. To simplify the proof, we assume that $\tau = \begin{pmatrix} \tau_1 & 0 \\ 0 & \tau_2 \end{pmatrix}$, where τ_1, τ_2 are Riemann matrices of Riemann surfaces M_1, M_2 of genus γ_1, γ_2, respectively. Then there exists a family of Riemann matrices given by

$$\tau(\epsilon) = \begin{pmatrix} \tau_1 & 0 \\ 0 & \tau_2 \end{pmatrix} - 2\pi i\epsilon \begin{pmatrix} 0 & X_1{}^tX_2 \\ X_2{}^tX_1 & 0 \end{pmatrix} + O(\epsilon^2), \tag{12.1}$$

where X_j is an element of $\mathbf{C}^{\gamma_j}$ determined by $\psi_1, \dots, \psi_{\gamma_j}$ (see, for example, [I-T]), so that the tangent vector at τ is given by $\begin{pmatrix} 0 & X_1{}^tX_2 \\ X_2{}^tX_1 & 0 \end{pmatrix}$. Taking many points corresponding to nodes, we note that tangent vectors span $\begin{pmatrix} 0 & Y \\ {}^tY & 0 \end{pmatrix}$, where Y

are (γ_1, γ_2)-matrices, while the assumption implies that the tangent vectors at τ by deformations of τ_1, τ_2 are $\begin{pmatrix} Z_1 & 0 \\ 0 & Z_2 \end{pmatrix}$, where Z_1 and Z_2 are symmetric matrices of size γ_1 and γ_2, respectively. Thus $\widetilde{L}$ is a normal vector at $P(\tau)$ of $P(H_\gamma)$. Lemma 12.1 completes the proof. □

Recall that there exist τ_0 and Φ for a positive definite symmetric matrix $\widetilde{L}$ of size 2γ such that $L_{11} = \mathrm{Re}\Phi$, $L_{12} = \mathrm{Re}(\Phi\tau_0)$, $L_{22} = \mathrm{Re}(\tau_0\overline{\Phi\tau_0})$. If $P(\tau_0)$ belongs to the closure of $P(RM)$, the only minimum point of the height function on the closure of $P(RM)$ corresponding to $\widetilde{L}$ is τ_0.

COROLLARY 12.5. *If τ_0 is not an element of the closure of RM and $\gamma = 4$, then the minimum points of the height function on the closure of $P(RM)$ in the direction of a positive definite symmetric matrix $\widetilde{L}$ are Riemann matrices and the corresponding minimal surfaces are non-holomorphic and homotopically area-minimizing, that is, $\ell_1(\widetilde{L}) < \ell_2(\widetilde{L})$. In particular, the set of minimum points is compact and consists of one point for generic $\widetilde{L}$.*

Using the proof of Proposition 12.4, we can comment (see Corollary 12.6) on the following result proved by Arezzo and Micallef [A-M]:

THEOREM [A-M]. *The set of flat tori of real dimension 2γ which contain a stable minimal surfaces of genus γ not holomorphic with respect to any complex structure compatible with the metric, is open and dense in the space of flat tori of dimension 2γ.*

COROLLARY 12.6. *The set of $\widetilde{L}$ such that the torus with lattice vectors L corresponding to $\widetilde{L}$ admits a non-holomorphic, homotopically area minimizing minimal surfaces of genus γ contains an open and dense subset in S^2_R. In particular, if $\gamma = 4$, then equality holds.*

PROOF. We want to determine a necessary condition on $\widetilde{L}$ such that, on the closure of $P(RM)$, $E_{\widetilde{L}}$ has minimum points of non-Riemann matrices. To simplify our proof, we consider $\widetilde{L}$ on a submanifold N of H_γ defined by

$$\left\{\begin{pmatrix} \tau_1 & 0 \\ 0 & \tau_2 \end{pmatrix} : \tau_1,\ \tau_2 \text{ are Riemann matrices of size } \gamma_1,\ \gamma_2\right\}.$$

Then, by the proof of Proposition 12.4, $\widetilde{L}$ is orthogonal to tangent vectors of N and $\begin{pmatrix} 0 & Y \\ {}^tY & 0 \end{pmatrix}$. Thus we define a subbundle of the normal bundle of N in S^2_R defined by $\widetilde{L}$. Note that if $\gamma_2 = 0$, then the subbundle is defined by the induced bundle on $P(RM_{non-hyper})$, $P(RM_{hyper})$ the normal bundle of $\natural_{2\gamma}$ in $\mathbf{R}^{\gamma(2\gamma+1)}$. The real dimension of the total space is $(\gamma_1+\gamma_2)(2(\gamma_1+\gamma_2)+1) - 2\gamma_1\gamma_2$. So the image of the normal Gauss map for the subbundle is a set of measure zero in S^2_R. It is easy to see that we have at most countably many sets of measure zero. Thus the complement in PS^2_R is open and dense. Hence, for $\widetilde{L}$ in the open and dense set, the minimum points of $E_{\widetilde{L}}$ are Riemann matrices. The corresponding minimal surfaces are not holomorphic. Thus we have the required open dense subset. It is clear that equality holds if $\gamma = 4$ by Corollary 12.5. □

Furthermore, Arezzo has presented an interesting conjecture concerning the Torelli theorem for homotopically area-minimizing minimal surfaces:

AREZZO'S CONJECTURE. *Is a generalized homotopically area-minimizing map given in Theorem [Ar1] unique?*

Remark: Arezzo [Ar2] has obtained two full minimal surfaces of genus γ with the same periods in $T^{2\gamma}$. However one is hyperelliptic and hence is not area minimizing (nor even stable). From our viewpoint, the above problem can be restated as follows:

AREZZO'S CONJECTURE (FROM OUR VIEWPOINT). *Let $\widetilde{L}$ be a positive definite symmetric matrix. Then, is the minimum point for the height function corresponding to $\widetilde{L}$ on the closure of $P(RM)$ in $\mathbf{R}^{\gamma(2\gamma+1)}$ unique?*

Remark: By Corollary 12.5, if $\gamma = 4$, then, Arrezo's conjecture is true for a generic $\widetilde{L}$.

Since we would like to identify a minimal surface with its congruent minimal surfaces, we weaken Arezzo's conjecture as follows:

PROBLEM 6. *Let $\widetilde{L}$ be a positive definite symmetric matrix. Then, is the minimum point for the height function corresponding to $\widetilde{L}$ on the closure of $P(RM)$ in $\mathbf{R}^{\gamma(2\gamma+1)}$ unique up to $\widehat{G}(\widetilde{L})$?*

In general, we have:

LEMMA 12.7. *Let $\tau = X + iY \in H_\gamma$ and $\widetilde{L} = JP(\tau)^t J$, that is,*

$$\begin{pmatrix} Y^{-1} & Y^{-1}X \\ XY^{-1} & XY^{-1}X + Y \end{pmatrix}.$$

Then τ_0 and Φ (Lemma 12.1) corresponding to $\widetilde{L}$ are τ and Y^{-1}, respectively. Furthermore $I(\tau) = {}^t J\widehat{G}(\widetilde{L})J$ and $\widehat{G}(\widetilde{L}) = I(\tau)$ and hence $E_{\widetilde{L}}$ is $I(\tau)$-equivariant.

Unfortunately:

THEOREM 12.8. *The answer to Problem 6 is negative for $\gamma = 4$. In particular, Arezzo's conjecture fails.*

PROOF. In fact, we shall prove that the answer to Problem 6 is negative if $\gamma = 4$. Assume not. We consider the E_8-lattice. Then Buser and Sarnak [B-S] have proved that there exists $\tau_{E_8} \in H_4$ such that τ_{E_8} attains $\delta_8 = r_8$. The property we need is: τ_{E_8} does not belong to RM and is irreducible, and further, the automorphism group $I(\tau_{E_8})$ of the polarized abelian varieties corresponding to τ_{E_8} is of order 46080. Thus $\widetilde{L} = JP(\tau_{E_8})^t J$ gives an $I(\tau_{E_8})$-equivariant $E_{\widetilde{L}}$ by

Lemma 12.7. By our assumption and Corollary 12.5, the minimum points of $E_{\widetilde{L}}$ on the closure of $P(RM)$ are $\widehat{G}(\widetilde{L})$-orbit of an element τ belonging to $RM_{non-hyper}$ by Micallef's result. If τ is fixed by $\widehat{G}(\widetilde{L})$, then we have a non-hyperelliptic Riemann surface with the automorphism group of order $\frac{46080}{2}$, which is a contradiction(see [B-S]). Hence the number of elements of the orbit of τ is not less than 2 and is finite. So the minimum points must be isolated. Namely, by a small perturbation of $\widetilde{L}$, we obtain a critical point which is stable and is not a minimum and hence there exists a stable minimal surface which is not homotopically area-minimizing. Note that it is enough to use τ (in place of τ_{E_8}) which is not an element of the closure of RM and has sufficient automorphisms by Lemma 12.7. If we can prove the following Claim 1, then we obtain a contradiction, which completes the proof of Theorem 12.8.

CLAIM 1. *Under our assumption, all stable compact orientable branched minimal surfaces in Riemannian flat tori are homotopically area-minimizing.*

Proof of Claim 1. Let (M, f) be a Riemann surface of genus γ minimally mapped into a Riemannian flat torus T^n (with branch points). Then (M, f) may be non-hyperelliptic by Micallef's result and we have a positive semi-definite symmetric matrix $\widetilde{L}$ of size 2γ which is a normal vector of $P(RM)$ at $P(\tau)$ of (M, f). Let $\widetilde{L}'$ be the positive definite symmetric matrix corresponding to the Albanese map of M, which is also normal at $P(\tau)$. We set $\widetilde{L_s} = (1-s)\widetilde{L} + s\widetilde{L}'$, which is again a normal vector and positive definite except when $s = 0$. Thus we obtain a one parameter family of compact orientable minimal surfaces with fixed conformal structure. Suppose that M is stable in T^n, that is, the hessian of the height function corresponding to $\widetilde{L}$ is positive semi-definite. Using the fact that the hessian of the height function corresponding to $\widetilde{L}'$ is positive definite, so is the hessian corresponding to $\widetilde{L_s}$. Let χ_s be the minimal embedding $(s \neq 0)$ of M into the torus corresponding to $\widetilde{L_s}$. We shall prove that χ_s is homotopically area-minimizing for $s > 0$. If χ_0 is not homotopically area-minimizing, then $E_{\widetilde{L}}$ is not minimal at τ and, by a small perturbation, $E_{\widetilde{L_s}}$ for a small $s > 0$ is not minimal at τ. Thus χ_0 is homotopically area-minimizing. Let $0 < t < 1$. Then $\widetilde{L_s}$ on $[t, 1]$ gives a regular curve (τ_s, Φ_s) in the normal space $H_\gamma \times \mathbf{Her}$ such that $\tau_1 = \tau$ and $\Phi_1 = I_\gamma$. In this case, it is possible that $I(\tau_s) \neq \{I_{2\gamma}, -I_{2\gamma}\}$ occurs for some $t < s < 1$. It is well-known that $\{\tau \in H_\gamma \mid I(\tau) \neq \{I_{2\gamma}, -I_{2\gamma}\}\}$ are at most countably many analytic subvarieties in H_γ [Mu]. Thus we can construct a new curve $\widetilde{L}'_s = (\tau'_s, \Phi'_s)$ such that (τ'_s, Φ'_s) is close to (τ_s, Φ_s) and $I(\tau'_s) = \{I_{2\gamma}, -I_{2\gamma}\}$, $\tau'_1 = \tau$, $\Phi'_1 = I_\gamma$ and $\tau'_t = \tau_t$, $\Phi'_t = \Phi_t$. Furthermore, using the strict positiveness of the hessian of the height function in the direction of $\widetilde{L_s}$ for $t \leq s \leq 1$ at τ, we obtain a strictly stable minimal surface χ'_s for $t \leq s \leq 1$ close to χ_s such that $\chi'_1 = \chi_1$ and $\chi'_t = \chi_t$. We set $I_t = \{s \in [t, 1] : \chi'_s$ is homotopically area-minimizing $\}$. It is enough to show that $I_t(\ni 1)$ is open and closed. Openness: Let $s \in I_t$. Then the height function on the closure of the space of Riemann matrices corresponding to $\widetilde{L'_s}$ has a unique minimum, because $\widehat{G}(\widetilde{L'_s}) \subset I(\tau'_s) = \{I_{2\gamma}, -I_{2\gamma}\}$ and our assumption. So for a small perturbation of s, the minimum point of the height function corresponding to $\widetilde{L_s}$ only occurs in a small neighborhood of the critical point. In the small neighborhood, the critical point is unique. Closedness: This is easy. □

From the proof of Claim 1, we find that, for $\widetilde{L_s}$ close to $\widetilde{L}'$, χ_s is homotopically area-minimizing without the assumption, because the Albanese map corresponds to the unique minimum point of $E_{\widetilde{L}'}$. Using Theorem 8.6, Lemma 9.1 and Proposition 9.3, we have the following:

COROLLARY 12.9. *Let M be a non-hyperelliptic Riemann surface of genus $\gamma \geq 4$. Then there exists a full non-holomorphic homotopically area-minimizing minimal embedding of M in $T^{2\gamma}$*

Remark: Arezzo and Micallef [A-M] have constructed non-holomorphic, stable minimal immersions. In our proof, Problem 6 implies that stable minimal surface in Riemannian flat tori are homotopically area-minimizing. In spite of the negative answer to Problem 6, we ask:

PROBLEM 7. *Are all non-hyperelliptic stable compact orientable branched minimal surfaces in Riemannian flat tori homotopically area-minimizing?*

THEOREM 12.10. *The answer to Problem 7 is positive if and only if the following holds. Let $\widetilde{L}$ be a positive definite symmetric matrix of size 2γ such that there exist minimum points for the height function corresponding to $\widetilde{L}$ on the closure of $P(RM)$. Then if some minimum point is a Riemann matrix which has positive hessian, then it is unique. In particular, if the answer to Problem 7 is positive, then there are no two homotopic stable compact orientable branched minimal surfaces in a Riemannian flat torus such that one of them is immersed with only trivial Jacobi fields.*

PROOF. Assume that the answer to Problem 7 is positive for $\widetilde{L}$. Perturbing $\widetilde{L}$, we have strictly stable minimal surfaces smoothly dependent on the perturbation. If other minimum points exist, by choosing an appropriate perturbation, we can make the strictly stable minimal surface not homotopically area-minimizing, which is a contradiction. Conversely, by the same argument as in the proof of Claim 1, we can prove that the answer to Problem 7 is positive. □

If the minimum points of the height function in the direction of $JP(\tau_{E_8})^t J$ are a $\widehat{G}(\widetilde{L})$-orbit of an element of $RM_{non-hyper}$, then the answer to Problem 7 is also negative.

PROBLEM 8. *Determine the minimal surfaces corresponding to the above minimum points and their areas, that is, $\ell_2(\widetilde{L})$. Note that $\ell_1(\widetilde{L}) = 4$.*

If the answer to Problem 7 is positive, then the minimum point is unique by Theorem 12.10. An analogous problem to Problem 6 is:

PROBLEM 9. *Let $\widetilde{L}$ be a positive definite symmetric matrix. Then is the minimum point for the height function corresponding to $\widetilde{L}$ on the closure of $P(RM_{hyper})$ in $\mathbf{R}^{\gamma(2\gamma+1)}$ unique up to $\widehat{G}(\widetilde{L})$?*

For an element τ of the closure of RM_{hyper}, we have a result similar to (12.1) such that $\tau(\epsilon) \in RM_{hyper}$, however the nodes must be fixed by the hyperelliptic involution. Hence we have the following:

PROPOSITION 12.11. *Let τ be an element of the closure of RM_{hyper} such τ is not a Riemann matrix. If τ is a minimum point of a height function in the direction of a positive definite symmetric matrix $\widetilde{L}$ of size 2γ on the closure of $P(RM_{hyper})$ and the sizes of the Riemann matrices of the decomposition of τ are not greater than 2, then $\widetilde{L}$ is a normal vector of $P(H_\gamma)$ at $P(\tau)$ and hence τ is the unique minimum point.*

In particular, we have the following for $\gamma = 3$.

COROLLARY 12.12. *If τ_0 is not an element of the closure of RM_{hyper} and $\gamma = 3$, then the minimum points of the height function on the closure of $P(RM_{hyper})$ in the direction of a positive definite symmetric matrix $\widetilde{L}$ such that τ_0 is the minimum point of $E_{\widetilde{L}}$ are Riemann matrices and the corresponding minimal surfaces are non-holomorphic and hyperelliptically area-minimizing.*

Unfortunately:

THEOREM 12.13. *The answer to Problem 9 is negative for $\gamma = 3$.*

PROOF. Let C be the Klein curve. Then it is well-known that C is not hyperelliptic and is the unique curve of genus 3 admitting a group of 168 automorphisms, the largest possible number (see, for example, [Le]). Let τ be the Riemann matrix of C corresponding to an appropriate canonical homology basis. Then $I(\tau)$ has order 336. Let $\widetilde{L} = JP(\tau)^t J$. Then $G(\widetilde{L})$ has order 336 and the corresponding τ_0 is τ by Lemma 12.7. Thus by Corollary 12.12, minimum points belong to $P(RM_{hyper})$ whose cardinality is not less than 2, because there is no hyperelliptic Riemann surface with 336 automorphisms. If the answer to Problem 9 is positive, then there exists a non-hyperelliptically area-minimizing, hyperelliptically stable minimal surface in the torus corresponding to L. To prove Theorem 12.13, we need the following:

CLAIM 2. *If the answer to Problem 9 is positive, then hyperelliptically stable minimal surfaces are hyperelliptically area-minimizing.*

The method used in the proof of Claim 1 establishes Claim 2. □

It follows from Proposition 7.12 that we do not need the argument about the Klein curve. The same argument that we used in the proof of Claim 1 yields the following:

COROLLARY 12.14. *Let M be a hyperelliptic Riemann surface of genus $\gamma \geq 3$. Then there exists a full hyperelliptically area-minimizing embedding of M in $T^{2\gamma}$.*

We can pose a problem similar to Problem 8 for the above $\widetilde{L}$.

PROBLEM 10. *Determine hyperelliptically area-minimizing minimal surface corresponding to the minimum points of* $E_{\widetilde{L}}$ *on the closure of* $P(RM_{hyper})$ *for* $\widetilde{L} = JP(\tau)^t J$, *where* τ *is the Riemann matrix of the Klein curve,*

By Proposition 7.12, the answer to the following Problem is negative and Corollary 12.15 holds.

PROBLEM 11. *Are all hyperelliptically stable compact orientable branched minimal surfaces in Riemannian flat tori hyperelliptically area-minimizing?*

COROLLARY 12.15. *There exists a positive definite symmetric matrix* $\widetilde{L}$ *of size* 6 *such that there exists a non-minimal, critical point of the height function corresponding to* $\widetilde{L}$ *on* $P(RM_{hyper})$ *with strictly positive hessian.*

§13 Another solution of Problem 1.

Let M be a Riemann surface and $\{A_i, B_i\}$ a canonical homology basis. Collapsing A_γ, we obtain a Riemann surface $M(0)$ of genus $\gamma - 1$ with a node and a one parameter family $M(\epsilon)$ of Riemann surfaces which collapse to $M(0)$. Then the Riemann matrix $\tau(\epsilon)$ corresponding to $\{A_i, B_i\}$ is given by

$$\tau(\epsilon) = \begin{pmatrix} \tau_0 & q \\ {}^t q & \frac{1}{2\pi i}\log\epsilon + c_0 \end{pmatrix} + O(\epsilon),$$

where τ_0 is the Riemann matrix corresponding to $\{A_1, \dots, A_{\gamma-1}, B_1, \dots, B_{\gamma-1}\}$ of $M(0)$, $q \in \mathbf{C}^{\gamma-1}$, $c_0 \in \mathbf{C}$ (see, for example, [I-T]). We seek the condition that $\operatorname{tr}(P(\tau(\epsilon))\widetilde{L})$ is bounded from above. If $\widetilde{L}$ gives a torus, then the energy of the harmonic map of $M(\epsilon)$ to the torus corresponding to $\widetilde{L}$ is bounded from above. This situation was discussed in [S-Y], [S-U1] and was used in [Me]. We can recover their result when the ambient space is a Riemannian flat torus. Let $X(\epsilon), Y(\epsilon)$ be the real and imaginary part of $\tau(\epsilon)$ and $\epsilon = re^{i\theta}$.

LEMMA 13.1. $P(\tau(\epsilon))$ *is*

$$\begin{pmatrix} 0 & 0 & 0 & 0 \\ 0 & -\frac{1}{2\pi}\log r + \operatorname{Im} c_0 & 0 & 0 \\ 0 & 0 & 0 & 0 \\ 0 & 0 & 0 & 0 \end{pmatrix} + \begin{pmatrix} Y + XY^{-1}X & y + XY^{-1}x & -XY^{-1} & 0 \\ {}^t y + {}^t xY^{-1}X & {}^t xY^{-1}x & -{}^t xY^{-1} & 0 \\ -Y^{-1}X & -Y^{-1}x & Y^{-1} & 0 \\ 0 & 0 & 0 & 0 \end{pmatrix}$$

$+(1/\log r$ *terms)* $+O(\epsilon)$, *where* $x + iy = q$ *and* $\tau_0 = X + iY$. *In particular, if* M *is hyperelliptic and* A_γ *is invariant by the hyperelliptic involution, then* $M(\epsilon)$ *is hyperelliptic.*

COROLLARY 13.2. $\operatorname{tr}(P(\tau(\epsilon))\widetilde{L})$ *is bounded from above if and only if* $\ell_\gamma = 0$. *The limit of* $\operatorname{tr}(P(\tau(\epsilon))\widetilde{L})$ *is* $\operatorname{tr}(P(\tau_0)\widetilde{L}')$, *where* $\widetilde{L}'$ *corresponds to* $\{\ell_1, \dots, \ell_{\gamma-1}, \ell_{\gamma+1}, \dots, \ell_{2\gamma-1}\}$. *Thus we have a harmonic map of* $M(0)$ *to the torus spanned by* $\{\ell_1, \dots, \ell_{\gamma-1}, \ell_{\gamma+1}, \dots, \ell_{2\gamma-1}\}$.

We can give another solution of Problem 1.

COROLLARY 13.3. *For $\widetilde{L}$ corresponding to the Schwartz P-surface, the infimum of $E_{\widetilde{L}}$ on $P(RM_{hyper})$ is 0.*

PROOF. The Schwartz P-surface admits three null-homologous, simple closed curves without intersections which are invariant by the hyperelliptic involution. Using Corollary 13.2, we have a harmonic map of a sphere, which is a constant map. □

§14 The local Torelli problem for algebraic curves over C.

We shall study B for hyperelliptic minimal surfaces (see Section 7 for B).

PROPOSITION 14.1. *If all stable branched minimal immersions of a hyperelliptic Riemann surface (M, f) in Riemannian flat tori are holomorphic, then the range of B is $(\kappa(HT(R)))^{\perp}$ in H_γ.*

PROOF. The real dimension of Kerκ_* is $2\gamma - 4$ and the real dimension of the normal space is $\gamma^2 - 3\gamma + 2$. It follows from Theorem 7.10 that $B(\mathbf{J}X, \mathbf{J}Y) = -B(X, Y)$. Thus, let $e_1, \dots, e_{\gamma-2}, \mathbf{J}e_1, \dots, \mathbf{J}e_{\gamma-2}$ be a basis of Kerκ_*. Then the range of B is spanned by $B(e_i, e_j)(i \leq j)$, $B(\mathbf{J}e_i, e_j)(i \leq j)$ whose dimension is not greater than $\gamma^2 - 3\gamma + 2$. Assume that the dimension is less than $\gamma^2 - 3\gamma + 2$. Then we obtain a vector $\widetilde{L}$ orthogonal to the image of the range of B and the tangent space of $\kappa(HT(R))$ at $\kappa([(M, f)])$ which is in the tangent space of $P(H_\gamma)$. For $\widetilde{L_s}$ as in the proof of Claim 1, τ (the Riemann matrix of (M, f)) is a critical point of $E_{\widetilde{L_s}}|_{HT(R)}$ and hence we have a corresponding non-holomorphic hyperelliptic minimal surface if s is close to 1. The resulting hyperelliptic minimal surface satisfies $\lambda_1 = \cdots = \lambda_{\gamma-2} = 0$, where $\lambda_1, \dots, \lambda_{\gamma-2}$, are given in Theorem 7.10, which is stable if s is close to 1. This is a contradiction, so the dimension of the range of B is $\gamma^2 - 3\gamma + 2$. Furthermore $B(e_i, e_j)(i \leq j)$, $B(\mathbf{J}e_i, e_j)(i \leq j)$ are linearly independent. □

Hyperelliptic minimal surfaces satisfy $D^2E_\phi(\mu, \nu) = B(\mu, \nu)E_{\widetilde{L}}$ for $\mu, \nu \in$ Kerκ_*.

COROLLARY 14.2. *All stable branched minimal immersions of a hyperelliptic Riemann surface (M, f) into Riemannian flat tori are holomorphic if and only if the range of B at $[(M, f)]$ is full.*

By Micallef's result, B is full for any hyperelliptic Riemann surface. On the other hand, Oort and Steenbrink proved the marvelous result that the Torelli mapping of the moduli space of Riemann surfaces into the moduli space of principally polarized abelian varieties is a locally closed immersion [O-S]. This is equivalent to saying that, under the mapping $^{\#}\kappa$, the structure sheaf $O(U_\gamma/\vartheta)$ of U_γ/ϑ is isomorphic to $O(H_\gamma)/I$, where I is the local defining ideal of $\kappa(U_\gamma)$. Surprisingly:

COROLLARY 14.3. *The following statements are equivalent.*

1. *All stable hyperelliptic minimal surfaces in Riemannian flat tori are holomorphic.*

2. *The Torelli mapping of the moduli space of Riemann surfces into the moduli space of principally polarized abelian varieties is a locally closed immersion.*

PROOF. First of all, we assume (1). Let $U(\gamma)$ be the Torelli space and ${}^{\#}\kappa$ the holomorphic map of $U(\gamma)$ into H_γ. Then there is an involution ϑ of $U(\gamma)$ such that the hyperelliptic locus $\widetilde{H}$ is the set of fixed points and ${}^{\#}\kappa$ induces an injective map i of the analytic space $U(\gamma)/\vartheta$ into H_γ. For a proof of (2), it is enough to prove that i is a locally closed immersion. This idea was used in [O-S]. By (1) and Corollary 14.2, the range of B is full for any hyperelliptic point ($\in \widetilde{H}$). Naturally, we can define B in this case. We note that the fullness of B is equivalent to the existence of a complex local coordinate system $z^1, \dots, z^{3\gamma-3}$ for each point $p \in \widetilde{H}$ such that, at p, $\frac{\partial}{\partial z^1}, \dots, \frac{\partial}{\partial z^{2\gamma-1}}$ are tangent to $\widetilde{H}$ and $\frac{\partial}{\partial z^\alpha}{}^{\#}\kappa = 0$, $2\gamma \le \alpha \le 3\gamma - 3$ and further,

$$\frac{\partial^{\#}\kappa}{\partial z^1} \bigwedge \cdots \bigwedge \frac{\partial^{\#}\kappa}{\partial z^{2\gamma-1}} \bigwedge \frac{\partial^2 {}^{\#}\kappa}{\partial z^{2\gamma}\partial z^{2\gamma}} \bigwedge \cdots \bigwedge \frac{\partial^2 {}^{\#}\kappa}{\partial z^{\alpha}\partial z^{\beta}} \bigwedge \cdots \bigwedge \frac{\partial^2 {}^{\#}\kappa}{\partial z^{3\gamma-3}\partial z^{3\gamma-3}} \ne 0$$

for $2\gamma \le \alpha \le \beta \le 3\gamma - 3$, where we consider ${}^{\#}\kappa$ to have values in $\mathbf{C}^{\frac{\gamma(\gamma+1)}{2}}$.

We can find a new complex local coordinate system $(z^1, \dots, z^{2\gamma-1}, z^{2\gamma}, \dots, z^{3\gamma-3})$ at a hyperelliptic point p such that $p = (0, \dots, 0)$ and

$$\vartheta(z^1, \dots, z^{2\gamma-1}, z^{2\gamma}, \dots, z^{3\gamma-3}) = (z^1, \dots, z^{2\gamma-1}, -z^{2\gamma}, \dots, -z^{3\gamma-3})$$

and $(w^1, \dots, w^{\gamma(\gamma+1)/2})$ at ${}^{\#}\kappa(0, \dots, 0) = (0, \dots, 0)$ such that

$${}^{\#}\kappa(z^1, \dots, z^{2\gamma-1}, z^{2\gamma}, \dots, z^{3\gamma-3}) = (z^1, \dots, z^{2\gamma-1}, \dots, z^\alpha z^\beta, \dots),$$

$$2\gamma \le \alpha \le \beta \le 3\gamma - 3.$$

It is easy to obtain the above $(z^1, \dots, z^{2\gamma-1}, z^{2\gamma}, \dots, z^{3\gamma-3})$. Let $(\eta^1, \dots, \eta^{\gamma(\gamma+1)/2})$ be a complex local coordinate system at ${}^{\#}\kappa(0, \dots, 0) = (0, \dots, 0)$. The invariance of η^j by ϑ implies that the Taylor expansion is given by products of

$$z^1, \dots, z^{2\gamma-1}, \dots, z^\alpha z^\beta, \dots$$

where $2\gamma \le \alpha \le \beta \le 3\gamma - 3$. Then we define a holomorphic map of a neighborhood of the origin of $\mathbf{C}^{\gamma(\gamma+1)/2}$ into $\mathbf{C}^{\gamma(\gamma+1)/2}$ such that $(z^1, \dots, z^{2\gamma-1}, \dots, w^{\alpha\beta}, \dots) \mapsto (\eta^1, \dots, \eta^{\gamma(\gamma+1)/2})$, where $\eta^1 = z^1, \dots, \eta^{2\gamma-1} = z^{2\gamma-1}$ and η^j ($j \ge 2\gamma$) is a holomorphic function obtained by putting $w^{\alpha\beta}$ in place of the terms of $z^\alpha z^\beta$ of the Taylor expansion. The assumption on ${}^{\#}\kappa$ means that $(z^1, \dots, z^{2\gamma-1}, \dots, w^{\alpha\beta}, \dots)$ is a complex local coordinate system at ${}^{\#}\kappa(0, \dots, 0)$ as required. This complex local coordinate system shows that i is a locally closed immersion. Next, assume (2). Then i is a locally closed immersion. Thus it is easy to obtain the above coordinate systems at each point of $\widetilde{H}$, which implies the fullness of B. Hence we obtain (1) by Corollary 14.2. □

Thus we can understand the behaviour of RM in a neighborhood of a point of the hyperelliptic locus in H_γ from the complex geometric viewpoint. On the other hand, note that Problem 5 concerns the study of the differential-geometric property of $P(RM)$ in a neighborhood of a point of $P(RM_{hyper})$ in $P(H_\gamma)$. Furthermore:

COROLLARY 14.4. *Let (M, f) be a hyperelliptic Riemann surface of genus $\gamma \geq 3$. Then (M, f) can be embedded in a Riemannian flat 2γ-torus as a non-holomorphic, hyperelliptically area-minimizing minimal surface with $index_A = s$ and $nullity_A = 4\gamma - 4 - 2s$, where $s = 1, \ldots, \gamma - 2$.*

PROOF. For $N \in (\kappa(HT(R)))^{\perp}$, we obtain a symmetric bilinear $\mathbf{R}$-valued form $< B(\mu, \nu), N >$ of the space of symmetric bilinear forms C such that $C(\mathbf{J}\mu, \mathbf{J}\nu) = -C(\mu, \nu)$. This map is linear and onto by Corollary 14.2. Thus, choosing an appropriate normal vector of $\kappa(HT(R))$ in H_γ, we can make a hyperelliptic minimal surface such that $\lambda_1, \ldots, \lambda_s > 0$, $\lambda_{s+1} = \cdots = \lambda_{\gamma-2} = 0$ as in the proof of Proposition 14.1, which are hyperelliptically area-minimizing by Corollary 12.14.□

§15 Other problems.

We shall pose some other problems arising from the differential-geometric Schottky problem.

PROBLEM 12. *The null vectors of B give rise to the Jacobi fields on the corresponding hyperelliptic minimal surface. Study the geometric properties of these Jacobi fields.*

PROBLEM 13. *The embedding P of H_γ has an interesting geometric property. Ferus [Fe] determined isometric immersions of a Kähler manifold with $\sigma(JX, JY) = \sigma(X, Y)$ into Euclidean space. P satisfies half of these conditions. $\delta_{2\gamma}$ may be related to P. On the other hand, $\delta_{2\gamma}$ is related to the Seshadri constants of polarized abelian varieties [Ba], [La]. Thus it is very interesting to study the geometric properties of P.*

PROBLEM 14. *Is there a full minimal deformation of a homotopically area-minimizing minimal surface of genus γ in $T^{2\gamma}$?*

PROBLEM 15. *Nagano and Smith [N-S1] proved that two (hyperelliptic) minimal surfaces of genus 3 with the same conformal structure in T^3 are associate. What is the geometric meaning of their result from our viewpoint?*

PROBLEM 16. *For minimal surfaces of an indefinite torus, the nullity of the hessian of $E_{\tilde{L}}$ has still geometric meaning on deformations of minimal surfaces under the deformation of the indefinite torus. Thus it may be interesting to study the Jacobi operator of minimal surfaces in indefinite tori.*

References

[A] L.V. Ahlfors, *The complex analytic structure of the space of closed Riemann surfaces*, Analytic functions (Nevanlinna et al., eds.), Princeton Univ. Press, 1960, pp. 45-66.

[Ar1] C. Arezzo, *On minimal surfaces incompressible in homology and abelian varieties*, preprint (1996).

[Ar2] C. Arezzo, *On the Torelli theorem for minimal immersions*, Inter. J. Math. **10** (1999), 697-706.

[A-M] C. Arezzo and M.J. Micallef, *Stable minimal surfaces in flat tori*, preprint (1995).

[Ba] T. Bauer, *Seshadri constants and periods of polarized abelian varieties*, Math. Ann. **312** (1998), 607-623.

[Be] M. Beeson, *Some results on finiteness in Plateau's problem, part I*, Math. Z. **175** (1980), 103-123.

[B-T1] M.J. Besson and A.J. Tromba, *The cusp catastrophe of Thom in the bifurcation of minimal surfaces*, Manuscripta Mathematica **46** (1984), 273-308.

[B-T2] R. Böhme and A.J. Tromba, *The index theorem for classical minimal surfaces*, Ann. of Math. **113** (1981), 447-499.

[B-S] P. Buser and P. Sarnak, *On the period matrix of a Riemann surface of large genus (with an Appendix by J.H. Conway and N.J.A. Sloane)*, Invent. Math. **117** (1994), 27-56.

[E-L] J. Eells and K.D. Lemaire, *Deformations of metrics and associated harmonic maps*, Proc. Indian Acad. Sci. (Math. Sci.) **90** (1981), 33-45.

[F-K] H. Farkas and H. Kra, *Riemann Surfaces*, Springer, 1980.

[Fe] D. Ferus, *Symmetric submanifolds of Euclidean space*, Math. Ann. **247** (1980), 81-93.

[La] R. Lazarsfeld, *Lengths of periods and Seshadri constants of abelian varieties*, Math. Res. LetterS **3** (1996), 439-447.

[Gu] R.C. Gunning, *Some curves in abelian varieties*, Invent. Math. **66** (1982), 377-389.

[G-H] P. Griffith and J. Harris, *Principles of Algebraic Geometry*, Wiley, 1978.

[I-T] Y. Imayoshi and M. Taniguchi, *An Introduction to Teichmüller Spaces*, Springer-Verlag, Tokyo, 1992.

[J-J-P] J. Jost, X.L. Jost and X.W. Peng, *Bifurcation of minimal surfaces in Riemannian manifolds*, Trans. Amer. Math. Soc. **347** (1995), 49-62.

[K] N. Koiso, *Variation of harmonic mappings caused by a deformation of Riemannian metrics*, Hokkaido Math. J. **8** (1979), 199-213.

[Le] J. Lewittes, *Singular Riemann matrices*, Advances in the Theory of Riemann Surfaces, Ann. of Math. Studies, No. 66., Princeton Univ. Press, 1971, pp. 287-294.

[Me] W.H. Meeks III, *The theory of triple periodic minimal surfaces*, Indiana Univ. Math. J. **39** (1990), 877-936.

[Mi] M. Micallef, *Stable minimal surfaces in flat tori*, Contemporary Math. 49, 1986, pp. 73-78.

[M-M] M. Micallef and W.H. Meeks III, lecture at Oberwolfach.

[M-R] S. Montiel and A. Ros, *Schrödinger operator associated to a holomorphic map*, Global Diff. and Global Analysis, Lecture note in Math 1481, 1991, pp. 147-174.

[Mu] D. Mumford, *Tata Lectures on Theta 1*, Progress in Math. vol 28, Birkhäu-ser, 1983.

[N-S1] T. Nagano and B. Smith, *Periodic minimal surfaces*, Comm. Math. Helv. **53** (1978), 29-55.

[N-S2] T. Nagano and B. Smith, *Periodic minimal surfaces and Weyl groups*, Acta Math. **145** (1980), 1-27.

[N] S. Nayatani, *Morse index and gauss maps of complete minimal surfaces in Euclidean 3-space*, Comm. Math. Helv. **68** (1993), 511-537.

[O-S] F. Oort and J. Steenbrink, *The local Torelli problem for algebraic curves*, Journées de Géometrie Algébrique d'Angers Juillet 1979 (Reihard Bölling, eds.), Algebraic Geometry, Angers, 1979, Springer, Berlin, 1979.

[P-S] T. Poston and I. Stewart, *Catastrophe Theory and its Applications*, Pitman Publishing Limited, 1978.

[Ra] H.E. Rauch, *A transcendental view of the space of algebraic Riemann surfaces*, Bull. Amer. Math. Soc. **71** (1965), 1-39.

[Ro] M. Ross, *Schwarz' P and D surfaces are stable*, Diff. Geom. App. **2** (1992), 179-195.

[S-U1] J. Sacks and K. Uhlenbeck, *The existence of minimal immersions of 2-spheres*, Ann. of Math. **113** (1981), 1-24.

[S-U2] J. Sacks and K. Uhlenbeck, *Minimal immersions of closed Riemann surfaces*, Trans. Amer. Math. Soc. **271** (1982), 639-652.

[S-Y] R. Schoen and S.T. Yau, *The existence of incompressible minimal surfaces and the topology of three dimensional manifolds with non-negative scalar curvature*, Ann. of Math. **110** (1979), 127-142.

[Sc] R. Schoen, *The role of hamonic mapping in rigidity and deformations*, Proc. Osaka Int. Conf. 1991, pp. 179-200.

[Sh] T. Shiota, *Characterization of Jacobian varieties in terms of soliton equations*, Invent. Math. **83** (1986), 333-382.

[Si] J. Simons, *Minimal varieties in riemannian manifolds*, Ann. of Math. **88** (1968), 62-105.

[T1] A.J. Tromba, *Teichmüller Theory in Riemannian Geometry*, Birkhäuser Verlag, 1992.

[T2] A.J. Tromba, *A sufficient condition for a critical point of a functional to be a minimum and its application to Plateau's problem*, Math. Ann. **263** (1983), 303-312.

[W] B. White, *The space of minimal submanifolds for varying Riemannian metrics*, Indiana Univ. Math. J. **40** (1991), 161-200.

THE GRADUATE SCHOOL OF MATHEMATICS, NAGOYA UNIVERSITY, CHIKUSA-KU, FUROU-CHO, NAGOYA 464-8602, JAPAN

Current address: Department of Mechanical Engineering, Nagoya Institute of Technology, Gokisho-cho, Showa-ku, Nagoya 466-8555, JAPAN

E-mail address: ejiri@math.kyy.nitech.ac.jp

Contemporary Mathematics
Volume **308**, 2002

Surfaces in 3-space Possessing Nontrivial Deformations which Preserve the Shape Operator

E.V. Ferapontov

1. Introduction

In 1865 Bonnet initiated the study of the following two types of deformations of surfaces in Euclidean 3-space:

1. *isometric deformations preserving principal directions* [**3**];

2. *isometric deformations preserving principal curvatures* [**4**].

Deformations of type 1 characterise the so-called molding surfaces (see [**6**] for further discussion), while deformations of type 2 lead to constant mean curvature surfaces and a special class of Weingarten surfaces, which has been recently investigated in [**2**]. In 1933 Finikov and Gambier [**19**], [**18**] (see also [**7**]) introduced deformations of a third type which seem to be a natural counterpart of deformations 1, 2:

3. *Deformations preserving principal directions and principal curvatures or, equivalently, the shape operator (Weingarten operator).*

For short, we will call them S-deformations (notice that S-deformations are no longer isometric). Let M^2 be a surface in Euclidean space E^3 parametrized by coordinates R^1, R^2 of curvature lines. Let

$$G_{11}(dR^1)^2 + G_{22}(dR^2)^2 \tag{1}$$

be its third fundamental form (i.e. metric of Gaussian image, which is automatically of constant curvature 1). This condition can be written in the form

$$(\partial_2 a + a^2)G^{22} + \frac{a}{2}\partial_2 G^{22} + (\partial_1 b + b^2)G^{11} + \frac{b}{2}\partial_1 G^{11} + 1 = 0, \tag{2}$$

where $\partial_1 = \partial/\partial R^1$, $\partial_2 = \partial/\partial R^2$, $G^{11} = 1/G_{11}$, $G^{22} = 1/G_{22}$ and the coefficients a and b are defined by the formulae

$$\partial_2 G^{11} = -2aG^{11}, \ \partial_1 G^{22} = -2bG^{22}. \tag{3}$$

Notice that a and b are the Christoffel symbols of the Levi-Civita connection of the metric (1): $a = \Gamma^1_{12}$, $b = \Gamma^2_{21}$. We point out that for given a and b equations (2),

2000 *Mathematics Subject Classification.* Primary 53A85, 53Z05.

(3) are linear in G^{11} and G^{22}. Let k^1 and k^2 be the radii of principal curvatures of the surface M^2. In the coordinates of curvature lines the Codazzi equations take the form

$$\frac{\partial_2 k^1}{k^2-k^1}=a, \quad \frac{\partial_1 k^2}{k^1-k^2}=b. \tag{4}$$

Let us assume now that the surface M^2 possesses nontrivial S-deformations. Analytically, this means that for given k^1 and k^2 (and hence given a and b) the linear system (2), (3) is not uniquely solvable for G^{11} and G^{22}. The form of equations (2)-(4) leads to the following simple observations:

1. S-deformations occur in one-parameter (multi-parameter) families. Indeed, if (G^{11}, G^{22}) and $(\tilde G^{11}, \tilde G^{22})$ are two different solutions of (2), (3) (with a and b fixed) then $(\lambda G^{11}+(1-\lambda)\tilde G^{11},\ \lambda G^{22}+(1-\lambda)\tilde G^{22})$ is a solution as well. This is a consequence of the linearity of equations (2), (3) in G^{11} and G^{22}.

2. The property that a surface M^2 possesses nontrivial S-deformations depends entirely on the geometry of the orthogonal net on the sphere S^2 which is the Gauss image of its curvature lines. Indeed, equations (2), (3) depend on a and b only, so that any solution of the linear system (4) defines the radii of principal curvatures of an S-deformable surface. Surfaces corresponding to different solutions of (4) have one and the same spherical image of curvature lines and hence one and the same third fundamental form. Thus, they S-deform simultaneously. We refer to [**19**] for further comments.

In section 2 we list some of the known examples of S-deformable surfaces. These, in particular, include arbitrary quadrics, cyclids of Dupin, and conformal images of the surfaces of revolution (as well as all other surfaces with the same spherical image of curvature lines). Apart from several degenerate cases, these examples exhaust the list of surfaces which possess 3-parameter families of S-deformations (under certain assumptions the number 3 proves to be the maximal possible). Following [**17**], in section 3 we demonstrate the integrability of the Gauss-Codazzi equations governing S-deformable surfaces by explicitly constructing the Lax pair with a spectral parameter. Some results on multi-dimensional S-deformable hypersurfaces are discussed in section 4. It is pointed out that under certain "genericity" assumptions, n-dimensional quadrics are the only hypersurfaces which possess S-deformations depending on $n+1$ arbitrary constants. In sections 5-7 we draw parallels between S-deformable hypersurfaces and compatible Poisson brackets of hydrodynamic type. It turns out that coordinate hypersurfaces of the n-orthogonal coordinate systems arising in the theory of bi-Hamiltonian systems of hydrodynamic type are necessarily S-deformable.

2. Examples

This section contains a list of examples of surfaces in E^3 possessing nontrivial S-deformations. In our discussion we mainly follow [**19**].

Example 1. Molding surfaces of Monge correspond to the third fundamental form

$$\frac{(dR^1)^2}{1+c/\cos^2(R^1)}+\frac{\sin^2(R^1)(dR^2)^2}{\psi(R^2)}, \tag{5}$$

where c is a constant and $\psi(R^2)$ is an arbitrary function. The metric (5) has constant curvature 1 for any c, ψ. The coordinate curves R^2 = constant (R^1 = constant) represent the meridians (parallels) of the sphere S^2, respectively. It can be readily verified that the radii of principal curvatures k^1, k^2 satisfy the linear system

$$\partial_2 k^1=0, \quad \frac{\partial_1 k^2}{k^1-k^2}=\cot(R^1)$$

which does not explicitly depend on c, ψ. Thus, any surface with the third fundamental form (5) possesses S-deformations depending on one arbitrary constant and one arbitrary function. Geometrically, the molding surfaces of Monge are swept out by a planar curve while the plane rolls along a curvilinear cylinder.

Example 2. General molding surfaces are characterised by the third fundamental form

$$(dR^1)^2+\frac{\sin^2(R^1+\varphi(R^2))(dR^2)^2}{\psi(R^2)}, \tag{6}$$

where $\psi(R^2)$ and $\varphi(R^2)$ are arbitrary functions. The coordinate curves R^1 = constant represent a one-parameter family of great circles on S^2, the curves R^2 = constant are their orthogonal trajectories. It can be verified that the radii of principal curvatures k^1, k^2 satisfy the equations

$$\partial_2 k^1=0, \quad \frac{\partial_1 k^2}{k^1-k^2}=\cot(R^1+\varphi(R^2))$$

which do not explicitly depend on ψ. Thus, any surface with the third fundamental form (6) possesses S-deformations depending on one arbitrary function of one variable. A general molding surface is swept out by a planar curve while the plane rolls along a developable surface.

If a surface is not a molding surface (that is, both $\partial_2 k^1$ and $\partial_1 k^2$ are nonzero), the number of parameters on which S-deformations depend cannot exceed 3. The proof of this statement can be found in [**19**] (see also [**13**]). Examples presented below actually exhaust the list of surfaces which possess precisely 3-parameter families of S-deformations.

Example 3. Quadrics correspond to the third fundamental form

$$\frac{R^2-R^1}{4}\left(\frac{(dR^1)^2}{(R^1)^3+a(R^1)^2+bR^1+c}-\frac{(dR^2)^2}{(R^2)^3+a(R^2)^2+bR^2+c}\right) \tag{7}$$

where a, b, c are arbitrary constants. The coordinates R^1, R^2 are known as spherical-conical: the corresponding coordinate curves are intersections of the sphere with a confocal family of quadratic cones. Equations for the radii of principal curvatures k^1, k^2 take the form

$$\frac{\partial_2 k^1}{k^2-k^1}=\frac{1}{2(R^2-R^1)}, \quad \frac{\partial_1 k^2}{k^1-k^2}=\frac{1}{2(R^1-R^2)}$$

which does not explicitly depend on a, b, c. Thus, any surface with the third fundamental form (7) possesses S-deformations depending on three arbitrary constants. The case of quadrics corresponds to the choice

$$k^1 = \frac{1}{R^1\sqrt{R^1R^2}}, \quad k^2 = \frac{1}{R^2\sqrt{R^1R^2}}.$$

Example 4. Cyclids of Dupin correspond to the third fundamental form

$$\frac{1}{(R^1 - R^2)^2}\left(\frac{(dR^1)^2}{a(R^1)^2 + bR^1 + c} - \frac{(dR^2)^2}{a(R^2)^2 + bR^2 + c + 1}\right) \tag{8}$$

where a, b, c are arbitrary constants. The coordinate system R^1, R^2 consists of the two orthogonal pencils of circles on S^2. The radii of principal curvatures k^1, k^2 satisfy the linear system

$$\frac{\partial_2 k^1}{k^2 - k^1} = \frac{1}{R^1 - R^2}, \quad \frac{\partial_1 k^2}{k^1 - k^2} = \frac{1}{R^2 - R^1}$$

which does not explicitly depend on a, b, c. Thus, any surface with the third fundamental form (8) possesses S-deformations depending on three arbitrary constants. The case of cyclids of Dupin corresponds to the choice

$$k^1 = R^2, \ k^2 = R^1.$$

Example 5. Conformal transforms of surfaces of revolution correspond to the third fundamental form

$$\frac{1}{(R^1 - R^2)^2}\left(\frac{p^2(dR^1)^2}{a(R^1)^2 + bR^1 + c} - \frac{(p + p'(R^1 - R^2))^2(dR^2)^2}{a(R^2)^2 + bR^2 + c + p^2}\right) \tag{9}$$

where a, b, c are arbitrary constants and p is an arbitrary function of R^2. Here the curves $R^2 =$ constant represent a one-parameter family of circles on S^2. The radii of principal curvatures k^1, k^2 satisfy the system

$$\frac{\partial_2 k^1}{k^2 - k^1} = \frac{1}{R^1 - R^2} + \frac{p'}{p}, \quad \frac{\partial_1 k^2}{k^1 - k^2} = \frac{1}{R^2 - R^1} + \frac{p'}{p + p'(R^1 - R^2)} \tag{10}$$

which does not explicitly depend on a, b, c. Thus, any surface with the third fundamental form (9) possesses S-deformations depending on three arbitrary constants. The case of conformal images of surfaces of revolution corresponds to the choice

$$k^1 = \frac{1}{p}, \quad k^2 = \frac{1}{p + p'(R^1 - R^2)} \tag{11}$$

(see [**13**]). The choice $p = R^2$ and the subsequent change of variables $R^i \to 1/R^i$ brings us back to the case of cyclids of Dupin discussed above. Indeed, cyclids of Dupin are known to be conformal images of cylinders, cones and tori of revolution. Notice that equations (11) possess a more general class of solutions

$$k^1 = q, \ k^2 = q + \frac{qp'(R^1 - R^2)}{p + p'(R^1 - R^2)}$$

where q is another arbitrary function of R^2. Solution (11) corresponds to the choice $q = 1/p$. We refer to [**5**] where the Darboux-integrability of the system (10) was established.

Examples 3-5 provide a complete list of surfaces possessing 3-parameter families of S-deformations. We refer to [**13**] for the proof of this result. Notice that example

5 is missing from the classification proposed in [**19**] (although examples 3 and 4 are present).

The case of surfaces possessing 2-parameter families of S-deformations is not well enough understood at the moment. We present here just one example of that type.

Example 6. Surfaces possessing 2-parameter family of S-deformations. Consider the metric

$$\frac{R^1 - R^2}{(R^1 + R^2)^2}\left(\frac{(dR^1)^2}{a(R^1)^2 - R^1 + c} - \frac{(dR^2)^2}{a(R^2)^2 - R^2 + c}\right) \tag{12}$$

where a, c are arbitrary constants. The metric (12) has constant curvature 1 for any values of a, c. The coordinate system R^1, R^2 is the image of the ellipsoidal coordinate system on the plane E^2 under the stereographic projection $E^2 \to S^2$ (see [**19**]). The equations for the radii of principal curvature k^1, k^2

$$\frac{\partial_2 k^1}{k^2 - k^1} = \frac{1}{2(R^2 - R^1)} - \frac{1}{R^1 + R^2}, \quad \frac{\partial_1 k^2}{k^1 - k^2} = \frac{1}{2(R^1 - R^2)} - \frac{1}{R^1 + R^2}$$

do not explicitly depend on a, c. Thus, any surface with the third fundamental form (12) possesses S-deformations depending on two arbitrary constants. The particular case

$$k^1 = \frac{(R^1 + R^2)^2}{(R^1)^{5/2}(R^2)^{3/2}}, \quad k^2 = \frac{(R^1 + R^2)^2}{(R^1)^{3/2}(R^2)^{5/2}}$$

was discussed in detail in [**19**].

Example 7. Surfaces possessing 1-parameter family of S-deformations. Consider the metric

$$\frac{2}{\cosh^2(R^1 + R^2)}\left(\frac{(dR^1)^2}{1 + c} + \frac{(dR^2)^2}{1 - c}\right) \tag{13}$$

where c is an arbitrary constant. It has constant curvature 1 for any value of c. The radii of principal curvature k^1, k^2 satisfy the equations

$$\frac{\partial_2 k^1}{k^2 - k^1} = \frac{\partial_1 k^2}{k^1 - k^2} = -\tanh(R^1 + R^2)$$

which do not explicitly depend on c. Thus, any surface with the third fundamental form (13) possesses S-deformations which depend on one arbitrary constant. The choice

$$k^1 = -k^2 = \cosh^2(R^1 + R^2)$$

corresponds to a minimal surface. We refer to [**19**] for the geometry of this particular example. The discussion of surfaces possessing one-parameter families of S-deformations will be continued in the next section.

Some global aspects of S-deformations were discussed in [**33**], [**34**] where, in particular, the S-rigidity of ovaloids of revolution was established.

3. The Lax pair

In this section we discuss surfaces which possess 1-parameter families of S-deformations. Let $M^2 \in E^3$ be a surface parametrized by coordinates R^1, R^2 of curvature lines with the third fundamental form (1)

$$G_{11}(dR^1)^2 + G_{22}(dR^2)^2.$$

Let k^1, k^2 be the radii of principal curvature satisfying the Codazzi equations

$$\frac{\partial_2 k^1}{k^2 - k^1} = \partial_2 \ln \sqrt{G_{11}}, \quad \frac{\partial_1 k^2}{k^1 - k^2} = \partial_1 \ln \sqrt{G_{22}}. \tag{14}$$

Suppose there exists a flat metric

$$g_{11}(dR^1)^2 + g_{22}(dR^2)^2 \tag{15}$$

such that

$$G_{11} = g_{11}/\eta_1, \quad G_{22} = g_{22}/\eta_2, \tag{16}$$

where η_1, η_2 are functions of R^1, R^2, respectively. Under these assumptions the metric

$$\tilde{G}_{11}(dR^1)^2 + \tilde{G}_{22}(dR^2)^2 = \frac{g_{11}}{\lambda + \eta_1}(dR^1)^2 + \frac{g_{22}}{\lambda + \eta_2}(dR^2)^2 \tag{17}$$

has constant curvature 1 for any λ. Since equations (14) are still true if we replace G_{ii} by $\tilde{G}_{ii}$, we arrive at a 1-parameter family of surfaces M^2_λ with the third fundamental forms (17) (which depend on λ) and the principal curvatures k^1, k^2 (which are independent of λ). Hence, the shape operators of the surfaces M^2_λ coincide. The problem of the classification of surfaces which possess 1-parameter families of S-deformations is thus reduced to the classification of metrics (17) which have constant Gaussian curvature 1 for any λ. Any such metric generates an infinite family of S-deformable surfaces whose principal curvatures k^1, k^2 satisfy (14). In terms of the Lamé coefficients $H_1 = \sqrt{g_{11}}$, $H_2 = \sqrt{g_{22}}$ and the rotation coefficients $\beta_{12} = \partial_1 H_2/H_1$, $\beta_{21} = \partial_2 H_1/H_2$ our problem reduces to the nonlinear system

$$\begin{gathered} \partial_1 H_2 = \beta_{12} H_1, \quad \partial_2 H_1 = \beta_{21} H_2, \\ \partial_1 \beta_{12} + \partial_2 \beta_{21} = 0, \\ \eta_1 \partial_1 \beta_{12} + \eta_2 \partial_2 \beta_{21} + \tfrac{1}{2}\eta_1' \beta_{12} + \tfrac{1}{2}\eta_2' \beta_{21} + H_1 H_2 = 0, \end{gathered} \tag{18}$$

which possesses the Lax pair

$$\partial_1 \psi = \begin{pmatrix} 0 & -\sqrt{\frac{\lambda+\eta_2}{\lambda+\eta_1}}\beta_{21} & \frac{H_1}{\sqrt{\lambda+\eta_1}} \\ \sqrt{\frac{\lambda+\eta_2}{\lambda+\eta_1}}\beta_{21} & 0 & 0 \\ -\frac{H_1}{\sqrt{\lambda+\eta_1}} & 0 & 0 \end{pmatrix} \psi,$$

$$\partial_2 \psi = \begin{pmatrix} 0 & \sqrt{\frac{\lambda+\eta_1}{\lambda+\eta_2}}\beta_{12} & 0 \\ -\sqrt{\frac{\lambda+\eta_1}{\lambda+\eta_2}}\beta_{12} & 0 & \frac{H_2}{\sqrt{\lambda+\eta_2}} \\ 0 & -\frac{H_2}{\sqrt{\lambda+\eta_2}} & 0 \end{pmatrix} \psi.$$

Geometrically, this Lax pair governs infinitesimal displacements of the orthonormal frame of the orthogonal coordinate system on the unit sphere S^2, corresponding to the metric (17). In 2×2 matrices it takes the form

$$2\sqrt{\lambda+\eta_1}\,\partial_1\psi = \begin{pmatrix} i\sqrt{\lambda+\eta_2}\beta_{21} & H_1 \\ -H_1 & -i\sqrt{\lambda+\eta_2}\beta_{21} \end{pmatrix}\psi,$$

$$2\sqrt{\lambda+\eta_2}\,\partial_2\psi = i\begin{pmatrix} -\sqrt{\lambda+\eta_1}\beta_{12} & H_2 \\ H_2 & \sqrt{\lambda+\eta_1}\beta_{12} \end{pmatrix}\psi.$$

Example 8. The case of constant η_i. Let us assume $\eta_1 = -1/2,\ \eta_2 = 1/2$ in which case equations (18) take the form

$$\partial_1 H_2 = \beta_{12}H_1,\ \partial_2 H_1 = \beta_{21}H_2,$$

(19)

$$\partial_1\beta_{12} = H_1H_2,\ \partial_2\beta_{21} = -H_1H_2.$$

It is easily seen that system (19) possesses 2 integrals

$$\partial_1(\beta_{12}^2 - H_2^2) = 0,\ \partial_2(\beta_{21}^2 + H_1^2) = 0$$

implying

$$\beta_{12}^2 - H_2^2 = \mu_2(R^2),\ \beta_{21}^2 + H_1^2 = \mu_1(R^1),$$

where the functions μ_1, μ_2 can be reduced to ± 1 by virtue of the following symmetry of system (19):

$$\tilde{R}^1 = s_1(R^1),\ \tilde{H}_1 = H_1/s_1',\ \tilde{\beta}_{21} = \beta_{21}/s_1',$$

$$\tilde{R}^2 = s_2(R^2),\ \tilde{H}_2 = H_2/s_2',\ \tilde{\beta}_{12} = \beta_{12}/s_2'$$

(here $s_1(R^1),\ s_2(R^2)$ are arbitrary functions). Let us assume, for instance, that

$$\beta_{12}^2 - H_2^2 = 1,\ \beta_{21}^2 + H_1^2 = 1.$$

Introducing the parametrisation

$$H_1 = \sin\psi,\ H_2 = \sinh\varphi,\ \beta_{12} = \cosh\varphi,\ \beta_{21} = \cos\psi,$$

we rewrite (19) in the form

$$\partial_1\varphi = \sin\psi,\ \partial_2\psi = \sinh\varphi,$$

which implies the following integrable Monge-Ampere equations for φ, ψ:

$$\partial_1\partial_2\varphi = \sinh\varphi\sqrt{1-(\partial_1\varphi)^2},\ \partial_1\partial_2\psi = \sin\psi\sqrt{1+(\partial_2\psi)^2}.$$

4. Multidimensional hypersurfaces possessing nontrivial S-deformations

Let M^n be a hypersurface in E^{n+1} parametrised by coordinates R^i of curvature lines. We point out that the generic multidimensional hypersurface does not necessarily possess a curvature line parametrisation. However (at least for $n = 3$), such a parametrisation appears to be a necessary condition for the existence of nontrivial S-deformations. Let $k^i(R)$ be the radii of principal curvature of the hypersurface M^n and $G_{ii}\,(dR^i)^2$ the third fundamental form (which is automatically of constant curvature 1). These objects satisfy the Codazzi equations

$$\frac{\partial_j k^i}{k^j - k^i} = \partial_j \ln\sqrt{G_{ii}},\ i \neq j.$$

We will discuss just one important example.

Example 9. Hyperquadrics are characterized by the third fundamental form

$$\sum_{i=1}^{n} \frac{\prod_{k\neq i}(R^k - R^i)}{4P(R^i)} (dR^i)^2, \tag{20}$$

where $P(R) = \prod_{s=1}^{n+1}(R - a_s)$ is an arbitrary polynomial of order $n + 1$. The corresponding radii of principal curvature k^i satisfy the linear system

$$\frac{\partial_j k^i}{k^j - k^i} = \frac{1}{2(R^j - R^i)}, \ i \neq j,$$

which does not explicitly depend on a_s. Thus, any hypersurface M^n with the third fundamental form (20) possesses an $(n + 1)$-parameter family of S-deformations. The case of a hyperquadric corresponds to the choice

$$k^i = \frac{1}{R^i\sqrt{\prod_1^n R^k}}.$$

It can be demonstrated that quadrics are the only n-dimensional hypersurfaces which possess $(n + 1)$-parameter families of S-deformations and satisfy the "genericity" assumption $\partial_j k^i \neq 0$ for any $i \neq j$ (we recall that for $n = 2$ the condition $\partial_1 k^2 \neq 0$, $\partial_2 k^1 \neq 0$ forbids molding surfaces). Thus, unlike the case $n = 2$, the class of multidimensional hypersurfaces possessing the "maximal" number of S-deformations appears to be very restricted. It is likely that there exist examples of hypersurfaces M^n possessing k-parameter families of S-deformations for any intermediate $k = 1, 2, ..., n + 1$. Numerous examples of that type are provided by coordinate hypersurfaces of the n-orthogonal coordinate systems arising in the theory of multi-Hamiltonian equations of hydrodynamic type.

5. Compatible Poisson brackets of hydrodynamic type. Criterion of compatibility.

In 1983 Dubrovin and Novikov [**8**] introduced the Poisson brackets of hydrodynamic type

$$\{F, G\} = \int \frac{\delta F}{\delta u^i} A^{ij} \frac{\delta G}{\delta u^j}\, dx \tag{21}$$

defined by the Hamiltonian operators A^{ij} of the form

$$A^{ij} = g^{ij}\frac{d}{dx} + b^{ij}_k u^k_x, \ b^{ij}_k = -g^{is}\Gamma^j_{sk}. \tag{22}$$

They proved that in the nondegenerate case (det $g^{ij} \neq 0$) the bracket (21), (22) is skew-symmetric and satisfies the Jacobi identities if and only if the metric g^{ij} (with upper indices) is flat, and Γ^j_{sk} are the Christoffel symbols of the corresponding Levi-Civita connection.

Let us consider another Poisson bracket of hydrodynamic type defined on the same phase space by the Hamiltonian operator

$$\tilde{A}^{ij} = \tilde{g}^{ij}\frac{d}{dx} + \tilde{b}^{ij}_k u^k_x, \ \tilde{b}^{ij}_k = -\tilde{g}^{is}\tilde{\Gamma}^j_{sk}, \tag{23}$$

corresponding to a flat metric $\tilde{g}^{ij}$. Two Poisson brackets (Hamiltonian operators) are called compatible if their linear combinations $\tilde{A}^{ij} + \lambda A^{ij}$ are Hamiltonian as well. This requirement implies, in particular, that the metric $\tilde{g}^{ij} + \lambda g^{ij}$ must be flat for any λ (together with certain additional restrictions). Necessary and sufficient conditions for the compatibility were first formulated by Dubrovin [**11**], [**12**] (see [**25**], [**26**] for further discussion). Below we reformulate these conditions in terms of the operator $r^i_j = \tilde{g}^{is} g_{sj}$ (Theorem 1) which, in particular, imply the vanishing of the Nijenhuis tensor of the operator r^i_j:

$$N^i_{jk} = r^s_j \partial_s r^i_k - r^s_k \partial_s r^i_j - r^i_s(\partial_j r^s_k - \partial_k r^s_j) = 0$$

(see [**14**], [**26**]).

Examples of compatible Hamiltonian pairs arise naturally in the theory of Hamiltonian systems of hydrodynamic type — see for example [**1**], [**21**], [**22**], [**28**], [**29**], [**32**]. Dubrovin developed a deep theory for the particular class of compatible Poisson brackets arising in the framework of the associativity equations [**11**], [**12**]. Compatible Poisson brackets of hydrodynamic type can also be obtained as a result of Whitham averaging (dispersionless limit) from the local compatible Poisson brackets of integrable systems [**8**], [**9**], [**10**], [**37**], [**31**], [**16**], [**35**]. Some further examples and partial classification results can be found in [**13**], [**16**], [**28**], [**30**], [**25**], [**20**].

If the spectrum of r^i_j is simple, the vanishing of the Nijenhuis tensor implies the existence of a coordinate system where both metrics g^{ij} and $\tilde{g}^{ij}$ become diagonal. In these coordinates the compatibility conditions take the form of an integrable reduction of the Lamé equations. We present the corresponding Lax pairs in section 6. Another approach to the integrability of this system was recently proposed by Mokhov [**27**] by an appropriate modification of Zakharov's scheme [**38**].

Our main observation is a relationship between compatible Poisson brackets of hydrodynamic type and hypersurfaces $M^{n-1} \in E^n$ possessing nontrivial S-deformations. In section 7 we demonstrate that the n-orthogonal coordinate systems in E^n corresponding to flat metrics $\tilde{g}^{ij} + \lambda g^{ij}$ (rewritten in the diagonal coordinates) deform with respect to λ in such a way that the shape operators of coordinate hypersurfaces are preserved up to constant scaling factors.

Notice that the operator $r^i_j = \tilde{g}^{is} g_{sj}$ is automatically symmetric:

$$r^i_s g^{sj} = r^j_s g^{si}. \tag{24}$$

Hence $\tilde{g}^{ij} = r^i_s g^{sj} = r^j_s g^{si} = r^{ij}$. In what follows we use the first metric g^{ij} for raising and lowering the indices.

THEOREM 1. [**17**] *The Hamiltonian operators (22), (23) are compatible if and only if the following conditions are satisfied:*

1. The Nijenhuis tensor of r^i_j vanishes:

$$N^i_{jk} = r^s_j \partial_s r^i_k - r^s_k \partial_s r^i_j - r^i_s(\partial_j r^s_k - \partial_k r^s_j) = 0. \tag{25}$$

2. The metric coefficients $\tilde{g}^{ij} = r^{ij}$ satisfy the equations

$$\nabla^i \nabla^j r^{kl} + \nabla^k \nabla^l r^{ij} = \nabla^i \nabla^k r^{jl} + \nabla^j \nabla^l r^{ik}. \tag{26}$$

Here $\nabla^i = g^{is}\nabla_s$ *is the covariant derivative operator corresponding to the metric* g^{ij}. *The vanishing of the Nijenhuis tensor implies the following expression for the coefficients* $\tilde{b}^{ij}_k$ *in terms of* r^i_j*:*

$$2\tilde{b}^{ij}_k = \nabla^i r^j_k - \nabla^j r^i_k + \nabla_k r^{ij} + 2b^{sj}_k r^i_s \tag{27}$$

In a somewhat different form, necessary and sufficient conditions for compatibility were formulated in [**11**], [**12**], [**25**], [**26**].

Remark. The criterion for the compatibility of Hamiltonian operators of hydrodynamic type resembles that of finite-dimensional Poisson bivectors: two skew-symmetric Poisson bivectors ω^{ij} and $\tilde{\omega}^{ij}$ are compatible if and only if the Nijenhuis tensor of the corresponding recursion operator $r^i_j = \tilde{\omega}^{is}\omega_{sj}$ vanishes. We emphasize that in our situation the operator r^i_j does not coincide with the recursion operator.

Remark. If the spectrum of r^i_j is simple the condition (26) is redundant: it is automatically satisfied by virtue of (25) and the flatness of both metrics g and $\tilde{g}$. This was the motivation to drop condition (26) in the compatibility criterion formulated in [**14**]. However, in this general form the criterion proved to be incorrect: recently it was pointed out by Mokhov [**26**] that in the case when the spectrum of r^i_j is not simple the vanishing of the Nijenhuis tensor is no longer sufficient for compatibility.

6. Compatibility conditions in the diagonal form: the Lax pairs

If the spectrum of r^i_j is simple, the vanishing of the Nijenhuis tensor implies the existence of coordinates $R^1, \dots, R^n$ in which the objects r^i_j, g^{ij}, $\tilde{g}^{ij}$ become diagonal. Moreover, the i-th eigenvalue of r^i_j depends on the coordinate R^i only, so that $r^i_j = \mathrm{diag}(\eta_i)$, $g^{ij} = \mathrm{diag}(g^{ii})$, $\tilde{g}^{ij} = \mathrm{diag}(g^{ii}\eta_i)$ where η_i is a function of R^i. This is a generalization of the analogous observation by Dubrovin [**11**] in the particular case of compatible Poisson brackets originating from the theory of the associativity equations. Introducing the Lamé coefficients H_i and the rotation coefficients β_{ij} by the formulae

$$H_i = \sqrt{g_{ii}} = 1/\sqrt{g^{ii}}, \ \partial_i H_j = \beta_{ij} H_i, \tag{28}$$

we can rewrite the zero curvature conditions for the metric g in the form

$$\partial_k \beta_{ij} = \beta_{ik}\beta_{kj}, \tag{29}$$

$$\partial_i \beta_{ij} + \partial_j \beta_{ji} + \sum_{k \neq i,j} \beta_{ki}\beta_{kj} = 0. \tag{30}$$

The zero curvature condition for the metric $\tilde{g}$ imposes the additional constraint

$$\eta_i \partial_i \beta_{ij} + \eta_j \partial_j \beta_{ji} + \frac{1}{2}\eta_i{}' \beta_{ij} + \frac{1}{2}\eta_j{}' \beta_{ji} + \sum_{k \neq i,j} \eta_k \beta_{ki}\beta_{kj} = 0, \tag{31}$$

resulting from (30) after the substitution of the rotation coefficients $\tilde{\beta}_{ij} = \beta_{ij}\sqrt{\eta_i/\eta_j}$ of the metric $\tilde{g}$. As can be readily seen, equations (30) and (31) already imply the compatibility, so that in the diagonalisable case condition (26) of Theorem 1 is indeed superfluous. Solving equations (30), (31) for $\partial_i\beta_{ij}$, we can rewrite (29)–(31) in the form

$$\partial_k\beta_{ij} = \beta_{ik}\beta_{kj},$$

(32)

$$\partial_i\beta_{ij} = \tfrac{1}{2}\frac{\eta_i{}'}{\eta_j - \eta_i}\beta_{ij} + \tfrac{1}{2}\frac{\eta_j{}'}{\eta_j - \eta_i}\beta_{ji} + \textstyle\sum_{k\neq i,j}\frac{\eta_k - \eta_j}{\eta_j - \eta_i}\beta_{ki}\beta_{kj}.$$

It can be verified by a straightforward calculation that system (32) is compatible for any choice of the functions $\eta_i(R^i)$, and its general solution depends on $n(n-1)$ arbitrary functions of one variable (indeed, one can arbitrarily prescribe the value of β_{ij} on the j-th coordinate line). Under the additional "Egorov" assumption $\beta_{ij} = \beta_{ji}$, system (32) reduces to the one studied by Dubrovin in [**12**]. For $n \geq 3$ system (32) is essentially nonlinear. Its integrability follows from the Lax pair [**17**]

$$\partial_j\psi_i = \beta_{ij}\psi_j,\ \partial_i\psi_i = -\frac{\eta_i{}'}{2(\lambda+\eta_i)}\psi_i - \sum_{k\neq i}\frac{\lambda+\eta_k}{\lambda+\eta_i}\beta_{ki}\psi_k \tag{33}$$

with a spectral parameter λ (another demonstration of the integrability of system (33) has been proposed recently in [**27**] by an appropriate modification of Zakharov's approach [**38**]).

Remark. In fact, the Lax pair (33) is gauge-equivalent to the equations

$$(\tilde g^{ik} + \lambda g^{ik})\partial_k\partial_j\psi + (\tilde b^{ik}_j + \lambda b^{ik}_j)\partial_k\psi = 0$$

for the Casimirs $\int\psi dx$ of the Hamiltonian operator $\tilde A^{ij} + \lambda A^{ij}$.

After the gauge transformation $\psi_i = \varphi_i/\sqrt{\lambda+\eta_i}$ the Lax pair (33) assumes the manifestly skew-symmetric form

$$\partial_j\varphi_i = \sqrt{\frac{\lambda+\eta_i}{\lambda+\eta_j}}\beta_{ij}\varphi_j,\ \partial_i\varphi_i = -\sum_{k\neq i}\sqrt{\frac{\lambda+\eta_k}{\lambda+\eta_i}}\beta_{ki}\varphi_k. \tag{34}$$

Thus, we can introduce an orthonormal frame $\vec\varphi_1, ..., \vec\varphi_n$ in the Euclidean space E^n satisfying the equations

$$\partial_j\vec\varphi_i = \sqrt{\frac{\lambda+\eta_i}{\lambda+\eta_j}}\beta_{ij}\vec\varphi_j,\ \partial_i\vec\varphi_i = -\sum_{k\neq i}\sqrt{\frac{\lambda+\eta_k}{\lambda+\eta_i}}\beta_{ki}\vec\varphi_k,\ (\vec\varphi_i,\vec\varphi_j) = \delta_{ij}. \tag{35}$$

Let us introduce a vector $\vec r$ such that

$$\partial_i\vec r = \frac{H_i}{\sqrt{\lambda+\eta_i}}\vec\varphi_i$$

(the compatibility of these equations can be readily verified). In view of the formula

$$(\partial_i\vec r, \partial_j\vec r) = \frac{H_i^2}{\lambda+\eta_i}\delta_{ij}$$

the radius-vector $\vec r$ describes the n-orthogonal coordinate system in E^n corresponding to the flat metric

$$\sum_i \frac{H_i^2}{\lambda+\eta_i}(dR^i)^2.$$

Geometrically, the $\vec\varphi_i$ are the unit vectors along the coordinate lines of this n-orthogonal system.

Let us discuss in more detail the case $\eta_i = \text{constant} = c_i$, in which system (32) takes the form

$$\partial_k\beta_{ij} = \beta_{ik}\beta_{kj},$$

(36)

$$\partial_i\beta_{ij} = \sum\nolimits_{k\neq i,j} \frac{c_k - c_j}{c_j - c_i}\beta_{ki}\beta_{kj}.$$

One can verify that the quantity $P_i = \sum_{k\neq i}(c_k - c_i)\beta_{ki}^2$ is an integral of system (36), namely, $\partial_j P_i = 0$ for any $i \neq j$, so that P_i is a function of R^i. Utilising the obvious symmetry $R^i \to s_i(R^i)$, $\beta_{ki} \to \beta_{ki}/s_i'(R^i)$ of system (36), we can reduce P_i to ± 1 (if nonzero). Let us consider the simplest nontrivial case $n = 3$, $P_1 = P_2 = 1$, $P_3 = -1$:

$$P_1 = (c_2 - c_1)\beta_{21}^2 + (c_3 - c_1)\beta_{31}^2 = 1,$$

$$P_2 = (c_1 - c_2)\beta_{12}^2 + (c_3 - c_2)\beta_{32}^2 = 1,$$

$$P_3 = (c_1 - c_3)\beta_{13}^2 + (c_2 - c_3)\beta_{23}^2 = -1.$$

Assuming $c_3 > c_2 > c_1$ and introducing the parametrisation

$$\beta_{21} = \sin p/\sqrt{c_2 - c_1},\ \beta_{31} = \cos p/\sqrt{c_3 - c_1},$$

$$\beta_{12} = \sinh q/\sqrt{c_2 - c_1},\ \beta_{32} = \cosh q/\sqrt{c_3 - c_2},$$

$$\beta_{13} = \sin r/\sqrt{c_3 - c_1},\ \beta_{23} = \cos r/\sqrt{c_3 - c_2},$$

we readily rewrite (36) in the form

$$\partial_1 q = \mu_1 \cos p,\ \partial_1 r = -\mu_1 \sin p,$$

$$\partial_2 p = -\mu_2 \cosh q,\ \partial_2 r = \mu_2 \sinh q,$$

$$\partial_3 p = \mu_3 \cos r,\ \partial_3 q = \mu_3 \sin r,$$

where

$$\mu_1 = \sqrt{\frac{c_3 - c_2}{(c_2 - c_1)(c_3 - c_1)}},\ \mu_2 = \sqrt{\frac{c_3 - c_1}{(c_2 - c_1)(c_3 - c_2)}},\ \mu_3 = \sqrt{\frac{c_2 - c_1}{(c_3 - c_1)(c_3 - c_2)}}.$$

After rescaling, this system simplifies to

$$\partial_1 q = \cos p,\ \partial_1 r = -\sin p,$$

(37)

$$\partial_2 p = -\cosh q,\ \partial_2 r = \sinh q,$$

$$\partial_3 p = \cos r,\ \partial_3 q = \sin r.$$

Expressing p and r in the form $p = \arccos \partial_1 q$, $r = \arcsin \partial_3 q$, we can rewrite (37) as a triple of pairwise commuting Monge-Ampere equations

$$\partial_1\partial_2 q = \cosh q\sqrt{1 - \partial_1 q^2},$$

$$\partial_1\partial_3 q = -\sqrt{1 - \partial_1 q^2}\sqrt{1 - \partial_3 q^2},$$

$$\partial_2\partial_3 q = \sinh q\sqrt{1 - \partial_3 q^2}.$$

Similar triples of Monge-Ampere equations were obtained in [**15**] in the classification of quadruples of 3×3 hydrodynamic type systems which are closed under the Laplace transformations. However, there is no understanding of this coincidence at the moment. Notice that the coordinate surfaces of these 3-orthogonal coordinate systems are of the type discussed in example 8.

7. Deformations of n-orthogonal coordinate systems inducing rescalings of shape operators of the coordinate hypersurfaces

We have demonstrated in section 6 that the radius-vector $\vec{r}(R^1, ..., R^n)$ of the n-orthogonal coordinate system in E^n corresponding to the flat diagonal metric $\sum_i \frac{H_i^2}{\lambda+\eta_i}(dR^i)^2$ satisfies the equations

$$\partial_i \vec{r} = \frac{H_i}{\sqrt{\lambda+\eta_i}} \vec{\varphi}_i,$$

where the infinitesimal displacements of the orthonormal frame $\vec{\varphi}_i$ are governed by

$$\partial_j \vec{\varphi}_i = \sqrt{\frac{\lambda+\eta_i}{\lambda+\eta_j}} \beta_{ij} \vec{\varphi}_j, \ \partial_i \vec{\varphi}_i = -\sum_{k \neq i} \sqrt{\frac{\lambda+\eta_k}{\lambda+\eta_i}} \beta_{ki} \vec{\varphi}_k.$$

Since our formulae depend on the spectral parameter λ, we may speak of the "deformation" of the n-orthogonal coordinate system. To investigate this deformation in more detail, we fix a coordinate hypersurface $M^{n-1} \subset E^n$ (say, $R^n =$ constant). Its radius-vector $\vec{r}$ and the unit normal $\vec{\varphi}_n$ satisfy the Weingarten equations

$$\partial_i \vec{\varphi}_n = \frac{\beta_{ni}}{H_i} \sqrt{\lambda+\eta_n}\ \partial_i \vec{r}, \ i = 1, ..., n-1,$$

implying that

$$k^i = \frac{\beta_{ni}}{H_i} \sqrt{\lambda+\eta_n}$$

are principal curvatures of the hypersurface M^{n-1}. Since η_n is constant on M^{n-1}, our deformation preserves the shape operator of M^{n-1} up to a constant scaling factor $\sqrt{\lambda+\eta_n}$ (we point out that the curvature line parametrisation $R^1, ..., R^{n-1}$ is preserved by construction). Thus, compatible Poisson brackets of hydrodynamic type give rise to deformations of n-orthogonal systems in E^n which, up to scaling factors, preserve shape operators of the coordinate hypersurfaces. If we follow the evolution of a particular coordinate hypersurface M^{n-1}, this scaling factor can be eliminated by a homothetic transformation of the ambient space E^n, so that we arrive at the nontrivial deformation which preserves the shape operator. However, this scaling factor cannot be eliminated for all coordinate hypersurfaces simultaneously.

8. Concluding remarks

It seems to be an interesting and nontrivial problem to classify hypersurfaces $M^n \in E^{n+1}$ possessing S-deformations which depend on k essential parameters where k varies from 2 to n (the limiting cases $k = n+1$ and $k = 1$ reduce to hyperquadrics and generic S-deformable surfaces, respectively). Even the simplest nontrivial case $n = 2$, $k = 2$ is not understood at the moment (as I learned recently, some classification results in this direction will be published shortly by R. Bryant).

References

[1] M. Arik et al, Multi-Hamiltonian structure of the Born-Infeld equation, J. Math. Phys., 30 (1989) 1338–1344.

[2] A.I. Bobenko and U. Either, Bonnet surfaces and Painlevé equations, J. Reine Angew. Math. 499 (1998) 47-79.

[3] O. Bonnet, Theorie des surfaces applicables sur une surface donnee, J. de l'Ecole Polytech. 24 (1865) 209.

[4] O. Bonnet, Memoire sur la theorie des surfaces applicables, J. de l'Ecole Polytech. 25 (1867) 1-151.

[5] R.L. Bryant, On surfaces with prescribed shape operator, math.DG/0107083.

[6] R.L. Bryant, S.S. Chern and and P.A. Griffiths, Exterior differential systems, Proc. Beijing Symp. Diff. Geom. and Diff. Eq., Beijing,; N.Y. Vol. 1 (1982) 219-338.

[7] E. Cartan, Les systèmes différentiels extérieurs et leurs applications géométriques, Hermann, Paris, 1945, 214 pp.; 2nd ed., 1971, 210 pp.

[8] B.A. Dubrovin and S.P. Novikov, Hamiltonian formalism of one-dimensional systems of hydrodynamic type and the Bogolyubov-Whitham averaging method, Dokl. Akad. Nauk SSSR, 270, N.4 (1983) 781–785; (Soviet Math. Dokl., 27 (1983) 665–669).

[9] B.A. Dubrovin and S.P. Novikov, On Poisson brackets of hydrodynamic type, Dokl. Akad. Nauk SSSR, 279, N.2 (1984) 294–297; (Soviet Math. Dokl., 30 (1984) 651–654).

[10] B.A. Dubrovin and S.P. Novikov, Hydrodynamics of weakly deformed soliton lattices. Differential geometry and Hamiltonian theory, Usp. Mat. Nauk, 44, N.6 (1989) 29–98; (Russian Math. Surveys, 44:6 (1989) 35–124).

[11] B.A. Dubrovin, Differential geometry of the space of orbits of a Coxeter group, Preprint SISSA, Trieste, 1993, hep-th/9303152.

[12] B.A. Dubrovin, Geometry of 2D topological field theories, Lecture Notes in Math. 1620, Springer-Verlag (1996) 120-348.

[13] E.V. Ferapontov, Hamiltonian systems of hydrodynamic type and their realization on hypersurfaces of a pseudoeuclidean space, Geom. Sbornik, VINITI, 22 (1990) 59–96; (Soviet J. of Math., 55 (1991) 1970–1995).

[14] E.V. Ferapontov, Nonlocal Hamiltonian operators of hydrodynamic type: differential geometry and applications, Amer. Math. Soc. Transl. (2), 170 (1995) 33-58.

[15] E.V. Ferapontov, Laplace transformations of hydrodynamic type systems in Riemann invariants: periodic sequences, J. Phys. A: Math. Gen. 30 (1997) 6861-6878.

[16] E.V. Ferapontov and M.V. Pavlov, Quasiclassical limit of Coupled KdV equations. Riemann invariants and multi-Hamiltonian structure, Physica D 52 (1991) 211–219.

[17] E.V. Ferapontov, Compatible Poisson brackets of hydrodynamic type, J. Phys. A: Math. Gen. 34 (2001) 2377-2388.

[18] S.P. Finikoff, Surfaces dont les lignes de courbure se correspondent avec égalité des rayons de courbure principaux homologues, C.R. Acad. Sci., Paris 197 (1933) 984-986.

[19] S.P. Finikoff et B. Gambier, Surfaces dont les lignes de courbure se correspondent avec égalité des rayons de courbure principaux, Ann. Ecole Norm., III. Ser. 50 (1933) 319–370.

[20] A.P. Fordy and O.I. Mokhov, On a special class of compatible Poisson structures of hydrodynamic type, (2000).

[21] H. Gümral and Y. Nutku, Multi-Hamiltonian structure of equations of hydrodynamic type, J. Math. Phys., 31 (1990) 2606–2611.

[22] H. Gümral and Y. Nutku, Bi-Hamiltonian structures of d-Boussinesq and Benney-Lax equations, Phys. letters A: Math. Gen. 27 (1994) 193–200.

[23] S.P. Burtsev, V.E. Zakharov and A.V. Mikhailov, Inverse scattering method with variable spectral parameter, Teoreticheskaya i Matematicheskaya Fizika, 70, N3 (1987) 227-240.

[24] O.I. Mokhov and E.V. Ferapontov, Nonlocal Hamiltonian operators of hydrodynamic type associated with constant curvature metrics, Usp. Mat. Nauk, 45, N.3 (1990) 191–192.

[25] O.I. Mokhov, Compatible Poisson structures of hydrodynamic type and associativity equations, Proceedings of the Steklov Institute of Mathematics, 225 (1999) 269-284.

[26] O.I. Mokhov, Compatible and almost compatible pseudo-Riemannian metrics, arXiv.math.DG/0005051.

[27] O.I. Mokhov, On integrability of the equations for nonsingular pairs of compatible flat metrics, arXiv.math.DG/0005081.

[28] Y. Nutku, On a new class of completely integrable nonlinear wave equations. Multi-Hamiltonian structure, J. Math. Phys., 28 (1987) 2579–2585.

[29] P. Olver and Y. Nutku, Hamiltonian structures for systems of hyperbolic conservation laws, J. Math. Phys., 29 (1988) 1610–1619.

[30] M.V. Pavlov, Discrete symmetry and local Hamiltonian structures of systems of hydrodynamic type, Uspekhi Mat. Nauk 48 (1993), no. 6(294) 167–168.

[31] M.V. Pavlov, Multi-Hamiltonian structures of the Whitham equations. Dokl. Akad. Nauk 338 (1994), no. 2, 165–167; translation in Russian Acad. Sci. Dokl. Math. 50 (1995), no. 2, 220–223.

[32] M.V. Pavlov, Local Hamiltonian structures of Benney's system. Dokl. Akad. Nauk 338 (1994), no. 1, 33–34; translation in Phys. Dokl. 39 (1994), no. 9, 607–608.

[33] U. Simon, K. Voss, L. Vrancken and M. Wiehe, Surfaces with prescribed Weingarten-Operator, Preprint 2001.

[34] U. Simon, L. Vrancken, C.P. Wang and M. Wiehe, Intrinsic and extrinsic geometry of ovaloids and rigidity, Geometry and Topology of Submanifolds X, World Scientific, Singapore, 284-293, 2000.

[35] I.A.B. Strachan, Degenerate Frobenius manifolds and the bi-Hamiltonian structure of rational Lax equations, Journal of Mathematical Physics, 49, N10 (1999) 5058-5079.

[36] S.P. Tsarev, On Poisson brackets and one-dimensional Hamiltonian systems of hydrodynamic type, Dokl. Acad. Nauk SSSR, 282, N.3 (1985) 534–537; (Soviet Math. Dokl., 31 (1985) 488–491).

[37] S.P. Tsarev, The geometry of Hamiltonian systems of hydrodynamic type. The generalized hodograph method, Izv. Acad. Nauk SSSR, Ser. Mat. 54, N.5 (1990) 1048–1068 (Math. USSR Izvestiya 37, N.2 (1991) 397–419).

[38] V.E. Zakharov, Description of the n-orthogonal curvilinear coordinate systems and Hamiltonian integrable systems of hydrodynamic type, I: Integration of the Lamé equations, Duke Math. Journal 94, N1 (1998) 103-139.

DEPARTMENT OF MATHEMATICAL SCIENCES, LOUGHBOROUGH UNIVERSITY, LOUGHBOROUGH, LEICESTERSHIRE LE11 3TU, UK AND CENTRE FOR NONLINEAR STUDIES, LANDAU INSTITUTE OF THEORETICAL PHYSICS, ACADEMY OF SCIENCE OF RUSSIA, KOSYGINA 2, 117940 MOSCOW, GSP-1, RUSSIA

E-mail address: E.V.Ferapontov@lboro.ac.uk

Contemporary Mathematics
Volume **308**, 2002

Hamiltonian Stationary Lagrangian Surfaces in Hermitian Symmetric Spaces

Frédéric Hélein and Pascal Romon

Introduction

This paper is the third of a series on Hamiltonian stationary Lagrangian surfaces. We present here the most general theory, valid for any Hermitian symmetric target space, thus generalizing the $\mathbb{C}^2$ case analyzed in [**HR1**] and [**HR2**].

Hamiltonian stationary Lagrangian submanifolds are Lagrangian submanifolds of a given symplectic manifold M which are critical points of the volume functional with respect to a special class of infinitesimal variations preserving the Lagrangian constraint, namely the class of compactly supported Hamiltonian vector fields. This makes sense if M is not only symplectic but also Riemannian, in particular if M is a Kähler manifold. The Euler-Lagrange equation has then a particularly elegant formulation in terms of the so-called *Lagrangian angle* β, a $\mathbb{R}/2\pi\mathbb{Z}$-valued function defined along any Lagrangian submanifold. We may think of β as part of the Gauss map. A Lagrangian surface is Hamiltonian stationary if and only if β is harmonic.

This problem offers a nice generalization of the minimal surface problem. It shares even more similarities with the constant mean curvature surface problem, since the Lagrangian constraint could be thought as a replacement for the volume constraint. Other motivations for this problem come from isoperimetric type inequalities [**O**], incompressible elasticity [**Wo, HR1**] and the study of *special Lagrangian* surfaces [**HaL**] (these are Lagrangian surfaces in a Calabi-Aubin-Yau 4-manifold with *constant* Lagrangian angle function β). In [**ScWo**], R. Schoen and J. Wolfson gave the first analytical proof of the existence of special Lagrangian surfaces via the construction of a Hamiltonian stationary Lagrangian submanifold. These questions are also strongly motivated by string theory and branes theory, where special Lagrangian submanifolds play an important role.

In [**HR1**] we considered Hamiltonian stationary Lagrangian surfaces in $\mathbb{R}^4 \simeq \mathbb{C}^2$ and showed that this problem is completely integrable, like the famous KdV equation or, in differential geometry, the equation for harmonic maps from a surface to a symmetric space. Our formalism was similar to the harmonic map theory,

---

2000 *Mathematics Subject Classification.* Primary 53C55; Secondary 53C42, 53C25, 58E12.

except that the connection form was not of the form $\alpha_\lambda := \lambda^{-1}\alpha_1' + \alpha_0 + \lambda\alpha_1''$ but $\alpha_\lambda := \lambda^{-2}\alpha_2' + \lambda^{-1}\alpha_{-1} + \alpha_0 + \lambda\alpha_1 + \lambda^2\alpha_2''$, where λ is a complex (spectral) parameter. The equations being slightly more linear than in the harmonic maps context, we were able to simplify the integrable system theory and propose a Weierstrass type representation formula much simpler than the one constructed by J. Dorfmeister, F. Pedit and H. Y. Wu in [**DPW**]. In [**HR2**] we further simplified these formulae using quaternions and compared them with a similar formula due to B. G. Konopelchenko, suggesting a hidden structure of spinors. Using these formulae, H. Anciaux recently obtained an estimate which supports Oh's conjecture [**A**].

In this paper, we shall show that the integrable system structure — with a family of connections of the type $\alpha_\lambda := \lambda^{-2}\alpha_2' + \lambda^{-1}\alpha_{-1} + \alpha_0 + \lambda\alpha_1 + \lambda^2\alpha_2''$ — persists if one replaces the ambient space by any two-dimensional Hermitian symmetric space. It should be noted that such a space is automatically Kähler-Einstein, and comes in five types, namely : $\mathbb{R}^4 \simeq \mathbb{C}^2$, $\mathbb{CP}^2$ and its dual the complex hyperbolic space, $\mathbb{CP}^1 \times \mathbb{CP}^1$ and its dual. These manifolds share a crucial algebraic property: the existence of an order four automorphism τ. The specific properties of τ will be described below. As an application we show that the theory in [**DPW**] can be generalized (fully for compact spaces, partially with local versions for noncompact spaces as in [**H**]) and that conformal parametrizations of such surfaces are constructed using holomorphic data. There is no doubt that it is also possible to build a theory of finite type solutions (in particular for tori) using our equations, as was done in detail in [**HR1**]. However, unlike in the harmonic map setting, the choice of appropriate moving frames plays a key role in our theory. We call them *Lagrangian framings*. We show how they help to characterize the Lagrangian angle and describe their properties in great detail in section 1.2. Furthermore, the connection form α_λ is significantly different from the harmonic map case. We may finally observe that this connection mixes spinor-like quantities (α_{-1}, α_1) and non spinor-like ones (α_{-2}, α_0), and ask whether there is some supersymmetric interpretation of that.

By the same token, we can extend our description to Hamiltonian stationary Lagrangian cones in $\mathbb{C}^3$, since their links (intersection with the sphere S^5) are stationary Legendrian surfaces which project down to Hamiltonian stationary Lagrangian surfaces in $\mathbb{CP}^2$ by the Hopf fibration. We will give both a theoretical construction (in section 1.3) and a practical one (in section 4.3). Note that special Lagrangian cones in $\mathbb{C}^3$ with toric links have been constructed by M. Haskins in [**Has**].

1. Moving frames and Lagrangian lifts

1.1. Lagrangian framings of Lagrangian surfaces in a Kähler 4-manifold. Let M be a Kähler 4-manifold with complex structure J, $U(M)$ the principal $U(2)$-bundle of unitary frames on M and $P = U(M)/SU(2)$ the quotient bundle, indeed a principal $U(1)$-bundle associated to the canonical line bundle K on M, with the induced metric. Let $f : L \to M$ be an immersion of a surface into M. The surface is Lagrangian if $f^*\omega = 0$ where ω is the Kähler form; that amounts to saying that for any $p \in M$, the tangent plane T_pL at p to L is mapped by J to its (Riemannian) orthogonal complement or, in other words, that any Riemannian-orthonormal basis (e_1, e_2) of T_pL is actually Hermitian-orthonormal in TM.

Hence for any $p \in L$ the set of all Riemannian orthonormal bases of the tangent plane T_pL can be described as being $SO(2)_*(e_1, e_2)$ — i.e. some orbit of the action of $SO(2)$ — and this set is a subset of the set of all Hermitian-orthogonal bases at p, i.e. the orbit $U(2)_*(e_1, e_2)$. In between lies the orbit $SU(2)_*(e_1, e_2)$, which is what interests us.

DEFINITION 1.1. A (local) framing of the immersion f is a (local) section of the unitary frame bundle $f^*U(M)$. A framing F of f is called *Lagrangian*[1] if for any $p \in L$, $F(p)$ is equivalent mod $SU(2)$ to an orthonormal basis $(e_1(p), e_2(p))$ of T_pL.

Note that there may not exist global sections (Lagrangian or not) of $f^*U(M)$, except in coordinate charts. However it is clear that all local Lagrangian framings lift a single section of the quotient bundle P by the projection map and because of this uniqueness we obtain a globally defined section of P.

DEFINITION 1.2. The image of a Lagrangian framing by the quotient map $U(M) \to P$ is called a *tangent section* of the bundle P.

If M is Kähler-Einstein, we may describe Lagrangian framings more precisely. We follow here the exposition in [**Wo**] specialized to symmetric spaces. On the pull-back bundle f^*K of the canonical bundle, with the associated pull-back metric, the first Chern form is

$$f^*c_1 = f^*\mathrm{Ric} = Rf^*\omega = 0.$$

(Here Ric denotes the Ricci form and R the scalar curvature.) Hence f^*K is flat and we can construct (local) parallel sections. Any parallel unit section s_0 also defines a trivialization of f^*P, in which the tangent section is described by a unit complex number $e^{i\beta}$. In other words, if we pick any orthonormal framing e_1, e_2 of TL, then $e^{i\beta} = s_0(e_1, e_2)$. The real valued function β is called the Lagrangian angle[2]. A different choice of s_0 will only change β by a constant. A Lagrangian framing is a lift with the same Lagrangian angle as the tangent section. Finally we define the Maslov form as $\Theta = \frac{1}{\pi}d\beta$, and this is automatically closed. It is useful to have another expression of the Maslov form, namely $\Theta = \frac{1}{\pi}\iota_H\omega$ where H is the mean curvature vector field. Hence Lagrangian surfaces with constant Lagrangian angle are automatically minimal and vice-versa. If furthermore the canonical bundle on M is flat (as for $\mathbb{C}^2$ or more generally for any Calabi-Aubin-Yau manifold), there is a globally defined Lagrangian angle obtained by pulling back a global parallel unit section of K (e.g. $dz^1 \wedge dz^2$ for $\mathbb{C}^2$). In that particular setting, minimal Lagrangian surfaces are calibrated, and thus minimizing; they are called *special Lagrangian* (see [**Wo, HaL**]).

1.2. Symmetric space structure. Let us now consider a Hermitian symmetric space of the form $M = G/H$, where G is the group of unitary transformations

[1]This choice ought to be compared to Darboux framings in the Riemannian setting, for a submanifold N^n of M^m: a Darboux framing is an orthonormal framing such that the first n vectors span the tangent space of N, i.e. they are in the same class of the frame bundle modulo $SO(n) \times SO(m-n)$. Just as Darboux framings give insight into the Riemannian structure, Lagrangian framings yield valuable information on the Lagrangian immersion.

[2]We differ from Wolfson [**Wo**] in that β is defined mod 2π instead of mod 2.

of M and H the isotropy group at some point henceforth denoted by p_0. The symmetry around p_0 is induced by a group involution σ of G with $(G^\sigma)_0 \subset H \subset G^\sigma$. Abusing notation, we will also write σ for $d\sigma(e)$, its differential at the identity, acting on $T_eG = \mathfrak{g}$, and similarly, thinking of elements of G as matrices (which indeed they will be) we will identify $dg(p)$ with g.

The Lie algebra $\mathfrak{g}$ is the direct sum of two subspaces $\mathfrak{h} \oplus \mathfrak{m}$, eigenspaces of σ with respective eigenvalues $+1$ and -1 (so that $\mathfrak{h}$ is a Lie subalgebra, the Lie algebra of H). The subspace $\mathfrak{m}$ can be identified with $T_{p_0}M$. As a consequence, $\mathfrak{m}$ inherits a Hermitian structure. The reader should consult [**BRa**] for a more thorough introduction to homogeneous spaces from a geometer's point of view. Fixing a unitary frame (ϵ_1, ϵ_2) of $\mathfrak{m}$, any element $g \in G$ maps that reference frame at p_0 to another unitary frame (e_1, e_2) at $p = g \cdot p_0$ through its differential g $(= dg(e))$.

Suppose now that we can lift the map $f : L \to M$ to $F : L \to G$ (assume L is a contractible domain). Any choice of F yields a moving frame[3]. The choice is *a priori* wide and we have a gauge group $C^\infty(L, H)$ acting on lifts. In order for F to encode relevant information on the first derivatives of f, we only use Lagrangian lifts, i.e., Lagrangian framings in the sense of definition 1.1. The existence of such lifts will be clear from the analysis below. Consider for the moment any lift F and $\alpha = F^{-1}dF$ the associated (left) Maurer-Cartan form (the pull-back of the Maurer-Cartan form on G). Using the above symmetric splitting, we write $\alpha = \alpha_\mathfrak{h} + \alpha_\mathfrak{m}$; the $\mathfrak{m}$-valued 1-form $\alpha_\mathfrak{m}$ allows us to identify tangent vectors to $L \subset M$ with elements of $\mathfrak{m}$. However this identification is gauge dependent: if we replace F with Fh^{-1} where $h \in C^\infty(L, H)$, $\alpha_\mathfrak{m}$ changes to $\operatorname{Ad} h(\alpha_\mathfrak{m})$. Using a reference unitary frame (ϵ_1, ϵ_2), there exist at any $z \in L$ a unique element $a(z) \in GL(\mathfrak{m})$ such that

$$\alpha_\mathfrak{m}\big|_z = a(z)(\epsilon_1 dx + \epsilon_2 dy). \tag{1}$$

Without loss of generality, we assume that f is conformal (and that L is endowed with the induced conformal structure making it a Riemann surface; $z = x + iy$ is always a local holomorphic coordinate). Since (x, y) are conformal coordinates and L is Lagrangian, (1) can be written as

$$\alpha_\mathfrak{m}\big|_z = e^{\rho(z)} k(z)(\epsilon_1 dx + \epsilon_2 dy)\ , \quad k(z) \in U(\mathfrak{m}) \tag{2}$$

where $e^{\rho(z)}$ is the conformal factor[4].

To understand the notion of Lagrangian lift we need to delve deeper into the structure of G and $\mathfrak{g}$. First, as we shall see in later sections, there exists an order four automorphism τ acting on G, squaring to $\tau^2 = \sigma$. Thus we may split $\mathfrak{g}^\mathbb{C}$ as $\mathfrak{g}_0^\mathbb{C} \oplus \mathfrak{g}_2^\mathbb{C} \oplus \mathfrak{g}_{-1}^\mathbb{C} \oplus \mathfrak{g}_1^\mathbb{C}$, where $\mathfrak{g}_a^\mathbb{C}$ is the i^a-eigenspace of τ. Since $\tau^2 = \sigma$, $\mathfrak{h}^\mathbb{C} = \mathfrak{g}_0^\mathbb{C} \oplus \mathfrak{g}_2^\mathbb{C}$ and $\mathfrak{m}^\mathbb{C} = \mathfrak{g}_{-1}^\mathbb{C} \oplus \mathfrak{g}_1^\mathbb{C}$. Let G_0 be the subgroup of G fixed by τ, whose Lie algebra is of course $\mathfrak{g}_0$. Recall that the adjoint action of H on $\mathfrak{m}$ can be identified with the action of the linear isotropy $H^* = \{dh(p_0), h \in H\}$ on $T_{p_0}M$; in particular, since all elements of G are unitary transformations of M, $\operatorname{Ad} H$ can be identified with a subgroup of $U(\mathfrak{m})$ (and $\operatorname{ad}\mathfrak{h}$ with a subalgebra of $\mathfrak{u}(\mathfrak{m})$). Moreover, as we shall see later on, there is always an element $Y \in \mathfrak{h}$ such that $\exp(\frac{\pi}{2} \operatorname{ad} Y) = \operatorname{Ad}\exp(\frac{\pi}{2} Y)$

[3]The converse is not always true: one cannot always find a lift corresponding to a given moving frame; for instance one can if H is four dimensional (as for $\mathbb{CP}^2$) but not in general.

[4]If $\operatorname{Ad} H$ is big enough — namely transitive on orthonormal couples — then a suitable gauge change reduces $k(z)$ to the identity; such a lift is called *fundamental* (as for $\mathbb{C}^2$, see [**HR1**]). However that notion is dependent on the coordinate z, unlike the notion of *Lagrangian lift*.

is the complex structure J on $\mathfrak{m}$ and the automorphism τ can be chosen so that $\mathfrak{g}_2 = \mathbb{R}Y$ (and $\operatorname{Ad} G_0 \subset SU(\mathfrak{m})$).

In order to picture the moving frames, remember that the tangent bundle TM is canonically diffeomorphic to the subbundle $[\mathfrak{m}]$ of $M \times \mathfrak{g}$ with fiber $\operatorname{Ad} g(\mathfrak{m})$ over the point $g \cdot p_0$. The moving frame induced by F is $(\operatorname{Ad} F(\epsilon_1), \operatorname{Ad} F(\epsilon_2))$; and it is clear now that two lifts F, F' of f define the same moving frame mod $SU(2)$ if and only if they are gauge equivalent modulo the restricted gauge group $C^\infty(L, G_0)$. In these local coordinates a unitary framing $(e_1, e_2) \simeq \left(e^{-\rho}\frac{\partial f}{\partial x}, e^{-\rho}\frac{\partial f}{\partial y}\right)$ can be identified (in $[\mathfrak{m}]$) with

$$\operatorname{Ad} F\left(e^{-\rho}\alpha_{\mathfrak{m}}\left(\frac{\partial}{\partial x}\right), e^{-\rho}\alpha_{\mathfrak{m}}\left(\frac{\partial}{\partial y}\right)\right) \simeq \operatorname{Ad} F(z)\,(k(z)\epsilon_1, k(z)\epsilon_2)$$

using (2) — the later expressions being (inconspicuously) independent of the choice of the lift F. The tangent section is the class mod $SU(2)$ of (e_1, e_2). The lift F is Lagrangian if and only if its induced moving frame lies in the same class as (e_1, e_2). That requires exactly that k be special unitary ($\det_{\mathbb{C}} k(z) = 1$), which is a very simple condition to check. Such lifts do exist on contractible domains since a gauge change $F \to F\exp(-\theta Y)$ multiplies $\det_{\mathbb{C}} k$ by $\det_{\mathbb{C}} \operatorname{Ad}\exp(\theta Y) = e^{2i\theta}$; we may thus choose θ to force $k \in SU(\mathfrak{m})$. Notice that

(i) no integrability condition is involved in this process,
(ii) θ is defined mod π, so that the process can be applied to non simply-connected domains, by taking a suitable multiple covering (namely a double covering for each generator).

How does one read the Lagrangian angle now? First we have to build a parallel lift $\bar{F}$; recall that the covariant derivative is simply the flat differentiation followed by the projection on the bundle $[\mathfrak{m}]$, that we denote here by $\lfloor \ldots \rfloor_{[\mathfrak{m}]}$, then we have for $i = 1, 2$ (denoting $\bar{\alpha} = \bar{F}^{-1}d\bar{F}$)

$$\begin{aligned} \nabla \bar{e}_i &= \lfloor d(\operatorname{Ad}\bar{F}(\epsilon_i))\rfloor_{[\mathfrak{m}]} = \lfloor \operatorname{Ad}\bar{F}([\bar{\alpha}, \epsilon_i])\rfloor_{[\mathfrak{m}]} = \operatorname{Ad}\bar{F}(\lfloor([\bar{\alpha}, \epsilon_i])\rfloor_{\mathfrak{m}}) \\ &= \operatorname{Ad}\bar{F}([\bar{\alpha}_{\mathfrak{h}}, \epsilon_i]) = [\operatorname{Ad}\bar{F}(\bar{\alpha}_{\mathfrak{h}}), \bar{e}_i] \end{aligned}$$

so that the unitary frame $\bar{s} := (\bar{e}_1, \bar{e}_2)$ varies according to $\nabla \bar{s} = \zeta(s)$ with $\zeta : V \mapsto [\operatorname{Ad}\bar{F}(\bar{\alpha}_{\mathfrak{h}}), V]$. The class modulo $SU(2)$ of s varies following the class modulo $\mathfrak{su}(2)$ of ζ, i.e. it is the projection of ζ on the diagonal (central) part $\mathfrak{z}(\operatorname{Ad}\bar{F}(\mathfrak{m}))$ in the Lie algebra decomposition $\mathfrak{u}(\operatorname{Ad}\bar{F}(\mathfrak{m})) = \mathfrak{z}(\operatorname{Ad}\bar{F}(\mathfrak{m})) \oplus \mathfrak{su}(\operatorname{Ad}\bar{F}(\mathfrak{m}))$. Since

$$\lfloor\zeta\rfloor_{\mathfrak{z}(\operatorname{Ad}\bar{F}(\mathfrak{m}))} = \operatorname{Ad}\bar{F}\,\lfloor \operatorname{Ad}\bar{F}^{-1}\zeta \operatorname{Ad}\bar{F}\rfloor_{\mathfrak{z}(\mathfrak{m})} \operatorname{Ad}\bar{F}^{-1} = \operatorname{Ad}\bar{F}\,\lfloor \operatorname{ad}\bar{\alpha}_{\mathfrak{h}}\rfloor_{\mathfrak{z}(\mathfrak{m})} \operatorname{Ad}\bar{F}^{-1},$$

$\bar{F}$ is flat if and only if $\lfloor \operatorname{ad}\bar{\alpha}_{\mathfrak{h}}\rfloor_{\mathfrak{z}(\mathfrak{m})} = 0$, that is $\bar{\alpha}_2 = 0$, because the adjoint representation maps $\mathfrak{g}_2$ to $\mathfrak{z}(\mathfrak{m})$ and $\mathfrak{g}_0$ into $\mathfrak{su}(\mathfrak{m})$. We now have four ways of reading the Lagrangian angle:

1. $e^{i\beta}$ is the complex determinant of the tangent frame $(\operatorname{Ad}\bar{F}(z)(\bar{k}(z)\epsilon_1), \operatorname{Ad}\bar{F}(z)(\bar{k}(z)\epsilon_2))$ in the parallel frame $(\operatorname{Ad}\bar{F}(z)(\epsilon_1), \operatorname{Ad}\bar{F}(z)(\epsilon_2))$, namely is the complex determinant of $\bar{k}$ for any parallel framing of f^*P,
2. given a Lagrangian lift F, and $h = \bar{F}^{-1}F$ the (H-valued) gauge change, then $e^{i\beta} = \det_{\mathbb{C}} \operatorname{Ad} h$,

3. in the previous gauge change, write h as a (commutative) product $h_2 h_0$ in $G_0 G_2$, which is unique up to sign ($G_2 = \exp \mathfrak{g}_2$); then $h_2 = \exp(\beta Y/2)$,
4. finally, for a Lagrangian lift, it follows from the previous characterization (3) that

$$\begin{aligned}\frac{d\beta}{2} Y &= h_2^{-1} dh_2 = \lfloor \mathrm{Ad}\, h_0 (h^{-1}dh) - dh_0 h_0^{-1} \rfloor_{\mathfrak{g}_2} \\ &= \lfloor h^{-1}dh \rfloor_{\mathfrak{g}_2} = \lfloor \alpha - \mathrm{Ad}\, F^{-1}(\bar{\alpha}) \rfloor_{\mathfrak{g}_2} = \alpha_2.\end{aligned}$$

We shall actually exploit the last characterization. We conclude by noting that the Maslov form is exactly $\frac{2}{\pi}$ times the Y component of α_2.

So far we have not used at all the underlying complex structure of L. But decomposing the Maurer-Cartan form α into its $(1,0)$ and $(0,1)$ parts α' and α'' respectively yields an interesting condition for a lift F to be Lagrangian:

PROPOSITION 1.3. *A lift is Lagrangian if and only if the Maurer-Cartan form $\alpha = \alpha_2 + \alpha_0 + \alpha_{-1} + \alpha_1$ satisfies $\alpha''_{-1} = 0$ (which by the reality assumption implies $\alpha'_1 = 0$).*

PROOF. The equivalence rests upon the following simple fact: $\mathfrak{g}^{\mathbb{C}}_{-1}$ is *exactly* the orbit under $\mathbb{R}_+ \times SU(\mathfrak{m})$ of the vector $\epsilon = \frac{1}{2}(\epsilon_1 - i\epsilon_2) \in \mathfrak{m}^{\mathbb{C}}$ (for any choice of Hermitian basis (ϵ_1, ϵ_2)). A proof is given in [**HR1**]. Now writing $\alpha_{\mathfrak{m}} = e^{\rho} k(\epsilon dz + \bar{\epsilon} d\bar{z})$, we see that $\alpha'_{\mathfrak{m}} = e^{\rho} k \epsilon dz$ belongs to $\mathfrak{g}^{\mathbb{C}}_{-1}$ if and only if k is in $SU(\mathfrak{m})$. □

This type of condition should be compared with *primitivity* conditions [**BP**] or ω-maps [**Hi**]. However our "partial primitivity" differs in that it is a first order requirement on the immersion (namely that of being Lagrangian) and not an Euler-Lagrange condition, while primitive maps and ω-maps are automatically harmonic. Finally we remark that the assumption of conformality is not crucial until the very final step; indeed the concept of Lagrangian lift makes sense in any dimension.

1.3. Lagrangian surfaces in $\mathbb{CP}^2$ and Lagrangian cones in $\mathbb{C}^3$. We concern ourselves here with Lagrangian three dimensional cones centered at the origin in $\mathbb{C}^3$, and in particular their intersection with the unit sphere S^5, which is called the *link*. We assume henceforth that the cone is regular, i.e. the link M is a connected submanifold. Notice that our analysis applies as well to regular conical singularities in complex three dimensional manifolds. We recall here the known correspondence between Lagrangian cones and Lagrangian surfaces in $\mathbb{CP}^2$ (see for instance [**Re**]).

Lagrangian cones in $\mathbb{C}^3$ may be locally described by Lagrangian surfaces in $\mathbb{CP}^2$. This is done by the canonical projection map $P : \mathbb{C}^3 \setminus \{0\} \to \mathbb{CP}^2$. More precisely to any Lagrangian cone Σ^3 in $\mathbb{C}^3$ we may associate the Lagrangian surface $L := P(\Sigma^3)$ in $\mathbb{CP}^2$. Conversely this surface L describes Σ^3 up to a rigid motion in the sense that the set of Lagrangian cones of $\mathbb{C}^3$ which are mapped to L by P is $\{e^{i\alpha}\Sigma^3 / \alpha \in \mathbb{R}/2\pi\mathbb{Z}\}$. Indeed let $\langle .,. \rangle_H = \langle .,. \rangle_E - i\omega(.,.)$ be the standard Hermitian product in $\mathbb{C}^3$ ($\langle .,. \rangle_E$ is the Euclidean scalar product and $\omega(.,.)$ the symplectic form). Let $S^5 := \{Z \in \mathbb{C}^3 / \langle Z, Z \rangle_H = 1\}$ be the unit sphere in $\mathbb{C}^3$, and $\mathcal{L} := \Sigma \cap S^5$ the link, which fully determines Σ. At any point $p \in \mathcal{L} \subset \Sigma^3$ choose an orthonormal frame (e_1, e_2, e_3) of $T_p\Sigma^3$ such that $e_3 = p$. Then (e_1, e_2)

is an orthonormal basis of $T_p\mathcal{L}$. The Lagrangian constraint on Σ at the point p is $iT_p\Sigma \perp T_p\Sigma$; this is equivalent to the two conditions

a) $T_p\mathcal{L} \perp ip$ and
b) $ie_2 \perp e_1$.

The first condition a) means that $T_p\mathcal{L}$ is contained in Π_p, the 4-dimensional subspace of T_pS^5 orthogonal to ip. The collection $\Pi := (\Pi_p)_{p\in S^5}$ forms a distribution on S^5, orthogonal to the fibers of the Hopf fibration $\mathcal{H} : S^5 \to \mathbb{CP}^2$, and therefore named the *horizontal* distribution. Notice that each Π_p can also be identified with the orthogonal subspace to p in $\mathbb{C}^3$ for the Hermitian product. The restriction of $\langle .,. \rangle_H$ to Π_p is also Hermitian, which implies that the restriction of ω to Π_p is symplectic (non degenerate). Now the second condition b) just means that $T_p\mathcal{L}$ is a Lagrangian plane in Π_p. These two conditions actually mean that $\mathcal{L}$ is a *Legendrian* submanifold of the contact manifold $(S^5, \Pi, \omega_{|\Pi})$.

The horizontality condition a) on $\mathcal{L}$, $T_p\mathcal{L} \subset \Pi_p$ for all p, can be translated into the property that the restriction of the Hopf fibration $\mathcal{H}$ to $\mathcal{L}$ is an isometric covering map (and locally an isometric diffeomorphism onto its image). The second condition b) is then contained in the property that $\mathcal{H}(\mathcal{L})$ is a Lagrangian submanifold of $\mathbb{CP}^2$.

Conversely let us start from a contractible Lagrangian $L \subset \mathbb{CP}^2$ and let us try to construct first a Legendrian link lifting $\mathcal{L}$ in S^5 and then a Lagrangian cone in $\mathbb{C}^3$. We use the canonical connection ∇ on the Hopf fibration defined for any section $s : \mathbb{CP}^2 \to S^5$, any point $\underline{p} \in \mathbb{CP}^2$ and any $X \in T_{\underline{p}}\mathbb{CP}^2$ by

$$\nabla_X s(\underline{p}) := \langle ds_{\underline{p}}(X), s(\underline{p})\rangle_H s(\underline{p}).$$

The curvature of ∇ is given by $\nabla_X\nabla_Y s - \nabla_Y\nabla_X s - \nabla_{[X,Y]}s = 2is^\star\omega_{\mathbb{C}^3}(X,Y)s$ and hence vanishes along any Lagrangian surface. It follows that, if L is Lagrangian and if f denotes the immersion mapping $L \subset \mathbb{CP}^2$, then $f^\star\mathcal{H}$, the pull-back of the Hopf bundle by f is flat. Since L is also contractible we can construct a flat section $s : L \to S^5$ (unique up to the choice of the value of s at one point) and $\mathcal{L} := s(L)$ is just the Legendrian lift of L that we were looking for. Now this Legendrian surface spans a Lagrangian cone Σ^3 in $\mathbb{C}^3$.

Furthermore we observe that Σ^3 is Hamiltonian stationary if and only if L is so. We denote by $\Theta := dz^1 \wedge dz^2 \wedge dz^3$ the 3-form which helps us to define the Lagrangian angle function on Σ^3 by $e^{i\beta} = \Theta(e_1, e_2, e_3)$ for any orthonormal basis (e_1, e_2, e_3) of $T_p\Sigma^3$. The cone Σ^3 is Hamiltonian stationary if and only if $\Delta_{\Sigma^3}\beta = 0$. Since β obviously does not depend on the radius $r := \langle p,p\rangle_H^{1/2}$, this condition is equivalent to $\Delta_{\mathcal{L}}\beta = 0$ along $\mathcal{L}$. Now we need a parallel section θ of $f^\star K$, where K is the canonical bundle of $\mathbb{CP}^2$ and f denotes the immersion $L \subset \mathbb{CP}^2$. A very simple construction is the following: let $s : L \to \mathcal{L}$ be the lift mapping; it is a parallel section of $f^\star\mathcal{H}$ — because $\mathcal{L}$ is Legendrian — and hence the 2-form $\theta := s^\star(\iota_s\Theta)$ is parallel. We can thus use θ to construct the Lagrangian angle along L. Let us denote by $\underline{p}$ a point in L and $p = s(\underline{p})$ its lift in $\mathcal{L}$. The Lagrangian angle $\underline{\beta}$ at $\underline{p}$ is defined by $e^{i\underline{\beta}} = \theta(\underline{e}_1, \underline{e}_2)$, where $(\underline{e}_1, \underline{e}_2)$ is an orthonormal basis of $T_{\underline{p}}L$. But since $(p, s_\star\underline{e}_1, s_\star\underline{e}_2)$ is an orthonormal basis of $T_p\Sigma^3$, $\theta(\underline{e}_1, \underline{e}_2) = \Theta(s(\underline{p}), s_\star\underline{e}_1, s_\star\underline{e}_2) = e^{i\beta}$ and hence the two Lagrangian angles coincide. It follows that the harmonicity condition on β along $\mathcal{L}$ is equivalent to the harmonicity condition on $\underline{\beta}$ along L, i.e. the condition that L is Hamiltonian stationary.

2. Loop group formulation

2.1. A general formulation of the problem using a family of curvature free connections. In order to yield a well-defined Lagrangian immersion f on the contractible domain L, the Maurer-Cartan form α of a Lagrangian lift F needs only to satisfy the closedness condition (also called zero curvature equation, when thinking of $d+\alpha$ as a bundle connection): $d\alpha + \frac{1}{2}[\alpha\wedge\alpha] = 0$. This equation splits along the eigenspaces of τ to give:

$$\left\{\begin{array}{l} d\alpha_2 + [\alpha_0\wedge\alpha_2] + \frac{1}{2}[\alpha_{-1}\wedge\alpha_{-1}] + \frac{1}{2}[\alpha_1\wedge\alpha_1] = 0 \\ d\alpha_0 + \frac{1}{2}[\alpha_0\wedge\alpha_0] + \frac{1}{2}[\alpha_2\wedge\alpha_2] + [\alpha_{-1}\wedge\alpha_1] = 0 \\ d\alpha_{-1} + [\alpha_0\wedge\alpha_{-1}] + [\alpha_2\wedge\alpha_1] = 0 \\ d\alpha_1 + [\alpha_0\wedge\alpha_1] + [\alpha_2\wedge\alpha_{-1}] = 0\,. \end{array}\right.$$

However due to the commutation relations $[\mathfrak{g}_2, \mathfrak{g}_2] = [\mathfrak{g}_2, \mathfrak{g}_0] = 0$, and to the properties of our Lagrangian lift, we simplify that to:

$$\left\{\begin{array}{l} d\alpha_2 = 0 \\ d\alpha_0 + \frac{1}{2}[\alpha_0\wedge\alpha_0] + [\alpha'_{-1}\wedge\alpha''_1] = 0 \\ d\alpha'_{-1} + [\alpha_0\wedge\alpha'_{-1}] + [\alpha_2\wedge\alpha''_1] = 0 \\ d\alpha''_1 + [\alpha_0\wedge\alpha''_1] + [\alpha_2\wedge\alpha'_{-1}] = 0\,. \end{array}\right. \tag{3}$$

As noted in [**CM, Wo**], a Lagrangian surface is Hamiltonian stationary, i.e. stationary with respect to Hamiltonian deformations, if and only if β is harmonic, that is Θ is coclosed, in other words $d\star\alpha_2 = 0$. Furthermore this Lagrangian surface is minimal (special Lagrangian if $M=\mathbb{C}^2$) if and only if $\alpha_2 = 0$. A now familiar trick allows us to combine these two differential equations into one zero curvature equation, though formulated on a loop algebra.

PROPOSITION 2.1. *On a simply connected domain L, the $\mathfrak{g}$-valued one-form α is the Maurer-Cartan form of a weakly conformal Lagrangian immersion if and only if it is flat: $d\alpha + \frac{1}{2}[\alpha\wedge\alpha] = 0$, and partially primitive: $\alpha''_{-1} = \alpha'_1 = 0$. Furthermore the immersion is Hamiltonian stationary if and only if the* extended Maurer-Cartan form $\alpha_\lambda = \lambda^{-2}\alpha'_2 + \lambda^{-1}\alpha'_{-1} + \alpha_0 + \lambda\alpha''_1 + \lambda^2\alpha''_2$ *is flat for all $\lambda\in\mathbb{C}^*$ (or S^1):*

$$d\alpha_\lambda + \frac{1}{2}[\alpha_\lambda\wedge\alpha_\lambda] = 0 \tag{4}$$

with the minimal (special Lagrangian if $M=\mathbb{C}^2$) subcase characterized by $\alpha_2 = 0$.

PROOF. The only additional condition induced by (4) is the coclosedness condition on α_2. □

Equation (4) is a curvature free condition, or equivalently, the compatibility condition for the existence for any $\lambda\in S^1$ of a solution $F_\lambda : L\to G$ of the equation

$$dF_\lambda = F_\lambda\alpha_\lambda. \tag{5}$$

Moreover F_λ is unique provided that we know its value at some point $z_0\in L$. We may choose for instance $F_\lambda(z_0) = \mathbb{1}$. Is is not difficult to realize that each F_λ is a lift of a Hamiltonian stationary Lagrangian surface like F. We deduce the following:

COROLLARY 2.2. *Local Hamiltonian stationary Lagrangian surfaces come in families parametrized by $\lambda \in S^1$. (For global surfaces there are usually period problems.)*

2.2. Loop groups. In the light of Proposition 2.1 and Corollary 2.2 above it is natural to introduce the following loop groups — following techniques familiar since [**SW**] (see also [**U**]). We consider the set of maps from S^1 to G:

$$\Lambda G := \{S^1 \ni \lambda \mapsto \varphi_\lambda \in G\}.$$

We assume that these maps are bounded in the H^s topology for $s > 1/2$. Then ΛG is a group for the composition law $[\lambda \mapsto \varphi_\lambda].[\lambda \mapsto \psi_\lambda] = [\lambda \mapsto \varphi_\lambda \psi_\lambda]$. We also consider the twisted loop (sub)group

$$\Lambda G_\tau := \{S^1 \ni \lambda \mapsto \varphi_\lambda \in G \mid \tau(\varphi_\lambda) = \varphi_{i\lambda}\}.$$

These loop groups have Lie algebras which are respectively

$$\Lambda\mathfrak{g} := \{S^1 \ni \lambda \mapsto \xi_\lambda \in \mathfrak{g}\} \text{ and } \Lambda\mathfrak{g}_\tau := \{S^1 \ni \lambda \mapsto \xi_\lambda \in \mathfrak{g} \mid \tau(\xi_\lambda) = \xi_{i\lambda}\}.$$

A key observation is that α_λ can be seen as a 1-form with values in $\Lambda\mathfrak{g}_\tau$. More precisely, "partially primitive" extended 1-forms are exactly the 1-forms α_λ with values in $\Lambda\mathfrak{g}_\tau$ such that $\lambda^2\alpha_\lambda$ has a limit when λ goes to zero. Similarly, choosing $F_\lambda(z_0) = \mathbb{1}$, the family of maps $F_\lambda : L \to G$ solving (5) can be viewed as a map into ΛG_τ, called the *extended lift* of f. Such loop groups have already been considered in [**BP, DPW**] in the context of harmonic maps into a symmetric space or in [**HR1**] for Hamiltonian stationary Lagrangian surfaces in $\mathbb{C}^2$.

3. Weierstrass type representations

The above loop group formulation of the Hamiltonian stationary Lagrangian surface problem allows us to construct solutions using various methods coming from integrable systems theory. As an illustration, we will present here Weierstrass representations in the spirit of J. Dorfmeister, F. Pedit and H. Y. Wu [**DPW**]. See also [**HR1**].

3.1. Loop groups decompositions. At the base of this construction is the Iwasawa decomposition. Applied to G_0, the subgroup of G fixed by τ, it amounts to the existence of a solvable Lie subgroup B_{G_0} of $G_0^{\mathbb{C}}$ such that the following mapping

$$\begin{array}{ccc} G_0 \times B_{G_0} & \longrightarrow & G_0^{\mathbb{C}} \\ (g, b) & \longmapsto & gb \end{array}$$

is a diffeomorphism. We summarize this by $G_0^{\mathbb{C}} = G_0 B_{G_0}$. We need now an infinite dimensional extension of the Iwasawa decomposition to loop groups.

For that purpose we introduce the complexified versions of the above loop groups, $\Lambda G^{\mathbb{C}}$ and $\Lambda G_\tau^{\mathbb{C}}$, obtained by replacing G by its complexification $G^{\mathbb{C}}$, and their Lie algebras $\Lambda\mathfrak{g}^{\mathbb{C}}$ and $\Lambda\mathfrak{g}_\tau^{\mathbb{C}}$. We also introduce the subgroups

$$\Lambda^+ G_\tau^{\mathbb{C}} := \{[\lambda \mapsto \varphi_\lambda] \in \Lambda G_\tau^{\mathbb{C}} \text{ extending holomorphically in the disk } D^2\},$$

$$\Lambda^+_{B_{G_0}} G_\tau^{\mathbb{C}} := \{[\lambda \mapsto \varphi_\lambda] \in \Lambda^+ G_\tau^{\mathbb{C}} \mid \varphi_0 \in B_{G_0}\},$$

$$\Lambda^-_\star G_\tau^{\mathbb{C}} := \{[\lambda \mapsto \varphi_\lambda] \in \Lambda G_\tau^{\mathbb{C}} \text{ extending holomorphically in } S^2 \backslash D^2 \text{ and } \varphi_\infty = 1\}.$$

The two main tools are the following Lemmas, which are proved in [**DPW**] (the proofs are based on results in [**PrS**]).

LEMMA 3.1. *Assume that G is a compact Lie group. Let $\tau : G \to G$ be an order four automorphism of G and let G_0 be the subgroup of G fixed by τ. Suppose that the Iwasawa decomposition $G_0^{\mathbb{C}} = G_0 B_{G_0}$ holds. Then the mapping*

$$\begin{array}{ccc} \Lambda G_\tau \times \Lambda^+_{B_{G_0}} G_\tau^{\mathbb{C}} & \longrightarrow & \Lambda G_\tau^{\mathbb{C}} \\ (g_\lambda, b_\lambda) & \longmapsto & g_\lambda b_\lambda \end{array}$$

is a diffeomorphism. We write $\Lambda G_\tau^{\mathbb{C}} = \Lambda G_\tau . \Lambda^+_{B_{G_0}} G_\tau^{\mathbb{C}}$.

LEMMA 3.2. *Assume that G is a semisimple Lie group. Then there exists a dense open subset $\mathcal{C}$ of the connected component of the identity of $\Lambda G_\tau^{\mathbb{C}}$, called the* big cell, *such that the mapping*

$$\begin{array}{ccc} \Lambda_\star^- G_\tau^{\mathbb{C}} \times \Lambda^+ G_\tau^{\mathbb{C}} & \longrightarrow & \mathcal{C} \\ (\varphi_\lambda^-, \varphi_\lambda^+) & \longmapsto & \varphi_\lambda^- \varphi_\lambda^+ \end{array}$$

is a diffeomorphism. We write $\mathcal{C} = \Lambda_\star^- G_\tau^{\mathbb{C}} . \Lambda^+ G_\tau^{\mathbb{C}}$.

In some cases in this paper these results do not apply directly, either because the isometry group G is not compact (for $\mathbb{C}^2$, $\mathbb{CD}^2$ or the dual of $\mathbb{CP}^1 \times \mathbb{CP}^1$) or because this group is not semisimple (in the case of $\mathbb{C}^2$). However it is possible to extend the above Lemmas to these situations in two ways:

- for $G = U(2) \ltimes \mathbb{C}^2$, the properties stated in Lemmas 3.1 and 3.2 are true; this was proved in [**HR1**] by a direct construction.
- in all cases, in particular when G is not compact or not semi-simple, local versions of Lemmas 3.1 and 3.2 can be proved. In these versions, one replaces the loop groups by neighbourhoods of the identity. The proof of these results uses the inverse mapping theorem as in [**H**].

3.2. Solutions in terms of holomorphic data. Conformal immersions of Hamiltonian stationary Lagrangian surfaces are in correspondence with holomorphic data as defined below. We introduce

$$\Lambda_{-2,\infty}\mathfrak{g}_\tau^{\mathbb{C}} := \{[\lambda \mapsto \xi_\lambda] \in \Lambda\mathfrak{g}_\tau^{\mathbb{C}} \mid \xi_\lambda = \sum_{k=-2}^{\infty} \hat{\xi}_k \lambda^k\}.$$

DEFINITION 3.3. The set of *holomorphic potentials*, denoted $\mathcal{H}_{-2,\infty}(L)$, is the set of holomorphic 1-forms on L with values in $\Lambda_{-2,\infty}\mathfrak{g}_\tau^{\mathbb{C}}$. So any form μ_λ in $\mathcal{H}_{-2,\infty}(L)$ has the expression

$$\mu_\lambda = \sum_{k=-2}^{\infty} \hat{\mu}_k \lambda^k = \sum_{k=-2}^{\infty} \hat{\xi}_k(z) \lambda^k dz,$$

where for all z, $\sum_{k=-2}^{\infty} \hat{\xi}_k(z)\lambda^k \in \Lambda_{-2,\infty}\mathfrak{g}_\tau^{\mathbb{C}}$.

LEMMA 3.4. *Let $F_\lambda : L \to \Lambda G_\tau$ be the extended lift of a (conformal) Hamiltonian stationary Lagrangian immersion and assume that L is contractible. Then*

- *there exist a holomorphic map $H_\lambda : L \to \Lambda G_\tau^{\mathbb{C}}$ and a map $B_\lambda : L \to \Lambda^+_{B_{G_0}} G_\tau^{\mathbb{C}}$ such that $F_\lambda = H_\lambda B_\lambda$.*

- *the Maurer-Cartan form* $\mu_\lambda := (H_\lambda)^{-1}dH_\lambda$ *is a holomorphic potential.*

Proof. The existence of H_λ and B_λ relies on solving the equation

$$0 = \frac{\partial(F_\lambda(B_\lambda)^{-1})}{\partial\bar{z}} = F_\lambda\left(\alpha_\lambda\left(\frac{\partial}{\partial\bar{z}}\right) - (B_\lambda)^{-1}\frac{\partial B_\lambda}{\partial\bar{z}}\right)(B_\lambda)^{-1},$$

which is equivalent to

$$\frac{\partial B_\lambda}{\partial\bar{z}} = B_\lambda(\alpha_0 + \lambda\alpha_1 + \lambda^2\alpha_2)\left(\frac{\partial}{\partial\bar{z}}\right),$$

with the constraint that B_λ takes values in $\Lambda^+_{B_{G_0}}G^{\mathbb{C}}_\tau$. The proof of this result is left to the reader, following exactly the same arguments as in [**DPW**].

Conversely any holomorphic potential in $\mathcal{H}_{-2,\infty}(L)$ produces a Hamiltonian stationary Lagrangian immersion as follows.

THEOREM 3.5. *Let* $\mu_\lambda \in \mathcal{H}_{-2,\infty}(L)$, z_0 *a point in* L *and* H^0_λ *a constant in* $\Lambda G^{\mathbb{C}}_\tau$. *Then:*

- *There exists a unique holomorphic map* $H_\lambda : L \to \Lambda G^{\mathbb{C}}_\tau$, *such that* $dH_\lambda = H_\lambda\mu_\lambda$ *and* $H_\lambda(z_0) = H^0_\lambda$.
- *If the loop group decomposition* $\Lambda G^{\mathbb{C}}_\tau = \Lambda G_\tau.\Lambda^+_{B_{G_0}}G^{\mathbb{C}}_\tau$ *holds then we can apply it to* $H_\lambda(z)$ *for all values of* z. *It follows that there exists two maps* $F_\lambda : L \to \Lambda G_\tau$ *and* $B_\lambda : L \to \Lambda^+_{B_{G_0}}G^{\mathbb{C}}_\tau$ *such that* $H_\lambda(z) = F_\lambda(z)B_\lambda(z)$ *for all* $z \in L$. *Then* F_λ *is a lift of a (conformal) Hamiltonian stationary Lagrangian immersion.*

PROOF. Straightforward adaptation of the proof in [**DPW**]. □

3.3. Meromorphic potentials. The holomorphic potentials constructed in Lemma 3.4 are far from being unique. Moreover they involve in general infinitely many holomorphic maps. These defects can be remedied, provided we allow meromorphic potentials and under some hypotheses on G. We define

$$\Lambda_{-2,-1}\mathfrak{g}^{\mathbb{C}}_\tau := \{[\lambda \mapsto \varphi_\lambda] \in \Lambda\mathfrak{g}^{\mathbb{C}}_\tau \mid \xi_\lambda = \hat{\xi}_{-2}\lambda^{-2} + \hat{\xi}_{-1}\lambda^{-1}\}.$$

DEFINITION 3.6. The set of *meromorphic potentials*, denoted $\mathcal{M}_{-2,-1}(L)$, is the set of meromorphic 1-forms on L with coefficients in $\Lambda_{-2,-1}\mathfrak{g}^{\mathbb{C}}_\tau$. So any form μ_λ in $\mathcal{M}_{-2,-1}(L)$ has the expression

$$\mu_\lambda = \hat{\mu}_{-2}\lambda^{-2} + \hat{\mu}_{-2}\lambda^{-2} = (\hat{\xi}_{-2}(z)\lambda^{-2} + \hat{\xi}_{-1}(z)\lambda^{-1})dz,$$

where $\hat{\xi}_{-2}(z)\lambda^{-2} + \hat{\xi}_{-1}(z)\lambda^{-1} \in \Lambda_{-2,-1}\mathfrak{g}^{\mathbb{C}}_\tau$.

Then using the same methods as in [**DPW**], one can prove the following.

THEOREM 3.7. *Assume that the conclusion of Lemma 3.2 holds. Let* $F_\lambda : L \to \Lambda G_\tau$ *be the extended lift of a (conformal) Hamiltonian stationary Lagrangian immersion. Then there exists a finite subset* $\{a_1, \dots, a_p\}$ *of* L *such that*

- *there exists a holomorphic map* $F^-_\lambda : L \setminus \{a_1, \dots, a_p\} \to \Lambda^-_\star G^{\mathbb{C}}_\tau$ *and a map* $F^+_\lambda : L \setminus \{a_1, \dots, a_p\} \to \Lambda^+ G^{\mathbb{C}}_\tau$ *such that* $F_\lambda(z) = F^-_\lambda(z)F^+_\lambda(z)$ *for all* $z \in L \setminus \{a_1, \dots, a_p\}$.

- *$z \mapsto F_\lambda^-(z)$ extends to a meromorphic map on L*
- *the Maurer-Cartan form $\mu_\lambda := (F_\lambda^-)^{-1}dF_\lambda^-$ of F_λ^- is a meromorphic potential in $\mathcal{M}_{-2,-1}(L)$.*

PROOF. Straightforward adaptation of the proof in [**DPW**]. □

4. A list of cases

4.1. The Euclidean space. The case of $\mathbb{C}^2$ has been thoroughly studied in an earlier article [**HR1**] and includes an explicit description of all tori. We will only point out the — obvious — differences between $\mathbb{C}^2$ and the other Hermitian symmetric spaces in the light of our study. Since the group of isometries is the semi-direct product $G = U(2) \ltimes \mathbb{C}^2$, we have the additional commutation property $[\mathfrak{m}, \mathfrak{m}] = 0$. The equations in (3) then uncouple to yield two PDEs, one in $\mathfrak{h}$ (in α_2, α_0), and one in $\mathfrak{m}$ with parameters α_2, α_0. Moreover the nonlinearity has disappeared! One may go even further than the standard analysis of that case, since $d\alpha_0 + \frac{1}{2}[\alpha_0 \wedge \alpha_0] = 0$ implies the (local) existence of lifts gauging α_0 to zero (we call them spinor lifts). At that point the problem of finding surfaces is equivalent to solving two linear PDEs (plus integration). But we do not need to use the coclosedness condition, and the commutation property is also the key point for the linear Weierstrass representation of all Lagrangian surfaces in $\mathbb{C}^2$ — not only the stationary ones. Indeed we derive a Dirac-type equation characterizing all such surfaces (see [**HR2**]).

4.2. The complex projective plane. We write the projective plane as a symmetric space

$$\mathbb{CP}^2 = G/H = \frac{SU(3)}{S(U(2) \times U(1))}$$

where[5]

$$H = S(U(2) \times U(1)) = \left\{ \begin{pmatrix} A & 0 \\ 0 & \det A^{-1} \end{pmatrix}, \ A \in U(2) \right\}$$

with Lie algebra

$$\mathfrak{h} = \mathfrak{sl}(3, \mathbb{C}) \cap (\mathfrak{u}(2) \oplus \mathfrak{u}(1)) = \left\{ \begin{pmatrix} X & 0 \\ 0 & -\operatorname{tr} X \end{pmatrix}, \ X \in \mathfrak{u}(2) \right\}.$$

Here and in subsequent sections, $\tilde{X}$ will denote the conjugate of X with respect to the real form $\mathfrak{g} \subset \mathfrak{g}^{\mathbb{C}}$; in the $\mathfrak{su}(3)$ case, $\tilde{X} = -X^*$.

The quotient map is given simply by $SU(3) \to \mathbb{CP}^2$, $g \mapsto \mathbb{C}g\epsilon_3$ where $\epsilon_3 = (0, 0, 1)$. The natural involution σ acts on $SU(3)$ (and its differential on $\mathfrak{g} = \mathfrak{su}(3)$) by conjugation:

$$\sigma : \begin{pmatrix} A & u \\ -u^* & a \end{pmatrix} \mapsto \begin{pmatrix} \mathbb{1}_2 & 0 \\ 0 & -1 \end{pmatrix} \begin{pmatrix} A & u \\ -u^* & a \end{pmatrix} \begin{pmatrix} \mathbb{1}_2 & 0 \\ 0 & -1 \end{pmatrix} = \begin{pmatrix} A & -u \\ u^* & a \end{pmatrix}$$

---

[5]Notice that the adjoint action of $SU(3)$ is only almost effective, with a kernel consisting of three elements: the cube roots of the identity.

The Lie algebra $\mathfrak{g} = \mathfrak{su}(3)$ splits as the direct sum of the +1 eigenspace of σ, $\mathfrak{h}$, and the (-1)-eigenspace

$$\mathfrak{m} = \left\{ \begin{pmatrix} 0 & u \\ -u^* & 0 \end{pmatrix}, \ u \in \mathbb{C}^2 \right\} \simeq \mathbb{C}^2 .$$

The adjoint representation of H on $\mathfrak{m}$ is surjective and almost effective:

$$\mathrm{Ad}\begin{pmatrix} A & 0 \\ 0 & \det A^{-1} \end{pmatrix} \simeq \Big[u \mapsto (\det A) A u \Big]$$

and

$$\mathrm{ad}\begin{pmatrix} X & 0 \\ 0 & -\operatorname{tr} X \end{pmatrix} \simeq \Big[u \mapsto (X + \operatorname{tr} X \mathbb{1}) u \Big]$$

so that the complex structure is $\exp(\frac{\pi}{2}\,\mathrm{ad}\, Y)$ with

$$Y = \frac{i}{3}\begin{pmatrix} 1 & & \\ & 1 & \\ & & -2 \end{pmatrix} .$$

The order four automorphism acting on $G^{\mathbb{C}}$ is

$$\tau : g \mapsto \left(\begin{array}{cc|c} & 1 & \\ -1 & & \\ \hline & & 1 \end{array}\right) {}^t g^{-1} \left(\begin{array}{cc|c} & -1 & \\ 1 & & \\ \hline & & 1 \end{array}\right) . \tag{6}$$

Notice that on G itself ${}^t g^{-1} = \bar{g}$; hence its differential acting on $\mathfrak{g}^{\mathbb{C}}$ (still denoted τ) is:

$$\tau : X \mapsto - \left(\begin{array}{cc|c} & 1 & \\ -1 & & \\ \hline & & 1 \end{array}\right) {}^t X \left(\begin{array}{cc|c} & -1 & \\ 1 & & \\ \hline & & 1 \end{array}\right) .$$

We then have a direct sum $\mathfrak{g}^{\mathbb{C}} = \mathfrak{g}_0^{\mathbb{C}} \oplus \mathfrak{g}_2^{\mathbb{C}} \oplus \mathfrak{g}_{-1}^{\mathbb{C}} \oplus \mathfrak{g}_1^{\mathbb{C}}$ with $\mathfrak{g}_2^{\mathbb{C}} = \mathbb{C} Y$,

$$\mathfrak{g}_0 = \left\{ \begin{pmatrix} X & 0 \\ 0 & 0 \end{pmatrix} ; \ X \in \mathfrak{su}(2) \right\} , \quad \mathfrak{g}_{-1}^{\mathbb{C}} = \left\{ \left(\begin{array}{cc|c} & & -ia \\ & & b \\ \hline -ib & a & \end{array}\right) , \ a, b \in \mathbb{C} \right\}$$

and

$$\mathfrak{g}_1^{\mathbb{C}} = \widetilde{\mathfrak{g}_{-1}^{\mathbb{C}}} = \left\{ \left(\begin{array}{cc|c} & & ia \\ & & b \\ \hline ib & a & \end{array}\right) , \ a, b \in \mathbb{C} \right\} .$$

EXAMPLE 4.1. THE REAL PROJECTIVE PLANE $\mathbb{RP}^2$ is immersed minimally in $\mathbb{CP}^2$ (and its double cover is the only minimal Lagrangian sphere, up to unitary isometries, see [**Y**]). Choose the stereographic projection from the southern pole as conformal coordinate chart. The fundamental lift (real-valued of course) is:

$$F(z) = \frac{1}{1 + |z|^2} \begin{pmatrix} 1 - x^2 + y^2 & -2xy & 2x \\ -2xy & 1 + x^2 - y^2 & 2y \\ -2x & -2y & 1 - x^2 - y^2 \end{pmatrix}$$

The Maurer-Cartan form satisfies $\alpha_2 = 0$ and

$$
\begin{aligned}
(1+|z|^2)\alpha \;&=\; \underbrace{\left(\begin{array}{cc|c} & -i & \\ i & & \\ \hline & & \end{array}\right)(\bar z dz - z d\bar z)}_{\alpha_0} \\
&\quad + \underbrace{\left(\begin{array}{cc|c} & & 1 \\ & & -i \\ \hline -1 & i & \end{array}\right) dz}_{\alpha'_{-1}} + \underbrace{\left(\begin{array}{cc|c} & & 1 \\ & & i \\ \hline -1 & -i & \end{array}\right) d\bar z}_{\alpha''_1} .
\end{aligned}
$$

The associated family is merely a change of variable by rotation in the z-plane.

EXAMPLE 4.2. THE CLIFFORD TORUS is the quotient of the standard torus $S^1 \times S^1 \times S^1 \subset \mathbb{C}^3$ by the Hopf action; it can be conformally parametrized as $\mathbb{C}f$ where

$$
f(x+iy) = \frac{1}{\sqrt{3}} \begin{pmatrix} e^{2ix} \\ e^{i(y\sqrt{3}-x)} \\ e^{-i(x+y\sqrt{3})} \end{pmatrix}
$$

and the fundamental lift is

$$
F(z) = \frac{1}{\sqrt{6}} \begin{pmatrix} 2ie^{2ix} & 0 & \sqrt{2}e^{2ix} \\ -ie^{i(y\sqrt{3}-x)} & i\sqrt{3}e^{i(y\sqrt{3}-x)} & \sqrt{2}e^{i(y\sqrt{3}-x)} \\ -ie^{-i(x+y\sqrt{3})} & -i\sqrt{3}e^{-i(x+y\sqrt{3})} & \sqrt{2}e^{-i(x+y\sqrt{3})} \end{pmatrix}
$$

with Maurer-Cartan form

$$
\begin{aligned}
\alpha \;&=\; \underbrace{\left(\begin{array}{cc|c} i & -1 & \\ -1 & -i & \\ \hline & & \end{array}\right)\frac{dz}{2} + \left(\begin{array}{cc|c} i & 1 & \\ 1 & -i & \\ \hline & & \end{array}\right)\frac{d\bar z}{2}}_{\alpha_0} \\
&\quad + \underbrace{\left(\begin{array}{cc|c} & & 1 \\ & & -i \\ \hline -1 & i & \end{array}\right)\frac{dz}{\sqrt{2}}}_{\alpha'_{-1}} + \underbrace{\left(\begin{array}{cc|c} & & 1 \\ & & i \\ \hline -1 & -i & \end{array}\right)\frac{d\bar z}{\sqrt{2}}}_{\alpha''_1} .
\end{aligned}
$$

Note that $\alpha_2 = 0$, which agrees with the fact that the Clifford torus is minimal.

EXAMPLE 4.3. *Vacuum solutions* are obtained by taking extended lifts $F_\lambda = \exp(zM_\lambda + \bar z\widetilde{M_\lambda})$ where M_λ is a constant in ΛG_τ. Equation (3) amounts to $[M_\lambda, \widetilde{M_\lambda}] = 0$. For further simplification (M_λ being constant) we gauge the $\mathfrak{g}^{\mathbb{C}}_{-1}$ part to

$$
\frac{e^\rho}{2}\left(\begin{array}{cc|c} & & 1 \\ & & -i \\ \hline -1 & i & \end{array}\right).
$$

That yields a family parametrized by complex numbers b, c such that $e^{2\rho} = 8\,\mathrm{Im}(\bar b c)$ is positive :

$$
M_\lambda = \begin{pmatrix} b - i\lambda^{-2}a & c & \lambda^{-1}\frac{e^\rho}{2} \\ c & -b - i\lambda^{-2}a & -i\lambda^{-1}\frac{e^\rho}{2} \\ -\lambda^{-1}\frac{e^\rho}{2} & i\lambda^{-1}\frac{e^\rho}{2} & 2\lambda^{-2}ia \end{pmatrix}, \quad a = -\frac{\bar c + i\bar b}{3}
$$

Minimal conformal immersions correspond to $a = 0$ so $c = ib$ and $e^\rho = 2\sqrt{2}|b|$

$$M_\lambda = \begin{pmatrix} b & ib & \lambda^{-1}\sqrt{2}|b| \\ ib & -b & -i\lambda^{-1}\sqrt{2}|b| \\ -\lambda^{-1}\sqrt{2}|b| & i\lambda^{-1}\sqrt{2}|b| & 0 \end{pmatrix}$$

and for $b = \frac{i}{2}$ we recognize the Clifford torus above. Variations in $|b|$ amount to trivial scale changes, but changes in the argument of b yield different examples (not gauge-equivalent); these however may not be periodic.

4.3. Lagrangian cones in $\mathbb{C}^3$. We explain here how a similar formalism applies as well to Lagrangian cones in $\mathbb{C}^3$, and how this relates formally to the previous section. This relation is not one to one, but to each Lagrangian surface corresponds exactly a circle of cones, namely the orbit under the Hopf action of any member. To make it visually explicit, we will indicate by the symbol ˇ the corresponding quantity in S^5; for instance, the Legendrian map $\check{f} : L \to S^5$ projects down to a Lagrangian map $f : L \to \mathbb{CP}^2$.

We view now S^5 as the reductive (but not symmetric) space $\check{G}/\check{H} = U(3)/\check{H}$ where

$$\check{H} = \left\{ \begin{pmatrix} A & 0 \\ 0 & 1 \end{pmatrix}, \ A \in U(2) \right\}$$

is the isotropy group of ϵ_3. The quotient map is $g \mapsto g\epsilon_3$. We have the reductive splitting $\check{\mathfrak{g}} = \mathfrak{u}(3) = \check{\mathfrak{h}} \oplus \check{\mathfrak{m}}$ with

$$\check{\mathfrak{h}} = \left\{ \begin{pmatrix} X & 0 \\ 0 & 0 \end{pmatrix}, \ X \in \mathfrak{u}(2) \right\}, \ \check{\mathfrak{m}} = \left\{ \begin{pmatrix} 0 & u \\ -u^* & ia \end{pmatrix}, \ a \in \mathbb{R}, \ u \in \mathbb{C}^2 \right\}.$$

The same order four automorphism τ (formally) as in the previous section (see formula (6)) acts on $U(3)$ and splits the Lie algebra $\check{\mathfrak{g}}^{\mathbb{C}} = \mathfrak{gl}(3, \mathbb{C})$ into four eigenspaces: $\check{\mathfrak{g}}_0^{\mathbb{C}} \oplus \check{\mathfrak{g}}_2^{\mathbb{C}} \oplus \check{\mathfrak{g}}_{-1}^{\mathbb{C}} \oplus \check{\mathfrak{g}}_1^{\mathbb{C}}$:

$$\check{\mathfrak{g}}_2 = \mathbb{R}\check{Y} \oplus \mathbb{R}\check{Z} \text{ with } \check{Y} = \begin{pmatrix} i & & \\ & i & \\ & & 0 \end{pmatrix}, \ \check{Z} = \begin{pmatrix} 0 & & \\ & 0 & \\ & & i \end{pmatrix},$$

and $\check{\mathfrak{g}}_0 = \mathfrak{g}_0$, $\check{\mathfrak{g}}_{-1}^{\mathbb{C}} = \mathfrak{g}_{-1}^{\mathbb{C}}$, $\check{\mathfrak{g}}_1^{\mathbb{C}} = \mathfrak{g}_1^{\mathbb{C}}$. Comparing with 4.2, the only differences are that (i) $\check{\mathfrak{g}}_2$ is two dimensional and (ii) the complex structure changes from Y to $\check{Y}$. Note that the contact distribution is generated as the orbit under $U(3)$ of the subspace

$$\check{\mathfrak{p}} = \left\{ \begin{pmatrix} 0 & u \\ -u^* & 0 \end{pmatrix}, \ u \in \mathbb{C}^2 \right\} \simeq \mathbb{C}^2,$$

endowed with the complex structure $\operatorname{ad}\check{Y}$. Obviously $\check{\mathfrak{p}}^{\mathbb{C}} = \check{\mathfrak{g}}_{-1}^{\mathbb{C}} \oplus \check{\mathfrak{g}}_1^{\mathbb{C}}$ and $\check{\mathfrak{m}} = \check{\mathfrak{p}} \oplus \mathbb{R}\check{Z}$.

Consider a Lagrangian cone $\mathcal{C}$; its link is Legendrian, i.e., (i) its tangent bundle lies in the contact distribution Π ($T_xM \perp ix$) and (ii) the tangent space T_xM is Lagrangian in Π_x. Letting $\check{f} : L \to S^5$ be a conformal parametrization of M, conditions (i) and (ii) above amount to the existence of a (unique) fundamental lift $\check{F} \in U(3)$ such that:

$$d\check{f}_z = e^{\rho(z)}\check{F}(z)(\epsilon_1 dx + \epsilon_2 dy) \tag{7}$$

which can be rewritten in terms of $\check{\alpha} = \check{F}^{-1}d\check{F}$

$$\check{\alpha}\epsilon_3 = e^{\rho}(\epsilon_1 dx + \epsilon_2 dy) = e^{\rho}(\epsilon dz + \bar{\epsilon} d\bar{z}) \tag{8}$$

where

$$\epsilon_1 = \left(\begin{array}{cc|c} & & 1 \\ & & 0 \\ \hline -1 & 0 & \end{array}\right), \quad \epsilon_2 = \left(\begin{array}{cc|c} & & 0 \\ & & 1 \\ \hline 0 & -1 & \end{array}\right) \quad \text{and } \epsilon = \frac{\epsilon_1 - i\epsilon_2}{2}.$$

Recall that $\det_{\mathbb{C}} \check{F} = e^{i\beta}$ where β is the Lagrangian angle. Extending to the orbit under the gauge action of $C^\infty(L, \check{G}_0)$ with $\check{G}_0 = \check{H} \cap SU(3) \simeq SU(2)$, we define Lagrangian lifts by the property that $\check{\alpha}_{\mathfrak{m}} = \check{\alpha}'_{-1} + \check{\alpha}''_1$. Notice that the condition is more complicated here because we need to assume that the map is horizontal (i.e. lies in the contact distribution), which excludes components along $\check{Z}$.

We can now characterize Hamiltonian stationary Lagrangian cones intrinsically.

THEOREM 4.4. *A conical Lagrangian singularity (whose intersection with S^5 is conformally parametrized) is obtained by integrating a flat $\mathfrak{u}(3)$-valued 1-form ω satisfying $\check{\alpha}_{\mathfrak{m}} = \check{\alpha}'_{-1} + \check{\alpha}''_1$ (hence $\check{\alpha}_2$ lies in $\mathbb{C}\check{Y}$ since $\mathbb{R}\check{Z} = \check{\mathfrak{m}} \cap \check{\mathfrak{g}}_2$). Furthermore, the immersion is stationary Hamiltonian (respectively, special Lagrangian) if α_2 is coclosed (respectively, vanishes).*

Denoting by $\Lambda\mathfrak{u}(3)_{\tau,\mathfrak{p}}$ the subspace of the twisted loop-algebra where the Z-part vanishes, there is a bijective correspondence between this subspace and the loop algebra $\Lambda\mathfrak{su}(3)_\tau$, mapping flat extended connection forms to flat extended connections forms, which leaves all matrix coefficients unchanged but for the $\check{\mathfrak{g}}_2$ part, where the complex structure $\check{Y}$ is mapped to a complex structure Y:

$$\begin{pmatrix} a & & \\ & b & \\ & & 0 \end{pmatrix} \mapsto \frac{1}{3}\begin{pmatrix} 2a-b & & \\ & 2b-a & \\ & & 0 \end{pmatrix}.$$

Equivalently one can associate to the cone the projected Lagrangian surface in $\mathbb{CP}^2$, $f = \mathbb{C}\check{f}$. We claim that $F = e^{-i\beta/3}\check{F}$ is a Lagrangian lift of f. Clearly F lifts f since $\mathbb{C}F\epsilon_3 = f$, and $\det F = 1$. To prove that F is Lagrangian, consider its Maurer-Cartan form

$$\alpha = \check{\alpha} - i\frac{d\beta}{3}\mathbb{1} = \check{\alpha}_0 + \check{\alpha}'_{-1} + \check{\alpha}''_1 + \frac{d\beta}{2}Y$$

since $\frac{d\beta}{2}Y = \frac{d\beta}{2}\check{Y} - i\frac{d\beta}{3}\mathbb{1}$. Obviously $\alpha''_{-1} = 0$. Furthermore we see that the Lagrangian angle of the cone is equal to the Lagrangian angle of the surface in $\mathbb{CP}^2$. It may be noted that the fundamental lift of one is mapped thus to the fundamental lift of the other.

4.4. The complex hyperbolic plane. The non compact dual of $\mathbb{CP}^2$ is the complex hyperbolic space $\mathbb{CD}^2 = SU(2,1)/S(U(2) \times U(1))$ where

$$SU(2,1) = \left\{ g \in SL(3,\mathbb{C}),\ gBg^* = B = \begin{pmatrix} 1 & & \\ & 1 & \\ & & -1 \end{pmatrix} \right\}.$$

The same automorphism τ acts on M.

4.5. $\mathbb{CP}^1 \times \mathbb{CP}^1$**.** We consider the following Hermitian symmetric space:

$$\mathbb{CP}^1 \times \mathbb{CP}^1 = \frac{SU(2) \times SU(2)}{U(1) \times U(1)}$$

($G = SU(2) \times SU(2)$ will be written as block-diagonal four by four matrices) and define an order four automorphism $\tau : g \mapsto TgT^{-1}$ where

$$T = \left(\begin{array}{cc|cc} & & & -1 \\ & & 1 & \\ \hline & 1 & & \\ 1 & & & \end{array}\right).$$

Its differential at the identity diagonalizes on $\mathfrak{g}^{\mathbb{C}} = \mathfrak{sl}(2) \oplus \mathfrak{sl}(2)$ with eigenspaces $\mathfrak{g}_0^{\mathbb{C}} = \mathbb{C}X$, $\mathfrak{g}_2^{\mathbb{C}} = \mathbb{C}Y$ (as usual Y is the complex structure)

$$X = \frac{1}{2}\left(\begin{array}{cc|cc} -i & & & \\ & i & & \\ \hline & & i & \\ & & & -i \end{array}\right), \qquad Y = \frac{1}{2}\left(\begin{array}{cc|cc} -i & & & \\ & i & & \\ \hline & & -i & \\ & & & i \end{array}\right),$$

$$\mathfrak{g}_{-1}^{\mathbb{C}} = \left\{ \left(\begin{array}{cc|cc} & v & & \\ u & & & \\ \hline & & & iu \\ & & iv & \end{array}\right) u, v \in \mathbb{C} \right\}, \quad \mathfrak{g}_1^{\mathbb{C}} = \overline{\mathfrak{g}_{-1}^{\mathbb{C}}}.$$

4.6. The non compact dual of $\mathbb{CP}^1 \times \mathbb{CP}^1$**.** As expected, the situation is very close to its compact dual

$$M = G/H = \frac{SU(1,1) \times SU(1,1)}{U(1) \times U(1)}.$$

The automorphism has the same expression and so do the eigenspaces.

References

[A] Henri Anciaux, *An isoperimetric inequality for Hamiltonian stationary Lagrangian tori in* $\mathbb{C}^2$ *related to Oh's conjecture*, to appear in Math. Zeitschrift.

[BP] Francis E. Burstall, Franz Pedit, *Dressing orbits of harmonic maps*, Duke Math. J. 80 (1995), 353–382.

[BRa] Francis E. Burstall, John H. Rawnsley, *Twistor theory for Riemannian symmetric spaces*, Lecture Notes in Mathematics 1424, Springer-Verlag (1990).

[CM] Bang-Yen Chen, Jean-Marie Morvan, *Deformations of isotropic submanifolds in Kähler manifolds*, J. of Geometry and Physics 13 (1994), 79–104.

[DPW] Josef Dorfmeister, Franz Pedit, Hong-You Wu, *Weierstrass type representation of harmonic maps into symmetric spaces*, Comm. in Analysis and Geometry Vol 6, (1998), 633–668.

[HaL] Reese Harvey, H. Blaine Lawson, *Calibrated geometries*, Acta Mathematica 148 (1982), 47–157.

[Has] Mark Haskins: *Special Lagrangian cones*, arXiv:math.DG/0005164.

[H] Frédéric Hélein, *Willmore surfaces and loop groups*, J. Differential Geometry 50 (1998), 331–385.

[HR1] Frédéric Hélein, Pascal Romon, *Hamiltonian stationary Lagrangian surfaces in* $\mathbb{C}^2$, arXiv:math.DG/0009202, to appear in Comm. in Analysis and Geometry.

[HR2] Frédéric Hélein, Pascal Romon, *Weierstrass representation of Lagrangian surfaces in four-dimensional space using spinors and quaternions*, Commentarii Mathematici Helvetici 75 (2000), 668–680.

[Hi] Masanori Higaki, *Actions of loop groups on the space of harmonic maps into reductive homogeneous spaces*, J. Math. Sci. Univ. Tokyo 5 (1998), 401–421.

[O] Yong-Geun Oh, *Volume minimization of Lagrangian submanifolds under Hamiltonian deformations*, Math. Z. 212 (1993), 175–192.

[PrS] Andrew Pressley, Graeme Segal, *Loop Groups*, Oxford Mathematical Monographs, Clarendon Press, Oxford 1986.

[Re] Helmut Reckziegel, *Horizontal lifts of isometric immersions into the bundle space of a pseudo-Riemannian submersion*, Global Differential Geometry and Global Analysis (1984), Lecture Notes in Mathematics 1156, Springer Verlag (1985), 264–279, and *A correspondence between horizontal submanifolds of Sasakian manifolds and totally real submanifolds of Kählerian submanifolds*, Pap. Colloq., Hajduszoboszlo/Hung. 1984, Vol. 2, Colloq. Math. Soc. Janos Bolyai 46, 1063–1081 (1988).

[ScWo] Richard Schoen, Jon Wolfson, *Minimizing volume among Lagrangian submanifolds*, Proc. Sympos. Pure Math., 65, Amer. Math. Soc. (1999) and arXiv:math.DG/0008244.

[SW] Graeme Segal, George Wilson: *Loop groups and equations of KdV type*, Publ. Math. IHES 61(1985), 5–65, also published in *Surveys in Differential Geometry: Integrable Systems*, Vol. IV, C.L. Terng, K. Uhlenbeck (editors), International Press 1998.

[U] Karen Uhlenbeck, *Harmonic maps into Lie groups*, Journal of Differential Geometry 30 (1989), 1–50.

[Wo] Jon Wolfson, *Minimal Lagrangian diffeomorphisms and the Monge-Ampère equations*, J. Differential Geometry 46 (1997), 335–373.

[Y] Shing-Tung Yau, *Submanifolds with constant mean curvature I,II*, Amer. J. Math., 96 (1974), 346–366; ibid. 96 (1975), 76–100.

CMLA, ENS de Cachan, 61 avenue du Président Wilson, 94235 Cachan Cedex, FRANCE

E-mail address: helein@cmla.ens-cachan.fr

Université de Marne-la-Vallée, 5 bd Descartes, Champs-sur-Marne, 77454 Marne-la-Vallée cedex 2, FRANCE

Current address: CMLA, ENS de Cachan, 61 avenue du Président Wilson, 94235 Cachan Cedex, FRANCE

E-mail address: romon@math.univ-mlv.fr

Contemporary Mathematics
Volume **308**, 2002

Line Congruences and Integrable Systems

Hesheng Hu

Introduction

The theory of line congruences is an important part of classical differential geometry [1]. In recent years it was found that this theory is closely related to the modern theory of integrable systems of partial differential equations. In Euclidean space R^3, pseudo-spherical congruences play an important role in research on the sine-Gordon equation, which is one of the most famous soliton equations with wide applications [2-4]. There is another famous integrable system, the Toda equations, also closely related to differential geometry. In recent years the elliptic version of the 2-dimensional Toda equations has attracted many geometers. It is also important to study the geometrical aspects of the hyperbolic version of the Toda equations.

The present paper consists of two parts. In the first part, we give results on pseudo-spherical congruences in Minkowski space $R^{2,1}$ which is related to the sine-Gordon, sinh-Gordon, cosh-Gordon, sine-Laplace, sinh-Laplace and Liouville equations; these play an essential role in the construction of surfaces with constant Gaussian curvature. To obtain explicit solutions, we use Darboux transformations. This has the advantage that solutions can be constructed by using a purely algebraic algorithm and can be done successively to produce an infinite sequence of solutions. Meanwhile, a complete classification of pseudo-spherical congruences has been given [19,23]. In Minkowski space $R^{2,1}$, surfaces of constant mean curvature are parallel surfaces of surfaces of constant Gaussian curvature. Consequently, the pseudo-spherical congruences and Darboux transformation method can be used to construct surfaces of constant mean curvature [21].

In the second part, we will show that the Toda equations have a remarkable relationship with the Laplace sequence of surfaces in projective spaces. First, we point out that the Toda equations can be modified as **signed Toda equations**

2000 *Mathematics Subject Classification.* Primary 53A20, 35Q58.

Supported by the Chinese National Foundation of Natural Science, the Scientific Foundation of the Ministry of Education of China for Doctoral Programs, and the Research Foundation of the Science and Technology Commission of Shanghai.

[22]:

$$\frac{\partial^2 w_i}{\partial x \partial t} = -\alpha_{i-1} e^{w_{i-1}} + 2\alpha_i e^{w_i} - \alpha_{i+1} e^{w_{i+1}}, \quad \alpha_i = \pm 1 \quad (i = 1, 2, \cdots, n).$$

When all $\alpha_i = 1$, the above equations become the 2-dimensional hyperbolic Toda equations. It is shown that the Lax pair for n-periodic 2-dimensional Toda equations are just the fundamental equations of the n-periodic Laplace sequences of surfaces in the projective space $\mathbf{P}_{n-1}$. Moreover, we find that there are two types of n-periodic Laplace sequences of surfaces which are not equivalent in the sense of projective geometry. It is noted that type II occurs only for even n and was not mentioned by Darboux or other researchers in this field. We also use the Darboux transformation to construct explicit examples.

Part I. Pseudospherical congruence in $R^{2,1}$

1. Pseudospherical congruence in Euclidean space R^3. A line congruence in R^3 is a two parameter family of straight lines in R^3. A straight line in the family can be expressed by

$$Y(u, v, \lambda) = X(u, v) + \lambda \zeta(u, v). \tag{1}$$

If (u, v) take fixed values, and $\lambda \in (-\infty, \infty)$, (1) is a straight line in the congruence.

Let Σ be a line congruence with two focal surfaces S and S^*, i.e. all lines in Σ are the common tangent lines of S and S^*. Let $P \in S$, $P^* \in S^*$, and let $\overline{PP^*}$ be the common tangent line which belongs to Σ. Let the unit normal vectors of S and S^* at P and P^* be $\mathbf{n}$ and $\mathbf{n}^*$ respectively. This Σ is called a pseudo-spherical congruence if $d(PP^*) = l$ =constant $(\neq 0)$ and $\mathbf{n} \cdot \mathbf{n}^* = \cos\tau$ =constant $(\neq \pm 1)$.

In 1883, Bäcklund found the following important theorem.

Theorem. If Σ is a pseudospherical congruence then the two surfaces S and S^* are both surfaces of negative constant Gaussian curvature $K = -\sin^2\tau \,/\, l^2$.

Let $S(u, v)$ be a known surface of constant curvature -1 and (u, v) the Chebyshev coordinate based on the lines of curvature, i.e. the curves u =constant and v =constant are two systems of lines of curvature of S, and the first fundamental form of S is

$$ds^2 = \cos^2 \frac{\alpha}{2} du^2 + \sin^2 \frac{\alpha}{2} dv^2. \tag{2}$$

Here $\alpha = \alpha(u, v)$ is a solution of the sine-Gordon equation

$$\alpha_{uu} - \alpha_{vv} = \sin\alpha. \tag{3}$$

From $S(u, v)$ we can use the Bäcklund theorem to construct another focal surface $S^*(u, v)$ with the same constant curvature. The point $\mathbf{r}^* \in S^*$ can be expressed as

$$\mathbf{r}^* = \mathbf{r} + l(\cos\frac{\alpha_1}{2}\mathbf{e}_1 + \sin\frac{\alpha_1}{2}\mathbf{e}_2), \tag{4}$$

where $\mathbf{e}_1, \mathbf{e}_2$ are unit tangent vectors of the u-curve and v-curve respectively, and α_1 is another solution of the sine-Gordon equation satisfying

$$\begin{aligned} (\alpha_1 + \alpha)_\zeta &= 2\beta \sin \tfrac{\alpha_1 - \alpha}{2} \\ (\alpha_1 - \alpha)_\eta &= \tfrac{2}{\beta} \sin \tfrac{\alpha_1 + \alpha}{2}. \end{aligned} \tag{5}$$

Here

$$\zeta = \frac{u+v}{2}, \quad \eta = \frac{u-v}{2}, \quad \beta = \frac{1-\cos\tau}{\sin\tau} \quad (\sin^2\tau = l^2). \tag{6}$$

2. Pseudo-spherical congruences in Minkowski space $R^{2,1}$. In Minkowski space $R^{2,1}$, the scalar product of two vectors $a = (a_1, a_2, a_3)$ and $b = (b_1, b_2, b_3)$ is $a \cdot b = a_1b_1 + a_2b_2 - a_3b_3$ and $a^2 = a_1^2 + a_2^2 - a_3^2$. A vector is called space-like (respectively, time-like) if $a^2 > 0$ (respectively, $a^2 < 0$), and a surface is called space-like (respectively, time-like) if its normal vector is time-like (respectively, space-like).

The fundamental equations of a surface in the orthonormal frame $\{\mathbf{r}, \mathbf{e}_1, \mathbf{e}_2, \mathbf{n}\}$ are

$$\begin{aligned} d\mathbf{r} &= \omega^1\mathbf{e}_1 + \omega^2\mathbf{e}_2 \quad (\mathbf{e}_1^2 = 1, \mathbf{e}_2^2 = \pm 1) \\ d\mathbf{e}_a &= \omega_a^b\mathbf{e}_b + \omega_a^3\mathbf{n} \quad (\mathbf{n}^2 = \mp 1) \\ d\mathbf{n} &= \omega_3^a\mathbf{e}_a \quad (a, b = 1, 2) \end{aligned} \tag{7}$$

and their integrability conditions are

$$\begin{aligned} &\omega^a + \omega_b^a \wedge \omega^b = 0 \quad \omega_c^3 \wedge \omega^c = 0 \\ &d\omega_b^a + \omega_c^a \wedge \omega_b^c = -\omega_3^a \wedge \omega_b^3 \quad \text{Gauss equation} \\ &d\omega_a^3 + \omega_b^3 \wedge \omega_a^b = 0 \quad \text{Codazzi equation.} \end{aligned} \tag{8}$$

The Gaussian curvature K of the surface is defined by

$$d\omega_2^1 + \omega_c^1 \wedge \omega_2^c = \pm K\omega^1 \wedge \omega^2. \tag{9}$$

Here we should take "+" (respectively, "−") for a space-like (respectively, time-like) surface.

Let Σ be a line congruence in $R^{2,1}$ with two focal surfaces S and S^* and $\overline{PP^*}$ be lines in Σ which are common tangent lines of S and S^*, $P \in S$ and $P^* \in S^*$. If $d(PP^*) = l$ =constant and $\mathbf{n} \cdot \mathbf{n}^*$ =constant, where $\mathbf{n}, \mathbf{n}^*$ are the unit normal vectors of S and S^* at P and P^* respectively, then Σ is called a **pseudo-spherical congruence**.

Note that we can assume $\mathbf{n} \cdot \mathbf{n}^* > 0$, and when $0 < \mathbf{n} \cdot \mathbf{n}^* < 1$ (respectively, $\mathbf{n} \cdot \mathbf{n}^* > 1$), we write $\mathbf{n} \cdot \mathbf{n}^* = \cos\tau$ (respectively, $\cosh\tau$). We have the following theorem which corresponds to the Bäcklund theorem of the Euclidean case. Let Σ be a pseudo-spherical congruence in $R^{2,1}$, S and S^* be its focal surfaces. We have

THEOREM 1. *In $R^{2,1}$, there are four kinds of pseudo-spherical congruence.*

(1) Σ is space-like, S, S^* are space-like, and $K = K^* = \sinh^2\tau\,/\,l^2$.

(2) Σ is space-like, S, S^* are time-like, and $K = K^* = \sinh^2\tau\,/\,l^2$.

(3) Σ is space-like, S (respectively, S^*) is time-like, S^* (respectively, S) is space-like, and $K = K^* = -\cosh^2\tau\,/\,l^2$.

(4) Σ is time-like, S, S^* are time-like, and $K = K^* = \sinh^2\tau\,/\,l^2$.

The proof is similar to the Euclidean case [9,19,21].

3. Construction of surfaces of constant curvature and classification of pseudospherical congruences. The pseudo-spherical congruence can be used to construct all kinds of surfaces of constant curvature. As the most interesting cases we state the construction of the two kinds of surfaces of negative constant curvature by a space-like pseudo-spherical congruences (case (3) of Theorem 1) in detail.

For a time-like surface with $K = -1$, we can use Chebyshev coordinates, and the corresponding moving frame $\{\mathbf{r}, \mathbf{e_1}, \mathbf{e_2}, \mathbf{n}\}$ satisfies

$$\mathbf{e}_1^2 = 1, \quad \mathbf{e}_2^2 = -1, \quad \mathbf{n}^2 = 1, \tag{10}$$

$$\begin{aligned} &\omega^1 = \cos\frac{\beta}{2}du, \quad \omega^2 = \sin\frac{\beta}{2}dv \\ &\omega_2^1 = \omega_1^2 = \frac{1}{2}(-\beta_v du + \beta_u dv) \\ &\omega_1^3 = -\omega_3^1 = \sin\frac{\beta}{2}du, \quad \omega_2^3 = \omega_3^2 = \cos\frac{\beta}{2}dv \end{aligned} \tag{11}$$

where β is a solution of the sine-Laplace equation [7,8,10]

$$\Delta\beta = \frac{\partial^2\beta}{\partial u^2} + \frac{\partial^2\beta}{\partial v^2} = \sin\beta. \tag{12}$$

On the other hand, for a space-like surface of constant curvature $K = -1$ [5-8], by using Chebyshev coordinates and the corresponding moving frame $\{\mathbf{r}, \mathbf{e_1}, \mathbf{e_2}, \mathbf{n}\}$, we have

$$\begin{aligned} &\mathbf{e}_1^2 = 1, \quad \mathbf{e}_2^2 = 1, \quad \mathbf{n}^2 = -1, \\ &\omega^1 = \cosh\frac{\alpha}{2}du, \quad \omega^2 = \sinh\frac{\alpha}{2}dv \\ &\omega_2^1 = -\omega_1^2 = \frac{1}{2}(\alpha_v du - \alpha_u dv) \\ &\omega_1^3 = \omega_3^1 = \sinh\frac{\alpha}{2}du, \quad \omega_2^3 = \omega_3^2 = \cosh\frac{\alpha}{2}dv \end{aligned} \tag{13}$$

and α is a solution of the sinh-Laplace equation

$$\Delta\alpha = \sinh\alpha. \tag{14}$$

We start from a time-like surface S with $K = -1$: $\mathbf{r} = \mathbf{r}(u, v)$. Let S^* be the surface defined by

$$\mathbf{r}^*(u, v) = \mathbf{r} + l(\cosh\frac{\alpha_1}{2}\mathbf{e}_1 + \sinh\frac{\alpha_1}{2}\mathbf{e}_2) \tag{15}$$

which is space-like with $K = -1$ too. Here $l = \cosh\tau$ is a constant and α_1 is to be determined. Moreover, the space-like unit vector $\cosh\frac{\alpha_1}{2}\mathbf{e_1} + \sinh\frac{\alpha_1}{2}\mathbf{e_2}$ defines the direction of the line in the pseudospherical congruence Σ. It is required that S^* be another focal surface, $\overline{\mathbf{rr}^*}$ the common tangent to S and S^*, and $\mathbf{n}^* \cdot \mathbf{n} = \sinh\tau$. Here

$$\mathbf{n}^* = \cosh\tau(\sinh\frac{\alpha_1}{2}\mathbf{e}_1 + \cosh\frac{\alpha_1}{2}\mathbf{e}_2) + \sinh\tau\mathbf{n} \tag{16}$$

is the normal of S^*. The space-like congruence $\{\overline{\mathbf{rr}^*}\}$ is the pseudo-spherical congruence to be constructed and S, S^* are its focal surfaces.

Since $\mathbf{n}^*$ is the normal of S^*, we have $\mathbf{n}^* \cdot d\mathbf{r}^* = 0$. From (15) and (16) we have

$$(17)\qquad \begin{aligned} l\left(\tfrac{\alpha_{1u}}{2} - \tfrac{\beta_v}{2}\right) &= \sinh \tfrac{\alpha_1}{2} \cos \tfrac{\beta}{2} + \sinh \tau \cosh \tfrac{\alpha_1}{2} \sin \tfrac{\beta}{2}, \\ l\left(\tfrac{\alpha_{1v}}{2} + \tfrac{\beta_u}{2}\right) &= -\cosh \tfrac{\alpha_1}{2} \sin \tfrac{\beta}{2} + \sinh \tau \sinh \tfrac{\alpha_1}{2} \cos \tfrac{\beta}{2}. \end{aligned}$$

This is a system of equations for α_1 and so the problem reduces to solving (17) for α_1. The condition of integrability of (17) is $\Delta\beta = \sin\beta$ which is satisfied since S is time-like and of constant curvature -1, so the solution α_1 exists. Thus, the congruence $\{\overline{\mathbf{rr}^*}\}$ is pseudo-spherical, and from Theorem 1 we see that S^* is a space-like surface with $K = -1$. Moreover, from (17) we can prove that α_1 satisfies the sinh-Laplace equation. Equation (17) has appeared in [8,9,12].

Similarly, we can use pseudo-spherical congruences to construct time-like and space-like surfaces of positive constant curvature as follows:

(i) Let S be a space-like surface with $K = 1$. This is case (1) in Theorem 1. By using Chebyshev coordinates, we have

$$(18)\qquad \omega^1 = \cos\frac{\alpha}{2} du, \quad \omega^2 = \sin\frac{\alpha}{2} dv$$

and

$$(19)\qquad \alpha_{uu} - \alpha_{vv} = -\sin\alpha.$$

This surface S^* which is a space-like surface with $K = 1$ too has the expression

$$(20)\qquad \mathbf{r}^* = \mathbf{r} + l(\cos\frac{\alpha_1}{2}\mathbf{e}_1 + \sin\frac{\alpha_1}{2}\mathbf{e}_2).$$

The function α_1 is a solution of the system of equations

$$(21)\qquad \begin{aligned} l(\alpha_{1u} + \alpha_v) &= 2\sin\frac{\alpha_1}{2}\cos\frac{\alpha}{2} + 2\cosh\tau\cos\frac{\alpha_1}{2}\sin\frac{\alpha}{2} \\ l(\alpha_{1v} + \alpha_u) &= -2\cos\frac{\alpha_1}{2}\sin\frac{\alpha}{2} - 2\cosh\tau\sin\frac{\alpha_1}{2}\cos\frac{\alpha}{2} \\ l &= \sin\tau = \text{constant}. \end{aligned}$$

Here $\{\overline{\mathbf{rr}^*}\}$ is a space-like pseudo-spherical congruence and the construction of S^* reduces to solving (21). In fact, (21) is the condition for S^* to be another focal surface of the congruence.

We turn to cases (2) and (4) of Theorem 1; then the surface S is time-like and $K = 1$. Since the first fundamental form of S is indefinite, the principal curvatures k_1, k_2 may be imaginary. In the following discussion we consider the case when k_1, k_2 are real and $k_1 \neq k_2$. For the other cases (k_1 and k_2 are imaginary, or $k_1 = k_2$), the construction can be seen in a forthcoming paper [23].

(ii) Let S be a time-like surface with $K = 1$, and k_1, k_2 real and distinct. Using Chebyshev coordinates we have

$$(22)\qquad \omega^1 = \cosh\frac{\alpha}{2} du, \quad \omega^2 = \sinh\frac{\alpha}{2} dv$$

and

$$(23)\qquad \alpha_{uu} - \alpha_{vv} = -\sinh\alpha.$$

From Theorem 1 (case (2)), the time-like surface S^* with $K = 1$ has the expression

$$(24)\qquad \mathbf{r}^* = \mathbf{r} + l(\cosh\frac{\alpha_1}{2}\mathbf{e}_1 + \sinh\frac{\alpha_1}{2}\mathbf{e}_2).$$

Here α_1 is a solution of the system of equations

$$\begin{aligned} & l(\frac{\alpha_{1u}}{2}+\frac{\alpha_v}{2}) = \sinh\frac{\alpha_1}{2}\cosh\frac{\alpha}{2}+\cosh\tau\cosh\frac{\alpha_1}{2}\sinh\frac{\alpha}{2} \\ & l(\frac{\alpha_{1v}}{2}+\frac{\alpha_u}{2}) = -\cosh\frac{\alpha_1}{2}\sinh\frac{\alpha}{2}-\cosh\tau\sinh\frac{\alpha_1}{2}\cosh\frac{\alpha}{2} \\ & l = \sinh\tau = \text{constant} \end{aligned} \tag{25}$$

Here, $\{\overline{\mathbf{rr}^*}\}$ is a space-like pseudo-spherical congruence and the construction of S^* reduces to solving (25). The solutions exist since (23) holds.

(iii) We start with the same S as (ii). From Theorem 1 (case (4)) we can use a time-like pseudo-spherical congruence to construct time-like S^*, i.e.

$$\mathbf{r}^* = \mathbf{r} + l(\sinh\frac{\alpha_1}{2}\mathbf{e_1}+\cosh\frac{\alpha_1}{2}\mathbf{e_2}). \tag{26}$$

Here α_1 is a solution of system of equations.

$$\begin{aligned} & l(\frac{\alpha_{1u}}{2}+\frac{\alpha_v}{2}) = -\cosh\frac{\alpha_1}{2}\cosh\frac{\alpha}{2}-\cos\tau\sinh\frac{\alpha_1}{2}\sinh\frac{\alpha}{2} \\ & l(\frac{\alpha_{1v}}{2}+\frac{\alpha_u}{2}) = \sinh\frac{\alpha_1}{2}\sinh\frac{\alpha}{2}+\cos\tau\cosh\frac{\alpha_1}{2}\cosh\frac{\alpha}{2} \\ & l = \sin\tau. \end{aligned} \tag{27}$$

From the above statements, we have

THEOREM 2. *The four kinds of surfaces of constant curvature in $R^{2,1}$ can be constructed through pseudo-spherical congruences.*

4. Darboux transformations. As is well known, the solutions of completely integrable systems of partial differential equations such as (17), (21), (25) exist. In general it is not easy to find solutions with explicit expressions. However, the Darboux transformation provides an effective method to construct some explicit solutions.

We state the method for the most interesting case (3), which is related to the sine-Laplace equation and sinh-Laplace equation,

$$\Delta\alpha = \frac{\partial^2\alpha}{\partial u^2}+\frac{\partial^2\alpha}{\partial v^2} = \sin\alpha, \tag{28}$$

$$\Delta\alpha = \sinh\alpha. \tag{29}$$

We introduce complex coordinates of the (u, v) plane

$$\zeta = \frac{u+iv}{2}, \quad \bar{\zeta} = \frac{u-iv}{2}. \tag{30}$$

Then

$$\partial_\zeta = \frac{\partial}{\partial u} - i\frac{\partial}{\partial v}, \quad \partial_{\bar{\zeta}} = \frac{\partial}{\partial u} + i\frac{\partial}{\partial v}, \tag{31}$$

and equations (28), (29) can be written as

$$\partial^2_{\zeta\bar{\zeta}}\alpha = \sin\alpha, \tag{32}$$

$$\partial^2_{\zeta\bar{\zeta}}\alpha = \sinh\alpha. \tag{33}$$

If α is a complex function, equations (28) and (29) can be unified in the complexified sinh-Laplace equation

$$\Delta\alpha = \sinh\alpha. \tag{34}$$

In fact, when α is real, it reduces to the real sinh-Laplace equation. When α is purely imaginary and $\alpha = i\beta$ (β real), β satisfies the sine-Laplace equation.

The complexified sinh-Laplace equation has a Lax pair

$$\Phi_\zeta = \frac{\lambda}{2}\begin{bmatrix} 0 & -e^{-\alpha} \\ e^{\alpha} & 0 \end{bmatrix}\Phi, \quad \Phi_{\bar\zeta} = \frac{1}{2}\begin{bmatrix} -\alpha_{\bar\zeta} & \frac{1}{\lambda} \\ -\frac{1}{\lambda} & \alpha_{\bar\zeta} \end{bmatrix}\Phi, \tag{35}$$

i.e., (34) is the integrability condition of (35). Let $\Phi(\lambda)$ be a solution to (35) with $\det\Phi \neq 0$. If (α, Φ) is known, we use the Darboux transformation to produce new solutions of the complexified sinh-Laplace equation and its Lax pair. Let $\begin{bmatrix} h_1 \\ h_2 \end{bmatrix}$ be a column vector solution of the Lax pair with $\lambda = \lambda_1 \neq 0$, i.e.

$$\begin{bmatrix} h_1 \\ h_2 \end{bmatrix} = \Phi(\lambda_1)l \quad (l \neq 0, \text{ a constant column vector}) \tag{36}$$

Let

$$S = \frac{1}{\lambda_1}\begin{bmatrix} 0 & \frac{h_1}{h_2} \\ \frac{h_2}{h_1} & 0 \end{bmatrix}. \tag{37}$$

Then we have the following theorem:

THEOREM 3. *Let α be a solution of the complexified sinh-Laplace equation, and Φ a solution of its Lax pair. Then the transformation $(\alpha, \Phi) \to (\alpha_1, \Phi_1)$ with*

$$\Phi_1 = [I - \lambda S]\Phi, \tag{38}$$

$$e^{\alpha_1} = -\left(\frac{h_2}{h_1}\right)^2 e^{-\alpha}, \tag{39}$$

gives a new solution (α_1, Φ_1) of the sinh-Laplace equation and its Lax pair.

Proof. Substituting (38) into the Lax pair (35) in which α is replaced by α_1, we see that it is satisfied if (39) holds. The function α_1 satisfies the complexified sinh-Laplace equation (34), since (34) is the integrability condition of (35). □

It should be noted that we can apply Darboux transformations

$$(\alpha, \Phi) \to (\alpha_1, \Phi_1) \to (\alpha_2, \Phi_2) \to (\alpha_3, \Phi_3) \to \cdots \tag{40}$$

successively by using the same algebraic algorithm. In general, the DT's give complex solutions. However, we have the following theorem:

THEOREM 4. *If α is a real solution of the sinh-Laplace equation, there is a DT $\alpha \to \alpha_1$ such that $\alpha_1 = i\beta_1$ is purely imaginary, and β_1 is a real solution of the sine-Laplace equation. If α is purely imaginary, there is a DT $\alpha \to \alpha_1$ such that α_1 is a real solution of the sinh-Laplace equation.*

Proof. If α is real, from (39), α_1 is purely imaginary if and only if

$$\left|\frac{h_2}{h_1}\right|^2 e^{-\alpha} = 1. \tag{41}$$

Let

$$A = \left|\frac{h_2}{h_1}\right|^2 e^{-\alpha} - 1. \tag{42}$$

From the Lax pair (35), it can be seen that

$$\begin{aligned} h_{1\zeta} &= -\tfrac{\lambda_1}{2}e^{-\alpha}h_2, & h_{2\zeta} &= \tfrac{\lambda_1}{2}e^{\alpha}h_1, \\ h_{1\bar\zeta} &= -\tfrac{\alpha_{\bar\zeta}}{2}h_1 + \tfrac{1}{2\lambda_1}h_2, & h_{2\bar\zeta} &= -\tfrac{1}{2\lambda_1}h_1 + \tfrac{\alpha_{\bar\zeta}}{2}h_2. \end{aligned} \tag{43}$$

If $|\lambda_1| = 1$, for real α, from (43) it follows that

$$A_\zeta = \frac{\lambda_1}{2}\left(\frac{h_2}{h_1}e^{-\alpha} - \frac{\bar h_2}{\bar h_1}\right)A, \quad A_{\bar\zeta} = \frac{\lambda_1}{2}\left(\frac{h_2}{h_1} + \frac{\bar h_2}{\bar h_1}e^{-\alpha}\right)A. \tag{44}$$

Hence, if $A = 0$ at one point, then $A = 0$ everywhere.

If we choose $\begin{bmatrix} h_1 \\ h_2 \end{bmatrix}$ such that $\left|\frac{h_2}{h_1}\right|^2 e^{-\alpha} = 1$ at one point, then α_1 is purely imaginary. Let $\alpha_1 = i\beta_1$. We obtain the real solution β_1 of the sine-Laplace equation.

Similarly, if α is purely imaginary, α_1 is real if and only if

$$\overline{\left(\frac{h_2}{h_1}\right)^2} e^{-\alpha} = \left(\frac{h_2}{h_1}\right)^2 e^{\alpha} < 0 \quad (\overline{e^{-\alpha}} = e^{\alpha}) \tag{45}$$

or

$$B = \frac{\bar h_2 h_1}{\bar h_1 h_2}e^{\alpha} + 1 = 0. \tag{46}$$

If $|\lambda_1| = 1$, for pure imaginary α, by using (43) we see that

$$B_\zeta = \frac{\lambda_1}{2}\left(\frac{\bar h_1}{\bar h_2} + \frac{h_2}{h_1}e^{-\alpha}\right)B, \quad B_{\bar\zeta} = -\frac{1}{2\lambda_1}\left(\frac{h_2}{h_1} + \frac{\bar h_1}{\bar h_2}e^{\alpha}\right)B. \tag{47}$$

Hence, if we choose $\begin{bmatrix} h_1 \\ h_2 \end{bmatrix}$ such that $B = 0$ at one point, then $B = 0$ everywhere. Thus we obtain a solution α_1 of real sinh-Laplace equation. □

We can use the DT successively to obtain solutions of the sine-Laplace equation and sinh-Laplace equation alternatively.

$$\begin{aligned} &\alpha_0 \to i\beta_1 \to \alpha_2 \to i\beta_3 \to \alpha_4 \to \cdots & \Delta\alpha_j &= \sinh\alpha_j, \\ &i\beta_0 \to \alpha_1 \to i\beta_2 \to \alpha_3 \to i\beta_4 \to \cdots & \Delta\beta_j &= \sin\beta_j. \end{aligned} \tag{48}$$

Next we use the DT to solve (17) explicitly. Write equation (39) of the Darboux transformation in the form

$$e^{\alpha_1} = -e^{-i\beta}\left(\frac{h_2}{h_1}\right)^2. \tag{49}$$

Then we have

$$e^{(\alpha_1 + i\beta)/2} = i\frac{h_2}{h_1}. \tag{50}$$

Differentiating, and using (43), we obtain

$$(\alpha_1 + i\beta)_\zeta = 2i\lambda_1 \sinh \frac{i\beta - \alpha_1}{2},$$
$$(\alpha_1 - i\beta)_{\bar{\zeta}} = \frac{2i}{\lambda_1} \sinh \frac{i\beta + \alpha_1}{2}.$$

If $\lambda_1 = (i + \sinh\tau)/\cosh\tau$, it can be shown that α_1 in (49) gives exact solutions to equation (17). Thus, the DT's give explicit solutions of the problem of constructing space-like surfaces of negative constant curvature. The problem of constructing time-like surfaces of negative constant curvature can be solved similarly. Thus we have proved the following theorem:

THEOREM 5. *If* $(i\beta, \Phi) \to (\alpha_1, \Phi_1)$ *is a Darboux transformation with* $\lambda_1 = (i+\sinh\tau)/\cosh\tau$, *then (15) is a transformation of a time-like surface with* $K = -1$ *to a space-like surface with* $K = -1$. *Alternatively, a time-like surface of* $K = -1$ *can be obtained from a space-like surface of* $K = -1$ *by a Darboux transformation.*

Hence, we can use the pseudo-spherical congruence and Darboux transformations to construct the four kinds of surfaces of constant curvature explicitly. In conclusion, we give a full classification of these pseudo-spherical congruences in R^3 and $R^{2,1}$ in Table 1.

Part II. Laplace sequences of surfaces in projective space and 2-dimensional Toda equations

1. 2-dimensional signed Toda equations. The original Toda equations (periodic case) are

$$\frac{d^2\omega_i}{dt^2} = e^{\omega_{i-1}} - 2e^{\omega_i} + e^{\omega_{i+1}} \quad (i = 1, 2, \cdots, n;\ \omega_{n+1} = \omega_1). \tag{51}$$

The 2-dimensional Toda equations in $\mathbf{R}^{1+1}$ are a system of hyperbolic equations [11,13,15]

$$\frac{\partial^2\omega_i}{\partial x \partial t} = e^{\omega_{i-1}} - 2e^{\omega_i} + e^{\omega_{i+1}} \quad (i = 1, 2, \cdots, n), \tag{52}$$

or equivalently,

$$\frac{\partial^2\omega_i}{\partial x \partial t} = -e^{\omega_{i-1}} + 2e^{\omega_i} - e^{\omega_{i+1}} \quad (i = 1, 2, \cdots, n) \tag{52$'$}$$

since we can replace t by $-t$ in (52).

TABLE 1. Pseudo-spherical congruences in R^3 and $R^{2,1}$

Space	Surface	Congruence and BT	DT	Equation
1. $\mathbf{R}^3$	$\mathbf{r}$			
	$\mathbf{r}^*$	$\mathbf{r}^* = \mathbf{r} + l(\cos\frac{\alpha_1}{2}\mathbf{e}_1 + \sin\frac{\alpha_1}{2}\mathbf{e}_2)$	α	$\alpha_{uu} - \alpha_{vv} = \sin\alpha$
	$K = -1$	$\mathbf{e}_1^2 = \mathbf{e}_2^2 = 1,\ \mathbf{e}_1 \cdot \mathbf{e}_2 = 0$	$\downarrow\uparrow$	
	$I = \cos^2\frac{\alpha}{2}du^2 + \sin^2\frac{\alpha}{2}dv^2$		α_1	$\alpha_{1uu} - \alpha_{1vv} = \sin\alpha_1$
	$II = \cos\frac{\alpha}{2}\sin\frac{\alpha}{2}(du^2 - dv^2)$			
2. $\mathbf{R}^{2,1}$	$\mathbf{r}$			
	$\mathbf{r}^*$	$\mathbf{r}^* = \mathbf{r} + l(\cos\frac{\alpha_1}{2}\mathbf{e}_1 + \sin\frac{\alpha_1}{2}\mathbf{e}_2)$	α	$\alpha_{uu} - \alpha_{vv} = -\sin\alpha$
	$K = 1$ space-like	space-like congruence	$\downarrow\uparrow$	
	$I = \cos^2\frac{\alpha}{2}du^2 + \sin^2\frac{\alpha}{2}dv^2$	$\mathbf{e}_1^2 = \mathbf{e}_2^2 = 1,\ \mathbf{e}_1 \cdot \mathbf{e}_2 = 0$	α_1	$\alpha_{1uu} - \alpha_{1vv} = -\sin\alpha_1$
	$II = -\cos\frac{\alpha}{2}\sin\frac{\alpha}{2}(du^2 - dv^2)$			
3. $\mathbf{R}^{2,1}$	$\mathbf{r}$			
	$\mathbf{r}^*$	$\mathbf{r}^* = \mathbf{r} + l(\sinh\frac{\alpha_1}{2}\,\mathbf{e}_1 + \cosh\frac{\alpha_1}{2}\,\mathbf{e}_2)$	α	$\alpha_{uu} - \alpha_{vv} = \sinh\alpha$
	$K = 1$ time-like	time-like congruence	$\downarrow\uparrow$	
	principal curvatures $k_1 \neq k_2$, real	$\mathbf{r}^* = \mathbf{r} + l(\cosh\frac{\alpha_1}{2}\,\mathbf{e}_1 + \sinh\frac{\alpha_1}{2}\,\mathbf{e}_2)$	α_1	$\alpha_{1uu} - \alpha_{1vv} = \sinh\alpha_1$
	$I = \cosh^2\frac{\alpha}{2}du^2 - \sinh^2\frac{\alpha}{2}dv^2$	space-like congruence		
	or $\sinh^2\frac{\alpha}{2}du^2 - \cosh^2\frac{\alpha}{2}dv^2$	$\mathbf{e}_1^2 = -\mathbf{e}_2^2 = 1,\ \mathbf{e}_1 \cdot \mathbf{e}_2 = 0$		
	$II = \cosh\frac{\alpha}{2}\sinh\frac{\alpha}{2}(du^2 - dv^2)$			

TABLE 1. (continued)

Space	Surface	Congruence and BT	DT	Equation
4. $\mathbf{R}^{2,1}$	$\mathbf{r}$	$\mathbf{r}^* = \mathbf{r} + l(a\mathbf{e}_1 + b\mathbf{e}_2)$	ω	$\omega_{uv} = \frac{1}{2}e^{\omega}$
	$\mathbf{r}^*$	$abe^{\omega} = 1$	$\downarrow\uparrow$	
	$K = 1$ time-like	space-like congruence	ω_1	$\omega_{1uv} = -\frac{1}{2}e^{\omega_1}$
	$k_1 = k_2$	$abe^{\omega} = -1$		
	$I = -e^{\omega}dudv + dv^2$	time-like congruence		
	$II = -e^{\omega}dudv$	$\mathbf{e}_1^2 = \mathbf{e}_2^2 = 0, \mathbf{e}_1 \cdot \mathbf{e}_2 = \frac{e^{\omega}}{2}$		
5. $\mathbf{R}^{2,1}$	$\mathbf{r}$	$\mathbf{r}^* = \mathbf{r} + l(a\,\mathbf{e}_1 + b\,\mathbf{e}_2)$		
	$\mathbf{r}^*$	$abe^{\omega} = 1$	ω	$\omega_{uv} = \cosh\omega$
	$K = 1$ time-like	space-like congruence	$\downarrow\uparrow$	
	k_1, k_2 imaginary	$abe^{\omega} = -1$	ω_1	$\omega_{1uv} = \cosh\omega_1$
	$I = -du^2 - 2\sinh\omega dudv + dv^2$	time-like congruence		
	$II = -2\cosh\omega dudv$	$\mathbf{e}_1^2 = \mathbf{e}_2^2 = 0,\ \mathbf{e}_1 \cdot \mathbf{e}_2 = \frac{e^{\omega}}{2}$		
6. $\mathbf{R}^{2,1}$	$\mathbf{r}$			
	$K = -1$ time-like			
	$I = \cos^2\frac{\alpha}{2}du^2 - \sin^2\frac{\alpha}{2}dv^2$	$\mathbf{r}^* = \mathbf{r} + l(\cosh\frac{\alpha_1}{2}\,\mathbf{e}_1 + \sinh\frac{\alpha_1}{2}\,\mathbf{e}_2)$	α	$\Delta\alpha = \sin\alpha$
	$II = \cos\frac{\alpha}{2}\sin\frac{\alpha}{2}(du^2 - dv^2)$	$\mathbf{e}_1^2 = -\mathbf{e}_2^2 = 1,\ \mathbf{e}_1 \cdot \mathbf{e}_2 = 0$	$\downarrow\uparrow$	
	$\mathbf{r}^*$	$\mathbf{r} = \mathbf{r}^* + l(\cos\frac{\alpha}{2}\mathbf{e}'_1 + \sin\frac{\alpha}{2}\mathbf{e}'_2)$	α_1	$\Delta\alpha_1 = \sinh\alpha_1$
	$K = -1$ space-like	space-like congruence		
	$I = \cosh^2\frac{\alpha_1}{2}du^2 + \sinh^2\frac{\alpha}{2}dv^2$	$\mathbf{e}'_1 = \mathbf{e}'_2 = 1,\ \mathbf{e}'_2 \cdot \mathbf{e}'_2 = 0$		
	$II = \cosh\frac{\alpha_1}{2}\sinh\frac{\alpha_1}{2}(du^2 + dv^2)$			

The 2-dimensional Toda equations can be deduced from their Lax pair

$$(53)\qquad \begin{bmatrix} \psi_1 \\ \vdots \\ \psi_n \end{bmatrix}_t = \lambda \begin{bmatrix} 0 & 0 & \cdots & 0 & p_1 \\ p_2 & 0 & \cdots & 0 & 0 \\ 0 & p_3 & \cdots & 0 & 0 \\ \vdots & \vdots & \ddots & \vdots & \vdots \\ 0 & 0 & \cdots & p_n & 0 \end{bmatrix} \begin{bmatrix} \psi_1 \\ \vdots \\ \psi_n \end{bmatrix},$$

$$(54)\qquad \begin{bmatrix} \psi_1 \\ \vdots \\ \psi_n \end{bmatrix}_x = \begin{bmatrix} \sigma_1 & \frac{1}{\lambda} & 0 & \cdots & 0 \\ 0 & \sigma_2 & \frac{1}{\lambda} & \cdots & 0 \\ 0 & 0 & \sigma_3 & \cdots & 0 \\ \vdots & \vdots & \vdots & \ddots & \frac{1}{\lambda} \\ \frac{1}{\lambda} & 0 & 0 & \cdots & \sigma_n \end{bmatrix} \begin{bmatrix} \psi_1 \\ \vdots \\ \psi_n \end{bmatrix},$$

i.e. the integrability condition of (53) and (54) are

$$(55)\qquad \tfrac{\partial p_i}{\partial x} = p_i(\sigma_i - \sigma_{i-1}), \quad \tfrac{\partial \sigma_i}{\partial t} = p_i - p_{i+1}.$$

As is well-known, when $p_i > 0$, we can write $p_i = e^{\omega_i}$ and obtain the 2-dimensional Toda equation (52)$'$:

$$\frac{\partial^2 \omega_i}{\partial x \partial t} = -e^{\omega_{i-1}} + 2e^{\omega_i} - e^{\omega_{i+1}}.$$

In particular, if

$$(56)\qquad \begin{aligned} &p_1 = p_3 = \cdots = p_{2k-1} = e^{\omega}, \\ &p_2 = p_4 = \cdots = p_{2k} = e^{-\omega}, \qquad n = 2k, \end{aligned}$$

equations (52)$'$ reduce to the sinh-Gordon equation

$$(57)\qquad \omega_{xt} = 4\sinh\omega.$$

If x, t are replaced by ζ, $\bar{\zeta}$, the complex coordinates of the Euclidean plane, then (52) becomes the elliptic version of the Toda equations

$$(58)\qquad \frac{\partial^2 \omega_i}{\partial \zeta \partial \bar{\zeta}} = e^{\omega_{i-1}} - 2e^{\omega_i} + e^{\omega_{i+1}}.$$

In recent years 2-dimensional Toda equations have attracted much attention. (See [11,13-15].)

In studying Laplace sequences of surfaces in projective space (see §2), we find that the case with some $p_i < 0$ in (55) is also important. Hence we should put

$$(59)\qquad p_i = \alpha_i e^{\omega_i}.$$

Here $\alpha_i = -1$ (respectively, $\alpha_i = 1$) when $p_i < 0$ (respectively, $p_i > 0$) and equation (52)$'$ becomes

$$(60)\qquad \frac{\partial^2 \omega_i}{\partial t \partial x} = -\alpha_{i-1}e^{\omega_{i-1}} + 2\alpha_i e^{\omega_i} - \alpha_{i+1}e^{\omega_{i+1}}.$$

We call (60) the 2-dimensional signed Toda equation [22]. In particular, when

$$(61)\qquad \begin{aligned} &p_1 = p_3 = \cdots = p_{2k-1} = e^{-\omega}, \\ &p_2 = p_4 = \cdots = p_{2k} = -e^{\omega}, \end{aligned}$$

we have the cosh-Gordon equation

$$(62)\qquad \omega_{xt} = 4\cosh\omega.$$

REMARK 1. *For the 1-dimensional case, corresponding to (60) we have*

$$\frac{d^2\omega_i}{dt^2} = -\alpha_{i-1}e^{\omega_{i-1}} + 2\alpha_i e^{\omega_i} - \alpha_{i+1}e^{\omega_{i+1}}. \tag{63}$$

The physical interpretation of equations (63) is that the system consists of two kinds of particles, and the interaction forces between them are either attraction or repulsion. The values of the α_i's are determined by the distribution of the two kinds of particles and conversely.

REMARK 2. *The hyperbolic 2-dimensional Toda equations can be traced back to G. Darboux [16]. Starting from Laplace sequences of the following hyperbolic equations of 2nd order*

$$z_{xt} + a_i z_x + b_i z_t + c_i z = 0 \tag{64}$$

he derived the equations

$$\frac{\partial^2 \log h_i}{\partial x \partial t} = -h_{i-1} + 2h_i - h_{i+1} \quad (h_i = \frac{\partial a_i}{\partial t} + a_i b_i - c_i), \tag{65}$$

which coincide with (60) when $h_i = e^{\omega_i} > 0$.

2. Laplace sequences of surfaces in projective space $\mathbf{P}_{n-1}$. Let $\mathbf{P}_{n-1}$ be $n-1$ dimensional projective space, $(x_1, x_2, \cdots, x_n)$ the homogeneous coordinates of a point $N \in \mathbf{P}_{n-1}$ and

$$N = N(t,x) \quad (\text{i.e. } x_a = x_a(t,x),\ a = 1, 2, \cdots, n) \tag{66}$$

the equations of a surface of $\mathbf{P}_{n-1}$ in homogeneous coordinates. If a system of surfaces

$$N_i = N_i(t,x) \quad (i = 1, 2, \cdots, n) \tag{67}$$

satisfies

$$\begin{aligned} \tfrac{\partial N_i}{\partial t} &= \mu_i N_i + p_i N_{i-1} \quad (p_i \neq 0), \\ \tfrac{\partial N_i}{\partial x} &= \sigma_i N_i + q_i N_{i+1} \quad (q_i \neq 0), \end{aligned} \tag{68}$$

$(N_{n+1} = N_1,\ N_0 = N_n)$, then the system of surfaces constitute a Laplace sequence of period n. In fact, the line $\overline{N_i N_{i+1}}$ is a common tangent of the surfaces N_i and N_{i+1}. In other words, the surfaces N_i and N_{i+1} are the two focal surfaces of the line congruence $\{\overline{N_i N_{i+1}}\}$, or N_{i+1} (respectively, N_i) is the Laplace transform of N_i (respectively, N_{i+1}).

Periodic Laplace sequences of surfaces (LSS) are important in classical projective differential geometry and have been studied extensively [17,18,20,21]. In the present paper, we elucidate the relationship between LSS of period n in $\mathbf{P}_{n-1}$ and the Toda equations of period n. In contrast to Darboux, we consider this problem from the point of view of integrable systems.

In this section we first simplify the fundamental equation (68) of the Laplace sequences by multiplying the homogeneous coordinates of each surface $N_i = N_i(t,x)$ by a suitable factor. Let

$$\widetilde{N}_i = k_i(x,t) N_i, \qquad (k_i(x,t) \neq 0) \tag{69}$$

and take

$$k_l = k_i^0(x) e^{-\int \mu_i dt}. \tag{70}$$

Equation (68) reduces to

$$(68)' \qquad \begin{aligned} \frac{\partial \widetilde{N}_i}{\partial t} &= \widetilde{p}_i \widetilde{N}_{i-1}, \\ \frac{\partial \widetilde{N}_i}{\partial x} &= \widetilde{\sigma}_i \widetilde{N}_i + \widetilde{q}_i \widetilde{N}_{i+1}. \end{aligned}$$

From the integrability condition of $(68)'$, we have $\widetilde{q}_{i,t} = 0$. Hence $\widetilde{q}_i$ depends only on x. Furthermore, letting $N_i' = m_i(x)\widetilde{N}_i$ $(m_i(x) \neq 0)$, we have

$$(68)'' \qquad \begin{aligned} \partial_t N_i' &= p_i' N_{i-1}', \\ \partial_x N_i' &= \sigma_i' N_i' + q_i' N_{i+1}' \end{aligned}$$

where

$$(71) \qquad p_i' = \frac{m_i}{m_{i-1}} \widetilde{p}_i, \quad \sigma_i' = \frac{m_{i,x}}{m_i} + \widetilde{\sigma}_i, \quad q_i' = \frac{m_i}{m_{i+1}} \widetilde{q}_i, \\ (m_{n+1} = m_1, m_0 = m_n).$$

From the last equation, we see that

$$(72) \qquad q_1' q_2' \cdots q_n' = \widetilde{q}_1 \widetilde{q}_2 \cdots \widetilde{q}_n \equiv Q(x).$$

When n is odd, or n is is even and $Q > 0$, let

$$(73) \qquad q = Q^{1/n}.$$

If we choose $m_1 \neq 0$ and

$$(74) \qquad m_{i+1} = m_i \widetilde{q}_i q^{-1} \quad (i = 1, 2, \cdots, n),$$

then we have $m_{n+1} = m_1$ and

$$(75) \qquad q_i' = q \quad (i = 1, 2, \cdots, n).$$

When n is even and $Q < 0$, let

$$(76) \qquad q = (-Q)^{1/n}.$$

If we choose $m_1 \neq 0$, and

$$(77) \qquad m_{i+1} = m_i \widetilde{q}_i q^{-1} \quad (i = 1, 2, \cdots, n-1), \quad m_{n+1} = -m_n \widetilde{q}_n q^{-1},$$

then we have $m_{n+1} = m_1$, and

$$(78) \qquad q_i'(x) = q(x) \quad (i = 1, 2, \cdots, n-1), \quad q_n'(x) = -q(x).$$

By a transformation of the variable x, we can make $q = 1$ and obtain [22]:

THEOREM 6. *There are two types of Laplace sequences of surface with period n in projective space $\mathbf{P}_{n-1}$. Their fundamental equations can be written as*

$$(79) \qquad \text{Type I:} \quad \begin{cases} \frac{\partial N_i}{\partial t} = p_i N_{i-1}, & (n \text{ odd, or } n \text{ even and } Q > 0) \\ \frac{\partial N_i}{\partial x} = \sigma_i N_i + N_{i+1} & (i = 1, 2, \cdots, n) \end{cases}$$

and

$$(80) \qquad \text{Type II:} \quad \begin{cases} \frac{\partial N_i}{\partial t} = p_i N_{i-1}, & (n \text{ even and } Q < 0) \\ \frac{\partial N_i}{\partial x} = \sigma_i N_i + N_{i+1}, & (i = 1, 2, \cdots, n-1) \\ \frac{\partial N_n}{\partial x} = \sigma_n N_n - N_1. & \end{cases}$$

Type II occurs only when n is even.

REMARK 3. *For n even, sign(Q) is a projective invariant of LSS. This is the reason why there are two types of n periodic LSS for even n.*

We write

$$\Psi = \begin{bmatrix} N_1 \\ N_2 \\ \vdots \\ N_n \end{bmatrix}. \tag{81}$$

By the rescaling map $(t, x, \sigma_i, p_i) \to (\lambda t, \frac{x}{\lambda}, \lambda\sigma_i, p_i)$, we can introduce a spectral parameter in (79) and (80). Thus we obtain:

THEOREM 7. *The Lax pair of 2-dimensional Toda equations is just the fundamental equation (79) of the n periodic Laplace sequence of type I in $\mathbf{P}_{n-1}$.*

The Laplace sequence of a surface of type II corresponds to the Lax pair

$$\begin{bmatrix} \psi_1 \\ \vdots \\ \psi_n \end{bmatrix}_t = \lambda \begin{bmatrix} 0 & 0 & \cdots & 0 & p_1 \\ p_2 & 0 & \cdots & 0 & 0 \\ 0 & p_3 & \cdots & 0 & 0 \\ \vdots & \vdots & \ddots & \vdots & \vdots \\ 0 & 0 & \cdots & p_n & 0 \end{bmatrix} \begin{bmatrix} \psi_1 \\ \vdots \\ \psi_n \end{bmatrix}, \tag{82}$$

$$\begin{bmatrix} \psi_1 \\ \vdots \\ \psi_n \end{bmatrix}_x = \begin{bmatrix} \sigma_1 & \frac{1}{\lambda} & 0 & \cdots & 0 \\ 0 & \sigma_2 & \frac{1}{\lambda} & \cdots & 0 \\ 0 & 0 & \sigma_3 & \cdots & 0 \\ \vdots & \vdots & \vdots & \ddots & \frac{1}{\lambda} \\ -\frac{1}{\lambda} & 0 & 0 & \cdots & \sigma_n \end{bmatrix} \begin{bmatrix} \psi_1 \\ \vdots \\ \psi_n \end{bmatrix}. \tag{83}$$

Equation (83) differs slightly from (54). However, the integrability condition of (82), (83) is the signed Toda equations (60) if we set

$$p_a = \alpha_a e^{\omega_a} \qquad (a = 2, \cdots, n) \tag{84}$$

and

$$-p_1 = \alpha_1 e^{\omega_1} \tag{85}$$

where

$$\alpha_a = \operatorname{sign}(p_a), \qquad \alpha_1 = \operatorname{sign}(-p_1). \tag{86}$$

REMARK 4. *If the LSS is non-periodic, there are infinitely many surfaces. The nonperiodic LSS corresponds to an infinite number of Toda equations and the Lax pair contains matrices of infinite order.*

In the following, we put $\lambda = 1$.

3. The construction of Laplace sequence of type I. We use Darboux transformations to construct Laplace sequences of type I [21]. Let us recall the Darboux transformation of the 2-dimensional Toda lattice [11]. Let (Ψ, p, σ) be a known solution of (53)–(55) where Ψ is a column vector. If Ψ^0 is a known solution (with the same p, σ), then

$$\Psi'_a = \Psi_a - \frac{\Psi^0_a}{\Psi^0_{a-1}} \Psi_{a-1}, \tag{87}$$

$$p'_a = p_a - \left[\frac{\Psi^0_a}{\Psi^0_{a-1}}\right]_t, \tag{88}$$

$$\sigma'_a = \sigma_a + \frac{\Psi^0_{a-1}}{\Psi^0_a} - \frac{\Psi^0_a}{\Psi^0_{a-1}} \tag{89}$$

is a solution of (53)–(55) too. The transformation $(\Psi, p, \sigma) \to (\Psi', p', \sigma')$ is called a Darboux transformation (DT). This was obtained in [11] .

Let

$$\Psi = (N_1, \cdots, N_n) \tag{90}$$

be a known LSS of period n in $\mathbf{P}_{n-1}$ such that (79) is satisfied. Applying the DT, we have

$$\Psi' = (N'_1, \cdots, N'_n) \tag{91}$$

with

$$N'_a = N_a - \frac{\Psi^0_a}{\Psi^0_{a-1}} N_{a-1} \tag{92}$$

where

$$\Psi^0_a = N_a l^a \quad (l^a \text{ is a constant column vector}). \tag{93}$$

Then Ψ' is a new Laplace sequence of surfaces of period n. It should be noted that N'_a lies on the line $\overline{N_a N_{a-1}}$.

By using this method we can construct a series of periodic LSS explicitly. We write down the procedure briefly for $n = 4$. Take the trivial solution of the 2-dimensional Toda equation

$$\sigma_a = 0, \quad p_a = 1. \tag{94}$$

Solving the fundamental equations, we obtain the Laplace sequence of period 4:

$$\begin{array}{ccccc} N_1 & \cosh u & \sinh u & \cos v & \sin v \\ N_2 & \sinh u & \cosh u & \sin v & -\cos v \\ N_3 & \cosh u & \sinh u & -\cos v & -\sin v \\ N_4 & \sinh u & \cosh u & -\sin v & \cos v \end{array} \tag{95}$$

Here $u = t + x$, $v = t - x$. It is seen that N_1 and N_3 generate the surface

$$S_1 : x_1^2 - x_2^2 = x_3^2 + x_4^2 \quad \text{or} \quad x^2 + y^2 + z^2 = 1 \tag{96}$$

if we use the inhomogeneous coordinates $x = x_4/x_1$, $y = x_2/x_1$, $z = x_3/x_1$. Similarly, N_2 and N_4 generate the surface

$$S_2 : x_2^2 - x_1^2 = x_3^2 + x_4^2 \quad \text{or} \quad y^2 - x^2 - z^2 = 1. \tag{97}$$

By using the DT, we obtain a new Laplace sequence of period 4:

$$N_1' = N_1 - \frac{\Psi_1^0}{\Psi_4^0} N_4, \quad N_2' = N_2 - \frac{\Psi_2^0}{\Psi_1^0} N_1, \tag{98}$$
$$N_3' = N_3 - \frac{\Psi_3^0}{\Psi_2^0} N_2, \quad N_4' = N_4 - \frac{\Psi_4^0}{\Psi_3^0} N_3.$$

Here Ψ_a^0 is a linear combination of the four columns in (95). Then $N_1' \in \overline{N_1 N_4}$, $N_2' \in \overline{N_2 N_1}$, $N_3' \in \overline{N_3 N_2}$, $N_4' \in \overline{N_4 N_3}$. It can be proved that $N_1', \cdots, N_4'$ generate algebraic varieties. We can apply DT's successively to produce an infinite sequence of 4-periodic Laplace sequences of surfaces in $\mathbf{P}_3$.

An illustration of a typical quadrilateral together with S_1 and S_2 can be found in [22], with N_1, N_3 on the surface $x^2 + y^2 + z^2 = 1$, and N_2, N_4 on the surface $y^2 - x^2 - z^2 = 1$.

4. Laplace sequences of type II. For Laplace sequences of surfaces in $\mathbf{P}_{n-1}$ of type II, the Lax equations are slightly different from those of type I. We shall derive the formulae of the DT.

Let (Ψ, p, σ) be a solution of the Lax equations

$$\begin{bmatrix} \psi_1 \\ \vdots \\ \psi_n \end{bmatrix}_t = \begin{bmatrix} 0 & 0 & \cdots & 0 & p_1 \\ p_2 & 0 & \cdots & 0 & 0 \\ 0 & p_3 & \cdots & 0 & 0 \\ \vdots & \vdots & \ddots & \vdots & \vdots \\ 0 & 0 & \cdots & p_n & 0 \end{bmatrix} \begin{bmatrix} \psi_1 \\ \vdots \\ \psi_n \end{bmatrix}, \tag{99}$$
$$\begin{bmatrix} \psi_1 \\ \vdots \\ \psi_n \end{bmatrix}_x = \begin{bmatrix} \sigma_1 & 1 & 0 & \cdots & 0 \\ 0 & \sigma_2 & 1 & \cdots & 0 \\ 0 & 0 & \sigma_3 & \cdots & 0 \\ \vdots & \vdots & \vdots & \ddots & 1 \\ -1 & 0 & 0 & \cdots & \sigma_n \end{bmatrix} \begin{bmatrix} \psi_1 \\ \vdots \\ \psi_n \end{bmatrix}.$$

It can be proved that we can still use formula (87) to construct a DT. Moreover, (88) and (89) hold as well except for

$$\sigma_1' = \sigma_1 + \frac{\Psi_2^0}{\Psi_1^0} + \frac{\Psi_1^0}{\Psi_n^0},$$
$$\sigma_{n-1}' = \sigma_{n-1} + \frac{\Psi_n^0}{\Psi_{n-1}^0} - \frac{\Psi_{n-1}^0}{\Psi_{n-2}^0}, \tag{100}$$
$$\sigma_n' = \sigma_n - \frac{\Psi_1^0}{\Psi_n^0} - \frac{\Psi_n^0}{\Psi_{n-1}^0}.$$

For the trivial solution we have

$$\sigma_1 = \sigma_2 = \sigma_3 = \sigma_4 = 0, \quad p_1 = -1, \quad p_2 = p_3 = p_4 = 1. \tag{101}$$

Solving the fundamental equations we obtain

(102)

$$\begin{array}{lllll}
N_1 & e^v \cos u & e^v \sin u & e^{-v} \cos v & e^{-v} \sin v \\
N_2 & \frac{e^v}{\sqrt{2}}(\cos u - \sin u) & \frac{e^v}{\sqrt{2}}(\cos u + \sin u) & -\frac{e^{-v}}{\sqrt{2}}(\cos u + \sin u) & \frac{e^{-v}}{\sqrt{2}}(\cos u - \sin u) \\
N_3 & -e^v \sin u & e^v \cos u & e^{-v} \sin u & -e^{-v} \cos u \\
N_4 & -\frac{e^v}{\sqrt{2}}(\sin u + \cos u) & \frac{e^v}{\sqrt{2}}(\cos u - \sin u) & \frac{e^{-v}}{\sqrt{2}}(\cos u - \sin u) & \frac{e^{-v}}{\sqrt{2}}(\cos u + \sin u)
\end{array}$$

Here $u = (x-t)/\sqrt{2}$, $v = (x+t)/\sqrt{2}$, N_1 and N_3 generate the surface $S_1 : x_1x_4 = x_2x_3$ or $x = yz$, N_2 and N_4 generate the surface $S_2 : x_1x_3 = -x_2x_4$ or $z = -xy$. An illustration of a typical quadrilateral N_1, N_2, N_3, N_4 together with the surfaces S_1 and S_2 can be found in [22], with N_1, N_3 on the surface $x = yz$, and N_2, N_4 on the surface $z = -xy$.

By using formula (87), we obtain

$$N'_a = N_a - \frac{\Psi^0_a}{\Psi^0_{a-1}} N_{a-1} \tag{103}$$

as in the case of type I. Hence the DT gives a new Laplace sequence of period 4. The four surfaces are algebraic surfaces too.

REMARK 5. *The surfaces (96), (97) which appeared in the LSS of type I do not contain straight lines, while the surfaces S_1 and S_2 which appeared in LSS of type II contain two systems of straight lines. This reflects the fact that the LSS of type I and those of type II are non-equivalent in real projective geometry.*

References

[1] G. Darboux, Sur une proposition relative aux équations linéaires, C.R. Acad. Sci., Paris **94** (1882), 1456.

[2] L.P. Eisenhart, A Treatise on the Differential Geometry of Curves and Surfaces, Dover, New York, 1960.

[3] C.H. Gu, On the Bäcklund transformations for the generalized hierarchies of compound MKdV-SG equation, Lett. Math. Phys. **11** (1986), 31.

[4] M. Heyerhoff, The history of the early period of soliton theory, In: D. Wójcik and J. Cieśliński (eds.), Nonlinearity and Geometry, Polish Scientific Publishers, Warsaw, 1998, pp. 13–23.

[5] H.S. Hu, The construction of hyperbolic surfaces in 3-dimensional Minkowski space and sinh-Laplace equation, Acta Math. Sinica New Ser. **1** (1985), 79–86.

[6] H.S. Hu, Soliton and differential geometry, In: Gu Chaohao (ed.), Soliton Theory and its Applications, Springer-Verlag, Berlin and Zhejiang Science and Technology Publishing House, 1995,pp. 297–336.

[7] H.S. Hu, Sine-Laplace equation, sinh-Laplace equation and harmonic maps, Manuscripta Math. **40** (1982), 205.

[8] H.S. Hu, The geometry of sine-Laplace, sinh-Laplace equations, In: N. Hu (ed.), Proc. of the Third Marcel Grossman Meeting on General Relativity 1982 Shanghai, Science Press, Beijing, 1983, pp. 1073–1076.

[9] Y.Z. Huang, Bäcklund theorems in 3-dimensional Minkowski space and their higher-dimensional generalization, Acta Math. Sinica **29** (1986), 684 (in Chinese).

[10] G. Leibbrandt, Exact solutions of the elliptic sine equation in two space dimensions with the application to the Josephson effect, Phys. Rev. B **15** (1977), 3353– 3361.

[11] V.B. Matveev and M.A. Salle, Darboux Transformations and Solitons, Spring-Verlag, Berlin, 1991.

[12] C. Rogers and W.R. Shadwick, Bäcklund Transformations and their Applications, Academic Press, New York, 1982.
[13] M. Toda, Nonlinear Waves and Solitons, KTK Scientific Publishers, Tokyo and Kluwer Academic Publishers, Holland, 1989.
[14] A.P. Fordy and J.C. Wood (eds.), Harmonic Maps and Integrable Systems, Vieweg, 1994.
[15] M.A. Guest, Harmonic Maps, Loop Groups, and Integrable Systems, London Mathematical Society Student Texts 38, Cambridge University Press, 1997.
[16] G. Darboux, Leçons sur la Theorie Générale des Surfaces, Vol. 2, 2nd ed. Gauthier-Villars, Paris, 1915.
[17] S. Finikoff, Transformation T des congruences de droites, Annali della R. Scuola Normale Superiore di Pisa, **2** (1933), 59–88.
[18] B. Su, On certain periodic sequences of Laplace of period four in ordinary space, Science Reports of the Töhoku Imp. Univ. (1) **25** (1956), 227–256.
[19] H.S. Hu, Darboux transformations between $\Delta\alpha = \sinh\alpha$ and $\Delta\alpha = \sin\alpha$ and the application to pseudo-spherical congruences in $\mathbf{R}^{2,1}$, Lett. Math. Phys. **48** (1999), 187–195.
[20] H.S. Hu, Darboux transformations of Su-chain, Proc. Symposium in honor of Prof. Su Buchin, World Scientific (1993), 108–113.
[21] C.H. Gu, H.S. Hu and Z.X. Zhou, Darboux Transformation in Soliton Theory and its Geometric Applications, Shanghai Scientific and Technical Publishers, 1999.
[22] H.S. Hu, Laplace sequences of surfaces in projective space and 2-dimensional Toda equations, Lett. Math. Phys. **57** (2001), 19–32.
[23] C.H. Gu, H.S. Hu and J. Inoguchi, On time-like surface of positive constant Gaussian curvature with imaginary principal curvature, J. Geom. Phys. **41** (2002), 296–311.

E-mail address: `guch@fudan.ac.cn`

INSTITUTE OF MATHEMATICS, FUDAN UNIVERSITY, SHANGHAI 200433, P.R. CHINA

Contemporary Mathematics
Volume **308**, 2002

Factorizations of Harmonic Maps of Surfaces into Lie Groups by Singular Dressing Actions

Xiaoxiang Jiao

1. Introduction

In recent years, many mathematicians have studied harmonic maps into Lie groups, and obtained important results. In [1] Uhlenbeck studied certain harmonic maps into Lie groups in terms of integrable systems, and classified them by their uniton numbers. She constructed new harmonic maps of finite uniton number from known ones by non-degenerate Backlund transforms, i.e., the dressing action, and by adding a uniton, and proved that harmonic maps of two-spheres into unitary groups can be factorized into products of flag factors by adding unitons. Wood showed in [8] that all harmonic maps of two-spheres into unitary groups can be factorized into the product of maps into complex Grassmann manifolds called based flag factors. Dong and Shen gave local factorization theorems for certain harmonic maps in [6], by a different method to that of [8]. Factorizations display the process of obtaining harmonic maps from constant or holomorphic maps by adding flag factors, which are obtained by solving a system of differential equations.

In [4] Guest and Ohnita discussed the properties of loop group actions, and proved that the natural action of the loop group on the extended solutions of finite uniton number is the same as the group action introduced by Uhlenbeck in [1]. Bergvelt and Guest studied the singularities of the group action in [2]. The limiting group action does not exist in general at every point of the domain, but when there are a finite number of singular points, these are removable, and a new extended solution is obtained. They called this the singular dressing action or the modified completion procedure, and showed in [2] that a harmonic map of finite uniton number into complex projective space can be reduced to a holomorphic map, and then to a constant map, by this procedure. They asked whether any harmonic map of finite uniton number can be reduced to a constant map in the same way.

In this paper we discuss some properties of singular dressing actions, and answer a version of this question (Theorem 4.5).

2000 *Mathematics Subject Classification.* Primary 53C20.

The author would like to thank Professor M. Guest for his suggestions and also to acknowledge the support of the National Natural Science Foundation of China (grant numbers 10001033, 10071084) and the MSJ-IRI.

2. Preliminaries

Let $G = U(N)$ be the unitary group with Lie algebra $g = u(N)$, and let Ω be a simply connected domain in $R^2 \cup \{\infty\}$ with complex coordinate z. If $s : \Omega \to U(N)$ is a smooth map, we take $A = \frac{1}{2}s^{-1}ds = A_z dz + A_{\overline{z}} d\overline{z}$. It is known that s is a harmonic map if and only if

$$\partial A_{\overline{z}} = [A_{\overline{z}}, A_z],\ \overline{\partial} A_z = [A_z, A_{\overline{z}}], \tag{1}$$

where $\partial = \partial/\partial z$ and $\overline{\partial} = \partial/\partial \overline{z}$. The associated Lax equation is

$$\overline{\partial} E_\lambda = (1-\lambda)E_\lambda A_{\overline{z}},\ \partial E_\lambda = (1-\lambda^{-1})E_\lambda A_z, \tag{2}$$

A solution E_λ of (2) is called an *extended solution* associated to the harmonic map s if $s = QE_{-1}$ for some constant $Q \in U(N)$. It can be normalized such that

$$E_1 = I,\ E_\lambda E^*_{\sigma(\lambda)} = I,$$

where $\sigma(\lambda) = 1/\overline{\lambda}$ and $\lambda \in C^* = C\backslash\{0\}$. Moreover, $E : \Omega \times C^* \longrightarrow G^C$ is analytic and holomorphic in $\lambda \in C^*$. Note that every E_λ may be expanded in a Laurent series $E_\lambda = \sum_{\alpha=-\infty}^{+\infty} \lambda^\alpha T_\alpha$, where $T_\alpha : \Omega \longrightarrow gl(N, C)$. If E_λ is a polynomial in λ, i.e., $E_\lambda = \sum_{\alpha=0}^{n} \lambda^\alpha T_\alpha$, then s is called a *harmonic map of finite uniton number,* E_λ is called an *n-uniton extended solution* associated to s , and n is said to be a *uniton number* of s .

If s is a harmonic map of finite uniton number, then there is a unique n-uniton extended solution of the form $E_\lambda = \sum_{\alpha=0}^{n} \lambda^\alpha T_\alpha$ with $V_0(E_\lambda) = C^N$, where

$$V_0(E_\lambda) = \text{span}\{v : v = T_0(q)w,\ q \in \Omega,\ w \in C^N\}.$$

This n is called the *minimal uniton number* of s . Such an extended solution is called a *standard extended solution* associated to s in [6].

Let π be the Hermitian orthogonal projection of $\Omega \times C^N$ onto a k-dimensional subbundle η of $\Omega \times C^N$ and let a map $s : \Omega \longrightarrow U(N)$ be a harmonic map of finite uniton number with the n-uniton extended solution E_λ. The map π is called a *flag factor* or a *uniton* of s if $\widetilde{E}_\lambda = E_\lambda(\pi + \lambda\pi^\perp)$ is an extended solution or $\widetilde{s} = s(\pi - \pi^\perp)$ is a harmonic map, where $\pi^\perp = I - \pi$. The process of obtaining $\widetilde{E}_\lambda$ from E_λ is called *adding a uniton*. It is proved in [1] that π is a flag factor of s if and only if π satisfies the following system of differential equations: $\pi^\perp A_z \pi = 0,\ \pi^\perp(\overline{\partial}\pi + A_{\overline{z}}\pi) = 0$.

Lemma 2.1 ([8]) Let $s : \Omega \longrightarrow U(N)$ be a harmonic map of finite uniton number and π the Hermitian orthogonal projection onto a subbundle of $\Omega \times C^N$. Then π is a flag factor for s if and only if $\pi^\perp$ is a flag factor of $\widetilde{s} = s(\pi - \pi^\perp)$.

3. Singular dressing action

Consider the loop group $\Lambda G^C = \{f : S^1 \longrightarrow G^C \mid f \text{ is smooth }\}$ with Lie algebra $\Lambda g^C = \{\xi :\longrightarrow gl(N,\ C)\}$. A real form of ΛG^C is

$$\Lambda G = \{f \in \Lambda G^C \ :\ f(\lambda) \in G,\ \lambda \in S^1\}.$$

The based loop group of G is $\Lambda_u G = \{f \in \Lambda G \ :\ f(1) = I\}$. We may consider an extended solution as a complex analytic map $E : \Omega \longrightarrow \Lambda_u G$. Let $D = \{\lambda \in C \ :\ |\lambda| < 1\}$ be the interior of the unit disc. The space of loops extending holomorphically to D is denoted by $\Lambda^+ G^C$. Let $\xi \in \Lambda g^C$, with Fouries series

$\xi(\lambda) = \sum_{i=-\infty}^{+\infty} \lambda^i \xi_i$, where $\xi_i \in g^C$. Then the Lie subalgebras of Λg^C corresponding to ΛG, $\Lambda_u G$ and $\Lambda^+ G^C$ are, respectively,

$$\Lambda g = \{\xi \in \Lambda g^C \ : \ \xi_i^* = -\xi_{-i}\},$$

$$\Lambda_u g = \{\xi \in \Lambda g \ : \ \xi(1) = 0\},$$

$$\Lambda^+ g^C = \{\xi \in \Lambda g^C \ : \ \xi_i = 0 \ for \ i < 0\}.$$

Theorem 3.1 ([5]) The product map $\Lambda_u G \times \Lambda^+ G^C \longrightarrow \Lambda G^C$ is a diffeomorphism. In particular, $\Lambda G^C = \Lambda_u G \Lambda^+ G^C$, i.e., any $f \in \Lambda G^C$ may be written uniquely in the form $f = f_u f_+$, where $f_u \in \Lambda_u G$ and $f_+ \in \Lambda^+ G^C$.

From Theorem 3.1 we immediately obtain $\Lambda g^C = \Lambda_u g \bigoplus \Lambda^+ g^C$. Theorem 3.1 also shows that there is a natural action $f^\sharp E = (fE)_u$ of ΛG^C on $\Lambda_u G$, where $f \in \Lambda G^C$ and $E : \Omega \longrightarrow \Lambda_u G$. This is called the "dressing action" in [4] and [7]. If $R_f = (fE)_+^{-1} \in \Lambda^+ G^C$, then $f^\sharp E = fER_f$. When $f \in \Lambda_u G$, we have $f^\sharp E = fE$. When E_λ is an n-uniton extended solution, it was shown in [4] that the dressing action is essentially the same as the "the action of the group $A_R(S^2, G^C)$" introduced by Uhlenbeck in [1]. In [4] Guest and Ohnita showed that if E_λ is an n-uniton extended solution, then $f^\sharp E_\lambda$ is also an n-uniton extended solution for every $f \in \Lambda^+ G^C$.

Let $E_\lambda = \sum_{\alpha=0}^{n} \lambda^\alpha T_\alpha$ be a standard n-uniton extended solution and let

$$f(\lambda) = P + \xi_a(\lambda) P^\perp, \quad \text{where } \xi_a(\lambda) = \frac{(\lambda - a)(\bar{a} - 1)}{(\bar{a}\lambda - 1)(1 - a)}, \ a \neq 0, \infty,$$

and where P is the Hermitian orthogonal projection from C^N onto a subspace V. Such an $f(\lambda)$ is called a map of simplest type by Uhlenbeck in [1]. Then $R_f(\lambda) = \widetilde{P}(a) + \xi_a^{-1}(\lambda)\widetilde{P}^\perp(a)$, where $\widetilde{P}(a)$ is the Hermitian orthogonal projection onto a subbundle $\widetilde{V}$ of $\Omega \times C^N$ and $\widetilde{V} = (E_a)^* V$.

The limit $\lim_{a \to 0} \widetilde{P}(a)$ does not exist in general; it exists if and only if $\lim_{a \to 0} (E_a)^* V$ is a k-dimensional subbundle of $\Omega \times C^N$, i.e., $\lim_{a \to 0} (E_a(q))^* V$ is a k-dimensional subspace of C^N for all points $q \in \Omega$ (see [2]). Let us suppose that $\lim_{a \to 0} \widetilde{P}(a) = \widetilde{P}$ exists. Then we have

$$\lim_{a \to 0} f^\sharp E_\lambda = (P + \lambda P^\perp) E_\lambda (\widetilde{P} + \lambda^{-1} \widetilde{P}^\perp) = (\lambda^{-1} P + P^\perp) E_\lambda (\widetilde{P}^\perp + \lambda \widetilde{P}).$$

It was shown in [2] that $\widetilde{E}_\lambda = E_\lambda(\widetilde{P}^\perp + \lambda \widetilde{P})$ is a new extended solution and $\widetilde{P}^\perp$ is a flag factor of the harmonic map E_{-1}, away from points of the domain where the limit does not exist. When there are a finite number of singular points, the procedure of obtaining $\widetilde{E}_\lambda$ from E_λ is called the *singular dressing action* or the *modified completion procedure* in [2]. We shall give factorizations of harmonic maps of finite uniton number by using the singular dressing action in the next section.

4. Proof of the main theorem

If $s : \Omega \longrightarrow U(N)$ is a harmonic map of finite uniton number, s is said to be factorizable if it is the product of maps of Ω into complex Grassmann manifolds of the form

$$s = Q(\beta_0 - \beta_0^{\perp}) \cdots (\beta_k - \beta_k^{\perp}),$$

where k is a nonnegative integer, $Q \in U(N)$ is a constant matrix, and β_i is a flag factor of $s_i = Q(\beta_0 - \beta_0^{\perp}) \cdots (\beta_i - \beta_i^{\perp})$ for each $i = 0,\ 1,\ \cdots,\ k$.

Lemma 4.1 ([2]) Let V be a subspace of C^N and let $A_0,\ A_1,\ \cdots,\ A_n \in gl(N,\ C)$. Put $A = A_0 + aA_1 + \cdots + a^n A_n$, then

$$\{\lim_{a \to 0} A(Cv) \mid v \in V\} = A_0 V \cup A_1(\ker A_0 \cap V) \cup \cdots \cup A_n(\cap_{\alpha=0}^{n-1}(\ker A_\alpha \cap V)).$$

Hence $\lim_{a \to 0} AV$ is a subspace if and only if one term in this union contains all the other terms. Let $E_\lambda = \sum_{\alpha=0}^{n} \lambda^\alpha T_\alpha$ be an n-uniton extended solution. Assume that $\lim_{a \to 0} \widetilde{P}(a)$ exists and

$$\lim_{a \to 0} (E_a)^* V = \mathrm{Im}(T_0^* P).$$

Then π is the Hermitian orthogonal projection onto $\ker(PT_0)$, where $\widetilde{P}^{\perp} = \pi$, $\lim_{a \to 0} \widetilde{P}(a) = \widetilde{P}$, P is the Hermitian orthogonal projection onto V.

Lemma 4.2 ([1]) Let V be a subspace of C^N and P the Hermitian orthogonal projection onto V. Let $E_\lambda = \sum_{\alpha=0}^{n} \lambda^\alpha T_\alpha$ be an n-uniton extended solution and π the Hermitian orthogonal projection onto $\ker(PT_0)$. Then $\widetilde{E}_\lambda = E_\lambda(\pi + \lambda\pi^{\perp})$ is a new extended solution, and π is a flag factor of the harmonic map $s = QE_{-1}$.

Note: Since π is a flag factor of the harmonic map $s = QE_{-1}$, $\mathrm{rank}(\ker(PT_0))$ is constant off a set of isolated points.

Lemma 4.3 $\widetilde{E}_\lambda = E_\lambda(\pi + \lambda\pi^{\perp})$ is an n-uniton extended solution or $(n-1)$-uniton extended solution for any Hermitian projection $\pi : \Omega \times C^N \longrightarrow \ker(PT_0)$.

Proof: By the reality condition $E_\lambda E_{\sigma(\lambda)}^* = I$, we have $\mathrm{Im}\, T_0^* \subset \ker T_n$. Because of $\ker T_0 \subset \ker(PT_0)$, i.e., $\mathrm{Im}(T_0^* P) \subset \mathrm{Im}\, T_0^*$, $\pi^{\perp} \subset \ker T_n$. Hence $T_n \pi^{\perp} = 0$, it follows that

$$\widetilde{E}_\lambda = \sum_{\alpha=0}^{n} \lambda^\alpha \widetilde{T}_\alpha. \tag{3}$$

If $\widetilde{T}_0 = 0$ or $\widetilde{T}_n = 0$, then $\widetilde{E}_\lambda$ is an $(n-1)$-uniton extended solution. If $\widetilde{T}_0 \neq 0$ and $\widetilde{T}_n \neq 0$, then $\widetilde{E}_\lambda$ is an n-uniton extended solution. □

From $\widetilde{E}_\lambda = E_\lambda(\pi + \lambda\pi^{\perp})$ and (3), we see that $\widetilde{T}_0 = T_0\pi = T_0 - T_0\pi^{\perp}$. It is easy to see that $\mathrm{Im}\, \widetilde{T}_0 \subset \mathrm{Im}\, T_0, \mathrm{rank}\, \widetilde{T}_0 \leq \mathrm{rank}\, T_0$. Because of $\pi^{\perp}(C^N) = \mathrm{Im}(T_0^* P)$, we obtain

$$\mathrm{Im}(T_0 \pi^{\perp}) = \mathrm{Im}(T_0 T_0^* P). \tag{4}$$

Lemma 4.4 If $T_0 \neq 0$, then there exists a subspace V of $\mathrm{Im}\, T_0(q)$ for a point $q \in \Omega$ such that $\lim_{a \to 0} (E_a)^* V$ exists and $\mathrm{rank}\, \widetilde{T}_0 < \mathrm{rank}(T_0)$.

Proof: Since $E_a^* = T_0^* + \overline{a} T_1^* + \cdots + \overline{a}^n T_n^*$, $\lim_{a \to 0} \widetilde{P}(a)$ exists if and only if the set

$$T_0^* V \cup T_1^*(\ker T_0^* \cap V) \cup \cdots \cup T_n^*(\cap_{\alpha=0}^{n-1}(\ker T_\alpha^* \cap V)) \tag{5}$$

is a k-dimensional subbundle of $\Omega \times C^N$, i.e., if and only if one term in (5) contains all the other terms. If

$$T_0^* V \supset T_1^*(\ker T_0^* \cap V) \supset \cdots \supset T_n^*(\cap_{\alpha=0}^{n-1}(\ker T_\alpha^* \cap V)), \tag{6}$$

holds, then we have

$$\lim_{a \to 0} \widetilde{P}(a) = \pi^\perp : \Omega \times C^N \longrightarrow \operatorname{Im}(T_0^* P). \tag{7}$$

Let V be any non-trivial k-dimensional subspace of $\operatorname{Im} T_0(q)$ for a fixed point $q \in \Omega$. Because of $\ker T_0^* \bigoplus \operatorname{Im} T_0 = \Omega \times C^N$, we have

$$\ker T_0^* \cap V = \{0\}. \tag{8}$$

Hence (6) holds, i.e., $\lim_{a \to 0}(E_a)^* V$ exists for any k-dimensional subspace V for a fixed point $q \in \Omega$. Since $\operatorname{Im} T_0^* \cap \ker T_0 = \{0\}$, it follows from (8) that $T_0 T_0^* P(V) = T_0 T_0^*(V) \neq 0$ which, together with (4), yields $\operatorname{rank} \widetilde{T}_0 < \operatorname{rank} T_0$. □

From Lemma 4.4, we know that there exists a k-dimensional subspace V of $\operatorname{Im}(T_0(q))$ for some $1 \leq k \leq \dim(\operatorname{Im}(T_0(q)))$ such that

$$\lim_{a \to 0} \widetilde{P}(a) = \pi^\perp : \Omega \times C^N \longrightarrow \operatorname{Im}(T_0^* P),$$

i.e. $\pi : \Omega \times C^N \longrightarrow \ker(P T_0)$ gives a singular dressing action and $\operatorname{rank} \widetilde{T}_0 < \operatorname{rank} T_0$, where P is the Hermitian orthogonal projection onto V. By Lemma 4.4 we obtain the main theorem as follows:

Theorem 4.5 Let $s : \Omega \longrightarrow U(N)$ be a harmonic map with minimal uniton number n. Then s can be factorized into the product of maps of Ω into complex Grassmann manifolds of the form

$$s = Q(\beta_0 - \beta_0^\perp) \cdots (\beta_j - \beta_j^\perp)$$

in such a way that the flag factor $\beta_i^\perp$ of $s_i = Q(\beta_0 - \beta_0^\perp) \cdots (\beta_i - \beta_i^\perp)$ can be obtained by a singular dressing action for each $i = 0,\ 1,\ \cdots,\ j$, where $Q \in U(N)$ is a constant matrix.

Proof: Let $E_\lambda^{(0)} = E_\lambda = \sum_{\alpha=0}^n \lambda^\alpha T_\alpha^{(0)}$ be the standard extended solution of s and P_0 the Hermitian orthogonal projection onto any k_0-dimensional subspace V_0 of $\operatorname{Im}(T_0^{(0)}(q))$ for some $1 \leq k_0 \leq \dim(\operatorname{Im}(T_0^{(0)}(q)))$ at a point $q \in \Omega$. By Lemma 4.4, we can choose a k_0-dimensional subspace V_0 of $\operatorname{Im}(T_0^{(0)}(q))$ such that

$$\pi_0 : \Omega \times C^N \longrightarrow \ker(P_0 T_0^{(0)})$$

gives a singular dressing action and $\operatorname{rank}(T_0^{(1)}) < \operatorname{rank}(T_0^{(0)})$. Let $E_\lambda^{(1)} = E_\lambda^{(0)}(\pi_0 + \lambda \pi_0^\perp)$. Then by Lemma 4.3 we have either $E_\lambda^{(1)} = \lambda \sum_{\alpha=0}^{n-1} \lambda^\alpha T_\alpha^{(1)}$ (an $(n-1)$-uniton extended solution) or $E_\lambda^{(1)} = \sum_{\alpha=0}^n \lambda^\alpha T_\alpha^{(1)}$ ($\operatorname{rank}(T_0^{(1)}) \neq 0$). If $E_\lambda^{(1)} = \sum_{\alpha=0}^n \lambda^\alpha T_\alpha^{(1)}$ is an n-uniton extended solution, then we repeat the above step and choose a k_1-dimensional subspace V_1 of $\operatorname{Im}(T_0^{(1)}(q))$ such that

$$\pi_1 : \Omega \times C^N \longrightarrow \ker(P_1 T_0^{(1)})$$

gives a singular dressing action and $\operatorname{rank}(T_0^{(2)}) < \operatorname{rank}(T_0^{(1)})$, where P_1 is the Hermitian orthogonal projection onto a k_1-dimensional subspace V_1 of $\operatorname{Im}(T_0^{(1)}(q))$ for some $1 \leq k_1 \leq \dim(\operatorname{Im}(T_0^{(1)}(q)))$, $E_\lambda^{(2)} = \sum_{\alpha=0}^n \lambda^\alpha T_\alpha^{(2)}$.

Repeating this procedure, we have

$$0 \leq \cdots < \operatorname{rank}(T_0^{(i)}) < \cdots < \operatorname{rank}(T_0^{(1)}) < \operatorname{rank}(T_0^{(0)}) < N,$$

so $\operatorname{rank}(T_0^{(j_1)}) = 0$ for some j. At the same time $E_\lambda^{(0)}$ can certainly be reduced to an $(n-1)$-uniton extended solution, i.e.,

$$E_\lambda^{(j_1)} = E_\lambda^{(0)}(\pi_0 + \lambda\pi_0^\perp)\cdots(\pi_{j_1-1} + \lambda\pi_{j_1-1}^\perp)$$

is an $(n-1)$-uniton extended solution, and

$$\pi_i : \Omega \times C^N \longrightarrow \ker(P_i T_0^{(i)})$$

gives a singular dressing action and $\operatorname{rank}(T_0^{(i+1)}) < \operatorname{rank}(T_0^{(i)})$, where P_i is the Hermitian orthogonal projection onto a k_i-dimensional subspace V_i of $\operatorname{Im}(T_0^{(i)}(q))$ for some $1 \leq k_i \leq \dim(\operatorname{Im}(T_0^{(i)}(q)))$ and $E_\lambda^{(i)} = \sum_{\alpha=0}^n \lambda^\alpha T_0^{(i)}$ for $0 \leq i < j_1$. Since E_λ is a polynomial in λ, by repeating the above steps, it follows that there exists a positive integer j such that

$$E_\lambda^{(j)} = \lambda^n I, \ E_\lambda = \lambda^n(\pi_j + \lambda^{-1}\pi_j^\perp)\cdots(\pi_0 + \lambda^{-1}\pi_0^\perp).$$

Let $\beta_i = \pi_{n-i}^\perp$ $(i = 0,\ 1,\ \cdots,\ j)$, then

$$s = QE_{-1} = Q(\beta_0 - \beta_0^\perp)\cdots(\beta_j - \beta_j^\perp).$$

as required. □

Because every harmonic map of the two-sphere into a unitary group has finite uniton number, any harmonic map of the two-sphere into a unitary group has a factorization of the above type.

References

[1] K. Uhlenbeck, Harmonic maps into Lie groups (classical solutions of the chiral model), J. Diff. Geom., 30 (1989), 1–50.

[2] M.J. Bergvelt and M.A. Guest, Action of loop groups on harmonic maps, Trans. Amer. Math. Soc., 326 (2) (1991), 861–886.

[3] F.E. Burstall and M.A. Guest, Harmonic two-spheres in compact symmetric spaces, revisited, Math. Ann., 309 (1997), 541–572.

[4] M.A. Guest and Y. Ohnita, Group actions and deformations for harmonic maps, J. Math. Soc. Japan, 45 (4) (1993), 671–704.

[5] A.N. Pressley and G. Segal, Loop Groups, Oxford Math. Monographs, Clarendon Press, Oxford, 1986.

[6] Y.X. Dong and Y.B. Shen, Factorization and uniton numbers for harmonic maps into the unitary groups U(N), Sci. in China, Ser. A, 39 (1996), 589–597.

[7] J. Dorfmeister, F. Pedit and H. Wu, Weierstrass type representation of harmonic maps into symmetric spaces, Comm. in Anal. and Geom., 6 (4) (1998), 633–668.

[8] J.C. Wood, The explicit construction and parametrization of harmonic two-spheres in the unitary group, Proc. London Math. Soc., 58 (1989), 608–624.

Institute of Mathematics, Academy of Mathematics and System Sciences, Chinese Academy of Sciences, Beijing 100080, P.R. CHINA

Current address: Department of Mathematics, Graduate School, Chinese Academy of Sciences, Beijing 100039, P.R. CHINA

E-mail address: xxj@gscas.ac.cn

Contemporary Mathematics
Volume **308**, 2002

On Submersive p-Harmonic Morphisms and their Stability

Hong Jin and Xiaohuan Mo

1. Introduction

Let (M^m, g) and (N^n, h) be two Riemannian manifolds. Let $\phi : M \to N$ be a horizontally conformal submersion. When N is a surface (i.e. $n = 2$), J. Eells and P. Baird [**1**] showed that ϕ is harmonic if and only if ϕ has minimal fibres, which is equivalent to the fact that the parallel displacement defined by the horizontal distribution preserves the volume form of the fibres. When the target manifold is an arbitrary Riemannian manifold, P. Baird and S. Gudmundsson [**2**] proved that ϕ has minimal fibres if and only if ϕ is n-harmonic where $n = \dim N$. Combined with the recent result of E. Loubeau [**3**] that ϕ is a p-harmonic morphism if and only if ϕ is a horizontally conformal p-harmonic map, we have:

COROLLARY 1. *Let $\phi : (M^m, g) \to (N^n, h)$ be a horizontally conformal submersion. Then ϕ is an $n-$harmonic morphism if and only if the parallel displacement defined by the horizontal distribution preserves the volume of the fibres.*

A p-harmonic morphism is a map which pulls back p-harmonic functions on N to p-harmonic functions on M. In particular, it reduces to the usual concept of harmonic morphism when $n = 2$. Recently, R. Pantilie [**5**] developed the results of J. Eells and P. Baird further. He proved that a horizontally conformal submersion $\phi : (M, g) \to (N, h)$ with dilation λ is a harmonic morphism if and only if the parallel displacement defined by the horizontal distribution preserves the mass of the fibres, where the fibres are given mass density $\lambda^{2-n}\omega$, where ω is the volume form of the fibres. The first purpose of our article is to give a uniform treatment of the assertions mentioned above. We obtain:

THEOREM 2. *Let $\phi : (M^m, g) \to (N^n, h)$ be a horizontally conformal submersion with dilation λ. Then ϕ is a p-harmonic morphism if and only if the parallel displacement defined by the horizontal distribution preserves the mass of the fibres, where the fibres are given mass density $\lambda^{p-n}\omega$ and ω denotes the volume form of the fibres.*

2000 *Mathematics Subject Classification.* Primary 58E20, 53C42.
This work is supported by the National Natural Science Foundation of China (10171002) and the Foundation of the School of Mathematical Sciences, Peking University.

The second purpose of this note is to discuss the relation between stability of ϕ as an n-harmonic map and stability of its fibres as minimal submanifolds of M. We shall prove the following generalization of S. Montaldo's result **[4]** (which is the case n=2).

THEOREM 3. *Let $\phi : (M^m, g) \to (N^n, h)$ be a horizontally conformal submersion between two compact Riemannian manifolds. If its fibres are all stable minimal submanifolds in M, ϕ is a stable n-harmonic morphism.*

2. Preliminaries

Let (M, g) and (N, h) be two Riemannian manifolds. Let $\phi : M \to N$ be a smooth map. For $p \geq 1$,we define the p-energy of ϕ by

$$E_p(\phi) = \frac{1}{p}\int_M |d\phi|^p\, dx$$

where $|d\phi|$ is the Hilbert-Schmidt norm of $d\phi$ and dx the volume form on M. The map ϕ is called p-harmonic if and only if it is a critical point of $E_p(\phi)$. When $p = 2$, this reduces to the definition of harmonic map.

LEMMA 4. **[2]** *A map $\phi : (M, g) \to (N, h)$ is p-harmonic if and only if it satisfies the Euler-Lagrange equation*

$$\tau_p(\phi) := |d\phi|^{p-2}\left[\tau(\phi) + d\phi\left(\mathrm{grad}\left(\log|d\phi|^{p-2}\right)\right)\right] \equiv 0 \tag{1}$$

where $\tau(\phi)$ is the tension field of ϕ. We call $\tau_p(\phi)$ the p-tension field of ϕ.

We assume that $\phi : (M^m, g) \to (N^n, h)$ is a submersion once and for all. Then for $x \in M$ let $V_x = \ker d\phi_x \subset T_xM$ and $H_x = V_x^{\perp} \subset T_xM$. Let $V := \{V_x\,|x \in M\}$ and $H := \{H_x\,|x \in M\}$ be the two distributions on M; these are called the vertical and horizontal distributions defined by ϕ. By the suffices V and H we denote also the orthogonal projections of TM onto V and H respectively. We denote the pull-back of the metric tensor h on N by ϕ^*h. Then for $x \in M$, $X, Y \in T_xM$, $(\phi^*h)_x(X, Y) = h_{\phi(x)}(d\phi(X), d\phi(Y))$. We define a section A of the vector bundle $\mathrm{End}(TM, TM)$ such that for $x \in M$, $A_x = d\phi^*_{\phi(x)} \circ d\phi_x$. Hence A_x is a self-adjoint operator and its rank equals n, and we have, for all $x \in M$ and $X, Y \in T_xM$, $g(A_xX, Y) = \phi^*h(X, Y)$. Let $\lambda_1^2, \lambda_2^2, \cdots, \lambda_n^2$ denote the non-zero eigenvalues of A.

LEMMA 5. **[4]** *If M is compact, then*

(1) $e(\phi) := \frac{1}{2}|d\phi|^2 = \frac{1}{2}\left(\lambda_1^2 + \lambda_2^2 + \cdots + \lambda_n^2\right)$

(2) $E_p(\phi) = \frac{1}{p}\int_M \left(\lambda_1^2 + \lambda_2^2 + \cdots + \lambda_n^2\right)^{\frac{p}{2}} dx$

(3) $\int_N \mathrm{vol}\left(\phi^{-1}(y)\right) dy = \int_M \lambda_1\lambda_2\cdots\lambda_n dx$.

PROOF. (1) and (2) are obvious. For (3), we decompose the volume form dx of M into the exterior product of its horizontal part dx^H and vertical part dx^V: $dx = dx^H \wedge dx^V$. If we use dy to denote the volume form of N, we have $dy = \sqrt{\det A}\, dx^H = \lambda_1\lambda_2\cdots\lambda_n dx^H$. So $\int_N \mathrm{vol}\left(\phi^{-1}(y)\right) dy = \int_N \left(\int_{\phi^{-1}(y)} dx^V\right) dy = \int_M \lambda_1\lambda_2\cdots\lambda_n dx^H dx^V = \int_M \lambda_1\lambda_2\cdots\lambda_n dx$. $\square$

As a generalization of the concept of Riemannian submersion, we have the following definition.

DEFINITION 6. **[2]** *Let* $\phi : (M^m, g) \to (N^n, h)$ *be a submersion between two Riemannian manifolds . If there exists a function* $\lambda : M \to R^+$ *such that* $\lambda^2 g(X,Y) = \phi^* h(X,Y)$ *for all* $X, Y \in H_x$ *and* $x \in M$, *then* ϕ *is said to be horizontally conformal and* λ *is called the dilation of* ϕ.

LEMMA 7. **[4]** ϕ *is horizontally conformal if and only if* $\lambda_1^2 = \lambda_2^2 = \cdots = \lambda_n^2$ *for all* $x \in M$.

DEFINITION 8. **[6]** *Let* V *be a distribution on a Riemannian manifold* (M^m, g) *with codimension* n *and* $H = V^\perp$ *its orthogonal distribution. The second fundamental form of* V *is defined by* ${}^V B(X,Y) := \frac{1}{2}(\nabla_X Y + \nabla_Y X)_H$ *for all* $X,\ Y \in V$ *where the suffix* H *indicates the orthogonal projection of* TM *onto* H *and* ∇ *denotes Levi-Civita connection on* M.

It is clear that the trace of ${}^V B$ can be expressed by

$$\operatorname{trace}\left({}^V B\right) := \sum_{k=1}^{m-n} {}^V B(X_k, X_k)$$

for any orthonormal basis $X_1,\ X_2, \cdots, X_{m-n}$ of V.

LEMMA 9. **[1]** *Let* ϕ *be a horizontally conformal submersion with dilation* λ. *Then it satisfies the fundamental equation*

$$\tau(\phi) + \operatorname{trace}\left({}^V B\right) + (n-2)\left(\operatorname{grad}(\log\lambda)\right)_H = 0 \tag{2}$$

where the suffix H *indicates the orthogonal projection of* TM *onto* H.

LEMMA 10. **[2]** *Let* $\phi : (M^m, g) \to (N^n, h)$ *be a horizontally conformal submersion. Then* ϕ *is* n*-harmonic if and only if all the fibres of* ϕ *are minimal submanifolds of* M, *where* $n = \dim\ N$.

DEFINITION 11. *Let* ϕ *be a* p*-harmonic map. We call* ϕ *a stable* p*-harmonic map or* E_p-*stable if*

$$\frac{d^2}{dt^2} E_p(\phi_t)\,|_{t=0} \geq 0$$

for all smooth variations $\phi_t : M \times (-\varepsilon, +\varepsilon) \to N$ *such that* $\phi_0 = \phi$.

3. Proof of theorem 2

PROPOSITION 12. *Let* $\phi : (M^m, g) \to (N^n, h)$ *be a horizontally conformal submersion with dilation* λ . *Then*

$$\frac{\tau_p(\phi)}{(n\lambda^2)^{\frac{p-2}{2}}} + \operatorname{trace}\left({}^V B\right) + (n-p)\left(\operatorname{grad}(\log\lambda)\right)_H = 0 \tag{3}$$

where $\tau_p(\phi)$ *is the* p*-tension field of* ϕ *and the suffix* H *indicates the orthogonal projection of* TM *onto* H.

PROOF. Because ϕ is horizontally conformal, we know $|d\phi|^2 = n\lambda^2$. It is easy to complete the proof from Lemma 4 and Lemma 9. $\square$

Let ω denote the local volume form of the vertical distribution V of ϕ. Let X be a horizontal vector field (a section of the vector bundle H on M) and let $\mathcal{L}_X$ denote the Lie derivation operation along X. Let $\mathcal{V}^*\eta$ denote the restriction of a differential form η on M to the subbundle V of TM. We have the following statement which is equivalent to Proposition 12.

PROPOSITION 13. *Let $\phi : (M^m, g) \to (N^n, h)$ be a horizontally conformal submersion with dilation λ. Then for any horizontal vector field X*

$$\mathcal{V}^*\left(\mathcal{L}_X\left(\lambda^{p-n}\omega\right)\right) = n^{\frac{2-p}{2}}\left\langle X, \tau_p(\phi)\right\rangle \lambda^{2-n}\omega \tag{4}$$

where $\langle X, Y\rangle$ denotes $g(X, Y)$. We call $\lambda^{2-n}\omega$ the mass density of the fibres.

PROOF. This is a straightforward computation.

$$\begin{aligned}
\mathcal{V}^*\left(\mathcal{L}_X\left(\lambda^{p-n}\omega\right)\right) &= X\left(\lambda^{p-n}\right)\omega + \lambda^{p-n}\left(\mathcal{V}^*\mathcal{L}_X\omega\right) \\
&= \left\{X\left(\lambda^{p-n}\right) - \langle X, \lambda^{p-n}\,\mathrm{trace}\left({}^V B\right)\rangle\right\}\omega \\
&= \langle X, \mathrm{grad}\,\lambda^{p-n} - \lambda^{p-n}\,\mathrm{trace}\left({}^V B\right)\rangle\omega \\
&= \langle X, (\mathrm{grad}(\log\lambda^{p-n}))_H - \mathrm{trace}\left({}^V B\right)\rangle\lambda^{p-n}\omega \\
&= \frac{1}{(n\lambda^2)^{\frac{p}{2}-1}}\left\langle X, \tau_p(\phi)\right\rangle\lambda^{p-n}\omega \\
&= n^{\frac{2-p}{2}}\langle X, \tau_p(\phi)\rangle\lambda^{2-n}\omega.
\end{aligned}$$

The second equality follows from Lemma 14. □

LEMMA 14. *With the same notation as before, we have*

$$\mathcal{V}^*\mathcal{L}_X\omega = \langle X, \mathrm{trace}\left({}^V B\right)\rangle\omega. \tag{5}$$

PROOF. Let $X_1, X_2, \cdots, X_m$ be a local orthonormal frame field on M such that $X_1, X_2, \cdots, X_n$ are horizontal vector fields and $X_{n+1}, X_{n+2}, \cdots, X_m$ are vertical vector fields. Let $\theta_1, \theta_2, \cdots, \theta_m$ be a local co-frame field dual to $X_1, X_2, \cdots, X_m$. Then $\omega = \theta_{n+1} \wedge \theta_{n+2} \wedge \cdots \wedge \theta_m$. Let X be a horizontal vector field. Then we have $\langle \mathcal{L}_X\theta_\alpha, X_\alpha\rangle = -\langle\theta_\alpha, \mathcal{L}_X X_\alpha\rangle = \langle\theta_\alpha, [X_\alpha, X]\rangle = \langle X_\alpha, \nabla_{X_\alpha}X + \nabla_X X_\alpha\rangle = \langle\nabla_{X_\alpha}X, X_\alpha\rangle$ and

$$\begin{aligned}
\mathcal{V}^*\mathcal{L}_X\omega &= \mathcal{L}_X\left(\theta_{n+1}\wedge\theta_{n+2}\wedge\cdots\wedge\theta_m\right) \\
&= \sum_{\alpha=n+1}^{m}\theta_{n+1}\wedge\theta_{n+2}\wedge\cdots\wedge\mathcal{L}_X\theta_\alpha\wedge\cdots\wedge\theta_m \\
&= \left(\sum_{\alpha=n+1}^{m}\langle\nabla_{X_\alpha}X, X_\alpha\rangle\right)\omega \\
&= -\left(\sum_{\alpha=n+1}^{m}\langle\nabla_{X_\alpha}X_\alpha, X\rangle\right)\omega \\
&= -\langle X, \mathrm{trace}\left({}^V B\right)\rangle\omega
\end{aligned}$$

as required. □

PROOF OF THEOREM 2. If ϕ is a horizontally conformal submersion and p-harmonic, then $\tau_p(\phi) = 0$. From Proposition 13, $\mathcal{L}_X(\lambda^{p-n}\omega) = 0$ holds for any horizontal vector field X. On the other hand, if ϕ is a horizontally conformal submersion and $\mathcal{L}_X\left(\lambda^{p-n}\omega\right)=0$ for any horizontal vector field X, then we obtain $\tau_p(\phi)=0$ from the fact that $\tau_p(\phi)$ is also a horizontal vector field on M. □

4. Proof of theorem 3

If M is a compact manifold and $\phi:(M^m,g)\to(N^n,h)$ is a horizontally conformal n-harmonic submersion between two compact Riemannian manifolds, then all fibres of ϕ are minimal submanifolds of M, by Lemma 10. Let $\phi_t: M\times(-\varepsilon,+\varepsilon)\to N$, $\phi_0=\phi$ be an arbitrary variation of ϕ. When t is small, ϕ_t is a submersion and $\phi_t^{-1}(y)$ is a submanifold of M for all $y\in N$. We can consider $\phi_t^{-1}(y)$ as a variation of $\phi_0^{-1}(y)$. So we have

$$\frac{d}{dt}\operatorname{vol}\left(\phi^{-1}(y)\right)|_{t=0}=0. \tag{6}$$

PROOF OF THEOREM 3. From Lemma 5 we have

$$\begin{aligned} E_p(\phi_t) &= \frac{1}{n}\int_M [\left(\lambda_1^{\phi_t}\right)^2+\left(\lambda_2^{\phi_t}\right)^2+\cdots+\left(\lambda_n^{\phi_t}\right)^2]^{\frac{n}{2}}dx \\ &\geq \frac{1}{n}\int_M\left[n\left(\lambda_1^{\phi_t}\lambda_2^{\phi_t}\cdots\lambda_n^{\phi_t}\right)^{\frac{2}{n}}\right]^{\frac{n}{2}}dx \\ &= n^{\frac{n-2}{2}}\int_M \lambda_1^{\phi_t}\lambda_2^{\phi_t}\cdots\lambda_n^{\phi_t}dx \\ &= n^{\frac{n-2}{2}}\int_N \operatorname{vol}\left(\phi^{-1}(y)\right)dy. \end{aligned}$$

Let $f(t)=E_n(\phi_t)$, $g(t)=n^{\frac{n-2}{2}}\int_N\operatorname{vol}\left(\phi^{-1}(y)\right)dy$. Then $f(t)\geq g(t)$. When $t=0$, $\phi_t=\phi_0=\phi$ is horizontally conformal, so $\lambda_1=\lambda_2=\cdots=\lambda_n$ and the inequality above becomes an equality. Then we have $f(0)=g(0)$ and $f'(0)=g'(0)$ from the fact that ϕ is $n-$harmonic and its fibres are all minimal submanifolds of M. So $f''(0)\geq g''(0)$. But $g''(0)\geq 0$ as all fibres of ϕ are stable minimal submanifolds, so we can conclude that $f''(0)\geq 0$ and ϕ is E_n-stable. □

References

[1] P. Baird and J. Eells, A conservation law for harmonic maps, Lec. Notes in Math., **894** (1981) 1–25.

[2] P. Baird and S. Gudmundsson, p-harmonic maps and minimal submersions, Math. Ann., **294** (1992) 611–624.

[3] E. Loubeau, On p-harmonic morphisms, Diff. Geom. and Appl. **12** (2000), 219–229.

[4] S. Montaldo, Stability of harmonic morphisms to a surface, Internat. J. Math. **9** (1998) 865–875.

[5] R. Pantilie, Harmonic morphisms with one-dimensional fibres, Internat. J. Math. **10** (1999) 457–501.

[6] J.C. Wood, Harmonic morphisms, foliations and Gauss maps, Complex Differential Geometry and Non-linear Differential Equations, Contemp. Math. **49** (1986) 145–183

E-mail address: moxh@pku.edu.cn

LMAM, SCHOOL OF MATHEMATICAL SCIENCE, PEKING UNIVERSITY, BEIJING 100871, P.R. CHINA

Contemporary Mathematics
Volume **308**, 2002

On Kähler-Liouville Manifolds

Kazuyoshi Kiyohara

1. Introduction

In [**K1**] we introduced the notion of Kähler-Liouville manifolds and studied their global structures. The definition was as follows. Let M be a complete Kähler manifold, with $\dim_{\mathbb{C}} M = n$, and let $\mathcal{F}$ be an n-dimensional real vector space of functions on the (real) cotangent bundle T^*M. We say $(M, \mathcal{F})$ is a Kähler-Liouville manifold if the following conditions are satisfied:

1. The Hamiltonian E of the geodesic flow belongs to $\mathcal{F}$;
2. $\{F, H\} = 0$ for any $F, H \in \mathcal{F}$;
3. $F_p = F|_{T_p^*M}$ is a Hermitian form for any $p \in M$ and $F \in \mathcal{F}$;
4. F_p $(F \in \mathcal{F})$ are simultaneously normalizable for any $p \in M$;
5. $\dim \{F_p \mid F \in \mathcal{F}\} = n$ for some $p \in M$.

Our purpose there was to present a certain class of Kähler manifolds whose geodesic flows are integrable in the sense of symplectic geometry. Strictly speaking, however, the definition itself was insufficient to deduce the integrability of the geodesic flow. In fact, we assumed a certain nondegeneracy condition in order to obtain the integrability. Kähler-Liouville manifolds satisfying that condition were called of type (A) (see Section 2). In the compact case the structures of such manifolds were investigated in full detail. However, there was an incorrect statement in [**K1, Section 3**], and consequently we need another type of nondegeneracy condition (called properness) to save the results in the compact case (see [**K2**]). Thus, for instance, it was shown that compact, proper Kähler-Liouville manifolds of type (A) possess the structure of toric varieties in a natural way. Thanks to the general theory for toric varieties, we know a lot about the global structure of such manifolds.

Thus it becomes interesting to consider how (proper) Kähler-Liouville manifolds can fail to be of type (A), since the definition of Kähler-Liouville manifold itself is quite natural in relation to (real) Liouville manifolds (cf. [**K1, Part 1**]). The purpose of this paper is to study global structures of such manifolds, primarily in the compact case. Our main results may be stated as follows.

2000 *Mathematics Subject Classification.* Primary 53D25 53C22; Secondary 37J35, 70H06.

1. *A finite, partially ordered set $\mathcal{A}$ is naturally associated with any n-dimensional Kähler-Liouville manifold $(M, \mathcal{F})$, which is the quotient of the set of indices $\{1, \ldots, n\}$ by a certain equivalence relation. It has the property that for each $\alpha \in \mathcal{A}$ the subset $\{\beta \in \mathcal{A} \mid \beta \preceq \alpha\}$ is totally ordered.*

For each $\alpha \in \mathcal{A}$ we denote by $|\alpha|$ the number of indices in the equivalence class α. It turns out that $(M, \mathcal{F})$ is of type (A) if and only if $|\alpha| \geq 2$ for every maximal element $\alpha \in \mathcal{A}$. Now let $(M, \mathcal{F})$ be a compact, proper Kähler-Liouville manifold.

2. *There exists a natural holomorphic fiber bundle $\rho : M \to M^b$, where (the universal cover of) M^b is a product of one-dimensional Kähler manifolds, and the typical fiber is a compact, proper Kähler-Liouville manifold. This fiber is not necessarily of type (A), but has similar properties, i.e., the geodesic flow is integrable with the aid of a commutative Lie algebra of infinitesimal automorphisms, and with this Lie algebra it becomes a toric variety. The partially ordered set associated to the fiber is naturally identified with the subset $\mathcal{A}^f$ of $\mathcal{A}$, which is obtained by deleting all maximal elements $\alpha \in \mathcal{A}$ such that $|\alpha| = 1$. If $(M, \mathcal{F})$ is of type (A), then M^b is a point.*

Here the metric on the base manifold M^b is the one that is naturally obtained from $(M, \mathcal{F})$ (ρ-image of a certain element in $\mathcal{F}$), though ρ is not necessarily a Riemannian submersion. Also, the structure of a Kähler-Liouville manifold on the fiber is the one that is obtained by restricting the metric on M and the first integrals in $\mathcal{F}$ to a fiber.

We say that a compact, proper Kähler-Liouville manifold is of type (B) if it is not of type (A), and if the corresponding base manifold M^b described above is a product of projective lines $\mathbb{C}P^1$ and each line admits an effective action of $\mathbb{C}^\times$ whose subgroup $U(1)$ preserves the given metric on the $\mathbb{C}P^1$. In this case it will turn out that the action of the product of $(\mathbb{C}^\times)$'s on the base manifold can be lifted to M, and consequently M itself becomes a toric variety, and the geodesic flow becomes integrable. Namely:

3. *Any Kähler-Liouville manifold of type (B) admits the structure of a toric variety and with this the geodesic flow becomes integrable as in the case of type (A).*

In contrast with the case of type (A), the toric variety structure is not necessarily unique in the case of type (B). That is, if a factor $\mathbb{C}P^1$ is of constant curvature, then the choice of $U(1)$ is not unique. We also have the following result, which implies that compact, proper Kähler-Liouville manifolds of type (A) and (B) form a closed subclass in some sense.

4. *The fiber in result 2 is a compact, proper Kähler-Liouville manifold of type (A) or type (B).*

Result 2 indicates that there are two extreme cases for compact, proper Kähler-Liouville manifolds: One is the case of type (A) (the base is a point); the other is the case of a product of one-dimensional compact Kähler manifolds (the fiber is a point). In the latter case, the geodesic flow is generally not integrable. Since our primary concern is the integrability of geodesic flows, we may say that any compact, proper Kähler-Liouville manifold has a natural bundle structure whose fiber is the good factor and whose base is the bad factor.

The organization of this paper is as follows. In Section 2 we briefly review the major results obtained in [**K1, Part 2**], as well as the definitions of type (A) and properness. In Sections 3 and 4 we study local and global structures of Kähler-Liouville manifolds respectively. In the latter we assume properness of Kähler-Liouville manifolds for simplicity, though some of the results there can be obtained without this assumption. In Section 5 we study the compact case, and prove the results cited above. Many arguments needed here are similar or parallel to those in [**K1**]. In such cases we shall omit details and only indicate the corresponding places in [**K1**].

2. Review of the type (A) case

Let us first recall the definition of type (A). Let $(M, \mathcal{F})$ be a Kähler-Liouville manifold, with $\dim_{\mathbb{C}} M = n$, and let I be the complex structure. Define M^0 as the set of points $p \in M$ where condition (5) in the definition of Kähler-Liouville manifold holds. Then it turns out that M^0 is an open and dense subset of M.

Now, take a point $p \in M^0$ and a basis $F_1, \dots, F_n \in \mathcal{F}$. Then, by definition, there are an orthonormal frame $V_1, \dots, V_n$, $IV_1, \dots, IV_n$ of TM and functions f_{ij} $(1 \le i, j \le n)$ on M around p such that F_i can be written as

$$F_i = \sum_{j=1}^{n} f_{ij} \left(V_j^2 + (IV_j)^2\right) \qquad (1 \le i \le n).$$

Here V_j^2, $(IV_j)^2$ are symmetric tensor products, identified with functions on T^*M. Then it turns out that $\dim \operatorname{span} \{(df_{ij})_p \mid 1 \le i, j \le n\} \le n$. Thus we reach the following definition: $(M, \mathcal{F})$ is called *of type (A)* if

$$\dim \operatorname{span} \{(df_{ij})_p\} = n \ \text{ for some point } p \in M^0$$

Put $(a_{ij}) = (f_{ij})^{-1}$. Then we have $d\,(a_{ik}/a_{il})\,|_{D_j} = 0$ for any $j \neq i$ and any k, l, where D_j is the distribution spanned by V_j and IV_j. Therefore we can take functions $a_i > 0$ so that $d\,(a_{ik}/a_i)\,|_{D_j} = 0$ for any $j \neq i$ and any k. The function a_i can be taken as one of the functions $|a_{ik}|$ $(k = 1, \dots, n)$ that does not vanish at p (cf. [**K1, Proposition 1.1**]). Then we have another expression of the definition of type (A), which is used in [**K1, (1.9)**]: $(M, \mathcal{F})$ is called of type (A) if there is a point $p \in M^0$ such that

(2.1) for any i there is some $j (\neq i)$ such that $da_j|_{D_i} \neq 0$ at p.

Remark: There is a mistake in statement (1.9) in [**K1, p. 85**]; the condition $(\neq i)$ for j is omitted. The statement above is the correct one.

Next, we will give the definition of properness. A Kähler-Liouville manifold is called proper if for each $p \in M - M^0$ and each $F \in \mathcal{F} - \{0\}$ satisfying $F_p = 0$, there is some $\lambda \in T_p^* M$ such that $dF \neq 0$ at λ. Note that this definition of properness is formally identical with that for Liouville manifolds (see [**K1, Part 1**]).

Now let us give a quick overview of the results obtained in [**K1**] subject to the correction in [**K2**]. Suppose $(M, \mathcal{F})$ is of type (A). Let M^1 be the set of points $p \in M^0$ where condition (2.1) above is satisfied. Then

- M^1 is open, connected, and dense in M.

For any $p \in M^1$ we define a binary relation $\preceq$ on the set of indices $\{1, \ldots, n\}$ as follows: $i \preceq j \iff i \neq j$ and $da_j|_{D_i} \neq 0$ at p or $i = j$. It turns out that $i \preceq j$ and $j \preceq k$ imply $i \preceq k$. Hence, writing $i \sim j$ if $i \preceq j$ and $j \preceq i$, we have an equivalence relation $\sim$ and a partially ordered set $\mathcal{A} = \{1, \ldots, n\}/\sim$, whose order is also denoted by $\preceq$. Then we have the following result.

- The indices $\{1, \ldots, n\}$ (the numbering of the D_i's) are globally defined on M^0. Also, the relation $\preceq$ is constant and thus $\mathcal{A}$ is constant on M^1.

For each $\alpha \in \mathcal{A}$ let $|\alpha|$ denote the number of indices contained in the equivalence class α. So we have $\sum_{\alpha \in \mathcal{A}} |\alpha| = n$.

- $\mathcal{A}$ has the following properties: (1) For any $\alpha \in \mathcal{A}$, the subset $\{\beta \in \mathcal{A} \mid \beta \preceq \alpha\}$ is totally ordered; (2) If $\alpha \in \mathcal{A}$ is a maximal element, then $|\alpha| \geq 2$.

- A commutative Lie algebra $\mathfrak{k}$ of infinitesimal automorphisms of the Kähler manifold M is naturally associated to this structure, satisfying $\{Y, F\} = 0$ ($Y \in \mathfrak{k}$, $F \in \mathcal{F}$). In particular, the geodesic flow of M is integrable, with first integrals in $\mathfrak{k}$ and $\mathcal{F}$.

- Suppose $(M, \mathcal{F})$ is compact and proper. Put $\mathfrak{g} = \mathfrak{k} + I\mathfrak{k}$. Then the Lie transformation group G generated by $\mathfrak{g}$ is isomorphic to $(\mathbb{C}^\times)^n$, and with this action M becomes a toric variety. We call such a toric variety a *toric variety of KL-A type.*

- For any toric variety of KL-A type, there are (many) compact Kähler-Liouville manifolds of type (A) such that the associated toric variety is isomorphic to the given one.

3. Local structures on M^1

Let $(M, \mathcal{F})$ be an n-dimensional Kähler-Liouville manifold that is not necessarily of type (A). Let M^0 be as before and let $p \in M^0$. As explained in the previous section, we take an orthonormal frame $V_1, \ldots, V_n, IV_1, \ldots, IV_n$, subbundles D_i (defined around p), and positive functions a_i. Then, arranging the numbering suitably, we have a number m such that

$$\begin{aligned} 1 \leq i \leq m &\Rightarrow da_j|_{D_i} = 0 \text{ at } p \text{ for any } j \neq i \\ m+1 \leq i \leq n &\Rightarrow da_j|_{D_i} \neq 0 \text{ at } p \text{ for some } j \neq i. \end{aligned} \tag{3.1}$$

The number m may depend on the point p. It is zero somewhere if $(M, \mathcal{F})$ is of type (A). Otherwise, $m \geq 1$ everywhere. Let n_0 be the minimum of the number m on M^0, and put

$$M^1 = \{p \in M^0 \mid m = n_0 \text{ at } p\}.$$

Since the assignment $p \mapsto m$ is upper semicontinuous, it follows that M^1 is an open subset of M^0. Also, the definition coincides with the previous one in the case of type (A).

Now let $p \in M^1$, and suppose that the numbering of D_i's is taken so that (3.1) holds for $m = n_0$. Then it is clear that (3.1) also holds on a neighborhood of p, because n_0 is minimal. Let a_{ij} (and a_i) be as in the previous section. Put $b_{ij} = a_{ij}/a_i$, $W_i = a_i^{-1/2} V_i$.

PROPOSITION 3.1. *If $1 \le i \le n_0$, then $da_{jk}|_{D_i} = 0$ for any j, k on a neighborhood of p. In particular, choosing a_i as one of the $|a_{ik}|$'s that are nonzero at p, we have $da_i|_{D_i} = 0$ and b_{ik} =constant for any k on a neighborhood of p $(1 \le i \le n_0)$.*

PROOF. Since $2E = \sum_j \left(V_j^2 + (IV_j)^2\right)$ belongs to $\mathcal{F}$, we have $\sum_j a_{jk}$ =constant for any k. Suppose $1 \le i \le n_0$. Then $da_j|_{D_i} = db_{jk}|_{D_i} = 0$ $(j \ne i$, any $k)$. Thus we have $da_{jk}|_{D_i} = 0$ for $j \ne i$, and hence $da_{ik}|_{D_i} = 0$ for any k. □

Put

$$D(i) = \sum_{j \ne i} D_j \quad (1 \le i \le n_0), \qquad D^f = \bigcap_{1 \le i \le n_0} D(i) = \sum_{j=n_0+1}^{n} D_j.$$

Note that D^f is well-defined over M^1. We also put $D^b = \sum_{i=1}^{n_0} D_i$.

PROPOSITION 3.2. *The subbundles $D(i)$ and D^f are completely integrable.*

PROOF. Since $\{F_j, F_k\} = 0$, it follows (see [**K1, p. 84**]) that

$$[V_j, V_k],\ [V_j, IV_k],\ [IV_j, IV_k] \in D_j + D_k \qquad (j \ne k).$$

Moreover, as proved in [**K1, p. 85**], the Kähler condition implies

$$[V_j, IV_j] \equiv \operatorname{sgrad}(\log a_j) \qquad \bmod D_j \tag{3.2}$$

for any j, where $\operatorname{sgrad} u = I(\operatorname{grad} u)$, $\operatorname{grad} u$ being the gradient vector field of the function u. Therefore for $i \le n_0$, the D_i-component of $[V_j, IV_j]$ is zero for any $j \ne i$. This shows that $D(i)$ is completely integrable. Since D^f is an intersection of $D(i)$'s, it is also completely integrable. □

As shown in [**K1, p. 86**] we can take V_j $(n_0 + 1 \le j \le n)$ so that

$$(IV_j)b_{jl} = (IV_j)a_i = 0 \qquad (\text{any } i \ne j, \text{ any } l).$$

For $i \le n_0$, there is no particular choice of V_i. Then the corresponding W_j and IW_j satisfy the following

PROPOSITION 3.3. (1) *If $n_0 + 1 \le j, k \le n$, $j \ne k$, then*

$$[W_j, W_k] = [W_j, IW_k] = [IW_j, IW_k] = 0.$$

(2) *If* $1 \le i \le n_0$, $n_0 + 1 \le j \le n$, *then*

$$[W_i, W_j] = \alpha_{ij} IW_i, \quad [IW_i, IW_j] = \beta_{ij} W_i$$
$$[W_i, IW_j] = -\beta_{ij} IW_i, \quad [IW_i, W_j] = -\alpha_{ij} W_i,$$

where α_{ij} *and* β_{ij} *are functions.*

In particular, we know that the subbundles

$$D_+^f = \operatorname{span}\{V_{n_0+1}, \dots, V_n\}, \quad D_-^f = \operatorname{span}\{IV_{n_0+1}, \dots, IV_n\}$$

are completely integrable. The following proposition is similar to Proposition 1.7 in [**K1, p. 87**].

PROPOSITION 3.4. *Suppose* $1 \le i \le n$, $n_0 + 1 \le j, k \le n$, $i \ne j \ne k \ne i$. *Then:*
(1) $W_j W_k \log a_i = 0$,
(2) $(W_k \log a_i)(W_j \log a_i) = (W_k \log a_j)(W_j \log a_i) + (W_j \log a_k)(W_k \log a_i)$,
(3) $W_k W_j \log a_k = (W_k \log a_j)(W_j \log a_k)$,
(4) $[W_k, IW_k]|_{D_k}$ *is of the form* $c_k IW_k$, *and if* $W_k \log a_i \ne 0$, *then*

$$c_k = W_k \log a_i - W_k \log a_k - \frac{W_k^2 \log a_i}{W_k \log a_i}\ .$$

As in [**K1, p. 88**], we define a binary relation $\preceq$ on the set $\{1, \dots, n\}$ at each point on a neighborhood U of $p \in M^1$:

$$i \preceq j \text{ at } q \iff i \ne j \text{ and } d \log a_j|_{D_i} \ne 0 \text{ at } q, \text{ or } i = j.$$

The following lemma is clear.

LEMMA 3.5. (1) *If* $1 \le i \le n_0$, *then* $i \not\preceq j$ *at* q *for any* $j \ne i$.
(2) *If* $n_0 + 1 \le j \le n$, *then there is some* $k \ne j$ *such that* $j \preceq k$ *at* q.

Writing $i \sim j$ if $i \preceq j$ and $j \preceq i$, we know by Proposition 3.4 that $\sim$ is an equivalence relation, and the quotient set $\mathcal{A} = \{1, \dots, n\}/\sim$ becomes a partially ordered set with respect to the inherited relation (denoted by the same symbol $\preceq$). We shall denote by $[i] \in \mathcal{A}$ the equivalence class of i. Also, let $|\alpha|$ denote the number of indices contained in the equivalence class $\alpha \in \mathcal{A}$. The following proposition is obtained in the same way as Proposition 1.9 in [**K1, p. 88**].

PROPOSITION 3.6. *Let* $\alpha, \beta, \gamma \in \mathcal{A}$.
(1) *If* $\alpha \preceq \gamma$ *and* $\beta \preceq \gamma$, *then* $\alpha \preceq \beta$ *or* $\beta \preceq \alpha$. *Namely, for any fixed* $\gamma \in \mathcal{A}$, *the set of* $\delta \in \mathcal{A}$ *such that* $\delta \preceq \gamma$ *is a totally ordered subset of* $\mathcal{A}$.
(2) *If* $\alpha = [i]$ *and* $1 \le i \le n_0$, *then* $|\alpha| = 1$ *and* α *is a maximal element.*
(3) *If* $\alpha = [j]$ *is a maximal element, and if* $j \ge n_0 + 1$, *then* $|\alpha| \ge 2$.

We put

$$\mathcal{A}^b = \{[i] \in \mathcal{A} \mid 1 \le i \le n_0\}, \qquad \mathcal{A}^f = \mathcal{A} - \mathcal{A}^b.$$

Shrinking the neighborhood U of the point p if necessary, we have the following

PROPOSITION 3.7. *The relation $\preceq$ on the set $\{1,\dots,n\}$ is constant on U. Therefore the partially ordered set $\mathcal{A}$ and the subset $\mathcal{A}^b$ are constant on U.*

The proof is similar to that of Proposition 1.11 in [**K1, p. 89**]. Proposition 1.12 in [**K1, p. 90**] is replaced by the following

PROPOSITION 3.8. *For any $p \in M^1$ there are a neighborhood U of p, functions h_j $(n_0+1 \le j \le n)$ on U, and constants $e_{\alpha\beta}$ $(\alpha, \beta \in \mathcal{A}, \alpha \prec \beta)$ satisfying the following four conditions:*

(1) $dh_j|_{D^f_-} = dh_j|_{D_k} = 0 \quad (k \ne j), \quad W_j h_j \ne 0$ *everywhere on U;*

(2) $e_{\alpha\beta} = e_{\alpha\gamma} \qquad (\alpha \prec \beta \preceq \gamma)$;

(3) $h_j \ne h_k$ $(j \sim k, j \ne k)$, $\quad h_j + e_{\alpha\beta} \ne 0$ $(j \in \alpha, \alpha \prec \beta)$ *everywhere on U;*

(4) *The functions a_i $(1 \le i \le n)$ can be taken in the form*

$$a_i^{-1} = \left| \prod_{j\in\alpha, j\ne i} (h_j - h_i) \prod_{\gamma\prec\alpha} \prod_{l\in\gamma} (h_l + e_{\gamma\alpha}) \right| \qquad (i \in \alpha).$$

Put

$$u_\alpha = \prod_{\gamma\prec\alpha} \prod_{l\in\gamma} (h_l + e_{\gamma\alpha}).$$

Then we have $a_i^{-1} = |u_\alpha|$ if $|\alpha| = 1$. Also, we have $u_\alpha = 1$ if α is minimal. Note that if $\{\tilde{h}_j\}$ and $\{\tilde{e}_{\alpha\beta}\}$ are other possible choices, then there are constants $c_\alpha \ne 0$ and d_α $(\alpha \in \mathcal{A}^f)$ such that $\tilde{h}_j = c_\alpha h_j - d_\alpha$, $\tilde{e}_{\alpha\beta} = c_\alpha e_{\alpha\beta} + d_\alpha$ $(j \in \alpha, \alpha \prec \beta)$.

Now we assume that each equivalence class $\alpha \in \mathcal{A}$ consists of successive numbers: $\alpha = \{s(\alpha), s(\alpha)+1, \dots, t(\alpha)\}$. Then, by taking a suitable basis $F_1, \dots, F_n$ of $\mathcal{F}$, the functions $\{b_{ij}\}$ can be described explicitly in terms of $\{h_i\}$ and $\{e_{\alpha\beta}\}$, by the same formulae as given in Proposition 1.13 in [**K1, p. 92**]. Note that $b_{il} = \delta_{il}$ for $1 \le i \le n_0$ in this situation. Computing the inverse of (a_{ij}), we have the following proposition. Let $n(\alpha)$ be the set of $\beta \in \mathcal{A}$ such that $\alpha \prec \beta$ and there is no γ between α and β.

PROPOSITION 3.9. *Suppose $i \in \alpha$. Then*

$$f_{ij} = \begin{cases} |u_\alpha| \mathcal{S}_{t(\alpha)-i}(h_l\,;\, l \in \alpha - \{j\}) & (j \in \alpha) \\ |u_\alpha| \sum_{k=0}^{t(\alpha)-i} e^k_{\alpha\beta} \mathcal{S}_{t(\alpha)-i-k}(h_l\,;\, l \in \alpha) & (j \in \gamma \succeq \beta, \beta \in n(\alpha)) \\ 0 & (\textit{otherwise}), \end{cases}$$

where $\mathcal{S}_k(h_l\,;\, l \in \alpha)$ stands for the elementary symmetric function of degree k $(0 \le k \le |\alpha|)$ with respect to $|\alpha|$ functions $\{h_l \mid l \in \alpha\}$. In particular, $f_{ij} = |u_\alpha|\delta_{ij}$ for $i \le n_0$, i.e., $F_i = |u_\alpha|\left(V_i^2 + (IV_i)^2\right)$ $(1 \le i \le n_0,\ \alpha = [i])$.

The following proposition can be proved in the same way as in [**K1, Section 2**].

PROPOSITION 3.10. *Let $M^{1,0}$ be a connected component of M^1. Then the subbundles D_i $(1 \le i \le n)$ and the binary relation $\preceq$ on $\{1,\dots,n\}$ (and hence the partially ordered set $\mathcal{A}$ and its subset $\mathcal{A}^b$) are globally defined on $M^{1,0}$. Moreover,* *the functions h_i $(i > n_0)$ and the constants $e_{\alpha\beta}$ are also globally defined on $M^{1,0}$. Namely, one may take $U = M^{1,0}$ in Proposition 3.8.*

Rearranging the ordering of indices in each $\alpha = \{s(\alpha), \dots, t(\alpha)\}$ if necessary, we may assume $h_{s(\alpha)} > h_{s(\alpha)+1} > \cdots > h_{t(\alpha)}$ $(\alpha \in \mathcal{A}^f)$ on $M^{1,0}$. Put $v_\alpha(\lambda) = u_\alpha \prod_{i\in\alpha}(h_i - \lambda)$ for each $\alpha \in \mathcal{A}^f$, and let $\mathfrak{k}$ be the vector space spanned by vector fields of the form $\operatorname{sgrad} v_\alpha(\lambda)$ $(\alpha \in \mathcal{A}^f,\ \lambda \in \mathbb{R})$.

PROPOSITION 3.11. *$\mathfrak{k}$ is an $(n - n_0)$-dimensional commutative Lie algebra of infinitesimal automorphisms of the Kähler manifold $M^{1,0}$, and it satisfies $\{Y, F\} = 0$ $(Y \in \mathfrak{k},\ F \in \mathcal{F})$.*

The proof is similar to that of Proposition 1.16 in [**K1**].

4. Global structures

Similar results to Theorem 3.1 and Proposition 3.13 in [**K1**] also hold in this case. But, as stated in the Introduction, some of those results are incorrect without an additional condition. So, from now on, we shall assume that *Kähler-Liouville manifolds are always proper* unless otherwise stated.

THEOREM 4.1. *$M - M^1$ is a locally finite union of closed, totally geodesic, complex hypersurfaces L. In particular M^1 is connected and dense in M. Moreover, the functions h_i $(i > n_0)$ can be continuously extended to the whole of M and the subbundles D_i $(1 \le i \le n)$ can be smoothly extended to M^0, and they possess the following properties:*

1. *$h_i + e_{\alpha\beta}$ $(i \in \alpha,\ \alpha \prec \beta)$ are everywhere nonzero on M;*
2. *$h_{s(\alpha)} > h_{s(\alpha)+1} > \cdots > h_{t(\alpha)}$ on M^0 $(\alpha \in \mathcal{A}^f)$;*
3. *h_i are C^∞ functions on M^0;*
4. *$\mathcal{S}_k(h_i\,;\, i \in \alpha)$ $(1 \le k \le |\alpha|)$ are C^∞ functions on M;*
5. *$D_\alpha = \sum_{i\in\alpha} D_i$ $(\alpha \in \mathcal{A})$ extend to the whole of M as C^∞ subbundles of TM;*
6. *The vector fields $Y \in \mathfrak{k}$ are globally defined and C^∞ on M;*
7. *For each hypersurface L there are $\alpha \in \mathcal{A}^f$ and $c \in \mathbb{R}$ such that a connected component of the set of zeros of* $\operatorname{sgrad} v_\alpha(c) \in \mathfrak{k}$ *coincides with L, and $v_\alpha(c)$ vanishes on L.*

PROPOSITION 4.2. *Let $q \in M - M^0$. Then the simultaneous eigenspaces D of the $\mathbb{C}$-linear endomorphisms $\{F_q^e \mid F \in \mathcal{F}\}$ are of dimension one or two. They can be smoothly extended to a neighborhood U of q as subbundles of TM in the following way: If* $\dim_{\mathbb{C}} D = 1$, *then $D = D_i$ on $U \cap M^0$ for some i; if* $\dim_{\mathbb{C}} D = 2$, *then there is $\alpha \in \mathcal{A}$ and $i, i+1 \in \alpha$ such that $D = D_i + D_{i+1}$ on $U \cap M^0$. In the first case, and if $i \in \alpha$, $|\alpha| \ge 2$, the function h_i is smooth around q, and*

$$h_j(q) \ne h_i(q) \qquad (j \in \alpha,\ j \ne i).$$

In the second case,

$$h_i(q) = h_{i+1}(q), \quad h_j(q) \ne h_i(q) \qquad (j \in \alpha,\ j \ne i, i+1),$$

and the functions h_i *and* h_{i+1} *are not differentiable at* q. *Moreover,* $h_i + h_{i+1}$ *and* $h_i h_{i+1}$ *are smooth functions around* q *such that* $d(h_i + h_{i+1}) \neq 0$ *and* $d((h_i - h_i(q))(h_{i+1} - h_i(q))) = 0$ *at* q.

The proofs are exactly the same as in [**K1, Section 3**] and its correction in [**K2**].

5. Compact case

In this section we assume M is compact (properness is also assumed). Put $\mathfrak{g} = \mathfrak{k} + I\mathfrak{k}$, and let K and G be the Lie transformation groups of M generated by $\mathfrak{k}$ and $\mathfrak{g}$ respectively.

THEOREM 5.1. *The Lie groups* K *and* G *are isomorphic to* $U(1)^{n-n_0}$ *and* $(\mathbb{C}^\times)^{n-n_0}$ *respectively.*

The proof is the same as that of Theorem 4.2 in [**K1**]. The crucial part is to show that $\mathfrak{k}$ is generated by elements $\operatorname{sgrad} v_\alpha(c)$ such that a hypersurface L contained in $M - M^1$ is given by $v_\alpha(c) = 0$. By Theorem 4.1 we have the global decomposition $TM = D^b \oplus D^f$. Let N be a maximal integral manifold of D^f in M.

PROPOSITION 5.2.
(1) $N \cap M^1$ *is connected, and* G *acts on* $N \cap M^1$ *simply transitively.*
(2) $N = \overline{N \cap M^1}$ *(the closure in* M*). In particular,* N *is closed.*

PROOF. (1) Recall that $M - M^1$ is a locally finite union of complex hypersurfaces. Since the normal bundle of each hypersurface L is contained in some D_α, $\alpha \in \mathcal{A}^f$, it follows that N is transversal to L. Therefore $N \cap M^1$ is connected. Since the vector fields in $\mathfrak{g}$ span the tangent space of N at each point in $N \cap M^1$, and since $N \cap M^1$ is preserved by G, it must be a G-orbit. Its isotropy subgroup is trivial, because G is abelian and the action is effective.

(2) Take a basis Y_i $(1 \le i \le n - n_0)$ of $\mathfrak{k}$ of the form

$$Y_i = \operatorname{sgrad} v_{\alpha_i}(c_i), \qquad \alpha_i \in \mathcal{A}^f, \quad c_i \in \mathbb{R},$$

where $v_{\alpha_i}(c_i) = 0$ represents some hypersurface contained in $M - M^1$. Let $\Phi : M \to \mathbb{R}^{n-n_0}$ be the map given by

$$p \mapsto (v_{\alpha_1}(c_1)(p), \ldots, v_{\alpha_{n-n_0}}(c_{n-n_0})(p)).$$

Then, for each $p \in N \cap M^1$, $\Phi^{-1}(\Phi(p)) \cap N$ is identical with the K-orbit passing through p.

Now take a point $p_0 \in \overline{N \cap M^1}$ and its neighborhood U in M. Let N_0 be the maximal integral manifold of D^f through p_0 in M. Let S be a small submanifold of M such that $\dim S = n_0$, $p_0 \in S$, and S is transversal to N_0 at p_0. We can take U and S in such a way that every integral manifold of D^f in U meets S at one point. Let $\rho : U \to S$ be such an assignment of intersection points. Namely, $\rho : U \to S$ is a fiber bundle whose fiber is an integral manifold of D^f in U. Let $q \in \rho^{-1}(p_0) \cap M^1$

and consider the K-orbit through q. Since it is tangent to the fiber, and since K is compact, we have $Kq \cap U \subset \rho^{-1}(p_0)$, provided U sufficiently small. Since $\Phi^{-1}(q)$ meets any fiber $\rho^{-1}(s)$ near $\rho^{-1}(p_0)$, and since those intersections are contained in different K-orbits from Kq, it follows that they are contained in different G-orbits from $Gq = N_0 \cap M^1$. Namely, we have $U \cap N_0 \cap M^1 = \rho^{-1}(p_0) \cap M^1$, provided U is sufficiently small. Now let $\{p_j\}$ be a sequence of points in $N \cap M^1$ converging to p_0. Then we have a sequence $\{q_j\}$ such that $\rho(p_j) = \rho(q_j)$, $q_j \in M^1$, q_j converge to $q_0 \in \rho^{-1}(p_0) \cap M^1$, and $\Phi(q_j) = \Phi(q_0)$ for any j. Since q_j are on the same K-orbit contained in N, so is the limit point q_0. Thus we have $\rho(q_j) = \rho(q_0)$ for large j, and $p_0 \in N$. □

The proposition above shows that every maximal integral manifold of D^f is compact. Therefore we have a holomorphic fiber bundle $\rho : M \to M^b$ whose fibers are maximal integral manifolds of D^f. The base manifold M^b is a compact complex manifold of dimension n_0.

THEOREM 5.3.

(1) *Each subbundle D_i ($1 \le i \le n_0$) is projectable with respect to ρ. Putting $\rho_*(D_i) = \tilde{D}_i$, we have a product structure, i.e., any $\tilde{D}_i$ and $\tilde{D}_i + \tilde{D}_j$ are integrable. In particular, the universal cover of M^b is the product of n_0 one-dimensional complex manifolds.*

(2) *For each $i \le n_0$, $F_i = \left|u_{[i]}\right| (V_i^2 + (IV_i)^2)$ is projectable, and defines a section $2\tilde{E}_i$ of $S^2\tilde{D}_i \subset S^2(TM^b)$ (S^2 denotes the symmetric product of two copies of a vector bundle).*

(3) *For each $q \in M^b$ let F' be the $S^2(D^f)$-component of $F \in \mathcal{F}$, and let $\mathcal{F}_q^f$ be the set of F' ($F \in \mathcal{F}$) restricted to $\rho^{-1}(q)$. Then $(\rho^{-1}(q), \mathcal{F}_q^f)$ is a compact, proper Kähler-Liouville manifold with respect to the induced Kähler metric from M, and the associated partially ordered set is naturally identified with $\mathcal{A}^f$. Moreover, the geodesic flow of $\rho^{-1}(q)$ is integrable with the first integrals in $\mathfrak{k}$ and $\mathcal{F}_q^f$.*

(4) *$(\rho^{-1}(q), \mathcal{F}_q^f)$ is a toric variety with respect to the action of G. This toric variety is of KL type.*

(5) *Let $\hat{X}$ denote the horizontal lift of a vector field X on M^b with respect to the horizontal distribution $D^b \subset TM$. Let $\hat{\psi}_t$ (respectively, ψ_t) be the flow generated by $\hat{X}$ (respectively, X). Then $\hat{\psi}_t$ gives the isomorphism of Kähler-Liouville manifolds $(\rho^{-1}(q), \mathcal{F}_q^f) \simeq (\rho^{-1}(\psi_t(q)), \mathcal{F}_{\psi_t(q)}^f)$.*

PROOF. By Proposition 3.2 we know that $D_i + D^f$ and $D_i + D_j + D^f$ are completely integrable for any $i \le n_0$. Thus we have (1). By Proposition 3.3 we also have

$$\{W_i^2 + (IW_i)^2, W_j\} = \{W_i^2 + (IW_i)^2, IW_j\} = 0 \qquad (1 \le i \le n_0,\ j > n_0).$$

Therefore $W_i^2 + (IW_i)^2 = \left|u_{[i]}\right| (V_i^2 + (IV_i)^2)$ is projectable, and (2) follows. (3) is easy. For (5) we use the formula in Proposition 3.3 (2). If $W_i = \sqrt{\left|u_{[i]}\right|}V_i$ ($i \le n_0$) is given as a horizontal lift of a vector field on M^b, then the flow generated by W_i

maps fibers to fibers. Therefore $\alpha_{ij} = \beta_{ij} = 0$ in this case. This shows

$$[\hat{X}, W_j] = [\hat{X}, IW_j] = \hat{X} h_j = 0 \qquad (n_0 + 1 \leq j \leq n),$$

and (5) follows. The proof of (4) is (rather long, but) exactly the same as Section 4 of [**K1**], so we omit it. The definition of "toric variety of KL type" is given at the beginning of Section 5 of [**K1**]. □

As mentioned in the Introduction, a compact, proper Kähler-Liouville manifold $(M, \mathcal{F})$ not of type (A) is called of type (B) if the base manifold M^b is the product of n_0 projective lines $\mathbb{C}P^1$, and each factor admits an effective, holomorphic $\mathbb{C}^\times$-action whose subgroup $U(1)$ preserves the given metric. Thus we have a $(\mathbb{C}^\times)^{n_0}$-action on M^b, which we call an admissible action.

THEOREM 5.4. *Let $(M, \mathcal{F})$ be of type (B), and suppose that an admissible action of $(\mathbb{C}^\times)^{n_0}$ on M^b is given. Then we have the following.*

1. *The infinitesimal action of $(\mathbb{C}^\times)^{n_0}$ on M^b can be lifted to M in a natural way.*
2. *The lifted action of $(\mathbb{C}^\times)^{n_0}$ together with the action of G makes M into a toric variety.*
3. *This toric variety is of KL type.*
4. *The Lie algebra of the maximal compact subgroup of $(\mathbb{C}^\times)^{n_0}$ together with $\mathfrak{k}$ and $\mathcal{F}$ makes the geodesic flow of M integrable.*

PROOF. Let Z_i be a vector field generating the $U(1)$-action on the i-th factor $\mathbb{C}P^1$ of M^b. Then there is a smooth function f_i on $\mathbb{C}P^1$ such that $\operatorname{sgrad} f_i = Z_i$ with respect to the given Kähler metric on $\mathbb{C}P^1$. Putting $\hat{f}_i = \rho^*(f_i)$, we have

$$\operatorname{sgrad}\left(u_{[i]}\hat{f}_i\right) = \hat{Z}_i + \hat{f}_i \operatorname{sgrad}\ u_{[i]}, \tag{5.1}$$

where $\hat{Z}_i$ is the horizontal lift of Z_i, and the skew gradient in the left hand side is the one with respect to the Kähler metric on M. Let W_j $(j \geq n_0 + 1)$ be as before, and choose W_i $(i \leq n_0)$ such that W_i is a horizontal lift of the unit vector field on the i-th factor $\mathbb{C}P^1$ proportional to $\operatorname{grad} f_i = -IZ_i$. Then the vector fields W_i are defined on an open and dense subset of M^1 where $d\hat{f}_i$ are not zero, and we have

$$[W_k, W_l] = [IW_k, IW_l] = [W_k, IW_l] = 0 \tag{5.2}$$

for any k, l $(k \neq l)$. We shall show that

$$\left[\operatorname{sgrad}\left(u_{[i]}\hat{f}_i\right), W_k\right] = \left[\operatorname{sgrad}\left(u_{[i]}\hat{f}_i\right), IW_k\right] = 0,\ 1 \leq i \leq n_0,\ 1 \leq k \leq n. \tag{5.3}$$

If the commutativity is shown, then by the same argument as Proposition 1.16 in [**K1**], we have (1) and (4). For $k \neq i$, (5.3) easily follows from (5.1) and (5.2). When $k = i$, the latter equality is also easy, and we have

$$\rho_*[\hat{Z}_i, W_i] = \left[Z_i, -|\operatorname{grad} f_i|^{-1} IZ_i\right] = 0.$$

We also have by formula (3.2)

$$[\hat{Z}_i, W_i] = [(\rho^* |\operatorname{grad} f_i|)\, IW_i, W_i] \equiv (\rho^* |\operatorname{grad} f_i|) \operatorname{sgrad} u_{[i]} \mod D_i.$$

Therefore we have

$$[\hat{Z}_i, W_i] = [(\rho^* |\mathrm{grad}\, f_i|) IW_i, W_i] = (\rho^* |\mathrm{grad}\, f_i|)\, \mathrm{sgrad}\, u_{[i]}.$$

Since $\rho^* |\mathrm{grad}\, f_i| = W_i \hat{f}_i$, (5.3) follows from (5.1) in this case.

Now let $\mathfrak{k}'$ be the commutative Lie algebra spanned by $\mathfrak{k}$ and vector fields $\mathrm{sgrad}\left(u_{[i]}\hat{f}_i\right)$ $(1 \le i \le n_0)$. Put $\mathfrak{g}' = \mathfrak{k}' + I\mathfrak{k}'$. Let K' and G' be the Lie transformation groups on M generated by $\mathfrak{k}'$ and $\mathfrak{g}'$ respectively. Let c_i^+ and c_i^- be, respectively, the maximum and minimum of the function $\hat{f}_i$, and let $L_i^\pm$ be the complex hypersurface determined by $u_{[i]}\left(\hat{f}_i - c_i^\pm\right)$. Put

$$(M^1)' = M^1 - \bigcup_{i=1}^{n_0} \left(L_i^+ \cup L_i^-\right).$$

Then we see that K' and G' are isomorphic to $U(1)^n$ and $(\mathbb{C}^\times)^n$ respectively as Lie groups, and that G' acts on M' simply transitively. Thus we can apply the same argument as Section 4 of [**K1**], and obtain (2) and (3). □

Finally, we shall prove result 4 from the Introduction. As shown in Theorem 5.3, each fiber $\rho^{-1}(q)$ $(q \in M^b)$ naturally possesses the structure of a Kähler-Liouville manifold, i.e., $(\rho^{-1}(q), \mathcal{F}_q^f)$.

THEOREM 5.5. *The compact, proper Kähler-Liouville manifold* $(\rho^{-1}(q), \mathcal{F}_q^f)$ *is of type (A) or type (B).*

PROOF. Suppose that it is not of type (A). Then there is some $\alpha \in \mathcal{A}^f$ such that α is maximal in $\mathcal{A}^f$ and $|\alpha| = 1$. The problem is to show that the corresponding base manifold is the product of projective lines. As we have stated in the proof of Theorem 5.3, the argument of Section 4 in [**K1**] is applicable to the Kähler-Liouville manifold $(\rho^{-1}(q), \mathcal{F}_q^f)$. Therefore, the resulting toric variety should be of KL type. Thus by [**K1, Proposition 5.5**] we have the desired result. □

References

[K1] K. Kiyohara, *Two classes of riemannian manifolds whose geodesic flows are integrable*, Mem. Amer. Math. Soc. **130/619** (1997).

[K2] K. Kiyohara, *Errata and addenda to "Two classes of riemannian manifolds whose geodesic flows are integrable"*, preprint.

DEPARTMENT OF MATHEMATICS, FACULTY OF SCIENCE, HOKKAIDO UNIVERSITY, SAPPORO 060-0810, JAPAN

E-mail address: kiyohara@math.sci.hokudai.ac.jp

Contemporary Mathematics
Volume **308**, 2002

Minimal Surfaces that Attain Equality in the Chern-Osserman Inequality

Masatoshi Kokubu, Masaaki Umehara, and Kotaro Yamada

The total curvature $\mathrm{TC}(M)$ of any complete minimal surface M in $\mathbb{R}^n$ has a value in $2\pi\mathbb{Z}$ and satisfies the following inequality called the *Chern-Osserman inequality* [**CO**]:

$$\mathrm{TC}(M) \leq 2\pi(\chi_M - m), \tag{1}$$

where χ_M denotes the Euler number of M and m is the number of ends of M.

In an earlier paper [**KTUY**], Takahashi and the authors generalized the theory of minimal surfaces in Euclidean n-space to that of surfaces with holomorphic Gauss map in a certain class of non-compact symmetric spaces, which also includes the theory of constant mean curvature one surfaces in hyperbolic 3-space. Moreover, a Chern-Osserman type inequality for such surfaces was obtained. Though the meaning of the equality condition is not yet clear, the authors have noticed that the equality condition of the Chern-Osserman inequality itself is not explained in the literature, except for the case $n = 3$, in spite of its importance. In the case $n = 3$, Jorge and Meeks [**JM**] have already shown that equality holds if and only if all of the ends are embedded.

In this paper, a simple geometric condition for minimal surfaces that attain equality in the Chern-Osserman inequality is given. The authors hope that it will be of interest. We shall prove the following:

Main Theorem. *A complete minimal surface in $\mathbb{R}^n$ attains equality in the Chern-Osserman inequality if and only if each end is asymptotic to a catenoid-type end or a planar end in some 3-dimensional subspace $\mathbb{R}^3$ in $\mathbb{R}^n$. When this is the case, all ends are embedded.*

For $n = 3$, according to Jorge-Meeks [**JM**] and Schoen [**S**], one can easily observe that embedded ends are all asymptotic to catenoids or planes (see the Appendix). So our theorem generalizes the result of Jorge-Meeks [**JM**]. For general n (> 3), we remark that the embeddedness of the ends is not a sufficient condition for equality in (1). For example, the embedded holomorphic curve $f\colon \mathbb{C}\setminus\{0\} \to \mathbb{C}^2$

2000 *Mathematics Subject Classification.* Primary 53A10; Secondary 53A07, 53C42.

defined by $f(z) = (z, 1/z^2)$ (considered as a complete minimal surface in $\mathbb{R}^4$) has total curvature -6π. So it does not satisfy equality in (1).

Preliminaries

We shall review the properties of minimal surfaces in $\mathbb{R}^n$ (cf. [**L**]). Let $f = (f^1, \dots, f^n)\colon M \to \mathbb{R}^n$ be a conformal minimal immersion of a Riemann surface M, where $n \geq 3$ is an integer. Then ∂f is a $\mathbb{C}^n$-valued holomorphic 1-form on M. We define the Gauss map $\nu\colon M \to \mathbb{CP}^{n-1}$ of f as

$$\nu := [\partial f] = \left[\frac{\partial f^1}{\partial z} : \frac{\partial f^2}{\partial z} : \cdots : \frac{\partial f^n}{\partial z}\right],$$

where z is a complex coordinate of M. Since f is conformal, we have

$$\langle \partial f, \partial f \rangle = \sum_{i=1}^{n} \left(\frac{\partial f^i}{\partial z}\right)^2 dz^2 = 0. \tag{2}$$

Thus, the Gauss map ν takes values in the complex quadric $\boldsymbol{Q}^{n-2} \subset \mathbb{CP}^{n-1}$.

We assume that f is complete and of finite total curvature. Under this assumption, the following properties are well-known:

- M is biholomorphic to a compact Riemann surface $\overline{M}$ punctured at finitely many points $\{p_1, \dots, p_m\}$. Each point p_j is called an *end.*
- The Gauss map ν can be extended holomorphically on $\overline{M}$, and the total curvature is given by $-2\pi d$ where d is the homology degree of $\nu(\overline{M})$ in $\mathbb{CP}^{n-1}$.
- For each end p_j, there exists a local complex coordinate z on $\overline{M}$ centered at p_j such that the first fundamental form ds^2 can be written as
$$ds^2 = |z|^{2\mu_j}\, dz\, d\bar{z} \qquad (\mu_j \in \mathbb{Z},\ \mu_j \leq -2).$$
We call μ_j the *order* of the metric ds^2 at the end p_j and write $\operatorname{ord}_{p_j} ds^2 = \mu_j$. Since $ds^2 = 2\langle \partial f, \overline{\partial f}\rangle$, μ_j coincides with the order of ∂f at the end p_j.

Definition. An end p_j of $f\colon M = \overline{M} \setminus \{p_1, \dots, p_m\} \to \mathbb{R}^n$ is said to be asymptotic to a *catenoid-type* (respectively, *planar*) end if there exists a piece of the catenoid (respectively, the plane)

$$f_0\colon \{|z - p_j| < \varepsilon\} \to \mathbb{R}^3 \subset \mathbb{R}^n$$

which is complete at p_j such that $|f(z) - f_0(z)| = O(|z - p_j|)$, that is,

$$\frac{|f(z) - f_0(z)|}{|z - p_j|}$$

is bounded on $\{|z - p_j| < \varepsilon\}$ for sufficiently small $\varepsilon > 0$.

Proof of the Main Theorem

The Chern-Osserman inequality follows from the fact that $\operatorname{ord}_{p_j} ds^2 \leq -2$ at each end p_j. Moreover, equality holds if and only if $\operatorname{ord}_{p_j} ds^2 = -2$ (see [**L**, pp. 135–136], for example). Thus the Main Theorem immediately follows from the following Lemma.

LEMMA 1. *Let* $f\colon \Delta^* \to \mathbb{R}^n$ *be a conformal minimal immersion of a punctured disc* $\Delta^* = \{z \in \mathbb{C} \mid 0 < |z| < 1\}$ *into* $\mathbb{R}^n$ *which is complete at the origin* 0*. Then* $\operatorname{ord}_0 ds^2 = -2$ *holds if and only if the end* 0 *is asymptotic to a catenoid-type end or a planar end in* $\mathbb{R}^3(\subset \mathbb{R}^n)$*. When this is the case, it is an embedded end.*

PROOF. Suppose that $\operatorname{ord}_0 ds^2 = -2$. This implies that the Laurent expansion of ∂f is given by

$$\text{(3)} \qquad \partial f = \left(\frac{1}{z^2}\boldsymbol{a}_{-2} + \frac{1}{z}\boldsymbol{a}_{-1} + \cdots\right) dz, \qquad \boldsymbol{a}_{-2} \in \mathbb{C}^n \setminus \{0\}, \quad \boldsymbol{a}_{-1} \in \mathbb{C}^n.$$

Here, since

$$f(z) = 2\operatorname{Re}\int_{z_0}^{z} \partial f \qquad \text{(where } z_0 \in M \text{ is a basepoint)}$$

is well-defined on Δ^*, the residue of ∂f must be real: $\boldsymbol{a}_{-1} \in \mathbb{R}^n$. Moreover, it follows from (2) that $\langle \boldsymbol{a}_{-2}, \boldsymbol{a}_{-2}\rangle = 0$ and $\langle \boldsymbol{a}_{-2}, \boldsymbol{a}_{-1}\rangle = 0$. Therefore we have

$$|\operatorname{Re}\boldsymbol{a}_{-2}| = |\operatorname{Im}\boldsymbol{a}_{-2}|, \qquad \langle \operatorname{Re}\boldsymbol{a}_{-2}, \operatorname{Im}\boldsymbol{a}_{-2}\rangle = 0,$$
$$\langle \operatorname{Re}\boldsymbol{a}_{-2}, \boldsymbol{a}_{-1}\rangle = 0, \qquad \langle \operatorname{Im}\boldsymbol{a}_{-2}, \boldsymbol{a}_{-1}\rangle = 0, \quad \text{and} \quad \operatorname{Im}\boldsymbol{a}_{-1} = 0.$$

Hence we can choose an orthonormal basis $\boldsymbol{e}_1, \dots, \boldsymbol{e}_n$ of $\mathbb{R}^n$ so that

$$\operatorname{Re}\boldsymbol{a}_{-2} = a\,\boldsymbol{e}_1, \quad \operatorname{Im}\boldsymbol{a}_{-2} = a\,\boldsymbol{e}_2, \quad \boldsymbol{a}_{-1} = b\,\boldsymbol{e}_3$$

for some real constants $a(\neq 0)$, b. With respect to this basis, we have

$$\partial f = \left(\frac{a}{z^2}(\boldsymbol{e}_1 + \sqrt{-1}\boldsymbol{e}_2) + \frac{b}{z}\boldsymbol{e}_3 + \cdots\right) dz, \quad a, b \in \mathbb{R}, (a \neq 0).$$

Then, using $z = re^{\sqrt{-1}\theta}$, we have

$$\text{(4)} \qquad f(z) = 2\operatorname{Re}\int_{z_0}^{z} \partial f = -\frac{2a\cos\theta}{r}\boldsymbol{e}_1 - \frac{2a\sin\theta}{r}\boldsymbol{e}_2 + 2b\log r\,\boldsymbol{e}_3 + O(r),$$

where z_0 is a basepoint. Here, we have dropped the constant terms in $f(z)$ by making a suitable parallel translation. By definition, formula (4) implies that the surface $f(\Delta^*)$ is asymptotic to the catenoid (respectively, the plane) for sufficiently small r if $b \neq 0$ (respectively, if $b = 0$).

Conversely, suppose that $\operatorname{ord}_0 ds^2 \neq -2$. This implies that $\operatorname{ord}_0 ds^2 = -k$ $(k \geq 3)$ and

$$\text{(5)} \qquad \partial f = \left(\frac{1}{z^k}\boldsymbol{a}_{-k} + \cdots + \frac{1}{z}\boldsymbol{a}_{-1} + \cdots\right) dz, \quad \boldsymbol{a}_{-k} \neq 0 \in \mathbb{C}^n, \ \boldsymbol{a}_{-1} \in \mathbb{R}^n.$$

It is obvious that the end is asymptotic neither to a catenoid-type end nor to a planar end.

We shall prove next that an end is embedded if it is asymptotic to a catenoid-type end or a planar end. Assume that the end is not embedded. Then there exist sequences $\{z_i\}$, $\{z_i'\}$ convergent to 0 such that $z_i \neq z_i'$ and $f(z_i) = f(z_i')$ for all i. Then by (4), there exists a positive constant C such that

$$\left|\frac{\cos\theta_i}{r_i} - \frac{\cos\theta_i'}{r_i'}\right| \leq C|r_i - r_i'|, \qquad \left|\frac{\sin\theta_i}{r_i} - \frac{\sin\theta_i'}{r_i'}\right| \leq C|r_i - r_i'|,$$

where $z_i = r_i e^{\sqrt{-1}\theta_i}$ and $z'_i = r'_i e^{\sqrt{-1}\theta'_i}$ $(i = 1, 2, \dots)$. With these estimates, we have

$$\left(\frac{1}{r_i} - \frac{1}{r'_i}\right)^2 \le \frac{1}{r_i{}^2} + \frac{1}{r'^2_i} - \frac{2}{r_i r'_i}\cos(\theta_i - \theta'_i)$$
$$= \left|\frac{\cos\theta_i}{r_i} - \frac{\cos\theta'_i}{r'_i}\right|^2 + \left|\frac{\sin\theta_i}{r_i} - \frac{\sin\theta'_i}{r'_i}\right|^2 \le 2C^2|r_i - r'_i|^2,$$

hence,

$$\frac{1}{(r_i r'_i)^2} \le 2C^2. \tag{6}$$

However the left hand side of (6) diverges to ∞ as $i \to \infty$, a contradiction. $\square$

Besides the Chern-Osserman inequality (1), the following inequalities for fully immersed complete minimal surfaces are known. (We say that the immersion f is *full* if the image $f(M)$ is not contained in any hyperplane of $\mathbb{R}^n$.)

Gackstatter [**G**] proved that

$$\mathrm{TC}(M) \le (2\chi_M + m - 1 - n)\pi.$$

On the other hand, Ejiri [**E**] proved the inequality

$$\mathrm{TC}(M) \le (\chi_M + m - 2n + 2l)\pi \tag{7}$$

if the Gauss image $\nu(M)$ is contained in an $(n-1-l)$-dimensional subspace of $\mathbb{CP}^{n-1}$. The Gauss map ν is said to be *degenerate* if $l > 0$.

We shall give a new example of a complete minimal surface which satisfies the equality both in the Chern-Osserman equality (1) and in the Ejiri inequality (7).

Example (Generalized Jorge-Meeks' surface). For $i = 0, 1, \dots, m-2$, we put

$$g_i(z) = \frac{z^i(1 - z^{2m-2i-2})}{(z^m - 1)^2}, \quad h_i(z) = \frac{\sqrt{-1}z^i(1 + z^{2m-2i-2})}{(z^m - 1)^2},$$

and define a conformal minimal immersion

$$f_m \colon M = \bigl(\mathbb{C} \cup \{\infty\}\bigr) \setminus \{z \,|\, z^m = 1\} \longrightarrow \mathbb{R}^{2m-1}$$

by

$$f_m := \mathrm{Re}\int_{z_0}^{z}\left(g_0, h_0, g_1, h_1, \dots, g_{m-1}, h_{m-1}, \frac{2\sqrt{m-1}\, z^{m-1}}{(z^m - 1)^2}\right) dz. \tag{8}$$

The immersion f_m is well-defined on M, since the integrand of (8) has real residue at each pole. This can be verified by similar computations to those in [**JM**].

Obviously, the genus of M is zero and the number of ends is m. It is easy to see that the induced metric is complete. Moreover, f_m has non-degenerate Gauss map because the components of the integrand of (8) are linearly independent over $\mathbb{C}$. This also implies that f_m is full. Since the degree of the Gauss map of f_m is $2(m-1)$, the total curvature $\mathrm{TC}(M)$ is equal to $-4(m-1)\pi$. Therefore it attains equality in the Chern-Osserman inequality. On the other hand, for f_m, the right hand side of (7) is $-4(m-1)\pi$ because the Gauss map of f_m is non-degenerate. Hence equality in (7) also holds.

Appendix: Embedded ends in $\mathbb{R}^3$

For the case $n = 3$, embeddedness of the end 0 in Lemma 1 implies $\operatorname{ord}_0 ds^2 = -2$, and consequently the end is asymptotic to a catenoid-type end or a planar end ([**JM**, Theorem 4] or [**S**, Proposition 1]). Here we shall give a simple proof of this fact, which combines aspects of the proofs of Jorge-Meeks and Schoen. The crucial point of the Jorge-Meeks proof is to show that the intersection of the end and the sphere of radius r centered at the origin converges to a finite covering of a great sphere as $r \to \infty$. According to Schoen [**S**], we can prove this via the Weierstrass representation directly.

Consider the Laurent expansion as (5) for $k \geq 2$. Without loss of generality, we may set $\boldsymbol{a}_{-k} = (a, \sqrt{-1}a, 0)$ ($a \in \mathbb{R} \setminus \{0\}$) because of (2). Integrating this, we have

$$f(re^{\sqrt{-1}\theta}) = \frac{1}{r^{k-1}}\left[\frac{2a}{1-k}\bigl(\cos(k-1)\theta, \sin(k-1)\theta, 0\bigr) + o(1)\right],$$

where $o(1)$ means a term which approaches 0 as $r \to 0$. Let S^2_R be the sphere in $\mathbb{R}^3$ with radius R centered at the origin, and consider the intersection of the surface and S^2_R:

$$E_R := \frac{1}{R}\left(S^2_R \cap f(\Delta^*)\right) \subset S^2_1,$$

which is normalized as a subset of the unit sphere.

Here, $f \in S^2_R$ if and only if

$$R^2 = f_1^2 + f_2^2 + f_3^2 = \frac{1}{r^{2k-2}}\left(\frac{4a^2}{(1-k)^2} + o(1)\right).$$

Then $r \to 0$ as $R \to \infty$ when $f(re^{\sqrt{-1}\theta}) \in S^2_R$ because $k \geq 2$. In particular, $\lim_{r\to\infty} R^2 r^{2k-2} = 4a^2/(1-k)^2$ holds. Then under the condition $f(z) \in S^2_R$, we have

$$\lim_{R\to\infty} \frac{1}{R} f(re^{\sqrt{-1}\theta}) = \pm\bigl(\cos(k-1)\theta, \sin(k-1)\theta, 0\bigr).$$

This implies that, for sufficiently large R, E_R is a closed curve in a neighborhood of the equator of S^2_1 with rotation index $|k-1|$, which is embedded if and only if $k = 2$. $\square$

Acknowledgement. We would like to thank Wayne Rossman for valuable comments.

References

[CO] S. Chern and R. Osserman, *Complete minimal surfaces in euclidean n-space,* J. Analyse Math., **19** (1967) 15–34.

[E] N. Ejiri, *Degenerate minimal surfaces of finite total curvature in R^N*, Kobe J. Math., **14** (1997), 11–22.

[G] F. Gackstatter, *Über die Dimension einer Minimalfläche und zur Ungleichung von St. Cohn-Vossen,* Arch. Rational Mech. Anal., **61** (1976), 141–152.

[JM] L.P.M. Jorge and W.H. Meeks III, *The topology of complete minimal surfaces of finite total curvature,* Topology, **22** (1983), 203–221.

[KTUY] M. Kokubu, M. Takahashi, M. Umehara and K. Yamada, *An analogue of minimal surface theory in* $\mathrm{SL}(n,\mathbb{C})/\mathrm{SU}(n)$, to appear in Trans. Amer. Math. Soc., math.DG/0008016.

[L] H. B. Lawson, Lectures on Minimal Submanifolds (Volume 1), Publish or Perish Inc., 1980.

[S] R. Schoen, *Uniqueness, symmetry and embeddedness of minimal surfaces,* J. Differential Geometry, **18** (1983), 791–809.

Department of Natural Science, School of Engineering, Tokyo Denki University, 2-2 Kanda-Nishiki-Cho, Chiyoda-Ku, Tokyo 101-8457, JAPAN

E-mail address: `kokubu@cck.dendai.ac.jp`

Department of Mathematics, Faculty of Science, Hiroshima University, Higashi-Hiroshima 739-8526, JAPAN

E-mail address: `umehara@math.sci.hiroshima-u.ac.jp`

Faculty of Mathematics, Kyushu University 36, Fukuoka 812-8581, JAPAN

E-mail address: `kotaro@math.kyushu-u.ac.jp`

Contemporary Mathematics
Volume **308**, 2002

Low Dimensional Manifolds Admitting Metrics with the Same Geodesics

Vladimir S. Matveev

1. Introduction

DEFINITION 1. *Let g be a Riemannian metric on a manifold M^n. A Riemannian metric $\bar{g}$ on M^n is called* **geodesically equivalent** *to g if any geodesic of $\bar{g}$, considered as an unparametrised curve, is a geodesic of g. The metrics are* **non-trivially geodesically equivalent**, *if they are geodesically equivalent and if, for any constant C, the metric $\bar{g}$ is not equal to the metric $C \cdot g$. The metrics are* **strictly non-proportional** *at a point of the manifold if the eigenvalues of one metric with respect to the other are different at that point.*

In this paper we will show that, modulo the Poincaré conjecture, any closed 3-manifold admitting non-trivial examples of geodesically equivalent metrics can be covered either by the sphere S^3 or by the product of the circle and of a surface—see Corollary 1 and Theorem 2.

1.1. History. The theory of geodesically equivalent metrics has an old and fascinating history that goes back to the works of Beltrami, Dini and Levi-Civita. Beltrami [**1**] was the first to observe that two different metrics can have the same geodesics. His example is quite simple: the metric g is the restriction of the Euclidean metric $dx^2 + dy^2 + dz^2$ to the sphere

$$S^2 = \{(x, y, z) \in R^3 : \; x^2 + y^2 + z^2 = 1\}.$$

The metric $\bar{g}$ is the pull-back l^*g, where the mapping $l : S^2 \to S^2$ is given by $l : v \mapsto A(v)/\|A(v)\|$, where A is an arbitrary linear nondegenerate transformation of R^3. The metrics g and $\bar{g}$ are geodesically equivalent. Indeed, the geodesics of the metric g are great circles (the intersections of planes through the origin with the sphere). The mapping A is linear and therefore takes planes to planes. Since the normalisation $w \mapsto w/\|w\|$ takes planes to their intersections with the sphere, the mapping l takes great circles to great circles. Thus, the metrics g and $\bar{g}$ are geodesically equivalent. Evidently, if A is not orthogonal, the metrics g and $\bar{g}$ are non-trivially geodesically equivalent.

2000 *Mathematics Subject Classification.* 35J37, 57N10, 53A20, 53C24, 53C15, 53D25.

A point $x \in M^n$ is called *stable*, if in some neighbourhood the number of different eigenvalues of one metric with respect to the other is constant.

Locally, near the stable points, a complete description of geodesically equivalent metrics has been given by Dini [**4**] for surfaces and Levi-Civita [**11**] for manifolds of arbitrary dimension. We will formulate the three-dimensional version of Levi-Civita's Theorem in Section 3. In particular, Levi-Civita's Theorem shows that the set of metrics admitting geodesic equivalence is quite rich: locally, in the nontrivial case, it is controlled by $n = \dim(M)$ parameters, each of the parameters being a function of one variable.

Later, geodesically equivalent metrics were considered by Weyl, E. Cartan and Eisenhart. Weyl studied geodesically equivalent metrics on the tensor level and found several tensor reformulations of geodesic equivalence. One of his most remarkable results is the construction of the projective Weyl tensor [**24**]: if two metrics are geodesically equivalent then their projective Weyl tensors coincide. E. Cartan [**3**] studied geodesic equivalence on the level of affine connections; he introduced the so-called projective connection, which allows the reconstruction of geodesics as unparametrised curves. In his book [**5**], Eisenhart systematically applied both methods and obtained a series of local results.

However, the global (when the manifold is closed or complete) behaviour of geodesically equivalent metrics is not understood. Global aspects of geodesic equivalence have been actively studied by French, Soviet and Japanese geometers. But, probably because of the influence of earlier researchers, all known global results require strong additional geometrical assumptions. Roughly speaking, one takes some geometric assumption written in the tensor form, combines it with one of the tensor reformulations of geodesic equivalence and deduces some new object with global geometric properties; see the review paper [**19**].

1.2. Results.

THEOREM 1. *Suppose M^3 is a connected closed manifold. Let the Riemannian metrics g and $\bar g$ on M^3 be geodesically equivalent and strictly non-proportional at least at one point. Then there exists a finite cover with fundamental group either $Z \times Z \times Z$ or Z or $\{e\}$. In the first case, the manifold M^3 is covered by the torus T^3.*

COROLLARY 1. *Suppose M^3 is a connected closed manifold. Let the Riemannian metrics g and $\bar g$ on M^3 be geodesically equivalent and strictly non-proportional at least at one point. Then, modulo the Poincaré conjecture, M^3 is finitely covered either by the torus or by the product $S^2 \times S^1$ or by the sphere S^3.*

THEOREM 2. *Let the Riemannian metrics g and $\bar g$ on a closed connected manifold M^3 be non-trivially geodesically equivalent, and suppose there is no point where they are strictly non-proportional. Then M^3 can be finitely covered either by the sphere S^3 or by the direct product of a closed surface and the circle.*

It is worth noting that the sphere S^3 admits non-trivially geodesically equivalent metrics: the Beltrami construction can easily be generalised for all dimensions. The direct product of the circle and of any surface also admits non-trivially geodesically equivalent metrics. In fact, for any two Riemannian manifolds (M_1, g_1) and

(M_2, g_2), the metrics $g_1 + g_2$ and $g_1 + 2g_2$ are non-trivially geodesically equivalent on the product $M_1 \times M_2$.

1.3. Integrable systems in the theory of geodesically equivalent metrics. The new technique that allows us to prove Theorems 1 and 2 came from the theory of integrable geodesic flows. The connection between geodesically equivalent metrics and integrable geodesic flows is established by the following theorem.

Let $g = (g_{ij})$ and $\bar{g} = (\bar{g}_{ij})$ be Riemannian metrics on M^n. Consider the (1,1)-tensor field L given by the formula

(1) $$L^i_j \stackrel{\text{def}}{=} \left(\frac{\det(\bar{g})}{\det(g)}\right)^{\frac{1}{n+1}} \bar{g}^{i\alpha} g_{\alpha j}.$$

THEOREM 3 ([**12, 23**]). *Let the Riemannian metrics g, $\bar{g}$ be geodesically equivalent. For any $t \in R$, consider the $(1,1)$-tensor field*

(2) $$S_t \stackrel{\text{def}}{=} \det(L - t\ \mathrm{Id})\,(L - t\ \mathrm{Id})^{-1}.$$

Let us identify the tangent and cotangent bundles of M^n by using g. Consider the standard Poisson structure on T^*M^n. Then, for any t_1, t_2, the functions

(3) $$I_{t_i} : TM^n \to R, \quad I_{t_i}(\xi) \stackrel{\text{def}}{=} g(S_{t_i}(\xi), \xi)$$

are commuting integrals for the geodesic flow of g.

REMARK 1. *Although $(L - t\ \mathrm{Id})^{-1}$ is not defined for t lying in the spectrum of L, the tensor field S_t, and, therefore, the function I_t, is well-defined for any t. Moreover, S_t is a polynomial (in t) of degree $n-1$ whose coefficients are (1,1)-tensor fields.*

In Section 2, we will show (Corollary 3) that if the metrics are strictly non-proportional at one point of a connected complete manifold then they are so at almost every point, so that the integrals are functionally independent almost everywhere, and the geodesic flow is Liouville-integrable [**12**].

1.4. Geodesically equivalent metrics on surfaces. For two-dimensional manifolds, Theorem 3 becomes:

THEOREM 4 ([**12**]). *The Riemannian metrics g and $\bar{g}$ on a surface are geodesically equivalent if and only if the function $I \stackrel{\text{def}}{=} (\det(g)/\det(\bar{g}))^{\frac{2}{3}}\, \bar{g}(\xi, \xi)$ is an integral of the geodesic flow of g.*

We see that the integral I is quadratic in velocities. Combining Theorem 4 with known facts about quadratically integrable geodesic flows on surfaces, we get the two-dimensional analogue of Theorem 1 and Corollary 1:

COROLLARY 2 ([**16**]). *Let the Riemannian metrics g and $\bar{g}$ on a closed surface M^2 be non-trivially geodesically equivalent. Then M^2 can be covered either by the sphere S^2 or by the torus T^2.*

The theory of quadratically integrable geodesic flows is quite well developed. The fundamental classification results of Igarashi, Kiyohara, Kolokoltsov, Bolsinov,

Fomenko, Matveev (see [**10, 9, 2**]) allow one to describe completely the pairs of geodesically equivalent metrics on closed or geodesically complete surfaces.

Here we prove Corollary 2 on the basis of Theorem 2.1 from [**10**], which, adapted to our situation, says that, if a geodesic flow on a closed surface of negative Euler characteristic admits an integral, quadratic in velocities, then the integral must be a linear combination of the Hamiltonian of the geodesic flow and the square of some integral which is linear in velocities. Evidently, the geodesic flow of any metric on a closed surface of negative Euler characteristic admits no integral which is linear in velocities. More precisely, it is known that an integral which is linear in velocities allows one to construct a vector field on the surface that preserves the metric. Then the vector field must preserve the complex structure generated by this metric and therefore it is (anti)holomorphic with respect to this complex structure. By Abel's Lemma, any nontrivial (anti)holomorphic vector field on a Riemann surface of negative Euler characteristic must have at least one pole; hence it is not smooth, which contradicts the assumptions. Thus, any integral of the geodesic flow on any closed surface of negative Euler characteristic which is quadratic in velocities is a constant multiple of the Hamiltonian and, finally, any two geodesically equivalent metrics on a closed surface of negative Euler characteristic are proportional. The corollary is proved.

1.5. Acknowledgements. I would like to use this opportunity to thank all participants of the conference "Integrable Systems in Differential Geometry" for their interest in this problem and Prof. K. Kiyohara and Prof. S. Saito for supporting my visit by their grants. I am grateful to Prof. V. Bangert, Prof. A. Bolsinov, Prof. M. Igarashi and Prof. P. Topalov for useful discussions and to Prof. S. Matveev for explaining to me the theory of Seifert manifolds. I thank the European Post-Doctoral Institute for partial financial support and Warwick University and the Isaac Newton Institute for Mathematical Sciences for hospitality. My research at INIMS has been supported by EPSRC grant GRK99015.

2. The eigenvalues of the tensor L

The main goal of this section is to prove the following theorem, which plays a key role in the proof of Theorem 1:

THEOREM 5. *Let (M^n, g) be a geodesically complete connected Riemannian manifold. Let the Riemannian metric $\bar g$ on M^n be geodesically equivalent to g. Denote by L the (1,1)-tensor field (1). At each point $x \in M^n$, denote by*

$$\lambda_1(x) \le \lambda_2(x) \le ... \le \lambda_n(x)$$

the eigenvalues of L at the point x. Then, for any $i \in \{1, ..., n-1\}$, for any $x, y \in M^n$:

1. *$\lambda_i(x) \le \lambda_{i+1}(y)$.*
2. *If $\lambda_i(x) < \lambda_{i+1}(x)$ then $\lambda_i(z) < \lambda_{i+1}(z)$ for almost every point $z \in M^n$.*
3. *If $\lambda_i(x) = \lambda_{i+1}(y)$ then there exists $z \in M^n$ such that $\lambda_i(z) = \lambda_{i+1}(z)$.*

COROLLARY 3 ([**15**]). *Let (M^n, g) be a geodesically complete connected Riemannian manifold. Let $\bar g$ be geodesically equivalent to g. Denote by L the (1,1)-tensor (1). If the eigenvalues of L are all different at one point of M^n then they are* *all different at almost every point of M^n, and the geodesic flow of g is Liouville-integrable.*

Proof: By definition, the tensor L is self-adjoint with respect to g. Then, for any $x \in M^n$, there exists a basis in T_xM^n such that the metric g is given by the matrix $\mathrm{diag}(1, 1, ..., 1)$ and the tensor L is given by the matrix $\mathrm{diag}(\lambda_1, \lambda_2, ..., \lambda_n)$. The tensor (2) reads:

$$S_t = \det(L - t\mathrm{Id})(L - t\mathrm{Id})^{(-1)} = \mathrm{diag}(\Pi_1(t), \Pi_2(t), ..., \Pi_n(t)),$$

where the polynomials $\Pi_i(t)$ are given by the formula

$$(4) \qquad \Pi_i(t) \stackrel{\text{def}}{=} (\lambda_1 - t)(\lambda_2 - t)...(\lambda_{i-1} - t)(\lambda_{i+1} - t)...(\lambda_{n-1} - t)(\lambda_n - t).$$

Then, for any fixed $\xi = (\xi_1, \xi_2, ..., \xi_n) \in T_xM^n$, the function (3) is the following polynomial in t:

$$(5) \qquad I_t = \xi_1^2\Pi_1(t) + \xi_2^2\Pi_2(t) + ... + \xi_n^2\Pi_n(t).$$

Consider the roots of this polynomial. From the proof of Lemma 1 below, it will be clear that they are real. We denote them by $t_1(x, \xi) \leq t_2(x, \xi) \leq ... \leq t_{n-1}(x, \xi)$.

LEMMA 1.

1. *For any $\xi \in T_xM^n$, $\lambda_i(x) \leq t_i(x, \xi) \leq \lambda_{i+1}(x)$. In particular, if $\lambda_i(x) = \lambda_{i+1}(x)$ then $t_i(x, \xi) = \lambda_i(x) = \lambda_{i+1}(x)$.*
2. *If $\lambda_i(x) < \lambda_{i+1}(x)$ then for any constant τ the Lebesgue measure of the set*

$$V_\tau \subset T_xM^n, \quad V_\tau \stackrel{\text{def}}{=} \{\xi \in T_xM^n : t_i(x, \xi) = \tau\},$$

is zero.

Proof of Lemma 1: Evidently, the coefficients of the polynomial I_t depend continuously on the eigenvalues λ_i and on the components ξ_i. Then it is sufficient to prove the first statement of the lemma assuming that the eigenvalues λ_i are all different and that the ξ_i are non-zero. For any $\alpha \neq i$, evidently we have $\Pi_\alpha(\lambda_i) \equiv 0$. Then $I_{\lambda_i} = \sum_{\alpha=1}^{n} \Pi_\alpha(\lambda_i)\xi_\alpha^2 = \Pi_i(\lambda_i)\xi_i^2$. Hence I_{λ_i} and $I_{\lambda_{i+1}}$ have different signs and therefore the open interval $]\lambda_i, \lambda_{i+1}[$ contains a root of the polynomial I_t. The degree of the polynomial I_t is $n - 1$; we have $n - 1$ disjoint intervals, and each of these intervals contains at least one root so that all roots are real and the i-th root lies between λ_i and λ_{i+1}. The first statement of the lemma is proved.

Let us prove the second statement of Lemma 1. Suppose $\lambda_i < \lambda_{i+1}$. Let $\lambda_i < \tau < \lambda_{i+1}$. Then the set V_τ consists of the points ξ where the function $I_\tau(x, \xi) \stackrel{\text{def}}{=} (I_t(x, \xi))|_{t=\tau}$ is zero; it is a nontrivial quadric in $T_xM^n \equiv R^n$ and its measure is zero.

Let τ be one of the endpoints of the interval $[\lambda_i, \lambda_{i+1}]$. Without loss of generality, we can suppose $\tau = \lambda_i$. Let k be the multiplicity of the eigenvalue λ_i. Then any coefficient $\Pi_\alpha(t)$ of the quadratic form (5) has a factor $(\lambda_i - t)^{k-1}$. Therefore,

$$\tilde{I}_t \stackrel{\text{def}}{=} \frac{I_t}{(\lambda_i - t)^{k-1}}$$

is a polynomial in t and $\tilde{I}_\tau$ is a nontrivial quadratic form. Evidently, for any point $\xi \in V_\tau$, we have $\tilde{I}_\tau(\xi) = 0$ so that the set V_τ is a subset of a nontrivial quadric in T_xM^n and its measure is zero. Lemma 1 is proved.

The first statement of Theorem 5 follows immediately from the first statement of Lemma 1: Let us join the points $x, y \in M^n$ by a geodesic $\gamma : R \to M^n$, $\gamma(0) = x$, $\gamma(1) = y$. Consider the one-parameter family of integrals $I_t(x, \xi)$ and the roots $t_1(x, \xi) \le t_2(x, \xi) \le ... \le t_{n-1}(x, \xi)$.

By Theorem 3, each root t_i is constant on each orbit $(\gamma, \dot\gamma)$ of the geodesic flow of g so that $t_i(\gamma(0), \dot\gamma(0)) = t_i(\gamma(1), \dot\gamma(1))$. Using Lemma 1, we obtain $\lambda_i(\gamma(0)) \le t_i(\gamma(0), \dot\gamma(0))$, and $t_i(\gamma(1), \dot\gamma(1)) \le \lambda_{i+1}(\gamma(1))$. Therefore $\lambda_i(\gamma(0)) \le \lambda_{i+1}(\gamma(1))$ and the first statement of Theorem 5 is proved.

Let us prove the second statement of Theorem 5. Suppose $\lambda_i(y) = \lambda_{i+1}(y)$ for any point y of some subset $V \subset M^n$. Then by the first statement of Theorem 5, the value of λ_i is a constant (independent of $y \in V$). Indeed, joining any two points $y_0, y_1 \in V$ by a geodesic, we have $\lambda_i(y_0) \le \lambda_{i+1}(y_1)$ and $\lambda_i(y_1) \le \lambda_{i+1}(y_0)$. Denote this constant by C. Let us join the point x with every point of V by all possible geodesics. Consider the set $V_C \subset T_xM^n$ of the initial velocity vectors (at the point x) of these geodesics. By the first statement of Lemma 1, for any geodesic γ passing through any point of V, the value $t_i(\gamma, \dot\gamma)$ is equal to C. By the second statement of Lemma 1, the measure of the set V_C is zero. Since the set V lies in the image of the exponential mapping of the set V_C, the measure of the set V is also zero. The second statement of Theorem 5 is proved.

Let us prove the third statement of Theorem 5. Let $\lambda_i(\gamma(0)) = \lambda_{i+1}(\gamma(1)) = \lambda$ for some $i \in \{1, ..., n-1\}$ and for some constant λ. We will assume that $\lambda_i(\gamma(0)) < \lambda_{i+1}(\gamma(0))$. Let us show that the geodesic γ consists of the points where λ_i or λ_{i+1} is equal to λ. If t_i is a multiple root of the polynomial $I_t(\gamma(0), \dot\gamma(0))$, or if there exists a point $z \in M^n$ such that $\lambda_{i-1}(z) = \lambda$, then the statement obviously follows from Lemma 1 and the first statement of Theorem 5. Suppose t_i is not a multiple root and $\lambda_{i-1}(z) < \lambda$ for any $z \in M^n$. Consider the function $I_\lambda : TM^n \to R$. Suppose the differential dI_λ is zero at some point $(z, \nu) \in TM^n$, $\nu \ne 0$. Let us show then that $\lambda_i(z)$ or $\lambda_{i+1}(z)$ is equal to λ. Indeed, consider the coordinate system such that the metric g at the point z is given by the diagonal matrix $\operatorname{diag}(1, 1, .., 1)$ and the mapping L is given by the diagonal matrix $\operatorname{diag}(\lambda_1, \lambda_2, .., \lambda_n)$. Then the restriction of the function I_λ to the tangent space T_zM^n is given by $\sum_{\alpha=1}^n \xi_\alpha^2 \Pi_\alpha(\lambda)$. The partial derivatives $\frac{\partial I_\lambda}{\partial \xi_\alpha}$ are $\frac{\partial I_\lambda}{\partial \xi_\alpha} = 2\Pi_\alpha(\lambda)\xi_\alpha$. Then λ is equal to one of the numbers $\lambda_1, ..., \lambda_n$; by assumption it can be equal to $\lambda_i(z)$ or $\lambda_{i+1}(z)$.

Now let us show that the differential dI_λ vanishes at every point $(\gamma(\tau), \dot\gamma(\tau))$. Evidently the differential of any integral is preserved by the geodesic flow so that it is sufficient to prove that the differential vanishes at the point $(\gamma(0), \dot\gamma(0))$. By Lemma 1, we have

$$\lambda = \lambda_i(\gamma(0)) \le t_i(\gamma(0), \dot\gamma(0)) = t_i(\gamma(1), \dot\gamma(1)) \le \lambda_{i+1}(\gamma(1)) = \lambda$$

so that λ is a root of the polynomial $I_t(\gamma(0), \dot\gamma(0))$ and $I_\lambda(\gamma(0), \dot\gamma(0)) = 0$. By assumption, the eigenvalue λ_i has multiplicity one in a small neighbourhood of $\gamma(0)$. Then it is a smooth function on this neighbourhood, and the function

$I_{\lambda_i(z)}(z,\nu)$ is also smooth on the tangent bundle to this neighbourhood. Consider the function $I_\lambda(z,\nu) - I_{\lambda_i(z)}(z,\nu)$. Its differential vanishes at the point $(\gamma(0),\dot\gamma(0))$. More precisely, by assumption, λ is a simple root of the polynomial $I_t(\gamma(0),\dot\gamma(0))$ so that in a neighbourhood of the point $\big(\lambda,(\gamma(0),\dot\gamma(0))\big) \in R\times TM^n$ the function $I_t(z,\nu)$ is a monotone function in t. But λ_i is no greater than λ. Thus the difference $I_\lambda(z,\nu) - I_{\lambda_i(z)}(z,\nu)$ is always non-positive or always non-negative in a small neighbourhood of $(\gamma(0),\dot\gamma(0))$. By assumption, $\lambda_i(\gamma(0)) = \lambda$ so that $I_\lambda(\gamma(0),\dot\gamma(0)) - I_{\lambda_i(\gamma(0))}(\gamma(0),\dot\gamma(0))$ is zero and therefore the function $I_\lambda(z,\nu) - I_{\lambda_i(z)}(z,\nu)$ has a local extremum at the point $(\gamma(0),\dot\gamma(0))$ and its differential vanishes at this point.

Now, the differential of the function $I_{\lambda_i(z)}(z,\nu)$ also vanishes at the point $(\gamma(0),\dot\gamma(0))$. More precisely, as we have shown in the proof of Lemma 1, the function I_{λ_i} is always non-positive or always non-negative. But $I_{\lambda_i(\gamma(0))}(\gamma(0),\dot\gamma(0))$ is equal to $I_\lambda(\gamma(0),\dot\gamma(0))$ and is zero. Then the point $(\gamma(0),\dot\gamma(0))$ is an extremum of the function I_{λ_i} and therefore the differential of I_{λ_i} vanishes at the point $(\gamma(0),\dot\gamma(0))$. Thus, the differential of the function I_λ is zero at the point $(\gamma(0),\dot\gamma(0))$. Hence it vanishes at each point of the curve $(\gamma(\tau),\dot\gamma(\tau))$. Then at any point τ either $\lambda_i(\gamma(\tau)) = \lambda$ or $\lambda_{i+1}(\gamma(\tau)) = \lambda$. Therefore, any point of the segment $[0,1]$ lies in one of the following sets:

$$\begin{aligned}\Gamma_0 &= \{\tau\in[0,1] : \lambda_i(\gamma(\tau)) = \lambda\},\\ \Gamma_1 &= \{\tau\in[0,1] : \lambda_{i+1}(\gamma(\tau)) = \lambda\}.\end{aligned}$$

The subsets Γ_0, Γ_1 are evidently closed and non-empty. They intersect, and at each point τ of the intersection we have $\lambda_i(\gamma(\tau)) = \lambda_{i+1}(\gamma(\tau)) = \lambda$. Theorem 5 is proved.

3. The Levi-Civita theorem in dimension three

Let $g,\bar g$ be Riemannian metrics on M^n. Here we formulate the classical Levi-Civita theorem which describes geodesically equivalent metrics near the stable points. We formulate the theorem for the three-dimensional case only.

As in Section 1, we consider the tensor L given by (1).

THEOREM 6 ([**11**]). *Let $g,\bar g$ be geodesically equivalent Riemannian metrics on M^3.*

1. *If at each point of some open set $U^3\subset M^3$ the number of different eigenvalues of one metric with respect to the other is precisely three, then any point $P_0\in U^3$ has a neighbourhood $V^3(P_0)\subset U^3$ with coordinates x_1,x_2,x_3 such that the metrics $g,\bar g$ are given by:*

$$\begin{aligned}ds_g^2 &= (\lambda_1(x_1)-\lambda_2(x_2))(\lambda_1(x_1)-\lambda_3(x_3))dx_1^2\\ &\quad+(\lambda_2(x_2)-\lambda_1(x_1))(\lambda_3(x_3)-\lambda_2(x_2))dx_2^2\\ &\quad+(\lambda_3(x_3)-\lambda_1(x_1))(\lambda_3(x_3)-\lambda_2(x_2))dx_3^2\end{aligned}$$

$$
\begin{aligned}
ds_{\bar g}^2 = &\frac{(\lambda_1(x_1)-\lambda_2(x_2))(\lambda_1(x_1)-\lambda_3(x_3))}{\lambda_1(x_1)^2\lambda_2(x_2)\lambda_3(x_3)}dx_1^2\\
&+\frac{(\lambda_2(x_2)-\lambda_1(x_1))(\lambda_3(x_3)-\lambda_2(x_2))}{\lambda_1(x_1)\lambda_2(x_2)^2\lambda_3(x_3)}dx_2^2\\
&+\frac{(\lambda_3(x_3)-\lambda_1(x_1))(\lambda_3(x_3)-\lambda_2(x_2))}{\lambda_1(x_1)\lambda_2(x_2)\lambda_3(x_3)^2}dx_3^2,
\end{aligned}
$$

where λ_i denotes a smooth function of the coordinate x_i.

2. *If at each point of some open set $U^3 \subset M^3$ the number of different eigenvalues of one metric with respect to the other is precisely two, then any point $P_0 \in U^3$ has a neighbourhood $V^3(P_0) \subset U^3$ with coordinates x_1, x_2, x_3 such that the metrics $g, \bar g$ are given by:*

$$
\begin{aligned}
ds_g^2 &= \pm(\lambda-\lambda_1(x_1))(dx_1^2 + B(x_2,x_3)(dx_2^2+dx_3^2))\\
ds_{\bar g}^2 &= \pm\left(\frac{1}{\lambda_1(x_1)}-\frac{1}{\lambda}\right)\left(\frac{dx_1^2}{\lambda_1(x_1)}+\frac{B(x_2,x_3)(dx_2^2+dx_3^2)}{\lambda}\right),
\end{aligned}
$$

where λ_1 is a smooth function of the coordinate x_1, B is a smooth function of the coordinates x_2, x_3 and λ is a constant.

4. Proof of Theorem 1

We assume that the manifold M^3 is closed, connected, and that the Riemannian metrics g and $\bar g$ on M^3 are geodesically equivalent and strictly non-proportional at least at one point. Our goal is to prove Theorem 1. To do this, we will show that:

1. The fundamental group $\pi_1(M^3)$ is virtually Abelian (i.e., has an Abelian subgroup of finite index) so that the manifold is finitely-covered by a manifold with fundamental group homeomorphic to Z^k.
2. If $k = 3$ then the metrics are strictly non-proportional at each point of the manifold and the manifold is covered by the torus.

REMARK 2. *By* [**6**], *the fundamental group of a compact 3-manifold cannot have a subgroup isomorphic to $Z \times Z \times Z \times Z$, so k is at most 3. By the theorem of Reidemeister (which has been formulated and proved in* [**7**] *and* [**22**]*), it follows that $Z \times Z$ cannot be a fundamental group of a closed 3-manifold. Thus, k is three or one or zero.*

The second statement has been announced in [**17**]. In view of Theorem 3, if the manifold and the metrics are real-analytic, the first statement follows from [**21**]. The first statement has been proved in [**18**], pages 347–349. We will repeat the proof with minor changes.

Proof of Theorem 1: Consider the tensor L given by (1) and its eigenvalues

$$\lambda_1(x) \le \lambda_2(x) \le \lambda_3(x).$$

By Theorem 5, there exist numbers $\tau_1, \tau_2 \in R$ such that, for any $x \in M^3$,

$$\lambda_1(x) \le \tau_1 \le \lambda_2(x) \le \tau_2 \le \lambda_3(x).$$

Consider the following subsets of M^3:

$$V_{--} \overset{\text{def}}{=} \{x \in M^3 : \ \lambda_1(x) \le \tau_1 < \lambda_2(x) \le \tau_2 < \lambda_3(x)\},$$
$$V_{+-} \overset{\text{def}}{=} \{x \in M^3 : \ \lambda_1(x) < \tau_1 \le \lambda_2(x) \le \tau_2 < \lambda_3(x)\},$$
$$V_{-+} \overset{\text{def}}{=} \{x \in M^3 : \ \lambda_1(x) \le \tau_1 < \lambda_2(x) < \tau_2 \le \lambda_3(x)\},$$
$$V_{++} \overset{\text{def}}{=} \{x \in M^3 : \ \lambda_1(x) < \tau_1 \le \lambda_2(x) < \tau_2 \le \lambda_3(x)\}.$$

By assumption, the eigenvalues $\lambda_1, \lambda_2, \lambda_3$ are globally defined continuous functions on M^3. Therefore, the sets $V_{--}, V_{-+}, V_{+-}, V_{++}$ are open. We assume that there is a point of the manifold where the metrics are not strictly non-proportional. Let us show that the fundamental group of any connected component V of any of these sets is virtually Abelian and its rank is at most two.

At each point of V, the eigenvalues of L are all different. They are smooth functions on V. Hence the functions

$$\Pi_1(\lambda_1) = (\lambda_1 - \lambda_2)(\lambda_1 - \lambda_3)$$
$$\Pi_2(\lambda_2) = (\lambda_2 - \lambda_1)(\lambda_2 - \lambda_3)$$
$$\Pi_3(\lambda_3) = (\lambda_3 - \lambda_1)(\lambda_3 - \lambda_2)$$

are smooth functions on V. At each point of V, consider the vectors v_1, v_2, v_3 satisfying the following conditions:

$$(6) \qquad \left\{ \begin{array}{rcl} L(v_i) & = & \lambda_i v_i \\ g(v_1, v_1) & = & (\lambda_1 - \lambda_2)(\lambda_1 - \lambda_3) = \Pi_1(\lambda_1) \\ g(v_2, v_2) & = & (\lambda_1 - \lambda_2)(\lambda_2 - \lambda_3) = -\Pi_2(\lambda_2) \\ g(v_3, v_3) & = & (\lambda_1 - \lambda_3)(\lambda_2 - \lambda_3) = \Pi_3(\lambda_3). \end{array} \right.$$

The only freedom we have is the signs of the vectors. Then we can globally define vector fields v_1, v_2, v_3 satisfying conditions (6) on the universal cover $\tilde{V}$ of V. We see that, in Levi-Civita's coordinates x_1, x_2, x_3 from Theorem 6, the vector fields v_i are equal to $\pm\frac{\partial}{\partial x_i}$. Then the vector fields v_i commute and for any $j \neq i$ the eigenvalue λ_j is constant on the integral curves of v_i.

We can globally define Levi-Civita's coordinates on the universal cover $\tilde{V}$: Choose an origin P_0. Join any point P with P_0 by a curve. Then the coordinate x_i is equal to the action of the vector field v_i along the curve. Evidently, the definition is independent of the curve. By definition of the set V, if the universal cover $\tilde{V}$ contains the points with the coordinates $(\bar{x}_1, \bar{x}_2, \bar{x}_3)$ and $(\hat{x}_1, \hat{x}_2, \hat{x}_3)$, then it contains the whole parallelopiped

$$\{(x_1, x_2, x_3) : \ \min(\bar{x}_i, \hat{x}_i) \le x_i \le \max(\bar{x}_i, \hat{x}_i), \ \ i = 1, 2, 3\}.$$

Then the coordinates uniquely define the point, and the universal cover $\tilde{V}$ is homeomorphic to the band

$$\{(x_1, x_2, x_3) \in R^3 : \ \alpha_1 < x_1 < \beta_1, \alpha_2 < x_2 < \beta_2, \alpha_3 < x_3 < \beta_3\},$$

where $\alpha_i, \beta_i \in R_\infty$ (real numbers or $\pm\infty$).

The fundamental group $\pi_1(V)$ acts on $\tilde{V}$. The action preserves the metric g and the tensor L. Therefore, if γ is an element of the fundamental group then for each vector field v_i either $\gamma(v_i) = v_i$ or $\gamma(v_i) = -v_i$. Consider the subgroup $H \subset \pi_1(V)$ consisting of the elements that preserve the directions of the vector fields. Since we have at most eight possibilities for choosing the signs of the vectors v_i, the subgroup H has finite index (at most eight) in the fundamental group $\pi_1(V)$. Evidently, in the coordinates x_1, x_2, x_3, the group H acts by parallel translations. Then the group H is Abelian (so that the fundamental group $\pi_1(V)$ is virtually Abelian) and torsion-free; since it has finite index in $\pi_1(V)$, its rank is greater than or equal to the first Betti number of V.

The group of all translations of $\tilde V$ is a subgroup of the group $(R^3, +)$. Therefore the group H is isomorphic to a discrete subgroup of the group $(R^3, +)$. Clearly, $k < 4$, since the group $(R^3, +)$ has no discrete subgroups isomorphic to $Z \times Z \times Z \times Z$. If $k = 3$ then the quotient space R^3/H is homeomorphic to the torus and, therefore, is compact; this quotient space naturally covers V so that V is also compact. Since V is open by definition, it coincides with the whole manifold M^3, so that in each point of M^3 the eigenvalues of L are all different, which contradicts the assumptions. Thus, the fundamental group of any of the connected components of the sets $V_{--}, V_{-+}, V_{+-}, V_{++}$ is almost Abelian and its rank is at most two. By Theorem 5, there exists a point $x \in M^3$ such that $\lambda_1(x) < \lambda_2(x) < \lambda_3(x)$. Denote by $V_{--}(x), V_{-+}(x), V_{+-}(x), V_{++}(x)$ the connected components of the sets $V_{--}, V_{-+}, V_{+-}, V_{++}$ containing the point x.

Let us show that each element of the fundamental group $\pi_1(M^3)$ can be realised on one of the connected components $V_{\pm\pm}(x)$. By the Hopf-Rinow theorem, any element of the fundamental group can be realised by a geodesic loop γ, $\gamma(0) = \gamma(1) = x$. We consider this geodesic loop as a curve $(\gamma, \dot\gamma)$ on TM^3. The values of the roots t_i of the polynomial $I_t(\gamma, \dot\gamma)$ are constant on the curve $(\gamma, \dot\gamma)$. If $t_1(\gamma, \dot\gamma) \neq \tau_1$ and $t_2(\gamma, \dot\gamma) \neq \tau_2$ then, by Lemma 1, the geodesic loop γ already lies in one of the sets $V_{\pm\pm}(x)$. Suppose $t_1(\gamma, \dot\gamma) = \tau_1$ or $t_2(\gamma, \dot\gamma) = \tau_2$. Let us slightly perturb the initial velocity vector $\dot\gamma(0)$ and consider a geodesic γ_ϵ such that $\gamma_\epsilon(0) = \gamma(0)$, $|\dot\gamma_\epsilon(0) - \dot\gamma(0)| = \epsilon << 1$ and $t_1(\gamma_\epsilon, \dot\gamma_\epsilon) \neq \tau_1$, $t_2(\gamma_\epsilon, \dot\gamma_\epsilon) \neq \tau_2$. By Lemma 1, the geodesic γ_ϵ lies in one of the sets $V_{\pm\pm}(x)$. If ϵ is small then the geodesic segment $\gamma_\epsilon(T)$, $0 \le T \le 1$, lies in a thin tubular neighbourhood of the geodesic loop γ and the point $\gamma_\epsilon(1)$ lies in a small disk neighbourhood of x and we can connect the points $\gamma_\epsilon(1)$ and x by a segment in this disk. Then the curve which is made from this segment and from the geodesic segment $\gamma_\epsilon(T)$, $0 \le T \le 1$, represents the same homotopy class as the geodesic loop γ, so that any element of the fundamental group can be realized on one of the sets $V_{\pm\pm}(x)$.

Finally, the first homology group $H_1(M^3; Z)$ of M^3 is the union of a finite number of subgroups. Each of these subgroups has rank less than 3; therefore the first Betti number $b_1(M^3)$ is less than 3. The fundamental group $\pi_1(M^3)$ is the union of a finite number of virtually Abelian subgroups; therefore it is virtually Abelian itself. In particular, if the first Betti number $b_1(M^3)$ is equal to 3 then the metrics are strictly non-proportional at each point of the manifold. Then the manifold can be covered by the torus. Indeed, on some finite cover $\tilde M^3$ of M^3, we can globally define vector fields v_1, v_2, v_3 satisfying conditions (6). The vector fields commute and never vanish. Then $\tilde M^3$ is homeomorphic to the torus. Theorem 1 is proved.

5. Proof of Theorem 2

We assume that the manifold M^3 is closed, that the Riemannian metrics g and $\bar{g}$ on M^3 are non-trivially geodesically equivalent, and that there is no point where the metrics are strictly non-proportional. Our goal is to prove that M^3 can be covered by the sphere or by the product of the circle and a closed surface.

Without loss of generality, by Theorem 5, we can assume that

$$\lambda_1(x) \leq \lambda_2(x) = \lambda_3(x) = \text{constant} \stackrel{\text{def}}{=} \lambda$$

at each point $x \in M^3$. The following two cases are possible:

Case 1: There exists $y \in M^3$ such that $\lambda_1(y) = \lambda_2(y) = \lambda_3(y)$.

Case 2. At each point x of the manifold, $\lambda_1(x) < \lambda_2(x) = \lambda_3(x)$.

Proof for Case 1: We will show that the fundamental group of the manifold is finite so that, modulo the Poincaré conjecture, the universal cover of the manifold is homeomorphic to the sphere S^3. Essentially, this fact has been proved in [**18**], page 351. We repeat this proof with minor changes.

As in Section 2, at each point $x \in M^3$, we can find a basis of the space T_xM^3 such that in this basis the metric g is given by the diagonal matrix $\operatorname{diag}(1,1,1)$ and the matrix L is given by the diagonal matrix $\operatorname{diag}(\lambda_1(x), \lambda_2(x), \lambda_3(x))$. In this basis, the polynomial $I_t(x,\xi)$ given by (3) reads

$$I_t = (\lambda - t)^2\xi_1^2 + (\lambda_1(x) - t)(\lambda - t)(\xi_2^2 + \xi_3^2).$$

Therefore, for any t, the function

$$\tilde{I}_t \stackrel{\text{def}}{=} \frac{I_t}{(\lambda - t)} = (\lambda - t)\xi_1^2 + (\lambda_1(x) - t)(\xi_2^2 + \xi_3^2)$$

is an integral for the geodesic flow of g. Substituting $t = \lambda$ we see that the function

$$\tilde{I}_\lambda = (\lambda_1(x) - \lambda)(\xi_2^2 + \xi_3^2)$$

is also an integral

Since $\lambda_1(y) = \lambda$, for any geodesic γ passing through the point y we have $\tilde{I}_\lambda(\gamma, \dot{\gamma}) \equiv 0$. Then, for any $x \in \gamma$ such that $\lambda_1(x) \neq \lambda$, the sum $\xi_2^2 + \xi_3^2$, where ξ_2 and ξ_3 are the second and the third components of the vector $\dot{\gamma}(x) = (\xi_1, \xi_2, \xi_3)$ in the "diagonal" coordinates at the point x, is zero. Therefore, the velocity vector $\dot{\gamma}(x)$ is an eigenvector of L with eigenvalue $\lambda_1(x)$. Then two geodesics passing through y can transversally intersect only in the points z where $\lambda_1(z) = \lambda$. There can be at most two of such points: if there exist three such points then almost any point of g is a transversal intersection of at least two geodesics passing through the points where $\lambda_1 = \lambda$ so that $\lambda_1 = \lambda$ at each point of M^3. Then the fundamental group of M^n is either trivial or isomorphic to Z_2. Hence, modulo the Poincaré conjecture, the manifold is covered by S^3.

Proof for Case 2: The eigenvalue λ_1 is a smooth function on the manifold. Denote by h the metric $1/(\lambda - \lambda_1(x))g$. On some finite cover of M^3, we can define a vector field v_1 satisfying

$$\begin{cases} L(v_1) &= \lambda_1 v_1 \\ h(v_1, v_1) &= 1. \end{cases}$$

In the Levi-Civita coordinates from Theorem 6, the metric h reads

$$ds_h^2 = dx_1^2 + B(x_2, x_3)(dx_2^2 + dx_3^2),$$

and the vector field v_1 is equal to $\pm\frac{\partial}{\partial x_1}$. We see that the flow of the vector field v_1 preserves the metric h, and that the distribution orthogonal to v_1 is integrable. The following lemma completes the proof of Theorem 2.

LEMMA 2. *Suppose a closed Riemannian 3-manifold (M^3, h) has a nowhere vanishing vector field v_1 such that the distribution orthogonal to v_1 is integrable, and the flow of the vector field preserves the metric h. Then M^3 can be covered by the direct product of the circle and a surface.*

Proof: Without loss of generality, we can assume that the integral curves of the vector field v_1 are closed. Indeed, let us show that we always can find a vector field v_1^{new} and a Riemannian metric h^{new} on M^3 such that they satisfy the conditions of the lemma and the integral curves of v_1^{new} are closed.

Consider the group $\mathfrak{G}$ of all isometries of (M^3, h). It is known that $\mathfrak{G}$ is a compact Lie group. The flow F of the vector field v_1 is a one-parameter subgroup of $\mathfrak{G}$. Its closure is a compact Lie subgroup of $\mathfrak{G}$. It is evidently commutative, so it is isomorphic (as a Lie group) to the torus. We can clearly choose a one-dimensional subgroup F^{new} of the torus in such a way that it is closed and its generator in the Lie algebra is close enough to the generator of F so that the corresponding vector field on the manifold never vanishes.

Denote by v_1^{new} the corresponding vector field on the manifold. More precisely, parametrise the subgroup F^{new} by $t \in (R \bmod 2\pi)$ such that $F_{t_1}^{new} \cdot F_{t_2}^{new} = F_{t_1+t_2}^{new}$. Define the vector field v_1^{new} at $x \in M^3$ to be $\frac{d}{dt}\left(F_t^{new}(x)\right)$. By construction, its flow preserves the metric h and the vector field v, and its integral curves are closed.

Let us define the new Riemannian metric h^{new} by the relations

$h^{new}(u_1, u_2) = h(u_1, u_2)$ for any u_1, u_2 such that $h(u_1, v_1) = h(u_2, v_1) = 0$

$h^{new}(v_1^{new}, u) = 0$ for any u such that $h(v_1, u) = 0$

$h^{new}(v_1^{new}, v_1^{new}) = 1.$

In other words, the metrics h and h^{new} coincide on the distribution orthogonal to v_1 in h; the distribution orthogonal to v_1 in h coincides with the distribution orthogonal to v_1^{new} in h^{new}, and the length of v_1^{new} in h^{new} is one. By construction, the new metric h^{new} is preserved by the vector field v_1^{new} and the distribution orthogonal to v_1^{new} in h^{new} is integrable.

Thus, we can assume that the integral curves of the vector field v_1 are closed. Hence we have a locally-free action of the circle on the manifold generated by shifts along the integral curves of the vector field v_1. Then the manifold M^3 is Seifert. The theory of Seifert manifolds is well developed (see, for example, [**20**]), in particular it is known that there exists a finite cover $\tilde{M}^3$ such that the action of the circle on the cover, generated by the vector field v_1, is free [**8**], so that $\tilde{M}^3$ has the natural structure of a fibration with fibre homeomorphic to the circle. Without loss of generality, we can assume that this circle (which can also be identified with the circle freely acting on $\tilde{M}^3$) is isomorphic to $R \bmod 2\pi$.

Let us prove that the fibration of $\tilde{M}^3$ is indeed trivial. Denote by B^2 the base of the fibration; B^2 is a closed orientable surface. Consider the standard handle decomposition of B^2 into two 2-disks D_s and D_l and bands $L_1, ..., L_{2n}$. Recall that two opposite sides of each band are called short; the other two sides are called long. For any band, its short sides are glued to some intervals on the border of D_s. Since the base is orientable, one side is glued in one direction and the other in the other. Similarly, the long sides of each band are glued to some intervals of D_l, one side is glued in one direction and the other in the other.

The existence of the integrable distribution orthogonal to the fibres allows us to define naturally a monodromy map $\mu : \pi(B^2, x) \to R \bmod 2\pi$: for any point X lying over the point $x \in B^2$ and for any closed smooth loop $\gamma : [0,1] \to B^2$, $\gamma(1) = \gamma(0) = x$, there exists a unique curve on M^3 such that this curve starts at X, is tangent to the distribution, and lies over the loop. Denote by X_1 the end point of the curve. There exists a unique $t \in R \bmod 2\pi$ such that $F_t(X) = X_1$. By definition, put $\mu(\gamma)$ equal to t. Since the distribution is integrable, the definition does not depend on the choice of γ in the homotopy class of the loop. Clearly, the monodromy depends on the metric h.

Let us consider $D_s \cup L_1 \cup ... \cup L_{2n} = B^2 \setminus \operatorname{Int} D_l$. Its fundamental group is isomorphic to the free group with $2n$ generators: any band gives us one generator. Hence, each band has monodromy $\phi_i \in R \bmod 2\pi$. Let us change the metric h to the new metric $\bar{h}$ as follows:

(1) In each band L_i, choose an arc a_i joining the long sides of the band.

(2) Cut M along the annulus A_i which is the inverse image of a_i under the fibre projection; there appear two annuli A_i', A_i''.

(3) Glue them again via rotation by ϕ_i.

Clearly, this cut+rotate+glue operation does not change the topological type of the manifold or the fibration. Since the action of the circle preserves the metric h, rotation by ϕ_i preserves the metric h and therefore we get another metric $\bar{h}$ on the inverse image of $B^2 \setminus \operatorname{Int} D_l$. By construction, the new metric and the vector field v_1 satisfy the conditions of the lemma, and the monodromies of the bands (in the new metrics) are zero.

Let us calculate the monodromy of the border m of the disk D_l. With respect to h it is evidently zero. It is zero also with respect to $\bar{h}$, since for each i the border m crosses the arc a_i in opposite directions. Thus, the cut+rotate+glue operation along A_i does not contribute to the monodromy of m. Thus, the monodromy corresponding to the metric $\bar{h}$ is identically zero, and therefore the fibration is trivial. Theorem 2 is proved.

6. Proof of Corollary 1

Our goal is to prove that, modulo the Poincaré conjecture, any closed 3-manifold with fundamental group Z is homeomorphic to $S^2 \times S^1$.

Recall that the Poincaré conjecture says that any closed 3-manifold with trivial fundamental group is homeomorphic to the sphere; its corollary is that any compact

3-manifold with boundary homeomorphic to the 2-sphere is homeomorphic to the ball, and that any compact 3-manifold with boundary homeomorphic to two copies of the 2-sphere is homeomorphic to the direct product of the 2-sphere and the interval.

By **[25]**, any closed 3-manifold with fundamental group homeomorphic to Z is reducible. That is, there exists a submanifold $S^2 \subset M^3$ homeomorphic to the 2-sphere such that the set $M^3 \setminus S^2$ is either connected or each of its connected components is not homeomorphic to the 3-ball. The second possibility cannot occur in our case: otherwise, since the sphere is simply-connected, the fundamental group of M^3 must be the free product of the fundamental groups of these components; therefore at least one of these components is simply-connected and therefore (modulo the Poincaré conjecture) is isomorphic to the 3-ball. Thus, there exists a submanifold S^2 homeomorphic to the 2-sphere which does not divide M^3 into two parts. Consider the generator of the fundamental group Z. It clearly intersects S^2 and the index of the intersection is ± 1. Then the fundamental group of $M^3 \setminus S^2$ is trivial and therefore $M^3 \setminus S^2$ is homeomorphic to the direct product of the 2-sphere and the interval. Hence M^3 is homeomorphic to the direct product of the 2-sphere and the circle. Corollary 1 is proved.

References

[1] E. Beltrami, *Resoluzione del problema: riportari i punti di una superficie sopra un piano in modo che le linee geodetische vengano rappresentante da linee rette*, Ann. Mat., **1** (1865), no. 7, 185–204.

[2] A.V. Bolsinov, V.S. Matveev, A.T. Fomenko, *Two-dimensional Riemannian metrics with an integrable geodesic flow. Local and global geometries,* Sb. Math. **189** (1998), no. 9-10, 1441–1466.

[3] E. Cartan, *Lecons sur la theorie des espaces a connexion projective. Redigees par P. Vincensini,* Paris: Gauthier-Villars., 1937.

[4] U. Dini, *Sopra un problema che si presenta nella theoria generale delle rappresetazioni geografice di una superficie su un'altra*, Ann. Mat., **3** (1869), 269–293.

[5] L.P. Eisenhart, *Riemannian Geometry,* (2nd printing) Princeton University Press, Princeton, N. J., 1949.

[6] D.B.A. Epstein, *Projective planes in* 3*-manifolds,* Proc. London Math. Soc. (3) **11** (1961), 469–484.

[7] D.B.A. Epstein, *Finite presentations of groups and* 3*-manifolds,* Quart. J. Math. Oxford Ser. (2) **12** (1961), 205–212.

[8] D.B.A. Epstein, *Periodic flows on three-manifolds,* Ann. of Math. (2) **95** (1972) 66–82.

[9] M. Igarashi, K. Kiyohara, K. Sugahara, *Noncompact Liouville surfaces,* J. Math. Soc. Japan **45** (1993), no. 3, 459–479.

[10] K. Kiyohara, *Compact Liouville surfaces*, J. Math. Soc. Japan **43** (1991), 555-591.

[11] T. Levi-Civita, *Sulle trasformazioni delle equazioni dinamiche*, Ann. di Mat., serie 2^a, **24** (1896), 255–300.

[12] V.S. Matveev, P.J. Topalov, *Trajectory equivalence and corresponding integrals*, Regular and Chaotic Dynamics, **3** (1998), no. 2, 30–45.

[13] V.S. Matveev, *Quantum integrability of the Beltrami-Laplace operator for geodesically equivalent metrics*, Russian Math. Doklady **61** (2000), no 2, 216–219.

[14] V.S. Matveev, P.J. Topalov, *Geodesic equivalence of metrics on surfaces, and their integrability,* Dokl. Akad. Nauk **367** (1999), no. 6, 736–738.

[15] V.S. Matveev, P.J. Topalov, *Quantum integrability for the Beltrami-Laplace operator as geodesic equivalence,* Math. Z. **238** (2001), 833–866.

[16] V.S. Matveev and P.J. Topalov, *Metric with ergodic geodesic flow is completely determined by unparameterized geodesics,* ERA-AMS, **6** (2000), 98–104.

[17] Vladimir S. Matveev, *Commuting operators and separation of variables for Laplacians of projectively equivalent metrics,* Let. Math. Phys., **119** (2001), 193–201.

[18] Vladimir S. Matveev, *Geschlossene hyperbolische 3-Mannigfaltigkeiten sind geodätisch starr,* Manuscripta Math. 105 (2001), no. 3, 343–352.

[19] J. Mikes, *Geodesic mappings of affine-connected and Riemannian spaces. Geometry, 2.,* J. Math. Sci. **78** (1996), no. 3, 311–333.

[20] P. Orlik, *Seifert manifolds,* Lecture Notes in Mathematics, **291**, Springer-Verlag, Berlin-New York, 1972.

[21] I.A. Taimanov, *Topological obstructions to the integrability of geodesic flow on nonsimply connected manifold,* Math. USSR-Izv., **30** (1988), no. 2, 403–409.

[22] C.B. Thomas, *On 3-manifolds with finite solvable fundamental group,* Invent. Math. **52** (1979), no. 2, 187–197.

[23] P.J. Topalov and V.S. Matveev, *Geodesic equivalence and integrability,* Preprint of Max-Planck-Institut f. Math. no. 74 (**1998**).

[24] H. Weyl, *Zur Infinitesimalgeometrie: Einordnung der projektiven und der konformen Auffasung,* Nachrichten von der K. Gesellschaft der Wissenschaften zu Göttingen, Mathematisch-Physikalische Klasse, 99-112, 1921. Can be found in "Selecta Hermann Weyl", Birkhaüser Verlag, 1956.

[25] J.H.C. Whitehead, *On finite cocycles and the sphere theorem,* Colloq. Math., **6** (1958), 271–281.

MATHEMATISCHES INSTITUT, UNIVERSITÄT FREIBURG, 79104 FREIBURG, GERMANY

E-mail address: matveev@email.mathematik.uni-freiburg.de

Contemporary Mathematics
Volume **308**, 2002

Harmonic Maps of Finite Type into Generalized Flag Manifolds, and Twistor Fibrations

Yoshihiro Ohnita and Seiichi Udagawa

Dedicated to the memory of Professor Nobuo Sasakura

Introduction

It is a fundamental problem to classify all harmonic maps of Riemann surfaces into compact symmetric spaces, especially when the Riemann surface is compact. In order to investigate this problem, it is important to identify good classes of such harmonic maps. Let $\varphi : M \longrightarrow G/K$ be a harmonic map of a Riemann surface into a compact symmetric space with involution σ. Then φ is said to be of *finite type* if it is adapted by some finite Laurent polynomial function ξ with values in a twisted loop algebra $\Lambda\mathfrak{g}_\sigma$ (see Section 1) which is a solution of the Lax type equation $d\xi = [\xi, \alpha_\lambda]$, where $\lambda \in S^1$ is a loop parameter and α_λ is defined from the Maurer-Cartan form pulled back by φ (see Proposition 3.1 or [BP],[Gu]).

It was shown in [BFPP] that any non-conformal harmonic map of a two-torus into G/K of rank 1 is of finite type (see [Hi], [FPPS] for the pioneering works). After that, Burstall ([Bu]) generalized the above construction to the case where the target manifold is a normal homogeneous space admitting a k-symmetric structure and proved that any weakly conformal non-isotropic harmonic map φ of a two-torus into a sphere or a complex projective space can be lifted to a *primitive map* of finite type into a certain k-symmetric space (see also [BPW] for previous work). The k-symmetric spaces are well-known as a generalization of symmetric spaces and are also important homogeneous spaces related to symmetric spaces as well as generalized flag manifolds. In general, we may fix the *canonical k-symmetric structure* τ on an arbitrary generalized flag manifold for some $k(\geq 2)$ (see Section 1.1).

The starting point of our study is the question: Is φ itself of finite type? This will be solved affirmatively in a more general setting: Given a primitive map ψ of

---

1991 *Mathematics Subject Classification.* Primary 58E20 ; Secondary 53C43, 53C35.
The authors wish to thank Professor F. Burstall for many valuable suggestions and comments, thanks to which the paper has been improved.

finite type of M into a generalized flag manifold G/H, we can conclude that $\varphi = p \circ \psi : M \longrightarrow G/K$ is also a harmonic map of finite type for some choice of $K \supset H$ with G/K being a compact Hermitian symmetric space, where $p : G/H \longrightarrow G/K$ is the homogeneous projection (see Theorem 2.1). The condition of choosing a closed subgroup K of G is rather mild and satisfied in the case of complex projective space above. The case of the sphere cannot be treated directly, but it can be done through a totally geodesic isometric immersion into complex projective space. Therefore, combining the results in [BFPP] and [Bu] stated above with Theorem 2.1 we can conclude that *any non-isotropic harmonic map of a two-torus into a sphere or a complex projective space is of finite type* (Corollary 2.3).

Next, we generalize Theorem 2.1 to the case where the domain manifold is a complex manifold. To do this we need the concept of pluriharmonic maps. We consider not only the concepts of harmonic maps of Riemann surfaces and symmetric spaces but also the wider concepts of pluriharmonic maps of complex manifolds and k-symmetric spaces. Actually, in order to study harmonic maps of Riemann surfaces into compact symmetric spaces, we will need to use the concepts of pluriharmonic maps and k-symmetric spaces. A definition and the construction of pluriharmonic maps of finite type into compact symmetric spaces were given in [BFPP]. However, the definition of pluriharmonic maps of finite type into k-symmetric spaces has not been given. Therefore, in Section 3, we provide a definition and construction. Moreover, in Section 3.3, we give a generalization of Theorem 2.1 under some restrictions (Theorem 3.4). Although it is not a complete generalization of Theorem 2.1, it has two interesting applications in Sections 4 and 5.

In Section 4, we prove that *any non-isotropic pluriharmonic map φ of a complex torus into a complex projective space is of finite type.* A certain factorization theorem also holds under some additional conditions (see Theorem 4.3). We first prove that φ can be lifted to a primitive (pluriharmonic) map of finite type and we conclude that φ is of finite type using Theorem 3.4.

In Section 5, we consider an equation of Lax type similar to that of Section 1 on a compact Riemann surface M_g of genus $g > 1$. This gives a 1-form with values in the twisted loop algebra which satisfies the Maurer-Cartan equation. It gives a primitive harmonic map ψ of M_g into a k-symmetric space G/H. Therefore, we call this map a primitive harmonic map of *generalized finite type.* It turns out that ψ factors as $\psi = \hat{f} \circ h$, where $\hat{f} : J(M_g) \longrightarrow G/H$ is a primitive pluriharmonic map of finite type from the Jacobian torus and $h : M_g \longrightarrow J(M_g)$ is the Abel map (Theorem 5.2). When we represent $J(M_g)$ as $\mathbf{C}^g/\Lambda$, $\hat{f}$ is actually induced from some primitive pluriharmonic map of finite type $f : \mathbf{C}^g \longrightarrow G/H$. First, we show that $\hat{f}$ is well-defined on $\mathrm{Im}(h) \subset \mathbf{C}^g/\Lambda$. To prove that $\hat{f}$ is actually well-defined on the whole $\mathbf{C}^g/\Lambda$, we need to extend the concept of primitive maps of generalized finite type to the case where the domain is a complex manifold. We use the generalized Abel map to prove the statement about $\hat{f}$ above. This proof will be given at the end of Section 6. Moreover, a restricted version of Theorem 2.1 for primitive maps of generalized finite type also holds (Corollary 5.3). The harmonic maps of generalized finite type give an important class of harmonic maps of M_g into a k-symmetric space. It is also important to study the periodicity of primitive pluriharmonic maps of finite type from $\mathbf{C}^n$ into a k-symmetric space. We will discuss this problem elsewhere.

1. k-symmetric spaces and primitive maps of finite type

In this section, we describe the canonical k-symmetric structure on a generalized flag manifold and the definition of primitive harmonic maps of finite type.

1.1. Canonical k-symmetric structures on generalized flag manifolds

Let G be a compact semisimple Lie group with Lie algebra $\mathfrak{g}$. Let H be the centralizer of a torus T_H of G and $\mathfrak{h}$ its Lie algebra. The coset space $N = G/H$ is called a generalized flag manifold. We have a reductive decomposition $\mathfrak{g} = \mathfrak{h} \oplus \mathfrak{p}$. Let $\mathfrak{t}$ be a Cartan subalgebra of $\mathfrak{g}$ which is contained in $\mathfrak{h}$. Then $\mathfrak{t}^{\mathbf{C}}$ is a Cartan subalgebra of $\mathfrak{g}^{\mathbf{C}}$ which is contained in $\mathfrak{h}^{\mathbf{C}}$. Let $\alpha \in \sqrt{-1}\mathfrak{t}^*$ be any root with respect to $\mathfrak{t}^{\mathbf{C}}$ and let $\mathfrak{g}^\alpha$ the corresponding root space. Let Δ be the set of all non-zero roots and $\Delta = \Delta^+ \cup \Delta^-, \Delta^+ \cap \Delta^- = \varnothing$ a decomposition into positive roots and negative roots. We have $\overline{\mathfrak{g}^\alpha} = \mathfrak{g}^{-\alpha}$, where the complex conjugation is taken with respect to the real form $\mathfrak{g}$. Set

$$\mathfrak{n} = \sum_{\alpha \in \Delta^+} \mathfrak{g}^\alpha, \quad \bar{\mathfrak{n}} = \sum_{\alpha \in \Delta^-} \mathfrak{g}^\alpha .$$

We have $\mathfrak{g}^{\mathbf{C}} = \mathfrak{t}^{\mathbf{C}} \oplus \mathfrak{n} \oplus \bar{\mathfrak{n}}$. Suppose that the rank of $\mathfrak{g}$ is r. Let $\{\alpha_1, \cdots, \alpha_r\}$ be simple roots. For $j = 1, \cdots, r$, we define $\xi_j \in \mathfrak{t}$ by $\alpha_i(\xi_j) = \sqrt{-1}\delta_{ij}$, $(1 \le i \le r)$. We define a set of natural numbers by $I = \{i \mid \mathfrak{g}^{\alpha_i} \subset \mathfrak{p}^{\mathbf{C}}\}$. Any $\alpha \in \Delta$ is expressed as $\alpha = \sum_{j=1}^r n_j \alpha_j$ for some integers $n_1, \cdots, n_r$. Put $n_I(\alpha) = \sum_{j \in I} n_j$ and

$$\xi_I = \sum_{j \in I} \xi_j, \quad k = \operatorname*{Max}_{\alpha \in \Delta^+} n_I(\alpha) + 1, \quad \tau = \operatorname{Ad} \exp(2\pi \xi_I / k).$$

Then $\tau : \mathfrak{g} \longrightarrow \mathfrak{g}$ is an inner automorphism of order k. If we denote by $\hat{\mathfrak{g}}^r$ the eigenspace of $\operatorname{ad}\xi_I$ with eigenvalue $\sqrt{-1}r$ $(r \in \mathbf{Z})$, then we see that

$$\mathfrak{g}_j = \sum_{r \in k\mathbf{Z}+j} \hat{\mathfrak{g}}^r$$

is an eigenspace of τ with eigenvalue $\exp(2\pi\sqrt{-1}j/k)$ for $j = 0, 1 \cdots, k-1$. It follows from the definition of the value k that $\mathfrak{g}_0$ coincides with the centralizer of ξ_I, which is nothing but $\mathfrak{h}^{\mathbf{C}}$. Then we have the following relations:

$$\mathfrak{g}_0 = \mathfrak{h}^{\mathbf{C}}, \quad \mathfrak{g}^{\mathbf{C}} = \sum_{j=0}^{k-1} \mathfrak{g}_j, \quad [\mathfrak{g}_i, \mathfrak{g}_j] \subset \mathfrak{g}_{i+j}, \quad \overline{\mathfrak{g}_j} = \mathfrak{g}_{-j},$$

where the indices are considered mod k. Thus, we may give a k-symmetric space structure on G/H (cf. [Ko]) and we call ξ_I and τ constructed above the *canonical element* and the *canonical k-symmetric structure*, respectively.

1.2. Primitive maps of finite type

Let G be a compact semisimple Lie group and G/H a compact homogeneous space of G. Suppose that G/H admits a k-symmetric structure τ. Generalized flag manifolds with the canonical k-symmetric structures described in Section 1.1 give such examples. For $\omega = \exp(2\pi\sqrt{-1}/k)$, we consider a twisted loop algebra

$$\Lambda\mathfrak{g}_\tau^{\mathbf{C}} = \{\xi : S^1 \longrightarrow \mathfrak{g}^{\mathbf{C}} \mid \tau(\xi(\lambda)) = \xi(\omega\lambda) \text{ for any } \lambda \in S^1\} .$$

Let $\Lambda\mathfrak{g}_\tau = \{\xi \in \Lambda\mathfrak{g}_\tau^{\mathbf{C}} \mid \xi : S^1 \longrightarrow \mathfrak{g}\}$ be the real form of $\Lambda\mathfrak{g}_\tau^{\mathbf{C}}$. Any element $\xi(\lambda)$ of $\Lambda\mathfrak{g}_\tau$ has a Fourier series $\xi(\lambda) = \sum \xi_j \lambda^j$. Let $d \equiv 1 \bmod k$. Define a finite dimensional subspace Λ_d of $\Lambda\mathfrak{g}_\tau$ by

$$\Lambda_d = \{\xi = \sum \xi_j \lambda^j \in \Lambda\mathfrak{g}_\tau \mid \xi_j = 0 \text{ for } \mid j \mid > d\} .$$

Let $\psi : M \longrightarrow G/H$ be a smooth map of a Riemann surface. Take a local lift $F : U \longrightarrow G$ with $\psi = \pi \circ F$, where $\pi : G \longrightarrow G/H$ is the coset projection. Denote by θ the Maurer-Cartan form of G. Then $F^*\theta$ has a decomposition

$$F^*\theta = \alpha_{\mathfrak{p}} + \alpha_{\mathfrak{h}} .$$

Let $\alpha_{\mathfrak{p}} = \alpha_{\mathfrak{p}}' + \alpha_{\mathfrak{p}}''$ be the decomposition into (1,0)-forms and (0,1)-forms of M. Then ψ is said to be *primitive* if $\alpha_{\mathfrak{p}}'$ is $\mathfrak{g}_1$-valued. If $k \geq 3$, then any primitive map is a harmonic map with respect to any normal homogeneous metric on G/H (any G-invariant metric if G/H is completely reductive (cf. [Bl]). If $k = 2$, all maps are primitive.

Consider a map $\xi : M \longrightarrow \Lambda\mathfrak{g}_\tau$. Then ξ_{d-1} is $\mathfrak{g}_0$-valued. Set

$$\mathfrak{n}_{\mathfrak{h}} = \mathfrak{n} \cap \mathfrak{h}^{\mathbf{C}} , \quad \overline{\mathfrak{n}_{\mathfrak{h}}} = \overline{\mathfrak{n}} \cap \mathfrak{h}^{\mathbf{C}} .$$

We fix an Iwasawa decomposition of $\mathfrak{h}^{\mathbf{C}}$:

$$\mathfrak{h}^{\mathbf{C}} = \mathfrak{h} \oplus \mathfrak{b} , \quad \mathfrak{b} = \sqrt{-1}\mathfrak{t} \oplus \overline{\mathfrak{n}_{\mathfrak{h}}} .$$

Thus, $(\xi_{d-1}dz)_{\mathfrak{h}}$ has an expression $(\xi_{d-1}dz)_{\mathfrak{h}} = r(\xi_{d-1})dz + \overline{r(\xi_{d-1})}\,\overline{dz}$, where z is a local complex coordinate for M and a map $r : \mathfrak{h}^{\mathbf{C}} \longrightarrow \mathfrak{h}^{\mathbf{C}}$ is given by $r(\eta) = \eta_{\mathfrak{n}_{\mathfrak{h}}} + \frac{1}{2}\eta_{\mathfrak{t}^{\mathbf{C}}}$ for $\eta \in \mathfrak{h}^{\mathbf{C}}$. Take an initial condition $\xi_0 \in \Lambda_d$ and let $\xi : M \longrightarrow \Lambda_d$ be a solution of the differential equation

$$\frac{\partial \xi}{\partial z} = [\xi, \lambda\xi_d + r(\xi_{d-1})] , \quad \xi(0) = \xi_0 .$$

Then $\alpha_\lambda = \lambda\xi_d dz + (\xi_{d-1}dz)_{\mathfrak{h}} + \lambda^{-1}\xi_{-d}\overline{dz}$ satisfies the Maurer-Cartan equation (cf. [BP]). Let $\widetilde{M}$ be the universal covering space of M and denote by $\pi_M : \widetilde{M} \longrightarrow M$ a covering projection. Since $\pi_M^*\alpha_\lambda$ also satisfies the Maurer-Cartan equation, there is a map $F_\lambda : \widetilde{M} \longrightarrow G$ with $\pi_M^*\alpha_\lambda = F_\lambda^*\theta$. In this case, $\pi \circ F_1$ is a primitive harmonic map. When $k = 2$ it is a harmonic map with respect to any G-invariant metric on G/H. Moreover, F_λ is called an *extended framing* of $\pi \circ F_1$. Given a primitive harmonic map $\psi : M \longrightarrow G/H$, we say that ψ is of *finite type* if $\psi \circ \pi_M$ has an extended framing F_λ with $\psi \circ \pi_M = \pi \circ F_1$.

REMARK. *We sometimes use the notation $F^{-1}dF$ in place of $F^*\theta$, even in the case where G is not a matrix group. This notation can be justified by embedding G into some unitary group using a faithful linear representation of G.*

2. Twistor fibrations giving harmonic maps of finite type into compact symmetric spaces

Let G/H be a generalized flag manifold with the canonical k-symmetric structure τ described in Section 1. We keep the notation of Section 1. We have $\mathfrak{h}^{\mathbf{C}} = \mathfrak{t}^{\mathbf{C}} \oplus \mathfrak{n}_{\mathfrak{h}} \oplus \overline{\mathfrak{n}_{\mathfrak{h}}}$. We equip G/H with a G-invariant metric induced by an

AdG-invariant inner product on $\mathfrak{g}$. In this case, we may choose $\mathfrak{p}$ as the orthogonal complement of $\mathfrak{h}$ in $\mathfrak{g}$ with respect to the AdG-invariant inner product. Thus, G/H is a normal homogeneous Riemannian manifold. We choose a closed subgroup K of G satisfying the following conditions:

$$(2.1)\quad \begin{cases} (1)\ H \subset K \text{ and } G/K \text{ is a compact Hermitian symmetric space,} \\ (2)\ \text{the canonical decomposition } \mathfrak{g} = \mathfrak{k} \oplus \mathfrak{m} \text{ is } \tau\text{-stable and orthogonal,} \\ (3)\ \mathfrak{g}_1 \cap \bar{\mathfrak{n}} \subset \mathfrak{m}^{\mathbf{C}}\ . \end{cases}$$

The projection $p : G/H \ni gH \longrightarrow gK \in G/K$ is called the *homogeneous projection.* Let $\psi : M \longrightarrow G/H$ be a primitive map of finite type. Then it is known ([Bl],[BP]) that $\varphi = p \circ \psi : M \longrightarrow G/K$ is a harmonic map with respect to any G-invariant metric induced from that on $\mathfrak{g}$. In this situation, we pose a question: Is $\varphi : M \longrightarrow G/K$ of finite type? We have an affirmative answer as follows.

THEOREM 2.1. *Let G/H be a generalized flag manifold with the canonical $k(>2)$-symmetric structure τ and a normal homogeneous metric. Let $\psi : M \longrightarrow G/H$ be a primitive map of finite type. Choose a closed subgroup K of G as in (2.1). Then $\varphi = p \circ \psi : M \longrightarrow G/K$ is a harmonic map of finite type with respect to the G-invariant metric induced from any* AdG*-invariant inner product on $\mathfrak{g}$.*

PROOF. Set $\mathfrak{n}_{\mathfrak{k}} = \mathfrak{n} \cap \mathfrak{k}^{\mathbf{C}}$. It follows from the choice of K that $(\mathfrak{g}_1)_{\bar{\mathfrak{n}}_{\mathfrak{k}}} = 0$ and $(\mathfrak{g}_{k-1})_{\mathfrak{n}_{\mathfrak{k}}} = 0$. We have a decomposition

$$\mathfrak{g}^{\mathbf{C}} = \sum_{j=1}^{k-1} (\mathfrak{g}_j)_{\mathfrak{m}^{\mathbf{C}}} + \sum_{j=1}^{k-2} (\mathfrak{g}_j)_{\mathfrak{n}_{\mathfrak{k}}} + \sum_{j=2}^{k-1} (\mathfrak{g}_j)_{\bar{\mathfrak{n}}_{\mathfrak{k}}} + \mathfrak{g}_0\ .$$

This decomposition is τ-stable and AdH-invariant. Set $I = \{\, i \mid \mathfrak{g}^{\alpha_i} \subset \mathfrak{p}^{\mathbf{C}}\}$ and $J = \{\, j \mid \mathfrak{g}^{\alpha_j} \subset \mathfrak{m}^{\mathbf{C}}\}$. Let ξ_I and ξ_J be the canonical elements corresponding to I and J, respectively. It follows from the definitions of ξ_I and ξ_J that

$$\mathrm{ad}\xi_I = \begin{cases} \sqrt{-1}j & \text{on } (\mathfrak{g}_j)_{\mathfrak{n}} \\ \sqrt{-1}(j-k) & \text{on } (\mathfrak{g}_j)_{\bar{\mathfrak{n}}} \\ 0 & \text{on } \mathfrak{g}_0 \end{cases}, \qquad \mathrm{ad}\xi_J = \begin{cases} \sqrt{-1} & \text{on } (\mathfrak{m}^{\mathbf{C}})_{\mathfrak{n}} \\ -\sqrt{-1} & \text{on } (\mathfrak{m}^{\mathbf{C}})_{\bar{\mathfrak{n}}} \\ 0 & \text{on } \mathfrak{k}^{\mathbf{C}} \end{cases}.$$

Define an element $\nu \in H$ by $\nu = \exp(\pi t\, \xi_J) \cdot \exp(-\frac{2\pi}{k} t\, \xi_I), t \in \mathbf{R}$. We then see that Ad$\nu$ has the following properties:

$$\mathrm{Ad}\nu = \begin{cases} \zeta^{\frac{k}{2}-j}\mathrm{Id} & \text{on } (\mathfrak{g}_j)_{\mathfrak{m}^{\mathbf{C}}} \quad (j = 1, \cdots, k-1), \\ \zeta^{-j}\mathrm{Id} & \text{on } (\mathfrak{g}_j)_{\mathfrak{n}_{\mathfrak{k}}} \quad (j = 1, \cdots, k-2), \\ \zeta^{k-j}\mathrm{Id} & \text{on } (\mathfrak{g}_j)_{\bar{\mathfrak{n}}_{\mathfrak{k}}} \quad (j = 2, \cdots, k-1), \\ \mathrm{Id} & \text{on } \mathfrak{g}_0, \end{cases}$$

where $\zeta = \exp(\frac{2\pi\sqrt{-1}}{k}t) \in S^1$. Let F_ζ be an extended framing of ψ. Then ψ is a primitive map of finite type if and only if there is a $Y(z,\zeta) = \sum_{j=-d}^{d} \xi_j \zeta^j \in \Lambda_d$ such that the following hold

$$dY = [Y, F_\zeta^{-1} dF_\zeta]\ , \quad F_\zeta^{-1} \frac{\partial F_\zeta}{\partial z} = \zeta \xi_d + r(\xi_{d-1})\ ,$$

where $d \equiv 1 \bmod k$ and $r(\xi_{d-1}) = (\xi_{d-1})_{\mathfrak{n}_{\mathfrak{h}}} + \frac{1}{2}(\xi_{d-1})_{\mathfrak{t}^{\mathbf{C}}}$. We call such a Y an *adapted polynomial Killing field.* Now, set $\lambda = \zeta^{k/2}$ and $\widetilde{F} = \mathrm{Ad}\nu \cdot F_\zeta$. We see that

$$\begin{aligned}(2.2) \qquad \widetilde{F}^{-1}\frac{\partial \widetilde{F}}{\partial z} &= \mathrm{Ad}\nu \cdot F_\zeta^{-1}\frac{\partial F_\zeta}{\partial z} \\ &= \mathrm{Ad}\nu(\zeta\xi_d + r(\xi_{d-1})) \\ &= \lambda(\xi_d)_{\mathfrak{m}^{\mathbf{C}}} + (\xi_d)_{\mathfrak{n}_{\mathfrak{k}}} + r(\xi_{d-1}) \in \lambda\mathfrak{m}^{\mathbf{C}} + \mathfrak{k}^{\mathbf{C}} .\end{aligned}$$

The involution σ of compact symmetric space G/K is given by $\sigma = \mathrm{Ad}\nu\,|_{\zeta=\omega} \cdot \tau$. We denote by ΛG_σ the corresponding twisted loop group, that is,

$$\Lambda G_\sigma = \{\gamma : S^1 \longrightarrow G \mid \sigma(\gamma(\lambda)) = \gamma(-\lambda)\} .$$

Denote by $\Lambda\mathfrak{g}_\sigma$ the Lie algebra of ΛG_σ. Since $\widetilde{F}$ is real, it follows from (2.2) that $\widetilde{F}$ has values in ΛG_σ and $\widetilde{F}^*\theta$ has values in $\Lambda\mathfrak{g}_\sigma$. We can write $\widetilde{F} = \widetilde{F}_\lambda$ and see that $\varphi = \widetilde{F}_1 \cdot K$, that is, $\widetilde{F}_\lambda$ is an extended framing of φ. Define an integer d' and a finite dimensional subspace $\Lambda_{d'}$ of $\Lambda\mathfrak{g}_\sigma$ by

$$d' = \frac{2}{k}(d-1) + 1, \quad \Lambda_{d'} = \{\xi = \sum \xi_j\lambda^j \in \Lambda\mathfrak{g}_\sigma \mid \xi_j = 0 \quad \text{for } |\, j\,| > d'\}.$$

Now Theorem 2.1 follows from Lemma 2.2 below. □

LEMMA 2.2. *Set $\widetilde{Y} = \mathrm{Ad}\nu \cdot Y$. Then the following hold:*

1. $d\widetilde{Y} = [\widetilde{Y}, \widetilde{F}^{-1}d\widetilde{F}]$,
2. $\widetilde{Y} : M \longrightarrow \Lambda_{d'}$ *and we may write* $\widetilde{Y} = \sum_{j=-d'}^{d'} \widetilde{\xi}_j\lambda^j$,
3. $\widetilde{F}^{-1}\frac{\partial \widetilde{F}}{\partial z} = \lambda\widetilde{\xi}_{d'} + \widetilde{r}(\widetilde{\xi}_{d'-1})$, *where* $\widetilde{r} : \mathfrak{k}^{\mathbf{C}} \longrightarrow \mathfrak{k}^{\mathbf{C}}$ *and* $\widetilde{r}(\widetilde{\xi}_{d'-1}) = (\widetilde{\xi}_{d'-1})_{\mathfrak{n}_{\mathfrak{k}}} + \frac{1}{2}(\widetilde{\xi}_{d'-1})_{\mathfrak{t}^{\mathbf{C}}}$.

PROOF OF LEMMA 2.2. (1) follows from the definitions of $\widetilde{F}$ and $\widetilde{Y}$. We prove (2). We may set $d = 1 + Nk$, where N is a non-negative integer. Then we have $d' = 2N + 1$. Recall that the definition of $\mathrm{Ad}\nu$ and the setting $\lambda = \zeta^{k/2}$. For $Y = \sum_{j=-d}^{d} \xi_j\zeta^j$, for example, we have

$$(2.3) \quad \mathrm{Ad}\nu \cdot (\xi_d\zeta^d) = \lambda^{d'}(\xi_d)_{\mathfrak{m}^{\mathbf{C}}} + \lambda^{d'-1}(\xi_d)_{\mathfrak{n}_{\mathfrak{k}}} , \ \mathrm{Ad}\nu \cdot (\xi_{d-1}\zeta^{d-1}) = \lambda^{d'-1}\xi_{d-1} .$$

In general, for $1 \leq j \leq k-1$ we have

$$(2.4) \qquad \mathrm{Ad}\nu \cdot (\xi_j\zeta^j) = \lambda(\xi_j)_{\mathfrak{m}^{\mathbf{C}}} + (\xi_j)_{\mathfrak{n}_{\mathfrak{k}}} + \lambda^2(\xi_j)_{\overline{\mathfrak{n}_{\mathfrak{k}}}} .$$

Moreover, since $\zeta^{j+N'k} = \zeta^j\lambda^{2N'}$, the constraint $j + N'k \leq d = 1 + Nk$ means that $N' \leq N$ if $j = 1$ and $N' \leq N - 1$ if $j \geqq 2$. Therefore we see that the highest order term of $\widetilde{Y}$ is $\widetilde{\xi}_{d'}$. The reality of $\widetilde{Y}$ implies that $\widetilde{Y}$ is of the form $\widetilde{Y} = \sum_{j=-d'}^{d'} \widetilde{\xi}_j\lambda^j$ with $\overline{\widetilde{\xi}_j} = \widetilde{\xi}_{-j}$. Moreover, (2.3) and (2.4) mean that $\widetilde{\xi}_j \in \mathfrak{m}^{\mathbf{C}}$ if j is odd and $\widetilde{\xi}_j \in \mathfrak{k}^{\mathbf{C}}$ if j is even. Thus $\widetilde{Y}$ has values in $\Lambda_{d'}$. Finally we prove (3). Considering (2.3) and (2.4), we obtain $\widetilde{\xi}_{d'} = (\xi_d)_{\mathfrak{m}^{\mathbf{C}}}$ and

$$\widetilde{r}(\widetilde{\xi}_{d'-1}) = (\widetilde{\xi}_{d'-1})_{\mathfrak{n}_{\mathfrak{k}}} + \frac{1}{2}(\widetilde{\xi}_{d'-1})_{\mathfrak{t}^{\mathbf{C}}} = (\xi_d)_{\mathfrak{n}_{\mathfrak{k}}} + (\xi_{d-1})_{\mathfrak{n}_{\mathfrak{h}}} + \frac{1}{2}(\xi_{d-1})_{\mathfrak{t}^{\mathbf{C}}} = (\xi_d)_{\mathfrak{n}_{\mathfrak{k}}} + r(\xi_{d-1}),$$

which, together with (2.2), yield (3). □

COROLLARY 2.3. *Any non-isotropic harmonic map φ of two-torus into sphere or complex projective space is of finite type.*

PROOF. When φ is non-conformal, this is known from [BFPP]. When φ is (weakly) conformal, it is proved in [Bu] that φ can be lifted to a primitive map ψ of finite type into a k-symmetric space G/H for some k. For the case of a complex projective space, it is known that G/H is a generalized flag manifold. It follows from Theorem 2.1 that $\varphi = p \circ \psi$ is a harmonic map of finite type, as the conditions in (2.1) are clearly satisfied. For the case of a n-dimensional sphere S^n, the bundle of isotropy flags over S^n, denoted by $F^r(S^n)$, is not always a generalized flag manifold. However, we can show that there is an embedding of $so(n+1)$ into $su(n+1)$ such that, for each $j = 0, 1, \cdots, k-1$, the $\mathfrak{g}_j$-subspace for k-symmetric structure on $F^r(S^n)$ with $k = 2r+2$ is mapped into the $\mathfrak{g}_j$-subspace for the canonical k-symmetric structure on the generalized flag manifold over $\mathbf{C}P^n$. We denote by $i : S^n \longrightarrow \mathbf{C}P^n$ the corresponding totally-geodesic immersion. We see that φ is non-isotropic harmonic (respectively of finite type) if and only if $i \circ \varphi$ is non-isotropic harmonic (respectively of finite type). We present below such an embedding of $so(n+1)$ into $su(n+1)$.

We use the notation in [Bu]. For $F^r(S^n) = G/H$, we have $\mathfrak{g} = \mathfrak{h} \oplus \mathfrak{p}$ and

$$\begin{cases} \mathfrak{g}_j = \displaystyle\sum_{\substack{s+t=j \\ 0 \leqq s < t \leqq r+1}} l_s l_t \oplus \sum_{\substack{-s+t=j \\ 0<s<t\leqq r+1}} \overline{l_s} l_t \oplus \sum_{\substack{s+t=k-j \\ 0<s<t\leqq r}} \overline{l_s l_t}, \\ \mathfrak{g}_{k-j} = \overline{\mathfrak{g}_j}, \quad (j = 1, 2 \cdots, r+1), \\ \mathfrak{h}^{\mathbf{C}} = \mathfrak{g}_0 = \displaystyle\sum_{j=1}^{r} \overline{l_j} l_j \oplus \Lambda^2 l_{r+1}, \\ \mathfrak{p}^{\mathbf{C}} = \displaystyle\sum_{0\leqq i<j\leqq r+1} l_i l_j \oplus \sum_{0\leqq i\neq j \leqq r+1} \overline{l_i} l_j \oplus \sum_{0\leqq i<j\leqq r+1} \overline{l_i l_j}, \end{cases}$$

where l_0 is the complexified line defined by a point of S^n with $\overline{l_0} = l_0$ and $\overline{l_{r+1}} = l_{r+1}$.

[**The case of** $n = 2m$] The standard embedding of $so(2m+1)$ into $su(2m+1)$ is described in [He], p573. However, it cannot be used for our purpose as it stands. Define a real $m \times m$-matrix J by

$$J = \begin{pmatrix} & & & & 1 \\ & & & 1 & \\ & & \cdot & & \\ & \cdot & & & \\ 1 & & & & \end{pmatrix}.$$

Then we have $J^2 = I$, where I is the $m \times m$-identity matrix. We set $J = 1$ when $m = 1$. Define $\widetilde{J}$ by

$$\widetilde{J} = \begin{pmatrix} 1 & & \\ & I & \\ & & J \end{pmatrix}.$$

Conjugating the image of the above embedding $so(2m+1)$ into $su(2m+1)$ by $\widetilde{J}$, we see that $so(2m+1)$ is isomorphic to the space of matrices of the form

$$\begin{pmatrix} 0 & v_0 & \overline{v}_0 J \\ -v_0^* & X+iY & \overline{Z}J \\ -J^t v_0 & JZ & JXJ-iJYJ \end{pmatrix} \in su(2m+1),$$

where $v_0 \in \mathbf{C}^m$, X is a real $m\times m$ skew-symmetric matrix, Y is a real $m \times m$ symmetric matrix and Z is a complex $m\times m$ skew-symmetric matrix. Write l_{r+1} as $l_{r+1} = \sum_{s=1}^{m-r} l_{r+1}^s \oplus \sum_{s=1}^{m-r} \overline{l_{r+1}^s}$. Then we order the rows by

$$l_0, l_1, \cdots, l_r, l_{r+1}^1, \cdots, l_{r+1}^{m-r}, \overline{l_{r+1}^{m-r}}, \cdots, \overline{l_{r+1}^1}, \overline{l_r}, \cdots, \overline{l_1}$$

and the columns by

$$l_0, \overline{l_1}, \cdots, \overline{l_r}, \overline{l_{r+1}^1}, \cdots, \overline{l_{r+1}^{m-r}}, l_{r+1}^{m-r}, \cdots, l_{r+1}^1, l_r, \cdots, l_1.$$

Note that $l_j l_j = 0, l_{r+1}^s l_{r+1}^s = 0, \overline{l_j l_j} = 0$ and $\overline{l_{r+1}^s l_{r+1}^s} = 0$ for $j = 0, 1, \cdots, r$ and $s = 1, 2, \cdots, m-r$. This correspondence means that the $\mathfrak{g}_1$-subspace for the $2(r+1)$-symmetric space $F^r(S^{2m})$ is mapped into the $\mathfrak{g}_1$-subspace for

$$SU(2m+1)/S(\underbrace{U(1)\times\cdots\times U(1)}_{(r+1)-\text{times}} \times U(2(m-r)) \times \underbrace{U(1)\times\cdots\times U(1)}_{r-\text{times}})$$

with the canonical $2(r+1)$-symmetric structure.

[**The case of** $n = 2m+1 (m \geqq 1)$] In this case, we give an embedding of $so(2m+2)$ into $su(2m+2)$ by

$$\begin{pmatrix} 0 & v_0 & x & \overline{v}_0 J \\ -v_0^* & X+iY & -u_0^* & \overline{Z}J \\ -x & u_0 & 0 & \overline{u}_0 J \\ -J^t v_0 & JZ & -J^t u_0 & JXJ-iJYJ \end{pmatrix} \in su(2m+2),$$

where $x \in \mathbf{R}$, $v_0, u_0 \in \mathbf{C}^m$, X is a real $m\times m$ skew-symmetric matrix, Y is a real $m \times m$ symmetric matrix and Z is a complex $m\times m$ skew-symmetric matrix. The correspondence is similar to the case of $n = 2m$. The only difference is the decomposition of l_{r+1}: $l_{r+1} = \sum_{s=1}^{m-r} l_{r+1}^s \oplus l_{\mathbf{R}} \oplus \sum_{s=1}^{m-r} \overline{l_{r+1}^s}$, where $l_{\mathbf{R}}$ is the real 1-dimensional subspace of l_{r+1}. Note that the entry 0 in the middle part of the above matrix corresponds to $l_{\mathbf{R}} l_{\mathbf{R}} = 0$. □

Examples. When $G = SU(n), H = S(U(r_0)\times U(r_1) \times \cdots \times U(r_{k-1}))$ with $\sum_{j=0}^{k-1} r_j = n$, G/H admits the canonical k-symmetric structure τ. Choose the closed subgroup $K = S(U(r_0) \times U(n-r_0))$ of G. Then, condition (2.1) is satisfied. We may define $\nu \in H$ by $\nu = \operatorname{diag}(\zeta^{x_0}, \cdots, \zeta^{x_0}, \zeta^{x_1}, \cdots, \zeta^{x_1}, \cdots, \zeta^{x_{k-1}}, \cdots, \zeta^{x_{k-1}})$, where $\operatorname{diag}(a_1, \cdots, a_n)$ is a diagonal matrix with diagonal entries $a_1, \cdots, a_n$, and $x_0 = 0, x_j = k/2 - j$ for $j = 1, \cdots, k-1$ and ζ^{x_i} is repeated r_i-times for $i = 0, 1, \cdots, k-1$. Theorem 2.1 ensures that any primitive map of finite type into G/H gives a harmonic map of finite type into G/K, which is a complex Grassmann manifold. Next, consider the case where $G = SU(4), H = S(U(1)\times U(1)\times U(1) \times U(1))$. A superconformal harmonic map $\varphi : T^2 \longrightarrow \mathbf{C}P^3$ can be lifted to a primitive map of finite type $\psi : T^2 \longrightarrow G/H$ (superconformal harmonic maps are related with the Toda field model, see [BPW] and [Mi]). Theorem 2.1 ensures that φ is itself of finite type. However, Theorem 2.1 says more. Choose $K = S(U(2) \times U(2))$. In this case, we define ν by $\nu = \mathrm{diag}(1, \zeta^{-1}, 1, \zeta^{-1})$. Then, since condition (2.1) is satisfied, we have a harmonic map of finite type $\hat{\varphi} : T^2 \longrightarrow G/K$ due to Theorem 2.1. This is a case where the first return map of the harmonic sequence is degenerate (see [Wd] for the definition of the first return map of a harmonic sequence). The non-degenerate case is treated in [Ud1] and [Ud2]. On the other hand, the non-degenerate case seems to be constructed from *spectral data*. It is known ([Mc1], [Mc2]) that any non-isotropic harmonic map $T^2 \longrightarrow \mathbf{C}P^n$ can be constructed from spectral data (see also [Ta]).

3. Primitive pluriharmonic maps of finite type and twistor fibrations

In this section, we generalize Theorem 2.1 to a higher dimensional domain manifold. For this, we need the concept of pluriharmonic maps of complex manifolds. We give the definition and construction of primitive pluriharmonic maps of finite type into k-symmetric spaces, which generalize those of [BFPP] and [BP].

3.1. Primitive pluriharmonic maps into k-symmetric spaces

Let G/H be a compact homogeneous space with a normal homogeneous metric g induced from an AdG-invariant inner product on $\mathfrak{g}$. We have a reductive decomposition $\mathfrak{g} = \mathfrak{h} \oplus \mathfrak{p}$, where $\mathfrak{p}$ is the orthogonal complement of $\mathfrak{h}$. Denote by $[\mathfrak{p}]$ the subbundle of $G/H \times \mathfrak{g}$ of which the fiber at $x = g \cdot o \in G/H$ is given by $\mathrm{Ad}g \cdot \mathfrak{p}$. Denote by $P_{[\mathfrak{p}]} : G/H \times \mathfrak{g} \longrightarrow [\mathfrak{p}]$ the orthogonal projection. Denote by ∇ the Levi-Civita connection on $(G/H, g)$. Define $\hat{\nabla}$ by $\hat{\nabla} \circ \beta = \beta \circ \nabla$, where $\beta : T(G/H) \longrightarrow [\mathfrak{p}]$ is the isomorphism given by the inverse of the map $[\mathfrak{p}] \ni \xi \longrightarrow \frac{d}{dt}|_{t=0} \exp t\xi \cdot x \in T_x(G/H)$ (cf. [BR]). We have

$$\hat{\nabla} = \nabla^c + \frac{1}{2} P_{[\mathfrak{p}]} \circ \mathrm{ad}\beta = d - \mathrm{ad}\beta + \frac{1}{2} P_{[\mathfrak{p}]} \circ \mathrm{ad}\beta \,,$$

where ∇^c is the canonical connection of $G/H \times \mathfrak{g}$.

Let $\psi : (M, J) \longrightarrow G/H$ be a smooth map of a complex manifold with complex structure J. Denote by ∇^ψ the pull-back connection on the pull-back bundle $\psi^{-1}T(G/H)$ over M and by d^{∇^ψ} the covariant exterior differentiation defined by ∇^ψ. Then ψ is said to be *pluriharmonic* if $(d^{\nabla^\psi}(d\psi \circ J))^{(1,1)} = 0$. This definition coincides with the usual one given in [OU]. By a simple examination, we see that ψ is pluriharmonic if and only if

$$d \circ \beta(d\psi \circ J)^{(1,1)} = 0 \,. \tag{3.1}$$

Take a local lift $F : U \longrightarrow G$ of ψ with $\psi = \pi \circ F$, where $\pi : G \longrightarrow G/H$ is the coset projection. Set $\alpha = F^*\theta$ and denote by $\alpha = \alpha_\mathfrak{h} + \alpha_\mathfrak{p}$ the decomposition with respect to $\mathfrak{g} = \mathfrak{h} \oplus \mathfrak{p}$. Denote by $\alpha_\mathfrak{p} = \alpha'_\mathfrak{p} + \alpha''_\mathfrak{p}$ and $\alpha_\mathfrak{h} = \alpha'_\mathfrak{h} + \alpha''_\mathfrak{h}$ the decompositions into (1,0)-form and (0,1)-form of $\alpha_\mathfrak{p}$ and $\alpha_\mathfrak{h}$, respectively. We have $\psi^*\beta = \mathrm{Ad}F \cdot \alpha_\mathfrak{p}$ and we see that (3.1) is equivalent to

$$\partial \alpha''_\mathfrak{p} - \overline{\partial} \alpha'_\mathfrak{p} + [\alpha'_\mathfrak{h} \wedge \alpha''_\mathfrak{p}] - [\alpha''_\mathfrak{h} \wedge \alpha'_\mathfrak{p}] = 0 \,. \tag{3.2}$$

Now, suppose that G/H admits a k-symmetric structure τ. As in Section 1 we have a decomposition $\mathfrak{g}^\mathbf{C} = \sum_{j=0}^{k-1} \mathfrak{g}_j$ into the eigenspaces of τ. As in the case of a Riemann surface as domain manifold, we call a map $\psi : M \longrightarrow G/H$ *primitive* if $\alpha'_{\mathfrak{p}}$ is $\mathfrak{g}_1$-valued. First of all, consider the case where $k = 2$. In this case, since $(G/H, g)$ has non-negative curvature operator, the pluriharmonicity of ψ yields

$$[\alpha'_{\mathfrak{p}} \wedge \alpha'_{\mathfrak{p}}] = [\alpha''_{\mathfrak{p}} \wedge \alpha''_{\mathfrak{p}}] = 0 \tag{3.3}$$

(cf. [OU], [OV]). Next, consider the case where $k \geq 3$. In this case, we suppose that ψ is a primitive map. Then, it follows from the vanishing of $\mathfrak{g}_2$-part and $\mathfrak{g}_{-2}$-part of the Maurer-Cartan equation of α that equation (3.3) also holds in this case. Further analysis of the Maurer-Cartan equation of α, together with (3.2) and (3.3), yields:

PROPOSITION 3.1.
1. *Let $\psi : M \longrightarrow G/H$ be a smooth map of a complex manifold. Then ψ is a primitive pluriharmonic map if and only if $\alpha'_{\mathfrak{p}}$ is $\mathfrak{g}_1$-valued and $\alpha_\lambda = \lambda\alpha'_{\mathfrak{p}} + \alpha_{\mathfrak{h}} + \lambda^{-1}\alpha''_{\mathfrak{p}}$ satisfies the Maurer-Cartan equation for each $\lambda \in S^1$.*
2. *If M is a simply-connected complex manifold and $\hat{\alpha}_\lambda$ defined by $\hat{\alpha}_\lambda = \lambda\hat{\alpha}'_{\mathfrak{p}} + \hat{\alpha}_{\mathfrak{h}} + \lambda^{-1}\hat{\alpha}''_{\mathfrak{p}} \in \Lambda\mathfrak{g}_\tau$ satisfies the Maurer-Cartan equation, then, for each $\lambda \in S^1$, there is a map $F_\lambda : M \longrightarrow G$ with $F_\lambda^*\theta = \hat{\alpha}_\lambda$ such that $\psi_\lambda = \pi \circ F_\lambda : M \longrightarrow G/H$ is a primitive pluriharmonic map.*
3. *If $k \geq 3$ and $\psi : M \longrightarrow G/H$ is a primitive map of a complex manifold, then ψ is pluriharmonic.*

3.2. Primitive pluriharmonic maps of finite type

Let G/H be a compact k-symmetric space with a normal homogeneous metric. Assume that the k-symmetric structure τ on G/H is an inner automorphism. Let $V : \mathfrak{g}^{\mathbf{C}} \longrightarrow \mathfrak{g}^{\mathbf{C}}$ be a $\mathrm{Ad}G^{\mathbf{C}}$-equivariant polynomial map.

DEFINITION. We say that V is *τ-compatible* if $V(\omega\tau) = \omega\tau \circ V$ holds, where ω is a primitive k-th root of 1.

Let $V_1, \cdots, V_n : \mathfrak{g}^{\mathbf{C}} \longrightarrow \mathfrak{g}^{\mathbf{C}}$ be τ-compatible, $\mathrm{Ad}G^{\mathbf{C}}$-equivariant polynomial maps. Then we have $d(V_j)_\xi([\xi, \eta]) = [V_j(\xi), \eta]$ for arbitrary $\xi, \eta \in \mathfrak{g}^{\mathbf{C}}$. In particular, we see that $[V_i(\xi), V_j(\xi)] = 0$ for $i, j = 1, \cdots, n$. Consider a subspace $\Lambda^{+1}\mathfrak{g}_\tau^{\mathbf{C}} = \{\eta \in \Lambda\mathfrak{g}_\tau^{\mathbf{C}} \mid \eta \text{ is of the form } \sum_{j\leq 1} \eta_j\lambda^j\}$, where $\lambda \in S^1$ is a loop parameter. For each $j = 1, \cdots, n$, let $V_j = V_j^{(1)} + \cdots + V_j^{(k_j)}$ be a decomposition into homogeneous polynomial maps. Denote by d_{ji} the homogeneous degree of $V_j^{(i)}$ for $1 \leq j \leq n, 1 \leq i \leq k_j$. Since τ is inner, we must have $d_{ji} \in k(\mathbf{N} \cup \{0\}) + 1$. Let $(z^1, \cdots, z^n)$ be a complex coordinate for $\mathbf{C}^n$. Choose an element $\xi_0 \in \Lambda_d$, where $d \equiv 1 \bmod k$, and define $\hat{\eta} \in \Lambda^{+1}\mathfrak{g}_\tau^{\mathbf{C}}$ by

$$\hat{\eta} = \sum_{j=1}^{n} \sum_{i=1}^{k_j} \lambda^{1-dd_{ji}} V_j^{(i)}(\xi_0) z^j .$$

Set $g = \exp(\hat{\eta})$. Then g has values in $\Lambda G_\tau^{\mathbf{C}}$. Let $H^{\mathbf{C}} = H \cdot B$ be the Iwasawa decomposition of $H^{\mathbf{C}}$ which corresponds to that of $\mathfrak{h}^{\mathbf{C}}$. Note that the multiplication map $\Lambda G_\tau \times \Lambda_- G_\tau^{\mathbf{C}} \longrightarrow \Lambda G_\tau^{\mathbf{C}}$ is a diffeomorphism onto (cf. [DPW]), where

$$\Lambda_- G_\tau^{\mathbf{C}} = \{\gamma(\lambda) \in \Lambda G_\tau^{\mathbf{C}} \mid \gamma(\lambda) \text{ extends holomorphically to } D_0, \gamma(0) \in B\}$$

and $D_0 = \{\lambda \in \mathbf{C} \mid \mid \lambda \mid > 1\}$. Then we may write $g = a \cdot b$, where $a : \mathbf{C}^n \longrightarrow \Lambda G_\tau$, $b : \mathbf{C}^n \longrightarrow \Lambda_- G_\tau^{\mathbf{C}}$. Since each $V_j^{(i)}$ is $\mathrm{Ad}G^{\mathbf{C}}$-equivariant, we see that $[V_j^{(i)}(\xi), V_k^{(l)}(\xi)] = 0$ for any $\xi \in \mathfrak{g}^{\mathbf{C}}$. Therefore we obtain

$$dg \cdot g^{-1} = d\hat{\eta} = \sum_{j=1}^{n} \sum_{i=1}^{k_j} \lambda^{1-dd_{ji}} V_j^{(i)}(\xi_0) dz^j \ .$$

On the other hand, using $g = a \cdot b$, we obtain $dg \cdot g^{-1} = da \cdot a^{-1} + \mathrm{Ad}a(db \cdot b^{-1})$. Hence, if we denote by $\pi_{\mathfrak{g}} : \Lambda\mathfrak{g}_\tau^{\mathbf{C}} \longrightarrow \Lambda\mathfrak{g}_\tau$ the projection with respect to the decomposition $\Lambda\mathfrak{g}_\tau^{\mathbf{C}} = \Lambda\mathfrak{g}_\tau \oplus \Lambda_-\mathfrak{g}_\tau^{\mathbf{C}}$, where

$$\Lambda_-\mathfrak{g}_\tau^{\mathbf{C}} = \{\xi \in \Lambda\mathfrak{g}_\tau^{\mathbf{C}} \mid \xi \text{ is of the form } \sum_{j \le 0} \xi_j \lambda^j, \xi_0 \in \mathfrak{b}\} \ ,$$

we obtain

$$a^{-1} da = \pi_{\mathfrak{g}}(\mathrm{Ad}a^{-1} \cdot \sum_{j=1}^{n} \sum_{i=1}^{k_j} \lambda^{1-dd_{ji}} V_j^{(i)}(\xi_0) dz^j) \ .$$

Set $\xi = \mathrm{Ad}a^{-1} \cdot \xi_0$. Then we see that $\xi : \mathbf{C}^n \longrightarrow \Lambda_d$ and $d\xi = [\xi, a^{-1}da]$. Moreover, we see that

$$\mathrm{Ad}a^{-1} \cdot d\hat{\eta} = \sum_{j=1}^{n} \sum_{i=1}^{k_j} \lambda^{1-dd_{ji}} V_j^{(i)}(\xi) dz^j \ , \tag{3.4}$$

where $\lambda^{1-dd_{ji}} V_j^{(i)}(\xi) \in \Lambda^{+1}\mathfrak{g}_\tau^{\mathbf{C}}$. In general, for $\eta = \sum_{j \le 1} \eta_j \lambda^j \in \Lambda^{+1}\mathfrak{g}_\tau^{\mathbf{C}}$, we have $\pi_{\mathfrak{g}}(\eta) = \lambda\eta_1 + (\eta_0)_{\mathfrak{h}} + \lambda^{-1}\overline{\eta_1}$. We denote by $V_j^{(i)}(\xi_d + \xi_{d-1})_1$ the sum of the linear terms with respect to ξ_{d-1} in $V_j^{(i)}(\xi_d + \xi_{d-1})$. It follows from (3.4) that

$$\begin{aligned} a^{-1}da &= \pi_{\mathfrak{g}}(\mathrm{Ad}a^{-1} \cdot d\hat{\eta}) \\ &= \lambda \sum_{j=1}^{n} V_j(\xi_d) dz^j + (\sum_{j=1}^{n} V_j(\xi_d + \xi_{d-1})_1 dz^j)_{\mathfrak{h}} + \lambda^{-1} \overline{\sum_{j=1}^{n} V_j(\xi_d) dz^j} \ . \end{aligned}$$

Since $a^{-1}da$ clearly satisfies the Maurer-Cartan equation, we have proved:

THEOREM 3.2. *For an initial condition $\xi_0 \in \Lambda_d$, let $\xi = \sum \xi_j \lambda^j : \mathbf{C}^n \longrightarrow \Lambda_d$ be a solution of the differential equations*

$$\frac{\partial \xi}{\partial z^j} = [\xi, \lambda V_j(\xi_d) + r(V_j(\xi_d + \xi_{d-1})_1)] \quad \text{for } j = 1, \cdots, n.$$

Then there is an extended framing $F_\lambda : \mathbf{C}^n \longrightarrow \Lambda G_\tau$ satisfying $F_\lambda^{-1} \frac{\partial F_\lambda}{\partial z^j} = \lambda V_j(\xi_d) + r(V_j(\xi_d + \xi_{d-1})_1)$ for each $j = 1, \cdots, n$ such that $\psi = \pi \circ F_1 : \mathbf{C}^n \longrightarrow G/H$ is a primitive pluriharmonic map.

We call ψ in Theorem 3.2 a primitive pluriharmonic map of *finite type*. In general, a primitive pluriharmonic map $\psi : M \longrightarrow G/H$ from a complex manifold M is said to be of finite type if $\psi \circ \pi_M$ has an extended framing $F_\lambda : \widetilde{M} \longrightarrow G$ obtained from an equation of Lax type as in Theorem 3.2 such that $\psi \circ \pi_M = \pi \circ F_1$, where $\pi_M : \widetilde{M} \longrightarrow M$ is the holomorphic universal covering map.

3.3. Twistor fibrations over compact symmetric spaces

Let G/H be a compact homogeneous space with a normal homogeneous metric induced from an AdG-invariant inner product on $\mathfrak{g}$. Suppose that G/H admits a $k(>2)$-symmetric structure τ. Let G/K be a compact symmetric space. Here, K is a closed subgroup of G such that $H \subset K$ and the canonical decomposition $\mathfrak{g} = \mathfrak{k} \oplus \mathfrak{m}$ is τ-stable and orthogonal with respect to AdG-invariant inner product on $\mathfrak{g}$. Let $p : G/H \longrightarrow G/K$ be a homogeneous projection. Given a primitive map $\psi : M \longrightarrow G/H$ of a complex manifold M, can we say that $p \circ \psi : M \longrightarrow G/K$ is pluriharmonic? We have the answer.

PROPOSITION 3.3. *If $\psi : M \longrightarrow G/H$ is primitive, then $p \circ \psi : M \longrightarrow G/K$ is pluriharmonic.*

PROOF. A framing F of ψ is also a framing for $p \circ \psi$. Set $\alpha = F^*\theta$. Equation (3.3), together with the Maurer-Cartan equation of α and the primitivity of ψ, yields

$$\bar\partial \alpha'_{\mathfrak{p}} + [\alpha'_{\mathfrak{p}} \wedge \alpha''_{\mathfrak{h}}] = 0 \,. \tag{3.5}$$

Let $\mathfrak{k} = \mathfrak{h} \oplus \mathfrak{l}$ be an orthogonal decomposition. We have $\alpha_{\mathfrak{p}} = \alpha_{\mathfrak{l}} + \alpha_{\mathfrak{m}}$. As in Theorem 3.7 in [BP], we have $[\alpha'_{\mathfrak{m}} \wedge \alpha''_{\mathfrak{l}}] = 0$. Therefore, by considering the $\mathfrak{m}$-part of (3.5) we obtain $\bar\partial \alpha'_{\mathfrak{m}} + [\alpha'_{\mathfrak{m}} \wedge \alpha''_{\mathfrak{k}}] = 0$, which, together with the $\mathfrak{m}$-part of the Maurer-Cartan equation of α, implies that (3.2) for G/K holds. Thus $p \circ \psi$ is pluriharmonic. □

We now prove a higher dimensional version of Theorem 2.1 under some restrictions.

THEOREM 3.4. *Let G/H be a generalized flag manifold with the canonical k-symmetric structure τ and let M an n-dimensional complex manifold. Let $V : \mathfrak{g}^{\mathbf{C}} \longrightarrow \mathfrak{g}^{\mathbf{C}}$ be a τ-compatible, $\mathrm{Ad}G^{\mathbf{C}}$-equivariant polynomial map. Set $V_j = c_j V$ for $c_j \in \mathbf{C}$ and for $j = 1, \cdots, n$. Let $\psi : M \longrightarrow G/H$ be a primitive pluriharmonic map of finite type arising from $V_1, \cdots, V_n$. Choose a closed subgroup K of G as in (2.1). Then $\varphi = p \circ \psi : M \longrightarrow G/K$ is a pluriharmonic map of finite type.*

PROOF. Since ψ is a primitive pluriharmonic map of finite type arising from $c_1 V, \cdots, c_n V$, we have $Y = \sum \xi_j \zeta^j \in \Lambda_d$, $d \equiv 1 \bmod k$, satisfying the following equations:

$$\begin{cases} dY = [Y, F_\zeta^{-1} dF_\zeta] \,, \\ F_\zeta^{-1} \dfrac{\partial F_\zeta}{\partial z^j} = \zeta c_j V(\xi_d) + r(c_j V(\xi_d + \xi_{d-1})_1) \,, \end{cases}$$

where $j = 1, \cdots, n$ and where we use the same notation as before. Let $V = V^{(1)} + \cdots + V^{(l)}$ be a decomposition into homogeneous polynomials. Denote by d_j the homogeneous degree of $V^{(j)}$ for $j = 1, \cdots, l$. Recall the projection $\pi_{\mathfrak{g}} : \Lambda\mathfrak{g}^{\mathbf{C}}_\tau \longrightarrow \Lambda\mathfrak{g}_\tau$ and the Iwasawa decomposition $\Lambda\mathfrak{g}^{\mathbf{C}}_\tau = \Lambda\mathfrak{g}_\tau \oplus \Lambda_-\mathfrak{g}^{\mathbf{C}}_\tau$. Define an Ad$G$-equivariant map $\widetilde{V}$ by $\widetilde{V} = \pi_{\mathfrak{g}} \circ \sum_{j=1}^{l} \zeta^{d-dd_j} V^{(j)}$. Set $\widetilde{Y} = \mathrm{Ad}\nu \cdot \widetilde{V}(Y)$ and $\widetilde{F} = \mathrm{Ad}\nu \cdot F_\zeta$ where ν is as defined in Section 2. Then

$$d\widetilde{Y} = \mathrm{Ad}\nu{\cdot}d\widetilde{V}_Y(dY) = \mathrm{Ad}\nu{\cdot}d\widetilde{V}_Y([Y, F_\zeta^{-1}dF_\zeta]) = \mathrm{Ad}\nu{\cdot}[\widetilde{V}(Y), F_\zeta^{-1}dF_\zeta] = [\widetilde{Y}, \widetilde{F}^{-1}d\widetilde{F}].$$

Moreover we see that

$$\widetilde{F}^{-1}\frac{\partial \widetilde{F}}{\partial z^j} = \lambda c_j V(\xi_d)_{\mathfrak{m}^{\mathbf{C}}} + c_j V(\xi_d)_{\mathfrak{n}_{\mathfrak{k}}} + r(c_j V(\xi_d + \xi_{d-1})_1) , \tag{3.6}$$

where $\lambda = \zeta^{k/2} \in S^1$. Hence we may write $\widetilde{F} = \widetilde{F}_\lambda$. If we set $d' = \frac{2}{k}(d-1)+1$ then, as in Section 2, we may express $\widetilde{Y}$ as $\widetilde{Y} = \sum_{j=-d'}^{d'} \widetilde{\xi}_j \lambda^j$ and we obtain

$$\widetilde{\xi}_{d'} = V(\xi_d)_{\mathfrak{m}^{\mathbf{C}}} , \quad \widetilde{\xi}_{d'-1} \equiv V(\xi_d)_{\mathfrak{n}_{\mathfrak{k}}} + V(\xi_d + \xi_{d-1})_1 \mod \overline{\mathfrak{n}_{\mathfrak{k}}} . \tag{3.7}$$

Let σ be the involution of G/K and $U_1, \cdots, U_n$ be σ-compatible, $\mathrm{Ad}G^{\mathbf{C}}$-equivariant polynomial maps defined by $U_j = c_j \mathrm{Id}$ for $j = 1, \cdots, n$. It follows from (3.7) that

$$\begin{aligned} U_j(\widetilde{\xi}_{d'}) &= c_j V(\xi_d)_{\mathfrak{m}^{\mathbf{C}}} , \\ \widetilde{r}(U_j(\widetilde{\xi}_{d'} + \widetilde{\xi}_{d'-1})_1) &= c_j V(\xi_d)_{\mathfrak{n}_{\mathfrak{k}}} + r(c_j V(\xi_d + \xi_{d-1})_1) , \end{aligned}$$

where $j = 1, \cdots, n$ and $\widetilde{r} : \mathfrak{k}^{\mathbf{C}} \longrightarrow \mathfrak{k}^{\mathbf{C}}$ is as in Section 2. These, together with (3.6), yield

$$\widetilde{F}^{-1}\frac{\partial \widetilde{F}}{\partial z^j} = \lambda U_j(\widetilde{\xi}_{d'}) + \widetilde{r}(U_j(\widetilde{\xi}_{d'} + \widetilde{\xi}_{d'-1})_1)$$

for $j = 1, \cdots, n$. Moreover we also have $\widetilde{Y} \in \Lambda_{d'} \subset \Lambda\mathfrak{g}_\sigma$. Thus, $\varphi = p \circ \psi$ is a pluriharmonic map of finite type. □

When $n = 1$ and $V_1 = \mathrm{Id}$, Theorem 3.4 is nothing but Theorem 2.1.

3.4. Twistor fibrations over r-symmetric spaces with $2 < r < k$

Is Theorem 3.4 true even if we replace a compact Hermitian symmetric space G/K by a r-symmetric space G/L for some r with $2 < r < k$? We obtain an affirmative answer by modifying slightly the proof of Theorem 3.4.

Let G/H be a generalized flag manifold with the canonical k-symmetric structure τ. Suppose that $k > 3$ and $\psi : M \longrightarrow G/H$ is a primitive pluriharmonic map of finite type as in Theorem 3.4. Choose a closed subgroup L of G which satisfies the following conditions:

(3.8) (1) $H \subset L$ and G/L is a generalized flag manifold with the canonical r-symmetric structure τ' for some $2 < r < k$, (2) the reductive decomposition $\mathfrak{g} = \mathfrak{l} \oplus \mathfrak{q}$ is τ-stable and orthogonal, (3) the eigenspace decomposition of $\mathfrak{g}^{\mathbf{C}}$ with respect to τ', $\mathfrak{g}^{\mathbf{C}} = \sum_{j=0}^{r-1} \mathfrak{q}_j$ with $\mathfrak{q}_0 = \mathfrak{l}^{\mathbf{C}}$, satisfies $\mathfrak{g}_1 \cap \overline{\mathfrak{n}} \subset \mathfrak{q}_1$ and $\mathfrak{g}_j \cap \mathfrak{q}_s = 0$ for $j = 1, \cdots, r-2$ and $s = j+1, \cdots, r-1$.

Set $I = \{ i \mid \mathfrak{g}^{\alpha_i} \subset \mathfrak{p}^{\mathbf{C}} \}$ and $J' = \{ j \mid \mathfrak{g}^{\alpha_j} \subset \mathfrak{q}^{\mathbf{C}} \}$. Let ξ_I and $\xi_{J'}$ be the canonical elements corresponding to I and J', respectively. Define an element $\nu' \in H$ by

$$\nu' = \exp(\frac{2\pi}{r} t\, \xi_{J'}) \cdot \exp(-\frac{2\pi}{k} t\, \xi_I) , \qquad t \in \mathbf{R}.$$

We then see that $\mathrm{Ad}\nu'$ is given by

$$\mathrm{Ad}\nu' = \begin{cases} \zeta^{s\frac{k}{r}-j}\mathrm{Id} & \text{on } \mathfrak{g}_j \cap \mathfrak{q}_s \quad (j=1,\cdots,k-1; s=1,\cdots,r-1), \\ \zeta^{-j}\mathrm{Id} & \text{on } (\mathfrak{g}_j)_{\mathfrak{n}_\mathfrak{l}} \quad (j=1,\cdots,k-2), \\ \zeta^{k-j}\mathrm{Id} & \text{on } (\mathfrak{g}_j)_{\overline{\mathfrak{n}_\mathfrak{l}}} \quad (j=2,\cdots,k-1), \\ \mathrm{Id} & \text{on } \mathfrak{g}_0, \end{cases}$$

where $\zeta = \exp(\frac{2\pi\sqrt{-1}}{k}t) \in S^1$ and $\mathfrak{n}_\mathfrak{l} = \mathfrak{n} \cap \mathfrak{l}^{\mathbf{C}}$. We set $d' = \frac{r}{k}(d-1)+1$ and $\lambda = \zeta^{\frac{k}{r}}$. Define $\widetilde{Y}$ and $\widetilde{F}$ as in the proof of Theorem 3.4. Then we have

$$\widetilde{\xi}_{d'} = V(\xi_d)_{\mathfrak{q}_1}, \quad \widetilde{\xi}_{d'-1} \equiv V(\xi_d)_{\mathfrak{n}_\mathfrak{l}} + V(\xi_d + \xi_{d-1})_1 \mod \overline{\mathfrak{n}_\mathfrak{l}}$$

and hence we obtain

$$\begin{aligned} \widetilde{F}^{-1}\frac{\partial \widetilde{F}}{\partial z^j} &= \lambda c_j V(\xi_d)_{\mathfrak{q}_1} + c_j V(\xi_d)_{\mathfrak{n}_\mathfrak{l}} + r(c_j V(\xi_d + \xi_{d-1})_1) \\ &= \lambda U_j(\widetilde{\xi}_{d'}) + \widetilde{r}(U_j(\widetilde{\xi}_{d'} + \widetilde{\xi}_{d'-1})_1), \end{aligned}$$

where $U_j = c_j\mathrm{Id}$ for $j = 1, \cdots, n$. Therefore we obtain:

THEOREM 3.5. *Let $\psi : M \longrightarrow G/H$ be a primitive pluriharmonic map of finite type as in Theorem 3.4. Choose a closed subgroup L of G as in (3.8). Then $p \circ \psi : M \longrightarrow G/L$ is a primitive pluriharmonic map of finite type, where $p : G/H \longrightarrow G/L$ is a homogeneous projection.*

We address some interesting applications of Theorem 3.4 in the next two sections.

4. Classification of non-isotropic pluriharmonic maps of a complex torus into complex projective space

4.1. Primitive pluriharmonic maps of G-finite type

Suppose that G/H is a compact homogeneous space admitting a k-symmetric structure τ. Denote by $\iota : G/H \longrightarrow G$ the Cartan embedding. Let $d \equiv 1 \mod k$. Denote by $\Omega\mathfrak{g} = \{\xi : S^1 \to \mathfrak{g} \mid \xi(1) = 0\}$ *the based loop algebra.* Set $\Omega_d = \Lambda_d \cap \Omega\mathfrak{g}$. By applying the same arguments as in [BFPP] and [Bu] we obtain:

PROPOSITION 4.1. *Let $d \equiv 1 \mod k$. Let $\widetilde{V}_1, \cdots, \widetilde{V}_n : \mathfrak{g}^{\mathbf{C}} \longrightarrow \mathfrak{g}^{\mathbf{C}}$ be τ-compatible, $\mathrm{Ad}G^{\mathbf{C}}$-equivariant polynomial maps. For an initial condition $\widetilde{\xi}_0 \in \Omega_d \cap \Lambda(\mathfrak{g}, \tau)$, let $\widetilde{\xi} : \mathbf{C}^n \longrightarrow \Omega_d$ be a solution of the differential equation*

$$\begin{cases} d\widetilde{\xi} = [\widetilde{\xi}, 2\sqrt{-1}(1-\zeta)\sum_{j=1}^{n}\widetilde{V}_j(\widetilde{\xi}_d)dz^j - 2\sqrt{-1}(1-\zeta^{-1})\sum_{j=1}^{n}\overline{\widetilde{V}_j(\widetilde{\xi}_d)}\,\overline{dz^j}], \\ \widetilde{\xi}(0) = \widetilde{\xi}_0 . \end{cases}$$

Let $\hat{F} : \mathbf{C}^n \longrightarrow G$ be a map satisfying $\hat{F}(0) = e$, $\hat{F}^*\theta = 2\sqrt{-1}(1-\omega)\sum \widetilde{V}_j(\widetilde{\xi}_d)dz^j - 2\sqrt{-1}(1-\omega^{-1})\sum \overline{\widetilde{V}_j(\widetilde{\xi}_d)}\,\overline{dz^j}$, where $\omega = \exp(2\pi\sqrt{-1}/k)$ and e is the identity element of G. Then $\hat{F}$ has its image in $\iota(G/H)$, $\widetilde{\xi}(x) \in \Lambda(\mathfrak{g}, \tau_{\hat{F}(x)})$ for any $x \in \mathbf{C}^n$ and $\hat{F} = \iota \circ \hat{\psi}$ for a primitive pluriharmonic map $\hat{\psi} : \mathbf{C}^n \longrightarrow G/H$.

We call $\hat{\psi}$ in Proposition 4.1 a primitive pluriharmonic map of *G-finite type.* We present now the Symes method for constructing all primitive pluriharmonic maps of G-finite type. Set $\Lambda G^{\mathbf{C}} = \{\gamma : S^1 \longrightarrow G^{\mathbf{C}}\}$ and define subgroups

$$\Omega G = \{\gamma \in \Lambda G^{\mathbf{C}} \mid \gamma : S^1 \longrightarrow G \text{ and } \gamma(1) = e\} ,$$
$$\Lambda_- G^{\mathbf{C}} = \{\gamma \in \Lambda G^{\mathbf{C}} \mid \gamma \text{ extends holomorphically to } D_0 \longrightarrow G^{\mathbf{C}}\} ,$$

where $D_0 = \{\lambda \in \mathbf{C} \mid \ \mid \lambda \mid > 1\}$. Then the multiplication map $\Omega G \times \Lambda_- G^{\mathbf{C}} \longrightarrow \Lambda G^{\mathbf{C}}$ is a diffeomorphism onto ([PS]). Choose an initial condition $\widetilde{\xi}_0 \in \Omega_d \cap \Lambda(\mathfrak{g}, \tau)$. Let $(z^1, \cdots, z^n)$ be a complex coordinate for $\mathbf{C}^n$. Denote by $\widetilde{V}_1, \cdots, \widetilde{V}_n : \mathfrak{g}^{\mathbf{C}} \longrightarrow \mathfrak{g}^{\mathbf{C}}$ τ-compatible, $\mathrm{Ad}G^{\mathbf{C}}$-equivariant polynomial maps. For each $j = 1, \cdots, n$, let $\widetilde{V}_j = \widetilde{V}_j^{(1)} + \cdots + \widetilde{V}_j^{(k_j)}$ be a decomposition into homogeneous polynomials and d_{ji} the homogeneous degree of $\widetilde{V}_j^{(i)}$. Define $\widetilde{g} : \mathbf{C}^n \longrightarrow \Lambda G^{\mathbf{C}}$ by

$$\widetilde{g}(\mathbf{z}) = \exp(2\sqrt{-1}\sum_{j=1}^{n}\sum_{i=1}^{k_j}(-1)^{d_{ji}+1}\lambda^{1-dd_{ji}}\widetilde{V}_j^{(i)}(\widetilde{\xi}_0)z^j) .$$

Then we may write $\widetilde{g} = \widetilde{a} \cdot \widetilde{b}$, where $\widetilde{a} : \mathbf{C}^n \longrightarrow \Omega G$, $\widetilde{b} : \mathbf{C}^n \longrightarrow \Lambda_- G^{\mathbf{C}}$. Let $\pi_{\Omega\mathfrak{g}} : \Lambda\mathfrak{g}^{\mathbf{C}} \longrightarrow \Omega\mathfrak{g}$ be the projection with respect to the decomposition $\Lambda\mathfrak{g}^{\mathbf{C}} = \Omega\mathfrak{g} \oplus \Lambda_-\mathfrak{g}^{\mathbf{C}}$. As in Section 3 we obtain

$$\widetilde{a}^{-1}d\widetilde{a} = \pi_{\Omega\mathfrak{g}}(2\sqrt{-1}\sum_{j=1}^{n}\sum_{i=1}^{k_j}(-1)^{d_{ji}+1}\lambda^{1-dd_{ji}}\widetilde{V}_j^{(i)}(\widetilde{\xi})dz^j) , \tag{4.1}$$

where $\widetilde{\xi} = Ad\widetilde{a}^{-1} \cdot \widetilde{\xi}_0$, so that $d\widetilde{\xi} = [\widetilde{\xi}, \widetilde{a}^{-1}d\widetilde{a}]$ holds. Since we can express $\widetilde{\xi}$ as

$$\widetilde{\xi} = \sum_{0<|j|\leq d} (1-\lambda^j)\widetilde{\xi}_j \ \in \Omega_d \subset \Omega\mathfrak{g} ,$$

we see from (4.1) that

$$\widetilde{a}^{-1}d\widetilde{a} = 2\sqrt{-1}(1-\lambda)\sum_{j=1}^{n}\widetilde{V}_j(\widetilde{\xi}_d)dz^j - 2\sqrt{-1}(1-\lambda^{-1})\sum_{j=1}^{n}\overline{\widetilde{V}_j(\widetilde{\xi}_d)}\,\overline{dz^j} .$$

Since $\widetilde{a}^{-1}d\widetilde{a}$ satisfies the Maurer-Cartan equation, if we set $\hat{F} = \widetilde{a}\mid_{\lambda=\omega}$, then we obtain all primitive pluriharmonic maps of G-finite type by Proposition 4.1, where $\omega = \exp(2\pi\sqrt{-1}/k)$. We call $\widetilde{a}$ an *extended solution* of $\hat{\psi}$.

Next we discuss the question: Does any primitive pluriharmonic map of G-finite type give a primitive pluriharmonic map of finite type in the sense of Section 3? We have an affirmative answer as well as in the case of Riemann surface ([BP]):

PROPOSITION 4.2. *Any primitive pluriharmonic map of G-finite type is of finite type.*

PROOF. If $\hat{\psi} : \mathbf{C}^n \longrightarrow G/H$ is a primitive pluriharmonic map of G-finite type, $\hat{F} = \iota \circ \hat{\psi} : \mathbf{C}^n \longrightarrow G$ has an extended solution $\widetilde{a} : \mathbf{C}^n \longrightarrow \Omega G$. The Symes method ensures that $\widetilde{a}$ is obtained by $\widetilde{g} = \exp(2\sqrt{-1}\sum_j\sum_i(-1)^{d_{ji}+1}\lambda^{1-dd_{ji}}\widetilde{V}_j^{(i)}(\widetilde{\xi}_0)z^j)$ for some initial condition $\widetilde{\xi}_0 \in \Omega_d \cap \Lambda(\mathfrak{g}, \tau)$ and the Iwasawa decomposition $\Lambda G^{\mathbf{C}} =$

$\Omega G \cdot \Lambda_- G^{\mathbf{C}}$. On the other hand, as $\widetilde{\xi}_0 \in \Lambda(\mathfrak{g}, \tau)$, we have $\widetilde{g} = a \cdot b$ according to the decomposition $\Lambda G^{\mathbf{C}}_\tau = \Lambda G_\tau \cdot \Lambda_- G^{\mathbf{C}}_\tau$ because $\widetilde{g} = \exp(\sum_j \sum_i \lambda^{1-dd_{ji}} V^{(i)}_j (\widetilde{\xi}_0) z^j)$ if we set $V^{(i)}_j = 2\sqrt{-1}(-1)^{d_{ji}+1} \widetilde{V}^{(i)}_j$. We denote by a_μ the evaluation at $\lambda = \mu$ of a. Similarly for $\widetilde{a}$. Now, a_1 is an extended framing of some primitive pluriharmonic map $\psi : \mathbf{C}^n \longrightarrow G/H$ of finite type, that is, $\psi = \pi \circ a_1$, where $\pi : G \longrightarrow G/H$ is the coset projection. On the other hand, since the decomposition $\widetilde{g} = \widetilde{a} \cdot \widetilde{b}$ is unique, we have $\widetilde{a} = a \cdot a_1^{-1}, \widetilde{b} = a_1 \cdot b$. Hence we have

$$\begin{aligned} a_\omega \cdot a_1^{-1} &= \widetilde{a}_\omega = \hat{F} = \iota \circ \hat{\psi} \,, \\ \iota \circ \pi \circ a_1 &= \tau(a_1) \cdot a_1^{-1} = a_\omega \cdot a_1^{-1} = \iota \circ \hat{\psi} \,. \end{aligned}$$

Therefore we may conclude that $\psi = \hat{\psi}$ and that $\hat{\psi}$ is of finite type in the sense of Section 3. □

4.2. Non-isotropic pluriharmonic maps into complex projective space

Denote by $\mathbf{C}P^m$ the m-dimensional complex projective space with the Fubini-Study metric. Consider a non-isotropic pluriharmonic map $\varphi : \mathbf{C}^n/\Lambda \longrightarrow \mathbf{C}P^m$ from an n-dimensional complex torus, where the lattice Λ is generated by $2n$-vectors $\mathbf{x}_1, \cdots, \mathbf{x}_{2n}$ of $\mathbf{C}^n$ which are linearly independent over $\mathbf{R}$. Any pluriharmonic map ψ of a complex manifold M into $\mathbf{C}P^m$ has the property that $\mathrm{rank}_{\mathbf{C}} \partial\psi \leq 1$ if ψ is neither holomorphic nor anti-holomorphic. Moreover, if ψ is non-constant then $\mathrm{rank}_{\mathbf{C}} \partial\psi = 1$ on a dense subset of M (cf. [OU]). Therefore, if ψ is non-isotropic then we have a pluriharmonic map sequence $\underline{\psi} = \underline{\psi_0} \to \underline{\psi_1} \to \cdots \underline{\psi_r} \to$ which never terminates, where $\underline{\psi_j} = \psi_j^* T_1$ for the tautological line bundle T_1 over $\mathbf{C}P^m$ and $\mathrm{rank}_{\mathbf{C}} \underline{\psi_j} = 1$ for $j = 0, 1, \cdots, r, \cdots$ (cf. [OU]). We choose r as the largest integer such that $\underline{\psi}$ is Hermitian orthogonal to $\underline{\psi_j}$ with respect to the Hermitian fiber metric on the trivial bundle $M \times \mathbf{C}^{m+1}$ for each $j = 1, 2, \cdots, r$. We call r the *isotropy order* of ψ. If ψ has isotropy order r then $\underline{\psi_{r+1}}$ is not perpendicular to $\underline{\psi}$ but perpendicular to $\underline{\psi_j}$ for each $j = 1, 2, \cdots, r$ (cf. [OU]).

We have the following theorem.

THEOREM 4.3. *Let $\varphi : \mathbf{C}^n/\Lambda \longrightarrow \mathbf{C}P^m$ be a non-isotropic pluriharmonic map. Then φ is of finite type and obtained from a solution $\xi : \mathbf{C}^n/\Lambda \longrightarrow \Lambda_{d'}$ of the differential equation*

$$\partial\xi = [\xi, \, \lambda \sum_{j=1}^{n} c_j \xi_{d'} dz^j + \sum_{j=1}^{n} r(c_j \xi_{d'-1}) dz^j] \,,$$

where d' is an odd integer and $c_j \in \mathbf{C}$ for $j = 1, \cdots, n$. Moreover, if there are two generators $\mathbf{x}_j, \mathbf{x}_k$ of Λ such that, for a map $\hat{C} : \mathbf{C}^n \ni {}^t(z^1, \cdots, z^n) \longrightarrow \sum_{j=1}^n c_j z^j \in \mathbf{C}$, $\hat{C}(\mathbf{x}_j)$ and $\hat{C}(\mathbf{x}_k)$ are linearly independent over $\mathbf{R}$ and any other generator of Λ is a linear combination of $\hat{C}(\mathbf{x}_j)$ and $\hat{C}(\mathbf{x}_k)$ over the integer ring $\mathbf{Z}$, then there is a holomorphic map $h : \mathbf{C}^n/\Lambda \longrightarrow \mathbf{C}/\Gamma$ and a non-isotropic harmonic map of finite type $\hat{\varphi} : \mathbf{C}/\Gamma \longrightarrow \mathbf{C}P^m$ such that $\varphi = \hat{\varphi} \circ h$.

Before proving Theorem 4.3, we need some lemmas.

LEMMA 4.4. *Let* $\psi : \mathbf{C}^n/\Lambda \longrightarrow G/H$ *be a primitive map into a compact* $k(\geq 2)$*-symmetric space with a normal homogeneous metric. Suppose that there is a non-vanishing holomorphic vector field* Z *on* $\mathbf{C}^n/\Lambda$ *such that* $\alpha'_{\mathfrak{p}}(Z)$ *is semisimple on a dense subset of* $\mathbf{C}^n/\Lambda$. *Then there is an adapted polynomial Killing field* $\widetilde{\xi} : \mathbf{C}^n/\Lambda \longrightarrow \Omega_d$, *where* $d \equiv 1 \bmod k$.

PROOF. Let P be an $\mathrm{Ad}H^{\mathbf{C}}$-invariant polynomial. Define $\hat{P}$ by $\hat{P}_{g\cdot o} = P \circ \mathrm{Ad}g^{-1} : [\mathfrak{p}^{\mathbf{C}}]_{g\cdot o} \longrightarrow \mathbf{C}$. Then $\hat{P}$ is a G-invariant polynomial. If we denote by F a local lift of ψ to G, then we have

$$\psi^*(\hat{P} \circ \beta) = \hat{P}_{F\cdot o}(\psi^*\beta) = P(\mathrm{Ad}F^{-1} \circ \mathrm{Ad}F \cdot \alpha_{\mathfrak{p}}) = P(\alpha_{\mathfrak{p}}).$$

Since $\psi^*(\hat{P} \circ \beta')$ is a holomorphic differential by Lemma 9.1 in [BFPP], it follows that $P(\alpha'_{\mathfrak{p}}(Z))$ is constant and lies on the common level set of all $\mathrm{Ad}H^{\mathbf{C}}$-invariant polynomials on $\mathfrak{g}_1$. It is known from [V] that each level set contains a single semisimple orbit which is a unique closed orbit. Therefore, if $\alpha'_{\mathfrak{p}}(Z)$ is semisimple on a dense subset of the domain then $\alpha'_{\mathfrak{p}}(Z)$ has values in a single semisimple $\mathrm{Ad}H^{\mathbf{C}}$-orbit in $\mathfrak{g}_1$. For the Cartan embedding $\iota : G/H \longrightarrow G$, we have $\widetilde{\alpha} := (\iota \circ \psi)^*\theta = (\omega - 1)\mathrm{Ad}F \cdot \alpha_{\mathfrak{p}}$, hence $\widetilde{\alpha}'(Z)$ has values in a single semisimple $\mathrm{Ad}G^{\mathbf{C}}$-orbit. Then it follows from Theorems 6.2 and 7.1 in [BFPP] that there is a polynomial Killing field $\widetilde{\xi} : \mathbf{C}^n/\Lambda \longrightarrow \Omega_d$ with $\widetilde{\xi}_d = \widetilde{\alpha}'(Z)$. □

Denote by $F^r(\mathbf{C}P^m)$ the generalized flag manifold $SU(m+1)/S(U(1) \times \cdots \times U(1) \times U(m+1-r))$ over $\mathbf{C}P^m$, where $U(1)$ is repeated r-times. We give $F^r(\mathbf{C}P^m)$ a normal homogeneous metric and the canonical $(r+1)$-symmetric structure τ. Let $p_1 : F^r(\mathbf{C}P^m) \longrightarrow \mathbf{C}P^m$ be the homogeneous projection which assigns a point $(l_0, l_1, \cdots, l_r) \in F^r(\mathbf{C}P^m)$ to a first element $l_0 \in \mathbf{C}P^m$. The following lemma can be proved by a direct calculation. See also the Appendix in [Mc1].

LEMMA 4.5. *When* $G/H = F^r(\mathbf{C}P^m)$, *any element* $\mathbf{x} \in \mathfrak{g}_1$ *satisfying* $[\mathbf{y}, \mathbf{x}] = \mathbf{0}$ *for a nonzero semisimple element* $\mathbf{y} \in \mathfrak{g}_1$ *is necessarily of the form* $\mathbf{x} = c\mathbf{y}$ *for some complex number* c.

PROOF OF THEOREM 4.3. Since φ is non-isotropic, if we denote by r the isotropy order of φ then we have a primitive map $\psi : \mathbf{C}^n/\Lambda \longrightarrow F^r(\mathbf{C}P^m)$ with $\varphi = p_1 \circ \psi$ (cf. [Bu]). Let $\underline{\varphi_0} = \underline{\varphi} \to \underline{\varphi_1} \to \cdots \to \underline{\varphi_r} \to \underline{\varphi_0} \oplus \underline{R}$ be the harmonic sequence of φ (cf. [Wd], [Ud1]). Then we have $\mathrm{Hom}(\underline{\varphi_j}, \underline{\varphi_{j+1}})$-valued holomorphic $(1,0)$-forms on $\mathbf{C}^n/\Lambda$, which are denoted by $A^{\varphi_j, \varphi_{j+1}}_{(1,0)} : \underline{\varphi_j} \to \underline{\varphi_{j+1}}$ for $j = 0, 1, \cdots, r$, where we may use the convention $\underline{\varphi_{r+1}} = \underline{\varphi_0}$ by the non-isotropicity of φ. We know that each $A^{\varphi_j, \varphi_{j+1}}_{(1,0)}$ has complex 1-dimensional image on a dense open subset of $\mathbf{C}^n/\Lambda$ (cf. [OU]). Let $(z^1, \cdots, z^n)$ be a (global) complex coordinate for $\mathbf{C}^n$. For each $j = 0, 1, \cdots, r$, there exists an integer k_j with $1 \leq k_j \leq n$ such that $A^{\varphi_j, \varphi_{j+1}}_{\frac{\partial}{\partial z^{k_j}}} \neq 0$ on a dense open subset U_j of $\mathbf{C}^n/\Lambda$. Set $A^j_i = A^{\varphi_j, \varphi_{j+1}}_{\frac{\partial}{\partial z^{k_i}}}$ for notational simplicity. Then the vector $(A^j_0, \cdots, A^j_r) \in \mathbf{C}^{r+1}$ is non-zero on U_j for $j = 0, 1, \cdots, r$. Since each A^j_i is holomorphic, if $\sum_{i=0}^r A^j_i \equiv 0$ on some open subset of U_j then $\sum_{i=0}^r A^j_i \equiv 0$ on U_j, which implies that the vector $(A^j_0, \cdots, A^j_r)$ is orthogonal to the constant vector $(1, \cdots, 1)$ on U_j. In general, we can choose a vector in $\mathbf{C}^{r+1}$ which is not orthogonal to any of these fixed $(r+1)$ vectors in $\mathbf{C}^{r+1}$. Therefore, there is a constant vector $(c^0, \cdots, c^r) \in \mathbf{C}^{r+1}$ such that for $Z = \sum_{i=0}^{r} c^i \frac{\partial}{\partial z^{k_i}}$ we have $A_Z^{\varphi_j, \varphi_{j+1}} \neq 0$ on some dense open subset of $\mathbf{C}^n/\Lambda$ for $j = 0, 1, \cdots, r$. In this way, we see that $\alpha'_{\mathfrak{p}}(Z)$ is semisimple on a dense open subset of the domain (cf. [Ud1]). By Lemma 4.4, we obtain an adapted polynomial Killing field $\widetilde{\xi} : \mathbf{C}^n/\Lambda \longrightarrow \Omega_d$, that is, $\widetilde{\xi}$ satisfying

$$
\left\{
\begin{array}{l}
d\widetilde{\xi} = [\widetilde{\xi}, A_\lambda] \,, \quad A_\lambda = (\dfrac{1-\lambda}{1-\omega})\widetilde{\alpha}' + (\dfrac{1-\lambda^{-1}}{1-\omega^{-1}})\widetilde{\alpha}'' \,, \\
\widetilde{\alpha}'(Z) = \widetilde{\xi}_d \; : \; \text{semisimple} \,.
\end{array}
\right.
\tag{4.2}
$$

It follows from (4.2) that $[\widetilde{\xi}_d, \widetilde{\alpha}'] = 0$ and hence $[\alpha'_{\mathfrak{p}}(Z), \alpha'_{\mathfrak{p}}(\frac{\partial}{\partial z^j})] = 0$ for $j = 1, \cdots, n$. This, together with Lemma 4.5, implies that

$$
\alpha'_{\mathfrak{p}}(\frac{\partial}{\partial z^j}) = f_j \, \alpha'_{\mathfrak{p}}(Z) \,, \quad \widetilde{\alpha}'(\frac{\partial}{\partial z^j}) = f_j \, \widetilde{\alpha}'(Z) \,,
$$

where $f_j : \mathbf{C}^n/\Lambda \longrightarrow \mathbf{C}$ for $j = 1, \cdots, n$. We want to show that each f_j is constant. Since ψ is primitive, it follows from (3.5) that

$$
\overline{\partial}_k \alpha'_{\mathfrak{p}}(\frac{\partial}{\partial z^j}) + [\alpha''_{\mathfrak{h}}(\overline{\frac{\partial}{\partial z^k}}), \alpha'_{\mathfrak{p}}(\frac{\partial}{\partial z^j})] = 0 \,, \quad j, k = 1, \cdots, n \,.
\tag{4.3}
$$

On the other hand, extracting the $[\mathfrak{p}]$-component of the (0,1)-part of λ^d-coefficient in (4.2), we obtain

$$
\overline{\partial}_k \alpha'_{\mathfrak{p}}(Z) + [\alpha''_{\mathfrak{h}}(\overline{\frac{\partial}{\partial z^k}}), \alpha'_{\mathfrak{p}}(Z)] = 0 \,.
\tag{4.4}
$$

Substituting $\alpha'_{\mathfrak{p}}(\frac{\partial}{\partial z^j}) = f_j \alpha'_{\mathfrak{p}}(Z)$ into (4.3) and using (4.4), we see that $\overline{\partial} f_j = 0$ and f_j is constant on $\mathbf{C}^n/\Lambda$ for $j = 1, \cdots, n$. Therefore, we have $\widetilde{\alpha}' = \sum_{j=1}^{n} f_j \widetilde{\xi}_d dz^j$. If we define $\widetilde{V}_j$ by $\widetilde{V}_j = \frac{f_j}{2\sqrt{-1}(1-\omega)}\mathrm{Id}$ for $j = 1, \cdots, n$, we see from Propositions 4.1 and (4.2) that ψ is a primitive map of finite type. From the proof of Proposition 4.2, we see that ψ is obtained from τ-compatible, $\mathrm{Ad}G^{\mathbf{C}}$-equivariant polynomial maps $V_j = c_j \mathrm{Id}$ with $c_j = f_j/(1-\omega)$ for $j = 1, \cdots, n$. Therefore we can use Theorem 3.4 to see that $\varphi = p_1 \circ \psi : \mathbf{C}^n/\Lambda \longrightarrow \mathbf{C}P^m$ is a pluriharmonic map of finite type. From the proof of Theorem 3.4 we see that φ is also obtained by $c_1 \mathrm{Id}, \cdots, c_n \mathrm{Id}$.
We now prove the second part of Theorem 4.3. Let $\pi_\Lambda : \mathbf{C}^n \ni \mathbf{z} \longrightarrow [\mathbf{z}] \in \mathbf{C}^n/\Lambda$ be the natural quotient map. By the assumptions on the generators of Λ we have

$$
\hat{C}(\mathbf{x}_s) = m_s \hat{C}(\mathbf{x}_j) + n_s \hat{C}(\mathbf{x}_k), \qquad (m_s, n_s \in \mathbf{Z})
$$

for each $s = 1, \cdots, n$. If we define a lattice by $\Gamma = \{m\hat{C}(\mathbf{x}_j) + n\hat{C}(\mathbf{x}_k) \mid m, n \in \mathbf{Z}\}$, then we see that $\mathbf{C}/\Gamma$ is a 1-dimensional complex torus. Moreover, we have $\hat{C}(\Lambda) = \Gamma$. If we denote by $\pi_\Gamma : \mathbf{C} \ni z \longrightarrow [z]_\Gamma \in \mathbf{C}/\Gamma$ the quotient map, we obtain the following commmutative diagram:

$$
\begin{array}{ccc}
\mathbf{C}^n \ni \mathbf{z} & \xrightarrow{\hat{C}} & \hat{C}(\mathbf{z}) \in \mathbf{C} \\
{\scriptstyle \pi_\Lambda} \big\downarrow & & \big\downarrow {\scriptstyle \pi_\Gamma} \\
\mathbf{C}^n/\Lambda \ni [\mathbf{z}] & \xrightarrow[h]{} & [\hat{C}(\mathbf{z})]_\Gamma \in \mathbf{C}/\Gamma
\end{array}
$$

where $h : \mathbf{C}^n/\Lambda \ni [\mathbf{z}] \longrightarrow [\hat{C}(\mathbf{z})]_\Gamma \in \mathbf{C}/\Gamma$ is a holomorphic map with $\pi_\Gamma \circ \hat{C} = h \circ \pi_\Lambda$. On the other hand, since $\varphi \circ \pi_\Lambda : \mathbf{C}^n \longrightarrow \mathbf{C}P^m$ is also a pluriharmonic map of finite type, it follows from the proof of Theorem 8.4 in [BFPP] that there is a harmonic map of finite type $\hat{\varphi} : \mathbf{C} \longrightarrow \mathbf{C}P^m$ such that $\varphi \circ \pi_\Lambda = \hat{\varphi} \circ \hat{C}$. Then, by the fact $\varphi \circ \pi_\Lambda(\mathbf{z} + \Lambda) = \varphi \circ \pi_\Lambda(\mathbf{z})$ we have $\hat{\varphi} \circ \hat{C}(\mathbf{z}) = \hat{\varphi}(\hat{C}(\mathbf{z}) + \Gamma)$. Therefore we see that $\varphi \circ \pi_\Lambda(\mathbf{z}) = \hat{\varphi} \circ \hat{C}(\mathbf{z}) = \hat{\varphi}([\hat{C}(\mathbf{z})]_\Gamma) = \hat{\varphi} \circ h \circ \pi_\Lambda(\mathbf{z})$, which implies that $\varphi = \hat{\varphi} \circ h$ and $\hat{\varphi}$ induces $\hat{\varphi} : \mathbf{C}/\Gamma \longrightarrow \mathbf{C}P^m$. □

5. Primitive harmonic maps from compact Riemann surfaces of higher genus

Let M_g be a Riemann surface of genus g and G/H a compact k-symmetric space with k-symmetric structure τ and a normal homogeneous metric. Let $V_1, \cdots, V_n : \mathfrak{g}^{\mathbf{C}} \longrightarrow \mathfrak{g}^{\mathbf{C}}$ be τ-compatible, $\mathrm{Ad}G^{\mathbf{C}}$-equivariant polynomial maps. Let $C_1, \cdots, C_n$ be holomorphic 1-forms on M_g. Denote by $\Lambda\mathfrak{g}_\tau$ and Λ_d the twisted loop algebra and a finite dimensional subspace as before, where $d \equiv 1 \bmod k$. Let $\xi : M_g \longrightarrow \Lambda_d$ be a solution of the following equation of Lax type

(5.1)
$$d\xi = [\xi, \lambda \sum_{j=1}^{n} V_j(\xi_d)C_j + (\sum_{j=1}^{n} V_j(\xi_d + \xi_{d-1})_1 C_j)_{\mathfrak{h}} + \lambda^{-1} \sum_{j=1}^{n} \overline{V_j(\xi_d)}\,\overline{C_j}] \,,$$

where we use the same notation as in Section 3. Set

(5.2)
$$\alpha(C)_\lambda = \lambda \sum_{j=1}^{n} V_j(\xi_d)C_j + (\sum_{j=1}^{n} V_j(\xi_d + \xi_{d-1})_1 C_j)_{\mathfrak{h}} + \lambda^{-1} \sum_{j=1}^{n} \overline{V_j(\xi_d)}\,\overline{C_j} \,,$$

where $C = (C_1, \cdots, C_n)$. Then we have the following theorem.

THEOREM 5.1. *$\alpha(C)_\lambda$ satisfies the Maurer-Cartan equation for each $\lambda \in S^1$.*

PROOF. It follows from (5.1) that

(5.3)
$$\begin{cases} \bar{\partial}\xi_d = [\xi_d, \sum_{j=1}^{n} \overline{r(V_j(\xi_d + \xi_{d-1})_1)}\,\overline{C_j}] \,, \\ \bar{\partial}\xi_{d-1} = [\xi_{d-1}, \sum_{j=1}^{n} \overline{r(V_j(\xi_d + \xi_{d-1})_1)}\,\overline{C_j}] + [\xi_d, \sum_{j=1}^{n} \overline{V_j(\xi_d)}\,\overline{C_j}] \,. \end{cases}$$

Using (5.3) we may show that $\alpha(C)_\lambda$ satisfies the Maurer-Cartan equation for each $\lambda \in S^1$. Since it is a direct calculation, we give only the essential point. For any $\eta \in \mathfrak{h}^{\mathbf{C}}$, we define a map $\mu : \mathfrak{h}^{\mathbf{C}} \longrightarrow \mathfrak{h}^{\mathbf{C}}$ by $\mu(\eta) = \eta_{\sqrt{-1}\mathfrak{t}} + \eta_{\mathfrak{n}_{\mathfrak{h}}} - \overline{\eta_{\mathfrak{n}_{\mathfrak{h}}}}$. Then we have $\mu(\eta) = r(\eta) - \overline{r(\eta)}$, where $r : \mathfrak{h}^{\mathbf{C}} \longrightarrow \mathfrak{h}^{\mathbf{C}}$ is as in Section 1.2. Since $\bar{\partial} r(\eta) - \overline{\bar{\partial} r(\eta)} = r(\bar{\partial}\eta) - \overline{r(\bar{\partial}\eta)}$, Theorem 5.1 follows from our fixed Iwasawa decomposition $\mathfrak{h}^{\mathbf{C}} = \mathfrak{h} \oplus \mathfrak{b}$. Here, note that, for $j = 1, \cdots, n$

$$\begin{aligned}\bar{\partial} V_j(\xi_d + \xi_{d-1})_1 =& [V_j(\xi_d + \xi_{d-1})_1, \sum_{k=1}^{n} \overline{r(V_k(\xi_d + \xi_{d-1})_1)}\, \overline{C_k}] \\ &+ [V_j(\xi_d), \sum_{k=1}^{n} \overline{V_k(\xi_d)}\, \overline{C_k}] \,.\end{aligned}$$

Alternatively, we may use $g = \exp(\sum_{j=1}^{n} \sum_{i=1}^{l_j} \lambda^{1-dd_{ji}} V_j^{(i)}(\xi_0) f_j)$, where $\xi_0 \in \Lambda_d$, d_{ji} is the degree of homogeneous polynomial $V_j^{(i)}$ in the decomposition $V_j = V_j^{(1)} + \cdots + V_j^{(l_j)}$ and f_j is a locally defined holomorphic function on a simply-connected domain U of M_g with $C_j \mid_U = \partial f_j$. As in Section 3.2, we see that $\alpha(C)_\lambda \mid_U$ satisfies the Maurer-Cartan equation. Since $\alpha(C)_\lambda$ is defined on M_g we conclude that $\alpha(C)_\lambda$ satisfies the Maurer-Cartan equation. □

Let $\pi_M : \widetilde{M}_g \longrightarrow M_g$ be a holomorphic map of the universal covering space $\widetilde{M}_g$.

DEFINITION. $\psi : M_g \longrightarrow G/H$ is said to be of *generalized finite type* if there is a solution $\xi : M_g \longrightarrow \Lambda_d$ of (5.1) with $\alpha(C)_\lambda$ in (5.2) such that $\psi \circ \pi_M$ has an extended framing $F_\lambda : \widetilde{M}_g \longrightarrow \Lambda G_\tau$ satisfying $F_\lambda^* \theta = \pi_M^* \alpha(C)_\lambda$.

In the following, we assume that M_g is a compact Riemann surface of genus $g > 1$ and examine how to obtain the solution of (5.1). We fix the bases $\{a_j\}_{j=1}^{g}$ and $\{b_j\}_{j=1}^{g}$ for a-cycle and b-cycle on M_g. Let $\omega_1, \cdots, \omega_g$ be the basis of holomorphic 1$-$forms on M_g which are normalized by the condition $\int_{a_j} \omega_i = \delta_{ij}$, for $i, j = 1, \cdots, g$. If we set $B_{ij} = \int_{b_j} \omega_i$ then $P = (I_g, B)$ is called a *period matrix* of M_g, where I_g is the identity matrix of order g. We represent P as $P = (\mathbf{e}_1, \cdots, \mathbf{e}_g, \mathbf{b}_1, \cdots, \mathbf{b}_g)$ by the column vectors which belong to $\mathbf{C}^g$. We define a lattice Λ in $\mathbf{C}^g$ by

$$\Lambda = \{\sum_{j=1}^{g} m_j \mathbf{e}_j + \sum_{j=1}^{g} n_j \mathbf{b}_j \mid m_j, n_j \in \mathbf{Z} \text{ for } j = 1, \cdots, g\} \,.$$

Then $J(M_g) = \mathbf{C}^g/\Lambda$ is called a Jacobian torus, which is an Abelian variety since B is symmetric and the imaginary part of B is positive definite (P satisfies Riemann's condition). We may express M_g as $M_g = \mathbf{H}/\Gamma$, where $\mathbf{H} = SL(2, \mathbf{R})/SO(2)$ and Γ is a discrete subgroup of $SL(2, \mathbf{R})$. Let $\pi_M : \mathbf{H} \longrightarrow M_g$ be a holomorphic quotient map. Set $\widetilde{\omega}_j = \pi_M^* \omega_j$ for $j = 1, \cdots, g$. Define a map $\widetilde{h} = {}^t(\widetilde{h}_1, \cdots, \widetilde{h}_g) : \mathbf{H} \longrightarrow \mathbf{C}^g$ by

$$(5.4) \qquad \begin{cases} d\widetilde{h}_j = \widetilde{\omega}_j \,, & j = 1, \cdots, g \,, \\ \widetilde{h}(x_0) = \mathbf{0} \,, \end{cases}$$

where $x_0 \in \mathbf{H}$ is a fixed point. We then have $\widetilde{h}(x) = {}^t(\int_{x_0}^{x} \widetilde{\omega}_1, \cdots, \int_{x_0}^{x} \widetilde{\omega}_g) \in \mathbf{C}^g$. On the other hand, we may define a map $h = {}^t(h_1, \cdots, h_g) : M_g \longrightarrow \mathbf{C}^g/\Lambda$ by

$$h(p) = (\int_{p_0}^{p} \omega_1, \cdots, \int_{p_0}^{p} \omega_g) \in \mathbf{C}^g/\Lambda \,,$$

which is well-defined and called the *Abel map*, where we choose $p_0 \in M_g$ so that $p_0 = \pi_M(x_0)$. In this case, we also have

$$\begin{cases} dh_j = \omega_j \;, \quad j = 1, \cdots, g \;, \\ h(p_0) = \mathbf{0} \bmod \Lambda \;. \end{cases} \tag{5.5}$$

Denote by $\pi_{\mathbf{C}} : \mathbf{C}^g \longrightarrow \mathbf{C}^g/\Lambda$ the quotient map. It then follows from (5.4) and (5.5) that

$$\begin{cases} d(\pi_{\mathbf{C}} \circ \widetilde{h}) = d(h \circ \pi_M) \bmod \Lambda \\ (\pi_{\mathbf{C}} \circ \widetilde{h})(x_0) = (h \circ \pi_M)(x_0) = \mathbf{0} \bmod \Lambda \;, \end{cases}$$

which imply

$$h \circ \pi_M = \pi_{\mathbf{C}} \circ \widetilde{h} \;. \tag{5.6}$$

We now express the holomorphic 1-form C_j as $C_j = \sum_{i=1}^g c_{ji}\omega_i$ with $c_{ji} \in \mathbf{C}$ for $i = 1, \cdots, g$ and $j = 1, \cdots, n$. Suppose that $\psi : M_g \longrightarrow G/H$ is a primitive map of generalized finite type arising from (5.1). Then we see that $\psi \circ \pi_M : \mathbf{H} \longrightarrow G/H$ is also a primitive map of generalized finite type arising from

$$d\widetilde{\xi} = [\widetilde{\xi}, \lambda \sum_{j=1}^n V_j(\widetilde{\xi}_d)\widetilde{C}_j + (\sum_{j=1}^n V_j(\widetilde{\xi}_d + \widetilde{\xi}_{d-1})_1 \widetilde{C}_j)_{\mathfrak{h}} + \lambda^{-1} \sum_{j=1}^n \overline{V_j(\widetilde{\xi}_d)\widetilde{C}_j}] \;,$$

where $\widetilde{\xi} = \pi_M^*\xi$ and $\widetilde{C}_j = \sum_{i=1}^g c_{ji}\widetilde{\omega}_i$. Set $\hat{V}_i = \sum_{j=1}^n c_{ji}V_j$ for $i = 1, \cdots, g$. Let $f : \mathbf{C}^g \longrightarrow G/H$ be a primitive map of finite type arising from

$$\begin{cases} d\eta = [\eta, \lambda \sum_{i=1}^g \hat{V}_i(\eta_d)dz^i + (\sum_{i=1}^g \hat{V}_i(\eta_d + \eta_{d-1})_1 dz^i)_{\mathfrak{h}} + \lambda^{-1} \sum_{i=1}^g \overline{\hat{V}_i(\eta_d)}\, \overline{dz^i}] \;, \\ \eta(0) = \widetilde{\xi}(x_0) \;. \end{cases} \tag{5.7}$$

It follows from $d\widetilde{h}_j = \widetilde{\omega}_j$ and $\sum_{i=1}^g \hat{V}_i \widetilde{\omega}_i = \sum_{j=1}^n V_j \widetilde{C}_j$ that

$$\begin{aligned} &d(\eta \circ \widetilde{h}) = \widetilde{h}^* d\eta \\ =&[\eta \circ \widetilde{h}, \lambda \sum_{i=1}^g \hat{V}_i(\eta_d \circ \widetilde{h})\widetilde{\omega}_i + (\sum_{i=1}^g \hat{V}_i(\eta_d \circ \widetilde{h} + \eta_{d-1} \circ \widetilde{h})_1 \widetilde{\omega}_i)_{\mathfrak{h}} + \lambda^{-1} \sum_{i=1}^g \overline{\hat{V}_i(\eta_d \circ \widetilde{h})}\, \overline{\widetilde{\omega}_i}] \\ =&[\eta \circ \widetilde{h}, \\ &\lambda \sum_{j=1}^n V_j(\eta_d \circ \widetilde{h})\widetilde{C}_j + (\sum_{j=1}^n V_j(\eta_d \circ \widetilde{h} + \eta_{d-1} \circ \widetilde{h})_1 \widetilde{C}_j)_{\mathfrak{h}} + \lambda^{-1} \sum_{j=1}^n \overline{V_j(\eta_d \circ \widetilde{h})}\, \overline{\widetilde{C}_j}] \;. \end{aligned}$$

Moreover we have $(\eta \circ \widetilde{h})(x_0) = \eta(0) = \widetilde{\xi}(x_0)$. It then follows from the uniqueness of the solution of an ordinary differential equation that $\eta \circ \widetilde{h} = \widetilde{\xi}$ and $f \circ \widetilde{h} = \psi \circ \pi_M$. Therefore, for any point $x \in \mathbf{H}$, the equation $\psi \circ \pi_M(\Gamma x) = \psi \circ \pi_M(x)$ means that $f \circ \widetilde{h}(\Gamma x) = f \circ \widetilde{h}(x)$. However, by a simple calculation we have $\widetilde{h}(\Gamma x) = \widetilde{h}(x) \bmod \Lambda$, which, together with (5.6), implies that f induces a map $\hat{f} : \mathrm{Im} h (\subset \mathbf{C}^g/\Lambda) \longrightarrow G/H$ such that

$$\psi \circ \pi_M(x) = f \circ \widetilde{h}(x) = f \circ \widetilde{h}(\Gamma x) = \hat{f} \circ \pi_{\mathbf{C}} \circ \widetilde{h}(x) = \hat{f} \circ h \circ \pi_M(x).$$

Therefore, we see that $\psi = \hat{f} \circ h$ holds on M_g. Since $f = \hat{f} \circ \pi_{\mathbf{C}}$ on $\mathrm{Im}(\widetilde{h})(\subset \mathbf{C}^g)$, we see that $\hat{f}$ is a primitive map of finite type. We mention finally that $\hat{f}$ is actually defined over $\mathbf{C}^g/\Lambda$, a proof of which will be given in Section 6. Thus, we have:

THEOREM 5.2. *Let $\psi : M_g \longrightarrow G/H$ be a primitive map of generalized finite type. Let $J(M_g) = \mathbf{C}^g/\Lambda$ be the Jacobian torus. Then there is a primitive map of finite type $\hat{f} : \mathbf{C}^g/\Lambda \longrightarrow G/H$ such that ψ is the composition of the map $\hat{f}$ with the Abel map $h : M_g \longrightarrow \mathbf{C}^g/\Lambda$.*

Consider the special case where $V_1, \cdots, V_n$ are given by $V_j = c_j V$ with $c_j \in \mathbf{C}$ for $j = 1, \cdots, n$. In this case, by Theorems 3.4 and 5.2 we have:

COROLLARY 5.3. *Let $\psi : M_g \longrightarrow G/H$ be a primitive map of generalized finite type arising from $V_1 = c_1 V, \cdots, V_n = c_n V$ for some τ-compatible, $\mathrm{Ad}G^{\mathbf{C}}$-equivariant polynomial map $V : \mathfrak{g}^{\mathbf{C}} \longrightarrow \mathfrak{g}^{\mathbf{C}}$. Choose a closed subgroup K of G as in (2.1). Let $p : G/H \longrightarrow G/K$ be the homogeneous projection. Then $p \circ \psi : M_g \longrightarrow G/K$ is a harmonic map of generalized finite type.*

Example. Define a semisimple matrix A by

$$A = \begin{pmatrix} \mathbf{0} & \pi J \\ i\pi I & \mathbf{0} \end{pmatrix}, \quad I = \begin{pmatrix} 1 & 0 \\ 0 & 1 \end{pmatrix}, \quad J = \begin{pmatrix} i & 0 \\ 0 & -i \end{pmatrix}$$

where i is the imaginary unit. Then we have $[A, \overline{A}] = 0$. Set $F_\lambda(z) = \exp(\lambda z A + \lambda^{-1}\overline{z}\overline{A})$, where $z = x + iy$ is a complex coordinate of $\mathbf{C}$. We then obtain

$$F_1(z) = \begin{pmatrix} \cos(2\pi x) & 0 & i\sin(2\pi x) & 0 \\ 0 & \cos(2\pi y) & 0 & \sin(2\pi y) \\ i\sin(2\pi x) & 0 & \cos(2\pi x) & 0 \\ 0 & -\sin(2\pi y) & 0 & \cos(2\pi y) \end{pmatrix} \in \mathrm{SU}(4),$$

(cf. [Ud1, p456]). We see that $F_1(z)$ is invariant under the translations $z \to z+1, z \to z+i$. For ${}^t(z^1, z^2) \in \mathbf{C}^2$ and for each $\lambda \in S^1$, we define F_λ^2 by $F_\lambda^2(z^1, z^2) = F_\lambda(z^1) \cdot F_\lambda(z^2)$. It follows from $[A, \overline{A}] = 0$ that $F_\lambda(z^1)$ and $F_\lambda(z^2)$ commute. We easily observe that F_1^2 is invariant under the translations $z^j \to z^j + 1, z^j \to z^j + i$ for $j = 1, 2$. Setting $\mathbf{e}_1 = {}^t(1,0), \mathbf{e}_2 = {}^t(0,1), \mathbf{b}_1 = {}^t(i,0), \mathbf{b}_2 = {}^t(0,i)$, we define a lattice Λ in $\mathbf{C}^2$ by $\Lambda = \{\sum_{j=1}^2 m_j \mathbf{e}_j + \sum_{j=1}^2 n_j \mathbf{b}_j \mid m_1, m_2, n_1, n_2 \in \mathbf{Z}\}$. Then F_1^2 induces a map $\hat{F}_1^2 : \mathbf{C}^2/\Lambda \longrightarrow \mathrm{SU}(4)$ and $\hat{f} = \pi \circ \hat{F}_1^2 : \mathbf{C}^2/\Lambda \longrightarrow G_2(\mathbf{C}^4)$. It is easy to see that $\hat{f}$ is a pluriharmonic map of finite type arising from the solution of Lax type equation $d\xi = [\xi, \lambda A(dz^1 + dz^2) + \lambda^{-1}\overline{A}(d\overline{z}^1 + d\overline{z}^2)]$. The pluriharmonic map $\hat{f}$ is also constructed from spectral data (see [Mc3]). The matrix $B = (\mathbf{b}_1, \mathbf{b}_2)$ satisfies Riemann's conditions ${}^tB = B$, $\mathrm{Im}(B) > 0$. By the first case of the Torelli's theorem (cf. [ACGH]), we see that there is a compact Riemann surface M_2 of genus 2 with period matrix given by $(I\ B)$ above. Therefore, the torus $\mathbf{C}^2/\Lambda$ above is a Jacobian torus of M_2. We have a harmonic map of generalized finite type $\varphi = \hat{f} \circ h : M_2 \longrightarrow G_2(\mathbf{C}^4)$, where h is the Abel map. Note that $G(p) = F_1(\int_{p_0}^p \omega_1) \cdot F_1(\int_{p_0}^p \omega_2)$ is a framing of φ, where $\{\omega_1, \omega_2\}$ is a normalized basis of $H^1(M_2, \mathbf{C})$.

6. Primitive maps of generalized finite type from compact Kähler manifolds

We shall extend the concept of maps of generalized finite type to the case where the domain is a complex manifold.

Let X be a complex manifold. Let $B_1, \cdots, B_n$ be closed $(1,0)$-forms on X. Consider a solution $\xi : X \longrightarrow \Lambda_d$ of the Lax equation $d\xi = [\xi, \alpha(B)_\lambda]$, where $\alpha(B)_\lambda$ is as in (5.2). This $\alpha(B)_\lambda$ also satisfies the Maurer-Cartan equation for each $\lambda \in S^1$ as that in Theorem 5.1. Let $\pi_X : \widetilde{X} \longrightarrow X$ be the holomorphic universal covering map.

DEFINITION. A map $\psi : X \longrightarrow G/H$ is said to be primitive map of *generalized finite type* if there is a solution $\xi : X \longrightarrow \Lambda_d$ as above such that $\psi \circ \pi_X$ has an extended framing $F_\lambda : \widetilde{X} \longrightarrow \Lambda G_\tau$ satisfying $F_\lambda^*\theta = \pi_X^*\alpha(B)_\lambda$.

Let X be a compact Kähler manifold. Denote by $H^{p,q}(X, \mathbf{C})$ the Dolbeault cohomology of the space of $\bar\partial-$closed (p,q)-forms on X. Set $h^{p,q} = \dim_{\mathbf{C}} H^{p,q}(X, \mathbf{C})$. Suppose that $h^{0,1} > 0$. In this case, the fundamental group $\pi_1(X, p_0)$ of X acts on the universal covering space $\widetilde{X}$ of X, where $p_0 \in X$ is a chosen fixed point. Denote by $\pi_X : \widetilde{X} \longrightarrow X$ the covering projecton. Define a map $\widetilde{h}_X : \widetilde{X} \longrightarrow H^{0,1}(X, \mathbf{C})^*$ by

$$< \widetilde{h}_X(x), \omega > = \int_{x_0}^{x} \omega \ , \qquad x \in \widetilde{X}, \ \ \omega \in H^{0,1}(X, \mathbf{C}).$$

Then $\widetilde{h}_X$ induces a map $\Phi : \pi_1(X, x_0) \longrightarrow H^{0,1}(X, \mathbf{C})^*$, which is a homomorphism. In fact, we have $\Phi(s) = \widetilde{h}_X(s \cdot x) - \widetilde{h}_X(x)$ for $s \in \pi_1(X, p_0), x \in \widetilde{X}$. The map Φ also gives a homomorphism from $H_1(X, \mathbf{Z})$ to $H^{0,1}(X, \mathbf{C})^*$ and it is zero on the torsion part, and injective on the free part, of $H_1(X, \mathbf{Z})$. Thus, $\Phi(H_1(X, \mathbf{Z}))$ defines a lattice of $H^{0,1}(X, \mathbf{C})^*$ and $H^{0,1}(X, \mathbf{C})^*/\Phi(H_1(X, \mathbf{Z}))$ becomes a complex torus, which is called the *Albanese torus* and denoted by $Alb(X)$. We may express it as $Alb(X) = \mathbf{C}^k/\Lambda$, where $k = h^{0,1}$. Using the basis $\omega_1, \cdots, \omega_k$ of $H^{1,0}(X, \mathbf{C})$, the map $\widetilde{h}_X : \widetilde{X} \longrightarrow \mathbf{C}^k$ is explicitly given by

$$\widetilde{h}_X(x) = (\int_{x_0}^{x} \pi_X^*\omega_1, \int_{x_0}^{x} \pi_X^*\omega_2, \cdots, \int_{x_0}^{x} \pi_X^*\omega_k)$$

(cf. [NS]). The Albanese map $h_X : X \longrightarrow \mathbf{C}^k/\Lambda$ is defined by (cf. [GH])

$$h_X(p) = (\int_{p_0}^{p} \omega_1, \int_{p_0}^{p} \omega_2, \cdots, \int_{p_0}^{p} \omega_k),$$

where $p_0 = \pi_X(x_0), p = \pi_X(x)$. Denote by $\pi_{\mathbf{C}} : \mathbf{C}^k \longrightarrow \mathbf{C}^k/\Lambda$ the natural quotient map. We then have $\pi_{\mathbf{C}} \circ \widetilde{h}_X = h_X \circ \pi_X$. By the proof similar to that of Theorem 5.2, we obtain:

PROPOSITION 6.1. *Let $\psi : X \longrightarrow G/H$ be a primitive map of generalized finite type. Let $Alb(X) = \mathbf{C}^k/\Lambda$ be the Albanese torus. Then there is a primitive map of* *finite type* $\hat{f} : \mathrm{Im}(h_X)(\subset \mathbf{C}^k/\Lambda) \longrightarrow G/H$ *such that* ψ *is a composition of the map* $\hat{f}$ *with the Albanese map* $h_X : X \longrightarrow \mathbf{C}^k/\Lambda$.

Finally, we give a proof of the fact that $\hat{f}$ from the proof of Theorem 5.2 is actually defined over $\mathbf{C}^g/\Lambda$. For this, we consider the case where $X = M_g \times \cdots \times M_g$, a Riemannian product of m-copies of a compact Riemann surface of genus g. Assume that $m \geq g$. Let $\pi_X = \pi_M \times \cdots \times \pi_M : \widetilde{M}_g \times \cdots \times \widetilde{M}_g \longrightarrow M_g \times \cdots \times M_g$ be a holomorphic covering projection. For simplicity, we denote it by $\pi_X : \widetilde{M}_g^m \longrightarrow M_g^m$. Choosing a fixed point $(p_{01}, \cdots, p_{0m}) \in M_g^m$ with $p_{01} = p_0$ as in Section 5, we define a generalized Abel map $h^m : M_g^m \longrightarrow \mathbf{C}^g/\Lambda$ by

$$h^m(p_1, \cdots, p_m) = (\sum_{j=1}^m \int_{p_{0j}}^{p_j} \omega_1, \sum_{j=1}^m \int_{p_{0j}}^{p_j} \omega_2, \cdots, \sum_{j=1}^m \int_{p_{0j}}^{p_j} \omega_g),$$

where $(p_1, \cdots, p_m) \in M_g^m$ and $\{\omega_1, \cdots, \omega_g\}$ is a basis of $H^{1,0}(M_g, \mathbf{C})$. By Abel's theorem, it is known that h^m is surjective. The corresponding map $\widetilde{h}^m : \widetilde{M}_g^m \longrightarrow \mathbf{C}^g$ is defined by

$$\widetilde{h}^m(x_1, \cdots, x_m) = (\sum_{j=1}^m \int_{x_{0j}}^{x_j} \widetilde{\omega}_1, \sum_{j=1}^m \int_{x_{0j}}^{x_j} \widetilde{\omega}_2, \cdots, \sum_{j=1}^m \int_{x_{0j}}^{x_j} \widetilde{\omega}_g),$$

where $\pi_X(x_1, \cdots, x_m) = (p_1, \cdots, p_m), \pi_X(x_{01}, \cdots, x_{0m}) = (p_{01}, \cdots, p_{0m})$. We also choose $x_{01} = x_0$ as in Section 5. Obviously, $\pi_{\mathbf{C}} \circ \widetilde{h}^m = h^m \circ \pi_X$.

Given a primitive map of generalized finite type $\psi : M_g \longrightarrow G/H$ as in Section 5, we have an extended framing F_λ such that $\psi \circ \pi_M = \pi \circ F_1$, where $\pi : G \longrightarrow G/H$ is the coset projection. We may assume that $F_\lambda(x_0) = e$. Consider the maps I, J defined by

$$\begin{aligned} I \ &: \ \widetilde{M}_g \ni x \longrightarrow (x, x_{02}, \cdots, x_{0m}) \in \widetilde{M}_g^m, \\ J \ &: \ M_g \ni p \longrightarrow (p, p_{02}, \cdots, p_{0m}) \in M_g^m. \end{aligned}$$

Then it follows that

$$\widetilde{h} = \widetilde{h}^m \circ I, \quad h = h^m \circ J, \quad \pi_X \circ I = J \circ \pi_M. \tag{6.1}$$

Denote by w^j a holomorphic coordinate for the j-th factor $\widetilde{M}_g$ in $\widetilde{M}_g^m$. Let $\widetilde{\psi}^m : \widetilde{M}_g^m \longrightarrow G/H$ be a primitive map of generalized finite type obtained by $\widetilde{\psi}^m = \pi \circ F_1^m$ for an extended framing F_λ^m which arises from the solution $\widetilde{\xi}^m : \widetilde{M}_g^m \longrightarrow \Lambda_d$ of Lax type equation $d\widetilde{\xi}^m = [\widetilde{\xi}^m, \widetilde{\alpha}_\lambda]$ with

$$\begin{cases} \widetilde{D}_j = \displaystyle\sum_{s=1}^m \widetilde{C}_j(\partial/\partial w^s) dw^s \ , \qquad (j = 1, \cdots, n), \\ \widetilde{\alpha}_\lambda = \lambda \displaystyle\sum_{j=1}^n V_j(\widetilde{\xi}_d^m)\widetilde{D}_j + (\sum_{j=1}^n V_j(\widetilde{\xi}_d^m + \widetilde{\xi}_{d-1}^m)_1 \widetilde{D}_j)_{\mathfrak{h}} + \lambda^{-1} \sum_{j=1}^n \overline{V_j(\widetilde{\xi}_d^m)\widetilde{D}_j} \, , \\ \widetilde{\xi}^m(x_{01}, \cdots, x_{0m}) = \widetilde{\xi}(x_0) \, . \end{cases} \tag{6.2}$$

LEMMA 6.2. *For arbitrary fixed points* $y_1, \cdots, y_m \in \widetilde{M}_g$ *and an arbitrary point* $x \in \widetilde{M}_g$ *considered as a point in the* r*-th factor of* $\widetilde{M}_g^m, (r = 1, \cdots, m)$, *we have*

$$F_\lambda^m(y_1, \cdots, y_{r-1}, x, y_{r+1}, \cdots, y_m) = F_\lambda^m(y_1, \cdots, y_{r-1}, x_0, y_{r+1}, \cdots, y_m) \cdot F_\lambda(x).$$

Consequently, there exists a primitive map of generalized finite type $\psi^m : M_g^m \longrightarrow G/H$ *such that* $\psi^m \circ \pi_X = \widetilde{\psi}^m = \pi \circ F_1^m$.

PROOF. For $r = 1, \cdots, m$, define a map $I^r_{y_1 \cdots y_m} : \widetilde{M}_g \longrightarrow \widetilde{M}_g^m$ by

$$I^r_{y_1 \cdots y_m}(x) = (y_1, \cdots, y_{r-1}, x, y_{r+1}, \cdots, y_m).$$

Choose a solution $\widetilde{\xi}'$ with initial condition $\widetilde{\xi}^m(y_1, \cdots, y_{r-1}, x_0, y_{r+1}, \cdots, y_m) = \widetilde{\xi}'(x_0)$ in (6.2). Then, since $I^r_{y_1 \cdots y_m}{}^* \widetilde{D}_j = \widetilde{C}_j, (j = 1, \cdots, n)$, we obtain $\widetilde{\xi}^m \circ I^r_{y_1 \cdots y_m} = \widetilde{\xi}'$. Since $\widetilde{\xi}^m = \mathrm{Ad}(F_\lambda^m(y_1, \cdots, x_0, \cdots, y_m)^{-1} \cdot F_\lambda^m)^{-1} \cdot \widetilde{\xi}'(x_0)$ and $\widetilde{\xi}' = \mathrm{Ad}F_\lambda^{-1} \cdot \widetilde{\xi}'(x_0)$(cf. Section 3.2), we have $F_\lambda^m(y_1, \cdots, x_0, \cdots, y_m)^{-1} \cdot F_\lambda^m \circ I^r_{y_1 \cdots y_m} = F_\lambda$. Hence we obtain the first statement. Since $\psi \circ \pi_M = \pi \circ F_1$, it follows from the first statement that $\widetilde{\psi}^m = \pi \circ F_1^m$ factors through M_g^m and induces a primitive map of generalized finite type $\psi^m : M_g^m \longrightarrow G/H$ such that $\widetilde{\psi}^m = \psi^m \circ \pi_X$. □

Let $f : \mathbf{C}^g \longrightarrow G/H$ be a primitive map of finite type given by (5.7). Since

$$\sum_{i=1}^{g} \hat{V}_i(\eta_d \circ \widetilde{h}^m)(\widetilde{h}^m)^* dz^i = \sum_{j=1}^{n} V_j(\eta_d \circ \widetilde{h}^m)\widetilde{D}_j ,$$

we obtain $\widetilde{\psi}^m = f \circ \widetilde{h}^m$. Thus, we have $\psi^m \circ \pi_X = f \circ \widetilde{h}^m$ by Lemma 6.2. By an argument similar to that in the proof of Theorem 5.2, we have

$$\psi^m \circ \pi_X = f \circ \widetilde{h}^m = \hat{f} \circ \pi_{\mathbf{C}} \circ \widetilde{h}^m = \hat{f} \circ h^m \circ \pi_X,$$

where $\hat{f} : \mathbf{C}^g/\Lambda \longrightarrow G/H$ is a primitive map of finite type induced from f by the surjectivity of h^m and consequently $\psi^m = \hat{f} \circ h^m$. Therefore, it follows from (6.1) that

$$\psi \circ \pi_M = f \circ \widetilde{h} = f \circ \widetilde{h}^m \circ I = \psi^m \circ \pi_X \circ I = \psi^m \circ J \circ \pi_M,$$

which implies that $\psi = \psi^m \circ J$ and $\psi = \hat{f} \circ h^m \circ J = \hat{f} \circ h$. This completes the proof of Theorem 5.2.

References

[ACGH] E. Arbarello, M. Cornalba, P.A. Griffiths and J. Harris, *Geometry of Algebraic Curves*, Springer-Verlag, 1985.

[Bl] M. Black, *Harmonic Maps into Homogeneous Spaces*, vol. 255, Pitman Res. Notes in Math., Longman, Harlow, 1991.

[BPW] J. Bolton, F. Pedit and L. Woodward, *Minimal surfaces and the affine Toda field model*, J. reine angew. Math. **459** (1995), 119–150.

[Bu] F.E. Burstall, *Harmonic tori in spheres and complex projective spaces*, J. reine angew. Math. **469** (1995), 149–177.

[BFPP] F.E. Burstall, D. Ferus, F. Pedit and U. Pinkall, *Harmonic tori in symmetric spaces and commuting Hamiltonian systems on loop algebras*, Ann. of Math. **138** (1993), 173–212.

[BP] F.E. Burstall and F. Pedit, *Harmonic maps via Adler-Kostant-Symes theory*, Harmonic Maps and Integrable Systems (A. Fordy and J.C. Wood, eds.), Vieweg, 1994, pp. 541–572.

[BR] F.E. Burstall and J.H. Rawnsley, *Twistor Theory for Riemannian Symmetric Spaces*, vol. 1424, Lecture Notes in Math., Springer-Berlin, 1990.

[DPW] J. Dorfmeister, F. Pedit and H. Wu, *Weierstrass type representation of harmonic maps into symmetric spaces*, Comm. Anal. Geom. **6** (1998), 633–667.

[FPPS] D. Ferus, F. Pedit, U. Pinkall and I. Sterling, *Minimal tori in S^4*, J. reine angew. Math. **429** (1992), 1–47.

[GH] P. Griffiths and J. Harris, *Principles of Algebraic Geometry*, John Wiley & Sons, 1978.

[Gu] M.A. Guest, *Harmonic Maps, Loop Groups, and Integrable Systems*, vol. 38, London Math. Soc., Cambridge Univ. Press, 1997.

[He] S. Helgason, *Differential Geometry, Lie Groups, and Symmetric Spaces*, Academic Press, London, 1978.

[Hi] N.J. Hitchin, *Harmonic maps from a 2-torus to the 3-sphere*, J. Differential Geom. **31** (1990), 627–710.

[Ko] O. Kowalski, *Generalized Symmetric Spaces*, vol. 805, Lecture Notes in Math., Springer, Berlin, 1980.

[Mc1] I. McIntosh, *A construction of all non-isotropic harmonic tori in complex projective space*, International J. Math. **6** (1995), 831–879.

[Mc2] I. McIntosh, *Two remarks on the construction of harmonic tori in $\mathbf{C}P^n$*, International J. Math. **7** (1996), 515–520.

[Mc3] I. McIntosh, *Harmonic tori and generalized Jacobi varieties*, Comm. Anal. Geom. **9 No. 2** (2001), 423–449.

[Mi] R. Miyaoka, *The family of isometric superconformal harmonic maps and the affine Toda equations*, J. reine angew. Math. **481** (1996), 1–25.

[MO] M. Mukai and Y. Ohnita, *Geometry of the moduli spaces of harmonic maps into Lie groups via gauge theory over Riemann surfaces*, International J. Math. **12 No. 3** (2001), 339–371.

[NS] T. Nagano and B. Smyth, *Minimal varieties and harmonic maps in tori*, Comment. Math. Helv. **50** (1975 pages 249–265).

[OU] Y. Ohnita and S. Udagawa, *Complex-analyticity of pluriharmonic maps and their constructions*, Prospects in Complex Geometry (Proceedings, Katata/ Kyoto, 1989), vol. 1468, Lect. Notes in Math., Springer-Verlag, 1990, pp. 371–407.

[OV] Y. Ohnita and G. Valli, *Pluriharmonic maps into compact Lie groups and factorization into unitons*, Proc. London Math. Soc. **61** (1990), 546–570.

[PS] A.N. Pressley and G.B. Segal, *Loop Groups*, Oxford Univ. Press, 1986.

[Ta] T. Taniguchi, *Non-isotropic harmonic tori in complex projective spaces and configurations of points on rational or elliptic curves*, Tôhoku Math. J. **52** (2000), 603–628.

[Ud1] S. Udagawa, *Harmonic maps from a two-torus into a complex Grassmann manifold*, International J. Math. **6** (1995), 447–459.

[Ud2] S. Udagawa, *Harmonic tori in quaternionic projective 3-spaces*, Proc. Amer. Math. Soc. **125** (1997), 275–285.

[Wd] J.C. Wood, *The explicit construction and parametrization of all harmonic maps from the two-sphere to a complex Grassmannian*, J. reine angew. Math. **386** (1988), 1–31.

Department of Mathematics, Graduate School of Science, Tokyo Metropolitan University, Minami-Ohsawa 1-1, Hachioji, Tokyo 192-0397, JAPAN

E-mail address: `ohnita@comp.metro-u.ac.jp`

Department of Mathematics, School of Medicine, Nihon University, Itabashi, Tokyo 173-0032, JAPAN

E-mail address: `sudagawa@med.nihon-u.ac.jp`

Contemporary Mathematics
Volume **308**, 2002

Submanifolds Associated to Grassmannian Systems

Joonsang Park

§1 Introduction

This is a short report on work done in collaboration with M. Brück, X. Du, and C. L. Terng [3].

The study of immersions of space forms into space forms is one of the most important and interesting problems in classical differential geometry. A well-known theorem of Hilbert states that a complete 2-dimensional Riemannian manifold of constant negative curvature, such as the hyperbolic space form $\mathbb{H}^2$, cannot be isometrically immersed into 3-dimensional Euclidean space $\mathbb{R}^3$ [9]. The natural generalization of the Hilbert theorem, that a complete n-dimensional hyperbolic space form $\mathbb{H}^n$ cannot be isometrically immersed into $\mathbb{R}^{2n-1}$, remains an open question. The local problem of isometric immersions of space forms in space forms was studied by Cartan [4]. He showed that there is no immersion of $\mathbb{H}^n$ in $\mathbb{R}^{2n-2}$, and he showed that there is a local immersion of $\mathbb{H}^n$ in $\mathbb{R}^{2n-1}$. Furthermore, he proved that such an immersion is nondegenerate and has flat normal bundle. Later in [12] and [13], Tenenblat and Terng studied immersions of $\mathbb{H}^n$ into $\mathbb{R}^{2n-1}$ and $\mathbb{R}^n$ into $\mathbb{S}^{2n-1}$, too.

Recently, Terng [14] studied local immersions of space forms in space forms with the same constant curvatures, or,

$$\mathbb{R}^n \text{ in } \mathbb{R}^{2n},\ \mathbb{S}^n \text{ in } \mathbb{S}^{2n},\ \mathbb{H}^n \text{ in } \mathbb{H}^{2n}.$$

She introduced a partial differential equation called an n-dimensional system on a Lie group or on a symmetric space and showed that by properly choosing symmetric spaces, the solutions of the n-dimensional systems on those symmetric spaces correspond to such local immersions. We will call these n-dimensional systems Grassmannian systems. The purpose of this work is to generalize n-dimensional systems to various submanifolds in space forms. The basic idea of this work is that the fundamental equations of such immersions can be expressed by flat connections on the ambient spaces N with values in the Lie algebras $\mathcal{G}$ of the isometry groups

---

2000 *Mathematics Subject Classification.* Primary 57N35.

This work was supported by Grant No. R01-2000-0004 from the Basic Research Program of the Korea Science & Engineering Foundation of N, and these connections correspond to Lax pairs on $\mathcal{G}$. Thus it is plausible that the methods of integrable systems can be applied to submanifolds in space forms.

In this paper, we study the following problems:

1. Find a local nondegenerate immersion $X : M^n \longrightarrow N^{n+k}(c)$ of a Riemannian manifold M of constant sectional curvature c into the space form $N^{n+k}(c)$, which has flat normal bundle. Specifically, we are interested in the cases

(i) M is a space form $N^n(c')$ with $c' \leq c$,

(ii) M is a surface with constant mean curvature or an isothermic surface in $\mathbb{R}^3$,

(iii) M is a surface with constant Gaussian curvature -1 in $\mathbb{R}^3$.

These will correspond to solutions of systems on real Grassmannians.

2. Construct a new immersion $\tilde{X}$ which has the same geometric type as X.

Classically, we have examples of such transformations, i.e., Bäcklund and Darboux transformations.

§2 The G/K systems

G/K systems are introduced in [14] as the first flows of integrable systems of evolution, where all the variables are on an equal footing as time variables and the first flows in each variable can be regarded as a first flow. We will review the definitions briefly. For more details, see [14].

Let G/K be a symmetric space of rank n, $\mathcal{G} = \mathcal{K} + \mathcal{P}$ the Cartan decomposition, and $\mathcal{A} \subset \mathcal{P}$ a maximal abelian subalgbra. Let $a_1, \ldots, a_n$ be a basis for $\mathcal{A}$ consisting of regular elements with respect to the $Ad(K)$-action on $\mathcal{P}$. Denote by $\mathcal{A}^\perp$ the orthogonal complement of $\mathcal{A}$ in $\mathcal{G}$ with respect to the Killing form. Then the G/K *system* for $v : \mathbb{R}^n \to \mathcal{P} \cap \mathcal{A}^\perp$ is

$$[a_i, v_{x_j}] - [a_j, v_{x_i}] = \big[\, [a_i, v], [a_j, v] \,\big], \qquad 1 \leq i \neq j \leq n, \tag{2.1}$$

where, $v_{x_i} = \frac{\partial v}{\partial x_i}$. This system is a Cauchy problem which can be solved for any generic data decaying rapidly along $(x_1, 0, \ldots, 0)$ (cf. [14]).

We can express G/K system in terms of a Lax pair and also in terms of a connection 1-form on the trivial principal bundle $\mathbb{R}^n \times \mathcal{G}$ over $\mathbb{R}^n$. To see this, we need the following proposition, which can be proved by direct computation.

PROPOSITION 2.1 ([14]). *The following statements are equivalent for a map* $v : \mathbb{R}^n \to \mathcal{P} \cap \mathcal{A}^\perp$*:*

(i) v *is a solution of the* G/K*-system (2.1),*

(ii)

$$[\frac{\partial}{\partial x_i} + \lambda a_i + [a_i, v], \frac{\partial}{\partial x_j} + \lambda a_j + [a_j, v]] = 0, \quad \textit{forall } \lambda \in \mathbb{C}, \tag{2.2}$$

(iii) θ_λ *is a flat* $\mathcal{G}_{\mathbb{C}} = \mathcal{G} \otimes \mathbb{C}$*-connection 1-form on* $\mathbb{R}^n$ *for all* $\lambda \in \mathbb{C}$*, where*

$$\theta_\lambda = \sum_{i=1}^{n} (a_i\lambda + [a_i, v])dx_i, \tag{2.3}$$

i.e., $d\theta_\lambda + \theta_\lambda \wedge \theta_\lambda = 0$,

(iv) there exists a G*-valued map* E_λ *such that* $E_\lambda^{-1}dE_\lambda = \theta_\lambda$.

The one-parameter family of flat connections (2.2) or (2.3) is called the *Lax pair* of the G/K-system. Suppose E is a trivialization of θ, that is, $E^{-1}dE = \theta$. Let $g : \mathbb{R}^n \to G$. The gauge transformation of E by g is defined as $g * E = Eg^{-1}$. This induces a new flat connection

$$(Eg^{-1})^{-1}d(Eg^{-1}) = g\theta g^{-1} - dgg^{-1}.$$

We call $g * \theta = g\theta g^{-1} - dgg^{-1}$ the gauge transformation of θ by g. The following proposition is immediate:

PROPOSITION 2.2. *If* θ *is a flat* $\mathcal{G}$*-valued connection 1-form and* $g : \mathbb{R}^n \to G$ *a smooth map, then the gauge transformation of* θ *by* g*,* $g * \theta = g\theta g^{-1} - dgg^{-1}$*, is also flat. Moreover, if* E *is a trivialization of* θ*, then* Eg^{-1} *is a trivialization of* $g * \theta$.

We can find a solution v of G/K system from the above kind of a family of flat connections θ_λ as follows:

PROPOSITION 2.3. *Let* $a_1, \cdots, a_n$ *be a basis of a maximal abelian subalgebra* $\mathcal{A}$ *in* $\mathcal{P}$*, and* $u_i : \mathbb{R}^n \to \mathcal{K}_{\mathcal{A}}^{\perp}$ *smooth maps for* $1 \leq i \leq n$*. If*

$$\theta_\lambda = \sum_{i=1}^{n} (a_i\lambda + u_i)dx_i$$

is a flat connection 1-form on $\mathbb{R}^n$ *for all* $\lambda \in \mathbb{C}$*, then there exists a unique map* $v : \mathbb{R}^n \to \mathcal{P} \cap \mathcal{A}^{\perp}$ *such that* $u_i = [a_i, v]$.

PROOF. We may assume that $a_1, \cdots, a_n$ are regular by choosing a regular basis $b_1, \cdots, b_n$ of $\mathcal{A}$ so that $b_j = \sum_i c_{ij}a_i$ and by changing coordinates $x_i = \sum_j c_{ij}y_j$. Since θ_λ is flat for all λ,

$$[a_i, u_j] = [a_j, u_i]. \tag{2.4}$$

Because $a_1, \cdots, a_n$ are regular, $\mathrm{ad}(a_j)$ maps $\mathcal{P} \cap \mathcal{A}^{\perp}$ isomorphically to $\mathcal{K} \cap \mathcal{A}^{\perp}$. Hence there exists a unique $v_j \in \mathcal{P} \cap \mathcal{A}^{\perp}$ such that $u_j = \mathrm{ad}(a_j)(v_j)$ for $1 \leq j \leq n$. Then (2.4) implies that $\mathrm{ad}(a_i)\mathrm{ad}(a_j)(v_j) = \mathrm{ad}(a_j)\mathrm{ad}(a_i)(v_i)$. Since $[a_i, a_j] = 0$, $\mathrm{ad}(a_i)\mathrm{ad}(a_j) = \mathrm{ad}(a_j)\mathrm{ad}(a_i)$. But $\mathrm{ad}(a_i)$ is injective on $\mathcal{P} \cap \mathcal{A}^{\perp}$ so that $v_i = v_j$, which will be denoted by v. □

§3 The $Gr_m(\mathbb{R}^{m+n})$ system

In this section, we assume

$$G/K = Gr_m(\mathbb{R}^{m+n}) = SO(m+n)/S(O(m) \times O(n)) \text{ with } m \geq n.$$

We shall write down the $Gr_m(\mathbb{R}^{m+n})$-systems explicitly.

Let $\mathcal{G} = so(m+n)$. The Cartan decomposition of G/K is $\mathcal{G} = \mathcal{K} + \mathcal{P}$, where

$$\mathcal{K} = so(m) \times so(n) = \left\{ \begin{pmatrix} Y_1 & 0 \\ 0 & Y_2 \end{pmatrix} \,\middle|\, Y_1 \in so(m),\, Y_2 \in so(n) \right\},$$
$$\mathcal{P} = \left\{ \begin{pmatrix} 0 & \xi \\ -\xi^t & 0 \end{pmatrix} \,\middle|\, \xi \in \mathcal{M}_{m\times n} \right\}.$$

Here $\mathcal{M}_{m\times n}$ is the set of $m \times n$ matrices. Note that

$$\mathcal{A} = \left\{ \begin{pmatrix} 0 & -D \\ D^t & 0 \end{pmatrix} \,\middle|\, D = (d_{ij}) \in \mathcal{M}_{m\times n},\, d_{ij} = 0 \text{ if } i \neq j \right\}$$

is a maximal abelian subalgebra in $\mathcal{P}$ and

$$\mathcal{P} \cap \mathcal{A}^\perp = \left\{ \begin{pmatrix} 0 & \xi \\ -\xi^t & 0 \end{pmatrix} \,\middle|\, \xi = (\xi_{ij}) \in \mathcal{M}_{m\times n},\, \xi_{ii} = 0 \text{ for } 1 \leq i \leq n \right\}.$$

Let

$$a_i = \begin{pmatrix} 0 & -D_i \\ D_i^t & 0 \end{pmatrix},$$

where D_i is the $m \times n$ matrix which has 1 in the (i,i)-th entry and zero otherwise. Then $a_1, \dots, a_n$ form a basis of $\mathcal{A}$. The Lax pair for this G/K-system is

$$\theta_\lambda = \sum_{i=1}^{n} \left\{ \lambda \begin{pmatrix} 0 & -D_i \\ D_i^t & 0 \end{pmatrix} + \begin{pmatrix} D_i\xi^t - \xi D_i^t & 0 \\ 0 & -\xi^t D_i + D_i^t \xi \end{pmatrix} \right\} dx_i$$
$$= \lambda \begin{pmatrix} 0 & 0 & -\delta \\ 0 & 0 & 0 \\ \delta & 0 & 0 \end{pmatrix} + \begin{pmatrix} \delta F^t - F\delta & \delta G^t & 0 \\ -G^t\delta & 0 & 0 \\ 0 & 0 & \delta F - F^t \delta \end{pmatrix}, \tag{3.1}$$

where $\delta = \operatorname{diag}(dx_1, \cdots, dx_n)$, $\xi = \binom{F}{G}$, $F = (f_{ij}) \in \mathcal{M}_{n\times n}$ with $f_{ii} = 0$ and $G \in \mathcal{M}_{(m-n)\times n}$. Put $gl(n)_* = \{(x_{ij}) \in gl(n) \mid x_{ii} = 0,\ 1 \leq i \leq n\}$, then $F \in gl(n)_*$. Since the solution v of this system is completely determined by F and G, we will say (F, G) is the solution of this system, too. This system will be related to flat submanifolds in the unit spheres and Euclidean spaces, and surfaces in $\mathbb{R}^3$ with constant Gaussian curvature -1.

Now, let $G/K = Gr_m(\mathbb{R}^{m+n+1})$ with $m \geq n+1$. The rank of G/K is $n+1$. So the corresponding G/K-system has $(n+1)$ independent variables. Below we consider a partial G/K-system of n-variables. Let $(m_1, m_2, m_3, m_4) = (n, m-n, n, 1)$. We partition a matrix A in $so(m+n+1)$ into 4×4 blocks $A = (A_{ij})$, where $A_{ij} \in \mathcal{M}_{m_i \times m_j}$. Let

$$a_i = \begin{pmatrix} 0 & 0 & -C_i & 0 \\ 0 & 0 & 0 & 0 \\ C_i & 0 & 0 & 0 \\ 0 & 0 & 0 & 0 \end{pmatrix},$$

where $C_i = \operatorname{diag}(0, \dots, 1, \dots, 0)$. Then the space $\mathcal{A}$ spanned by $a_1, \cdots, a_n$ is an n-dimensional abelian subspace of $\mathcal{P}$. Let $\mathcal{K}_\mathcal{A}$ and $\mathcal{G}_\mathcal{A}$ denote the centralizers of $\mathcal{A}$

in $\mathcal{K}$ and $\mathcal{G}$ respectively. Then $\mathcal{P} \cap \mathcal{G}_{\mathcal{A}}^{\perp}$ is the space of matrices of the form

$$v = \begin{pmatrix} 0 & 0 & F & b \\ 0 & 0 & G & 0 \\ -F^t & -G^t & 0 & 0 \\ -b^t & 0 & 0 & 0 \end{pmatrix},$$

where, $F \in gl(n)_*$, $b \in \mathcal{M}_{n\times 1}$, $G \in \mathcal{M}_{(m-n)\times n}$. The *partial* $Gr_m(\mathbb{R}^{m+n+1})$*-system* is given by the flat connections

$$\theta_\lambda = \lambda \begin{pmatrix} 0 & 0 & -\delta & 0 \\ 0 & 0 & 0 & 0 \\ \delta & 0 & 0 & 0 \\ 0 & 0 & 0 & 0 \end{pmatrix} + \begin{pmatrix} \delta F^t - F\delta & \delta G^t & 0 & 0 \\ -G\delta & 0 & 0 & 0 \\ 0 & 0 & \delta F - F^t\delta & \delta b \\ 0 & 0 & -b^t\delta & 0 \end{pmatrix}. \tag{3.2}$$

Again, we will say (F, G, b) is the solution of this partial system. This system will be associated to submanifolds in the sphere $\mathbb{S}^{m+n}$ with constant sectional curvature 1.

§4 The $Gr_m(\mathbb{R}^{m+n,1})$ system

In this section, we assume $m \geq n+1$ and

$$G/K = Gr_m(\mathbb{R}^{m+n,1}) = SO(m+n,1)/S(O(m) \times O(n,1)),$$

i.e., G/K is the rank $(n+1)$ pseudo-Riemannian symmetric space of all Lorentzian $(n+1)$-dimensional linear subspaces of the Lorentz space $\mathbb{R}^{m+n,1}$. Then $\mathcal{G} = so(m+n,1) = \{X \in gl(m+n+1) \mid X^t I_{m+n,1} + I_{m+n,1}X = 0\}$, $\sigma : \mathcal{G} \to \mathcal{G}$ is an involution defined by $\sigma(X) = I_{m,n+1}^{-1} X I_{m,n+1}$, where $I_{p,q} = \begin{pmatrix} I_p & 0 \\ 0 & -I_q \end{pmatrix}$. The Cartan decomposition is $\mathcal{G} = \mathcal{K} + \mathcal{P}$, where

$$\mathcal{K} = so(m) \times so(n,1) = \left\{ \begin{pmatrix} Y_1 & 0 \\ 0 & Y_2 \end{pmatrix} \middle| \, Y_1 \in so(m), Y_2 \in so(n,1) \right\},$$

$$\mathcal{P} = \left\{ \begin{pmatrix} 0 & \xi \\ -J\xi^t & 0 \end{pmatrix} \middle| \, \xi \in \mathcal{M}_{m\times(n+1)} \right\}.$$

Here $J = I_{n,1} = \mathrm{diag}(1, \cdots, 1, -1)$. It is easy to see that

$$\mathcal{A} = \left\{ \begin{pmatrix} 0 & -DJ \\ D^t & 0 \end{pmatrix} \middle| \, D = (d_{ij}) \in \mathcal{M}_{m\times(n+1)},\ d_{ij} = 0 \text{ if } i \neq j \right\}$$

is a maximal abelian subalgebra in $\mathcal{P}$. Let $a_i = \begin{pmatrix} 0 & -D_iJ \\ D_i^t & 0 \end{pmatrix}$, where $D_i \in \mathcal{M}_{m\times(n+1)}$ is the matrix defined as before. Then $a_1, \cdots, a_{n+1}$ form a basis of $\mathcal{A}$. The G/K-system for this symmetric space is described by the flat connections

$$\begin{aligned} \theta_\lambda &= \sum_i \begin{pmatrix} D_i\xi^t - \xi D_i^t & -\lambda D_i J \\ \lambda D_i^t & D_i^t\xi - J\xi^t D_i J \end{pmatrix} dx_i \\ &= \begin{pmatrix} \delta F^t - F\delta & \delta G^t & -\lambda\delta J \\ -G\delta & 0 & 0 \\ \lambda\delta & 0 & \delta F - JF^t\delta J \end{pmatrix}, \end{aligned} \tag{4.1}$$

where $\delta = \mathrm{diag}\,(dx_1, \cdots, dx_{n+1})$, $\xi = \binom{F}{G}$, $F \in gl(n+1)_*$ and $G \in \mathcal{M}_{(n+1)\times(m-n-1)}$.

Next we shall write down the partial $Gr_m(\mathbb{R}^{m+n,1})$-systems of n variables. Let

$a_i = \begin{pmatrix} 0 & 0 & -C_i & 0 \\ 0 & 0 & 0 & 0 \\ C_i & 0 & 0 & 0 \\ 0 & 0 & 0 & 0 \end{pmatrix}$, where $C_i = \mathrm{diag}(0, \ldots, 1, \ldots, 0)$, $i \leq n$, as before.

The partial $Gr_m(\mathbb{R}^{m+n,1})$-system is given by the Lax pair

$$\theta_\lambda = \lambda \begin{pmatrix} 0 & 0 & -\delta & 0 \\ 0 & 0 & 0 & 0 \\ \delta & 0 & 0 & 0 \\ 0 & 0 & 0 & 0 \end{pmatrix} + \begin{pmatrix} \delta F^t - F\delta & \delta G^t & 0 & 0 \\ -G\delta & 0 & 0 & 0 \\ 0 & 0 & \delta F - F^t\delta & \delta b \\ 0 & 0 & b^t\delta & 0 \end{pmatrix}. \tag{4.2}$$

$Gr_m(\mathbb{R}^{m+n,1})$-systems and the partial $Gr_m(\mathbb{R}^{m+n,1})$-systems will be related to submanifolds of constant sectional curvature -1 in the space form $\mathbb{H}^{m+n}$, isothermic surfaces, and surfaces with constant mean curvature in $\mathbb{R}^3$.

§5 Submanifolds

Suppose M is an n-dimensional submanifold in a Riemannian manifold N. If the normal bundle $\nu(M)$ of M is flat, then it is an elementary fact that the family $\{A_v | v \in \nu(M)_p\}$ of shape operators of M at p is a family of commuting self-adjoint operators on TM_p, and generically there is a smooth common eigenframe.

Definition. Let M^n be a submanifold in a Riemannian manifold N. The normal bundle $\nu(M)$ is called *non-degenerate* if the space of the shape operators $\{A_v | v \in \nu(M)_p\}$ has dimension n for all $p \in M$.

When M is a surface in $\mathbb{R}^3$, it is well-known that if $p \in M$ is not umbilic, then there exist local coordinates x_1, x_2 such that $\frac{\partial}{\partial x_1}, \frac{\partial}{\partial x_2}$ are eigenvectors for the shape operator A_{e_3}, where e_3 is a local unit normal vector field. In other words, the two fundamental forms are of the form

$$\mathrm{I} = a_1^2 dx_1^2 + a_2^2 dx_2^2, \quad \mathrm{II} = b_1 dx_1^2 + b_2 dx_2^2,$$

for some functions a_i, b_i. Such coordinates are called *line of curvature coordinates*. We generalize this notion to submanifolds below.

Definition. Let M be an n-dimensional submanifold of $\mathbb{R}^m$ with flat normal bundle, and $\{e_\alpha\} = \{e_{n+1}, \cdots, e_m\}$ a local orthonormal parallel normal frame. Local coordinates $x_1, \cdots, x_n$ are called *line of curvature coordinates* with respect to $\{e_\alpha\}$ if $\{\frac{\partial}{\partial x_i}\}$ is a common orthogonal eigenbasis for the shape operators A_v for all $v \in \nu(M)$, or equivalently, if the two fundamental forms of M are of the form

$$\mathrm{I} = \sum_{i=1}^{n} a_i^2 dx_i^2, \quad \mathrm{II} = \sum_{i=1,j=1}^{n,m-n} b_{ji} dx_i^2 e_{n+j},$$

for some smooth functions a_i and b_i. If in addition $a_1^2 + \cdots + a_n^2 = 1$ ($a_1^2 + \cdots + a_{n-1}^2 - a_n^2 = \pm 1$, respectively), then $x_1, \cdots, x_n$ are called *spherical (hyperbolic, respectively) line of curvature coordinates*.

Let M^n be a submanifold of $\mathbb{R}^{n+m}$ with flat normal bundle. Although generically there exist orthonormal tangent frame vectors e_i that are a common eigenbasis for all shape operators, we cannot always find coordinates $x_1, \cdots, x_n$ so that $\frac{\partial}{\partial x_i}$ is parallel to e_i for all i. However, n-dimensional submanifolds in $\mathbb{R}^{2n-1}$ with constant sectional curvature are known to have flat and non-degenerate normal bundle, and admit spherical line of curvature coordinates (cf. [4], [10]).

Now we describe submanifolds associated to Grassmannian systems.

THEOREM 5.1 (FLAT n-SUBMANIFOLDS IN $\mathbb{S}^{m+n-1}$). *Let X be a local isometric immersion of a flat n-dimensional submanifold in $\mathbb{S}^{n+m-1}$ with flat, non-degenerate normal bundle, and $m \geq n$. Let $e_{n+1} = X$, and fix a local parallel normal frame $e_{n+2}, \cdots, e_{n+m}$. Then:*

(i) There exist spherical line of curvature coordinates $(x_1, ..., x_n)$ and a smooth $O(m)$-valued map $A = (a_{ij})$ such that the first and second fundamental forms of X are given by

$$\mathrm{I} = \sum_{i=1}^{n} a_{1i}^2\, dx_i^2, \qquad \mathrm{II} = \sum_{i=1}^{n}\sum_{j=2}^{m} a_{1i}a_{ji} dx_i^2\, e_{n+j}.$$

(ii) Let $f_{ij} = (a_{1i})_{x_j}/a_{1j}$ for $i \neq j$, $f_{ii} = 0$ for all $1 \leq i, j \leq n$, and $F = (f_{ij}) \in gl(n)_*$. Then we can find a solution (F, G) of the $Gr_m(\mathbb{R}^{m+n})$-system for some G. In particular, the Gauss, Codazzi and Ricci equations for the immersion X are equivalent to the flatness of θ_λ in (3.1).

(iii) Conversely, if (F, G) is a solution of the $Gr_m(\mathbb{R}^{m+n})$-system, then we can find an isometric immersion X of flat n-submanifolds in $\mathbb{S}^{n+m-1}$ with flat and non-degenerate normal bundle such that the two fundamental forms are as in (i).

SKETCH OF THE PROOF. (i) can be proved the same way as for isometric immersions of n-submanifolds in $\mathbb{R}^{2n-1}$ with constant sectional curvature -1 (cf. [4], [10]). For (ii), let $e_i = \frac{1}{a_{1i}} X_{x_i}$, $1 \leq i \leq n$ and $E' = (e_{n+1}, \cdots, e_{n+m}, e_1, \cdots, e_n)$. Then $E'^{-1}dE'$ is a flat connection which gives the Gauss, Codazzi and Ricci equations for the immersion X. For $g = \begin{pmatrix} A & 0 \\ 0 & I \end{pmatrix}$, the gauge transformation of $E'^{-1}dE'$ by g^{-1}, $g^{-1} * E'^{-1}dE' = \theta_1$, gives the flat connection θ_λ as in (3.1). Conversely, choose A which makes the $(1,1)$-block of θ_λ in (3.1) zero. Then for $g = \begin{pmatrix} A & 0 \\ 0 & I \end{pmatrix}$, $g * \theta_1 = E'^{-1}dE'$ for some E' and the first column of E' is the desired X. □

THEOREM 5.2 (FLAT n-SUBMANIFOLDS IN $\mathbb{R}^{n+m}$). *Let $n \leq m$, and X a local isometric immersion of a flat n-dimensional submanifold in $\mathbb{R}^{m+n}$ with flat and non-degenerate normal bundle. Fix a local parallel normal frame $e_{n+1}, \cdots, e_{n+m}$. Then:*

(i) There exist line of curvature coordinates $(x_1, ..., x_n)$, a map $b = (b_1, \cdots, b_n)$, and a smooth $O(m)$-valued map $A = (a_{ij})$, such that the first and second fundamental forms of X are given by

$$\mathrm{I} = \sum_i b_i^2 dx_i^2, \quad \mathrm{II} = \sum_{i=1}^{n}\sum_{j=1}^{m} b_i a_{ji} dx_i^2 e_{n+j}.$$

(ii) *Let* $f_{ij} = (b_i)_{x_j}/b_j$ *for* $i \neq j$, $f_{ii} = 0$, *and* $F = (f_{ij})$. *Then* (F, G) *is a solution of the* $Gr_m(\mathbb{R}^{m+n})$*-system given by (3.1) for some* G.

(iii) *Conversely, if* (F, G) *is a solution of the* $Gr_m(\mathbb{R}^{m+n})$*-system, then there exists a local isometric immersion* X *of* $\mathbb{R}^n$ *in* $\mathbb{R}^{n+m}$ *such that the two fundamental forms are given by (i).*

(iv) X *lies in a hypersphere of radius 1 if and only if* $b = vA_1$ *for some constant unit vector* $v \in \mathcal{M}_{1\times m}$.

PROOF. The proof is similar to that of the previous theorem. □

THEOREM 5.3 (LOCAL ISOMETRIC IMMERSIONS OF $\mathbb{S}^n$ IN $\mathbb{S}^{n+m}$). *Suppose* X *is a local isometric immersion of* $\mathbb{S}^n$ *in* $\mathbb{S}^{n+m}$ *with flat and non-degenerate normal bundle, and* $\{e_\alpha\}$ *is a parallel normal frame. Then:*

(i) *There exist a local coordinate system* $(x_1, \ldots, x_n)$, *a smooth* $O(m)$*-valued map* $A = (a_{ij})$, *and an* $\mathcal{M}_{n\times 1}$*-valued map* $b = (b_1, \cdots, b_n)^t$, *such that the two fundamental forms of* X *are given by*

$$\mathrm{I} = \sum_i b_i^2\, dx_i^2, \quad \mathrm{II} = \sum_{i=1,j=1}^{n,m} a_{ji} b_i dx_i^2 e_{n+j}.$$

(ii) *Let* $f_{ij} = (b_i)_{x_j}/b_j$ *if* $i \neq j$, $f_{ii} = 0$ *for all* $1 \leq i \leq n$, *and* $F = (f_{ij})$. *Then* (F, G, b) *is a solution of the partial* $Gr_m(\mathbb{R}^{m+n})$*-system given by (3.2).*

(iii) *If* (F, G, b) *is a solution of the partial* $Gr_m(\mathbb{R}^{m+n})$*-system given by (3.2), then there exists a local isometric immersion* X *of* $\mathbb{S}^n$ *in* $\mathbb{S}^{n+m}$ *whose fundamental forms are as in (i).*

Let $\mathbb{R}^{k,1}$ denote the Lorentz space equipped with the non-degenerate bilinear form of index one: $\langle x, y\rangle_1 = x_1 y_1 + \cdots + x_k y_k - x_{k+1} y_{k+1}$. Denote by $\mathbb{H}^k$ the k-dimensional simply connected space form of sectional curvature -1 (i.e., a *hyperbolic k-space*). It is well-known that $\{x \in \mathbb{R}^{k,1} \mid \langle x, x\rangle_1 = -1, x_{k+1} > 0\}$ with the induced metric is isometric to $\mathbb{H}^k$.

The following proposition is obtained by direct computation:

PROPOSITION ([11]). *Let* $v_0 \in \mathbb{R}^{k,1}$ *be a constant non-zero vector,* $c \in \mathbb{R}$ *a constant, and* $N_{v,c}$ *the hypersurface defined by* $N_{v_0,c} = \{x \in \mathbb{H}^k \mid \langle x, v_0\rangle_1 = c\}$. *Then* $N_{v,c}$ *is a totally umbilic hypersurface of* $\mathbb{H}^k$ *with constant sectional curvature*

$$\frac{-\langle v_0, v_0\rangle_1}{c^2 + \langle v_0, v_0\rangle_1}.$$

THEOREM 5.4 (LOCAL ISOMETRIC IMMERSIONS OF $\mathbb{H}^n$ IN $\mathbb{H}^{n+m}$). *Let* X *be a local isometric immersion of* $\mathbb{H}^n$ *in* $\mathbb{H}^{n+m}$ *with flat and non-degenerate normal bundle, and* $\{e_\alpha\}$ *a parallel normal frame. Then:*

(i) *There exists a line of curvature coordinate system* $(x_1, \ldots, x_n)$, *a smooth* $O(m)$*-valued map* $A = (a_{ij})$, *and a smooth* $\mathcal{M}_{n\times 1}$*-valued map* $b = (b_1, \cdots, b_n)^t$, *such* *that the two fundamental forms of X are given by*

$$\mathrm{I} = \sum_i b_i^2\, dx_i^2, \quad \mathrm{II} = \sum_{i=1,j=1}^{n,m-n} a_{ji} b_i dx_i^2 e_{n+j}.$$

(ii) *Let $f_{ij} = (b_i)_{x_j}/b_j$ if $i \neq j$, $f_{ii} = 0$ for all $1 \leq i \leq n$, and $F = (f_{ij})$. Then (F, G, b) is a solution of the partial $Gr_m(\mathbb{R}^{m+n,1})$-system given by (4.2).*

(iii) *If (F, G, b) is a solution of the partial $Gr_m(\mathbb{R}^{m+n,1})$-system given by (4.2), then there exists a local isometric immersion X of $\mathbb{H}^n$ in $\mathbb{H}^{n+m}$ whose fundamental forms are given as in (i).*

(vi) *M^n lies in a totally umbilic hypersurface of $\mathbb{H}^{n+m}$ if and only if there is a constant vector $w \in \mathcal{M}_{m\times 1}$ such that $b = A_1^t w$. Moreover, M^n lies in a flat totally umbilic hypersurface if and only if $\|w\| = 1$.*

For the above theorems, reducing to zero the (1,1)-block in θ_λ of the G/K-system relates to geometry of submanifolds in a space form. Now, we consider the lower (2,2)-block in a similar way. We concentrate on $Gr_3(\mathbb{R}^5)$ and $Gr_3(\mathbb{R}^{4,1})$.

THEOREM 5.5. *The $Gr_3(\mathbb{R}^5)$-system is equivalent to the Gauss-Codazzi equations for a surface in $\mathbb{R}^3$ that admits spherical line of curvature coordinates.*

PROOF. Suppose M is a surface in $\mathbb{R}^3$, which admits a spherical line of curvature coordinate system (x_1, x_2) with respect to a normal frame e_3. Then there exist a function u and a map $G = (g_1, g_2)$ such that

$$\mathrm{I} = \cos^2 u\, dx_1^2 + \sin^2 u\, dx_2^2, \quad \mathrm{II} = g_1 \cos u\, dx_1^2 + g_2 \sin u\, dx_2^2.$$

The Gauss-Codazzi equations are the following system for u, g_1, g_2:

$$\begin{cases} u_{x_1x_1} - u_{x_2x_2} + g_1 g_2 = 0, \\ (g_2)_{x_1} = g_1 u_{x_1}, \\ (g_1)_{x_2} = -g_2 u_{x_2}. \end{cases}$$

Let $B = \begin{pmatrix} \cos u & \sin u \\ -\sin u & \cos u \end{pmatrix}$, $F = \begin{pmatrix} 0 & u_{x_1} \\ -u_{x_2} & 0 \end{pmatrix}$, $G = (g_1, g_2)$. Then (F, G) is a solution of the $Gr_3(\mathbb{R}^5)$-system. Conversely, if (F, G) is a solution of the $Gr_3(\mathbb{R}^5)$-system, then take $B = \begin{pmatrix} \cos u & \sin u \\ -\sin u & \cos u \end{pmatrix}$, where u satisfies $F = \begin{pmatrix} 0 & u_{x_1} \\ -u_{x_2} & 0 \end{pmatrix}$. Then the flatness of the (1,1)-block in the connection obtained by the gauge transformation on θ_λ in (3.1) by $\begin{pmatrix} I & 0 \\ 0 & B \end{pmatrix}$ gives the Gauss-Codazzi equations for a surface in $\mathbb{R}^3$ that admits spherical line of curvature coordinates. □

If we take $G = (\sin u, \cos u)$ as a special case, then the surface in Theorem 5.5 has constant Gaussian curvature -1 and the system is equivalent to the sine-Gordon equation: $u_{x_1x_1} - u_{x_2x_2} = \sin u \cos u$.

THEOREM 5.6. *The $Gr_3(\mathbb{R}^{4,1})$-system is equivalent to the Gauss-Codazzi equations for surfaces in $\mathbb{R}^3$ admitting hyperbolic line of curvature coordinates.*

PROOF. This is similar to the proof of Theorem 5.5. □

COROLLARY 5.7. *Let X_1 be an isometric immersion of a surface M_1 in $\mathbb{R}^3$ admitting hyperbolic line of curvature coordinates (x_1, x_2). Then there exists an immersion X_2 of a surface M_2 admitting the same hyperbolic line of curvature coordinates (x_1, x_2) such that the fundamental forms of X_1, X_2 are given respectively by*

$$\begin{cases} \mathrm{I}_1 = \cosh^2 u\, dx_1^2 + \sinh^2 u\, dx_2^2, \\ \mathrm{II}_1 = g_1 \cosh u\, dx_1^2 + g_2 \sinh u\, dx_2^2, \end{cases}$$
$$\begin{cases} \mathrm{I}_2 = \sinh^2 u\, dx_1^2 + \cosh^2 u\, dx_2^2, \\ \mathrm{II}_2 = -g_1 \sinh u\, dx_1^2 - g_2 \cosh u\, dx_2^2, \end{cases}$$

and the Gauss-Codazzi equations are

$$\begin{cases} u_{x_1x_1} + u_{x_2x_2} + g_1g_2 = 0, \\ (g_1)_{x_2} = u_{x_2}g_2, \\ (g_2)_{x_1} = u_{x_1}g_1. \end{cases} \tag{5.1}$$

Recall that a coordinate system (x, y) for a surface M in $\mathbb{R}^3$ is called *isothermic* if the two fundamental forms are of the form

$$\mathrm{I} = e^{2u}(dx_1^2 + dx_2^2), \quad \mathrm{II} = e^u(g_1 dx_1^2 + g_2 dx_2^2),$$

and such M is called an *isothermic surface.*

It is easy to prove the following fact:

COROLLARY 5.8. *Let $Y_1 = X_1 - X_2$ and $Y_2 = X_1 + X_2$, where (X_1, X_2) is as in Corollary 5.7. Then both Y_1 and Y_2 are isothermic.*

It was first noted by Cieśliński, Goldstein and Sym in [6] that the equation for isothermic surfaces in $\mathbb{R}^3$ has a Lax pair. There have been many papers ([1], [2], [5], [6], [8]) using techniques from soliton theory to study isothermic surfaces in $\mathbb{R}^3$.

The following proposition, which states that a CMC surface in $\mathbb{R}^3$ without umbilic points admits isothermic coordinates, is well-known (for a proof see [11]).

PROPOSITION. *Let M be a surface without umbilic points in space form $N^3(c)$, and k_1, k_2 the principal curvature functions. If M has constant mean curvature H and $k_1 > k_2$, then there exist conformal line of curvature coordinates x, y such that*

$$\mathrm{I} = \frac{2}{k_1 - k_2}\left(dx^2 + dy^2\right),$$

$$\mathrm{II} = \left(1 + \frac{H}{k_1 - k_2}\right) dx^2 + \left(-1 + \frac{H}{k_1 - k_2}\right) dy^2.$$

Moreover, if we write $e^{2u} = \frac{2}{k_1 - k_2}$, then the Gauss-Codazzi equation for M is

$$u_{xx} + u_{yy} = e^{-2u} - \left(\frac{H^2}{4} + c\right) e^{2u}. \tag{5.2}$$

§6 Transformations

In this section, we give an explicit construction of a new solution from a given one, using by the dressing action on the space of solutions of the Grassmannian systems.

First we review the construction of the dressing action. Recall that a symmetric space G/K is determined by a conjugate linear Lie algebra involution τ and a complex linear Lie algebra involution σ on the complexified Lie algebra $\mathcal{G}_{\mathbb{C}} = \mathcal{G} \otimes \mathbb{C}$ such that

(i) τ and σ commute,

(ii) $\mathcal{G}$ is the fixed point set of τ, and $\mathcal{K}$ and $\mathcal{P}$ are the $+1, -1$ eigenspaces of σ on $\mathcal{G}$ respectively, and $\mathcal{G} = \mathcal{K} + \mathcal{P}$ is the Cartan decomposition.

We still use τ and σ to denote the corresponding involutions on the group $G_{\mathbb{C}}$. A map $g : \mathbb{C} \to G_{\mathbb{C}}$ ($g : \mathbb{C} \to \mathcal{G}_{\mathbb{C}}$, respectively) is said to satisfy the *G/K-reality condition* if

$$\tau(g(\bar{\lambda})) = g(\lambda), \quad \sigma(g(-\lambda)) = g(\lambda). \tag{6.1}$$

Consider

$$\begin{aligned} G_+ &= \{g : \mathbb{C} \to G_{\mathbb{C}} \mid g \text{ is holomorphic and satisfies (6.1)}\}, \\ G_- &= \{g : \mathbb{S}^2 \to G_{\mathbb{C}} \mid g(\infty) = I,\ g \text{ is meromorphic, and satisfies (6.1)}\}. \end{aligned}$$

Let $g \in G_-$, v a solution of the G/K-system (2.1), and $E(x, \lambda)$ the trivialization of v with $E(0, \lambda) = I$, i.e.,

$$E^{-1} dE = \theta_\lambda, \quad E(0, \lambda) = I,$$

where θ_λ is defined by (2.3). Since θ_λ is holomorphic for $\lambda \in \mathbb{C}$, $E(x, \cdot) \in G_+$. It is known that (cf. [15]) $g(\lambda)E(x, \lambda)$ can be factored as

$$g(\lambda)E(x, \lambda) = \tilde{E}(x, \lambda)\tilde{g}(x, \lambda)$$

such that $\tilde{E}(x, \cdot) \in G_+$ and $\tilde{g}(x, \cdot) \in G_-$ for x in a neighborhood of the origin. Moreover, $\tilde{E}$ is the trivialization for a new solution $\tilde{v}$ of the U/K-system with $\tilde{E}(0, \lambda) = I$, which we denote by $\tilde{v} = g \sharp v$.

To construct an element $g \in G_-$ in our case, let w and z be unit vectors in $\mathbb{R}^m$ and $\mathbb{R}^n$ respectively, and π the Hermitian projection of $\mathbb{C}^{n+m}$ onto $\mathbb{C}\begin{pmatrix} w \\ iz \end{pmatrix}$, the complex linear subspace spanned by $\begin{pmatrix} w \\ iz \end{pmatrix}$. So $\pi = \frac{1}{2}\begin{pmatrix} ww^t & -iwz^t \\ izw^t & zz^t \end{pmatrix}$.

Let $s \in \mathbb{R}, s \neq 0$. Define

$$g_{s,\pi}(\lambda) = \left(\pi + \frac{\lambda - is}{\lambda + is}(I - \pi)\right)\left(\bar{\pi} + \frac{\lambda + is}{\lambda - is}(I - \bar{\pi})\right),$$

or,

$$g_{s,\pi}(\lambda) = I + \frac{2s}{\lambda^2 + s^2}\begin{pmatrix} -sww^t & \lambda wz^t \\ -\lambda zw^t & -szz^t \end{pmatrix}. \tag{6.2}$$

THEOREM 6.1. *Let v be a solution of the $Gr_m(\mathbb{R}^{m+n})$-system, and $E(x,\lambda)$ a frame of v such that $E(x,\lambda)$ is holomorphic for $\lambda \in \mathbb{C}$. Let $g_{s,\pi}$ be the map defined by (6.2). Let $\tilde{\pi}(x)$ denote the Hermitian projection onto $\mathbb{C}\begin{pmatrix} \tilde{w}(x) \\ i\tilde{z}(x) \end{pmatrix}$, where*

$$\begin{pmatrix} \tilde{w} \\ i\tilde{z} \end{pmatrix}(x) = E(x,-is)^{-1}\begin{pmatrix} w \\ iz \end{pmatrix}.$$

Let $\hat{w} = \tilde{w}/\|\tilde{w}\|$ and $\hat{z} = \tilde{z}/\|\tilde{z}\|$,

$$\tilde{E}(x,\lambda) = g_{s,\pi}(\lambda)E(x,\lambda)g_{s,\tilde{\pi}(x)}(\lambda)^{-1},$$

$$\tilde{\xi} = \xi - 2s(\hat{w}\hat{z}^t)_*,$$

where $(\eta_*)_{ij} = \eta_{ij}$ if $i \neq j$, and $(\eta_*)_{ii} = 0$ if $1 \leq i \leq n$. Then $\tilde{v}$ given by $\tilde{\xi}$ is a solution of (3.1), $\tilde{E}$ is a frame for $\tilde{\xi}$, and $\tilde{E}(x,\lambda)$ is holomorphic in $\lambda \in \mathbb{C}$.

PROOF. See [3]. □

The functions $\tilde{w}, \tilde{z}$ in Theorem 6.1 also satisfy the following differential equations:

$$d\begin{pmatrix} \tilde{w} \\ i\tilde{z} \end{pmatrix} = dE_{-is}^{-1}\begin{pmatrix} w \\ iz \end{pmatrix} = -E_{-is}^{-1}dE_{-is}\begin{pmatrix} \tilde{w} \\ i\tilde{z} \end{pmatrix} = -\theta_{-is}\begin{pmatrix} \tilde{w} \\ i\tilde{z} \end{pmatrix},$$

where θ_λ is defined by (3.1) and $E_\lambda(x) = E(x,\lambda)$. We shall write this down more explicitly:

$$\begin{cases} (\tilde{w}_i)_{x_i} = -\sum_{j\neq i}^{n} f_{ji}\tilde{w}_j - \sum_{j=1}^{m-n} g_{ji}\tilde{w}_{j+n} + s\tilde{z}_i & i \leq n, \\ (\tilde{w}_i)_{x_j} = f_{ij}\tilde{w}_j & i \leq n, j \neq i, \\ (\tilde{w}_i)_{x_j} = g_{ij}\tilde{w}_j & i > n, \\ (\tilde{z}_i)_{x_j} = f_{ji}\tilde{z}_j & j \neq i, \\ (\tilde{z}_i)_{x_i} = -\sum_{j\neq i}^{n} f_{ij}\tilde{z}_j + s\tilde{w}_i. \end{cases} \tag{6.3}$$

The converse is also true:

THEOREM 6.2. *Given $\xi : \mathbb{R}^n \to \mathcal{M}_{m\times n}$ with $\xi_{ii} = 0$ and a real number $s \neq 0$, we have:*

(i) System (6.3) is solvable for $\tilde{w}, \tilde{z}$ if and only if ξ is a solution of the $Gr_m(\mathbb{R}^{m+n})$-system (3.1).

(ii) Let ξ be a solution of (3.1), and $\tilde{w}, \tilde{z}$ a solution of (6.3) with initial condition $\tilde{w}(0) = w, \tilde{z}(0) = z$ for some unit vectors $w \in \mathbb{R}^m, z \in \mathbb{R}^n$. Set

$$\tilde{\xi} = \xi - 2s\frac{(\tilde{w}\tilde{z}^t)_*}{\|\tilde{w}\|\|\tilde{z}\|},$$

where $(Y_*)_{ij} = Y_{ij}$ if $i \neq j$, and $(Y_*)_{ii} = 0$ if $1 \leq i \leq n$. Then $\tilde{v}$ given by $\tilde{\xi}$ is a solution of (3.1), and is equal to $g_{s,\pi}\sharp v$, where π is the projection onto $\mathbb{C}\begin{pmatrix} w \\ iz \end{pmatrix}$.

Now we consider the case for the $Gr_m(\mathbb{R}^{m+n,1})$-system. Since the calculation and proofs are similar to those for the $Gr_m(\mathbb{R}^{m+n})$-system, we only state the results.

The $Gr_m(\mathbb{R}^{m+n,1})$-*reality condition* is

$$\begin{cases} \overline{g(\bar{\lambda})} = g(\lambda), \\ I_{m,n+1} g(\lambda) I_{m,n+1} = g(-\lambda), \\ g(\lambda)^t I_{n+m,1} g(\lambda)^t = I_{n+m,1}. \end{cases} \tag{6.4}$$

Let $\mathbb{C}^{n+m+1}$ be equipped with the bilinear form $< u, v >_1 = \sum_{i=1}^{n+m} \bar{u}_i v_i - \bar{u}_{n+m+1} v_{n+m+1}$. Let $w \in \mathbb{R}^m$ and $z \in \mathbb{R}^{n,1}$ be unit vectors and π the orthogonal projection of $\mathbb{C}^{n+m+1}$ onto the span of $\begin{pmatrix} w \\ iz \end{pmatrix}$ with respect to $< , >_1$. So $\pi = \frac{1}{2}\begin{pmatrix} ww^t & -iwz^t \\ izw^t & zz^t \end{pmatrix} I_{n+m,1}$. Given a non-zero $s \in \mathbb{R}$, define

$$q_{s,\pi}(\lambda) = \left(\pi + \frac{\lambda - is}{\lambda + is}(I - \pi)\right)\left(\bar{\pi} + \frac{\lambda + is}{\lambda - is}(I - \bar{\pi})\right),$$

or,

$$q_{s,\pi}(\lambda) = I + \frac{2s}{\lambda^2 + s^2}\begin{pmatrix} -sww^t & \lambda wz^t J \\ -\lambda zw^t & -szz^t J \end{pmatrix}, \tag{6.5}$$

where $J = I_{n,1} = \mathrm{diag}(1, \cdots, 1, -1)$. A direct computation implies that $q_{s,\pi}$ satisfies the reality condition (6.4). So $q_{s,\pi} \in G_-$.

THEOREM 6.3. *Suppose* $\begin{pmatrix} F \\ G \end{pmatrix}$ *is a solution of the* $Gr_m(\mathbb{R}^{m+n,1})$*-system,* θ_λ *its Lax pair as in (4.1), and* E *a frame for* $\begin{pmatrix} F \\ G \end{pmatrix}$. *Let* $q_{s,\pi}$ *be defined by (6.5),*

$$\begin{pmatrix} \tilde{w} \\ i\tilde{z} \end{pmatrix}(x) = E(x, -is)^{-1}\begin{pmatrix} w \\ iz \end{pmatrix},$$

and $\tilde{\pi}(x)$ *the orthogonal projection onto the span of* $\begin{pmatrix} \tilde{w} \\ i\tilde{z} \end{pmatrix}(x)$ *with respect to the bilinear form* $< , >_1$. *Set*

$$\tilde{E}(x, \lambda) = q_{s,\pi}(\lambda) E(x, \lambda) q^{-1}_{s,\tilde{\pi}(x)}(\lambda),$$

$$\begin{pmatrix} \tilde{F} \\ \tilde{G} \end{pmatrix} = \begin{pmatrix} F \\ G \end{pmatrix} - 2s(\hat{w}\hat{z}^t J)_*,$$

where $\hat{w}(x) = \frac{1}{||\tilde{w}(x)||_m}\tilde{w}(x)$ *and* $\hat{z}(x) = \frac{1}{||\tilde{z}(x)||_{n,1}}\tilde{z}(x)$, *and* $(y_*)_{ij} = y_{ij}$ *for* $i \neq j$ *and* $(y_*)_{ii} = 0$ *for all* i. *Then*

(i) $\begin{pmatrix} \tilde{F} \\ \tilde{G} \end{pmatrix}$ *is a new solution of (4.2),* *(ii)* $\tilde{E}$ *is a frame for* $\begin{pmatrix} \tilde{F} \\ \tilde{G} \end{pmatrix}$.

COROLLARY 6.4. *Let* $\xi = \begin{pmatrix} F \\ G \end{pmatrix}$ *be a solution of the* $Gr_m(\mathbb{R}^{m+n,1})$*-system,* E *a frame of* ξ, *and* $(\tilde{w}, \tilde{z})$ *be as in Theorem (6.1). Then* $(\tilde{w}, \tilde{z})$ *is a solution of the following system*

$$\begin{cases} (\tilde{w}_i)_{x_i} = -\sum_{j \neq i}^{n+1} f_{ji}\tilde{w}_j - \sum_{j=1}^{m-n-1} g_{ji}\tilde{w}_{j+n+1} + s\tilde{z}_i\epsilon_i, \; i \leq n+1 \\ (\tilde{w}_i)_{x_j} = f_{ij}\tilde{w}_j, \qquad i \leq n+1, j \neq i, \\ (\tilde{w}_i)_{x_j} = g_{ij}\tilde{w}_j, \qquad i > n+1, \\ (\tilde{z}_i)_{x_j} = f_{ji}\tilde{z}_j\epsilon_i\epsilon_j, \qquad j \neq i, \\ (\tilde{z}_i)_{x_i} = -\sum_{j \neq i}^{n+1} f_{ij}\tilde{z}_j + s\tilde{w}_i. \end{cases} \tag{6.6}$$

Since the case of the partial Grassmannian systems is exactly the same as above, we omit that case.

Hence, from an immersion X given as in Section 5, we can construct a new immersion $\tilde{X}$ which has the same geometric properties as X by applying the action of $g_{s,\pi}$ or $q_{s,\pi}$.

Now, we give geometric interpretations of the action of $g_{s,\pi}$ or $q_{s,\pi}$. The following definition is a generalization of the classical Ribaucour transformation. We denote by $\gamma_{x,\xi}$ the geodesic passing though x in the direction ξ.

DEFINITION ([7]). Let M^n and $\tilde{M}^n$ be submanifolds of the space form $N^m(c)$. A *sphere congruence* is a vector bundle isomorphism $P : \nu(M) \to \nu(\tilde{M})$ that covers a diffeomorphism $l : M \to \tilde{M}$ with the following properties: (i) If ξ is a parallel normal vector field of M, then $P \circ \xi \circ l^{-1}$ is a parallel normal field of $\tilde{M}$. (ii) For any nonzero vector $\xi \in \nu_x(M)$, the geodesics $\gamma_{x,\xi}$ and $\gamma_{l(x),P(\xi)}$ intersect at a point that is equidistant from x and $l(x)$.

DEFINITION ([7]). A sphere congruence $P : \nu(M) \to \nu(\tilde{M})$ that covers $l : M \to \tilde{M}$ is called a *Ribaucour Transformation* if it satisfies the following additional properties: (i) If e is an eigenvector of the shape operator A_ξ of M, then $l_*(e)$ is an eigenvector of the shape operator $A_{P(\xi)}$ of $\tilde{M}$. (ii) the geodesics $\gamma_{x,e}$ and $\gamma_{l(x),l_*(e)}$ intersect at a point equidistant from x and $l(x)$.

THEOREM 6.5. *Let an immersion X be given as in Section 5 and let a new immersion $\tilde{X}$ be obtained from X by applying the action of $g_{s,\pi}$ or $q_{s,\pi}$. Then there exists a Ribaucour transformation that covers the map $X(x) \mapsto \tilde{X}(x)$ up to rigid motion of the space form $N^k(c)$ containing X and $\tilde{X}$.*

Now,we give examples of Ribaucour transformations for the $Gr_3(\mathbb{R}^5)$-system and the $Gr_3(\mathbb{R}^{4,1})$-system.

THEOREM 6.6. *For $m = 3$ and $n = 2$, let $(F, \sin u, -\cos u)$ be a solution of (3.1) with $F = \begin{pmatrix} 0 & u_x \\ -u_y & 0 \end{pmatrix}$, $z = (z_1, z_2)^t$, $z = (w_1, w_2, w_3)^t$ real constant unit vectors, and constant $s \neq 0$ such that $sw_3 = z_1 \sin u(0,0) - z_2 \cos u(0,0)$. Let π denote the Hermitian projection onto $(w_1, w_2, w_3, iz_1, iz_2)^t$. Then the new solution obtained by applying $g_{s,\pi}$ is of the form $(\tilde{F}, \sin \tilde{u}, -\cos \tilde{u})$ for some $\tilde{u}$.*

EXAMPLE. Let $u = 0$ be the trivial solution of the sin-Gordon equation(SGE). We can use $g_{s,\pi}$ to produce a three-parameter family of new solutions of the SGE. Some of these examples are:

(1) If $0 < s = \sin c \leq 1$, then

$$\tilde{u} = 2\tan^{-1}\left(\frac{\sin c \ \sin(y \cos c)}{\cos c \ \cosh(x \sin c)}\right).$$

This is the breather solution for the SGE, and the corresponding curvature -1 surface in $\mathbb{R}^3$ is given by

$$\tilde{X} = -\begin{pmatrix} x \\ 0 \\ 0 \end{pmatrix} + r \begin{pmatrix} \sinh(x \sin c) \\ \cos y \cos(y \cos c) + \sec c \, \sin y \sin(y \cos c) \\ \sin y \cos(y \cos c) - \sec c \, \cos y \sin(y \cos c) \end{pmatrix},$$

where

$$\frac{r = 2\cosh(x \sin c)}{\sin c \, \{\cosh^2(x \sin c) + \tan^2 c \sin^2(y \cos c)\}.}$$

(2) If $s = \cosh c > 1$, then

$$\tilde{u} = 2\tan^{-1}\left(\frac{\cosh c \cosh(y \sinh c)}{\sinh c \sinh(x \cosh c)}\right),$$

and the corresponding -1 curvature surface in $\mathbb{R}^3$ is

$$\tilde{X} = -\begin{pmatrix} x \\ 0 \\ 0 \end{pmatrix} + r \begin{pmatrix} \cosh(x \cosh c) \\ \cos y \sinh(y \sinh c) + \operatorname{sech} c \sin y \cosh(y \sinh c) \\ \sin y \sinh(y \sinh c) - \operatorname{sech} c \cos y \cosh(y \sinh c) \end{pmatrix},$$

where

$$r = \frac{2\sinh(x \cosh c)}{\cosh c \, \{\sinh^2(x \cosh c) + \coth^2 c \, \sinh^2(y \sinh c)\}}.$$

THEOREM 6.7. *Let (Y_1, Y_2) be an isothermic pair in $\mathbb{R}^3$ such that Y_1 has constant mean curvature $2c$, no umbilical points, and is parametrized by the canonical isothermic coordinates. Let $s \neq 0$ be a constant, and let $w \in \mathbb{R}^3$, $z \in \mathbb{R}^{1,1}$ be constant unit vectors respectively, and $\tilde{Y} = (\tilde{Y_1}, \tilde{Y_2}) = q_{s,\pi} \cdot (Y_1, Y_2)$. If $sw_3 = ce^{u(0,0)}(z_1 + z_2) + e^{-u(0,0)}(z_1 - z_2)$, then $\tilde{Y_1}$ has constant mean curvature $2c$ and $\tilde{Y_2}$ has mean curvature -2.*

EXAMPLE. Let $Y_1 = \begin{pmatrix} -x \\ \sin y \\ 1-\cos y \end{pmatrix}$ and $Y_2 = \begin{pmatrix} -x \\ -\sin y \\ -1+\cos y \end{pmatrix}$. This pair of cylinders (Y_1, Y_2) is an isothermic pair. Let $(\tilde{Y_1}, \tilde{Y_2})$ be the new isothermic pair with constant mean curvatures. Then

$$\tilde{Y_1} = \begin{pmatrix} -x \\ \sin y \\ 1 - \cos y \end{pmatrix} + r_1 \begin{pmatrix} a \sinh(x \sinh c) \\ -\cosh c \, \cos y \cos(y \cosh c) - \sin y \sin(y \cosh c) \\ -\cosh c \, \sin y \cos(y \cosh c) + \cos y \sin(y \cosh c) \end{pmatrix},$$

$$\tilde{Y_2} = \begin{pmatrix} -x \\ -\sin y \\ -1 + \cos y \end{pmatrix} + r_2 \begin{pmatrix} a \sinh(x \sinh c) \\ -\cosh c \, \cos y \cos(y \cosh c) - \sin y \sin(y \cosh c) \\ -\cosh c \, \sin y \cos(y \cosh c) + \cos y \sin(y \cosh c) \end{pmatrix},$$

where $a = \{(z_1^2 - w_1^2)/(w_3^2 + w_2^2 \operatorname{sech}^2 c)\}^{\frac{1}{2}}$, and

$$r_1 = \frac{2}{\sinh c \, \{a \cosh(x \sinh c) + \sinh c \, \sin(y \cosh c)\}},$$

$$r_2 = \frac{2}{\sinh c \, \{a \cosh(x \sinh c) - \sinh c \, \sin(y \cosh c)\}}.$$

References

[1] F. Burstall, *Isothermic surfaces: conformal geometry, Clifford algebras and integrable systems*, preprint, math-dg/0003096.

[2] F. Burstall, U. Hertrich-Jeromin, F. Pedit, U. Pinkall,, *Curved flats and isothermic surfaces*, Math. Z. **225** (1997), 199–209.

[3] M. Brück, X. Du, J. Park, C. Terng, *The submanifold geometries associated to Grassmannian systems*, Mem. Amer. Math. Soc. **155** (2002), No. 735.

[4] E. Cartan, *Sur les varièt̀es de courbure constante d'un espace euclidien ou non-euclidien*, Bull. Soc. Math. France **47** (1920), 125–160.

[5] J. Cieśliński, *The Darboux-Bianchi transformation for isothermic surfaces*, Differential Geom. Appl. **7** (1997), 1–28.

[6] J. Cieśliński, , P. Goldstein, A. Sym, *Isothermic surfaces in* E^3 *as soliton surfaces*, Phys. Lett. A **205** (1995), 37–43.

[7] M. Dajczer, R. Tojeiro, *Commuting Codazzi tensors and the Ribaucour transformation for submanifolds*, preprint.

[8] D. Ferus, F. Pedit, *Isometric immersions of space forms and soliton theory*, Math. Ann. **305** (1996), 329–342.

[9] D. Hilbert, *Über Flächen von konstanter Gausscher Krümmung*, Trans. Amer. Math. Soc. **2** (1901), 89–99.

[10] J. Moore, *Isometric immersions of space forms in space forms*, Pacific Jour. Math. **40** (1972), 157–166.

[11] R. Palais, C. Terng, *Critical Point Theory and Submanifold Geometry*, Springer-Verlag, LNM 1353, 1988.

[12] K. Tenenblat, *Bäcklund's theorem for submanifolds of space forms and a generalized wave equation*, Boll. Soc. Brasil. Mat. **16** (1985), 67–92.

[13] C. Terng, *A higher dimensional generalization of the sine-Gordon equation and its soliton theory*, Ann. Math. **111** (1980), 491–510.

[14] C. Terng, *Soliton equations and differential geometry*, Jour. Diff. Geom. **45** (1997), 407–445.

[15] C. Terng, K. Uhlenbeck, *Bäcklund transformations and loop group actions*, Comm. Pure. Appl. Math. **53** (2000), 1–75.

Department of Mathematics, Dongguk University, Seoul 100-715, KOREA

E-mail address: jpark@dgu.ac.kr

Contemporary Mathematics
Volume **308**, 2002

Harmonic Cohomology Groups of Compact Symplectic Nilmanifolds

Yusuke Sakane and Takumi Yamada

1. Introduction

The purpose of this paper is to give a survey on harmonic cohomology groups of compact symplectic nilmanifolds. A Poisson manifold $(M, \mathbf{G})$ is a smooth manifold M endowed with a Poisson structure $\mathbf{G}$, that is, a skew-symmetric contravariant 2-tensor $\mathbf{G}$ on M satisfying $[\mathbf{G}, \mathbf{G}] = 0$, where $[\ ,\ ]$ denotes the Schouten-Nijenhuis bracket. Let $\Omega^k(M)$ denote the space of all k-forms on M. For a Poisson manifold M, Koszul [K] introduced a differential operator $d^* : \Omega^k(M) \to \Omega^{k-1}(M)$ defined by $d^* = [d, i(\mathbf{G})]$, where $i(\mathbf{G})$ denotes the contraction by $\mathbf{G}$. The operator d^* is called the Koszul differential. For a $2m$-dimensional symplectic manifold (M, ω), Brylinski [B] introduced the concept of symplectic harmonic form. By analogy with the star operator for an oriented Riemannian manifold, he defined the star operator $* : \Omega^k(M) \to \Omega^{2m-k}(M)$ so that Koszul differential satisfies $d^* = (-1)^k * d*$ on $\Omega^k(M)$. A form α is called a harmonic form if it satisfies $d\alpha = d^*\alpha = 0$. Let $\mathcal{H}^k_\omega(M)$ denote the space of all harmonic k-forms. We define the symplectic harmonic k-cohomology group $H^k_{\omega\text{-}hr}(M)$ of the symplectic manifold (M, ω) by $\mathcal{H}^k_\omega(M)/(B^k(M) \cap \mathcal{H}^k_\omega(M))$, which is a subspace of the de Rham cohomology group $H^k_{DR}(M)$. Brylinski conjectured that, for a compact symplectic manifold, every de Rham cohomology class contains a harmonic representative. He proved that this conjecture is true for compact Kähler manifolds. We denote by $L_\omega : \Omega^k(M) \to \Omega^{k+2}(M)$ the linear map defined by wedge product with ω and by $L_{[\omega]} : H^k_{DR}(M) \to H^{k+2}_{DR}(M)$ the induced homomorphism in de Rham cohomology groups. However, Mathieu proved the following theorem.

THEOREM 1.1 ([Ma]). *Let (M^{2m}, ω) be a symplectic manifold with dimension $2m$. Then the following two assertions are equivalent:*

(a) *For any $k \leq m$, the homomorphism $L_{[\omega]^k} : H^{m-k}_{DR}(M) \to H^{m+k}_{DR}(M)$ is surjective.*

(b) *Any de Rham cohomology class contains a harmonic cocycle.*

2000 *Mathematics Subject Classification.* Primary 53D05, 57R17.

Mathieu's Theorem is a generalization of the Hard Lefschetz Theorem for compact Kähler manifolds. He gave a 4-dimensional compact nilmanifold, the Kodaira-Thurston surface, as a counter-example of Brylinski's conjecture. Mathieu [Ma] also proved that, for $i = 0, 1, 2$, $H^i_{\omega\text{-}hr}(M) = H^i_{DR}(M)$. Yan ([Yn]) gave a simpler, elegant proof of Mathieu's Theorem by using a special type of infinite dimensional $\mathfrak{sl}(2)$-representation theory.

In connection with the study of harmonic forms, we are interested in the following question raised by B. Khesin and D. McDuff (see Yan [Yn]).

Question : On which compact manifolds M does there exist a family ω_t of symplectic forms such that the dimension of $H^k_{\omega_t\text{-}hr}(M)$ varies for some k?

This question was considered by Yan [Yn] for 4-manifolds and he constructed compact 4-manifolds M which have a family ω_t of symplectic forms such that the dimension of $H^3_{\omega_t\text{-}hr}(M)$ varies. Yan also observed that for compact 4-dimensional nilmanifolds the dimension of $H^3_{\omega_t\text{-}hr}(M)$ is independent of the symplectic form. Benson and Gordon [BG] proved that the Hard Lefschetz Theorem fails for any symplectic structure on a non-toral nilmanifold, hence a non-toral nilmanifold does not admit any Kähler structure. We now have the following question.

Question : On which compact nilmanifolds M does there exist a family ω_t of symplectic forms such that the dimension of $H^k_{\omega_t\text{-}hr}(M)$ variess?

For 6-dimensional compact nilmanifolds, this question was considered independently by one of us [Ym] and Ibáñez, Rudyak, Tralle and Ugarte [IRTU1]. The latter group proved that there exist at least five 6-dimensional nilmanifolds M with a family ω_t of symplectic form such that the dimension of $H^k_{\omega_t\text{-}hr}(M)$ varies, by computing harmonic cohomology groups $H^4_{\omega_t\text{-}hr}(M)$ and $H^5_{\omega_t\text{-}hr}(M)$. In [IRTU1] a compact manifold M is called flexible if M possesses a continuous family of symplectic forms ω_t ($t \in [a, b]$) such that $\dim H^k_{\omega_a\text{-}hr}(M) \neq \dim H^k_{\omega_b\text{-}hr}(M)$ for some k. Note that if there exists a diffeomorphism $f : M_1 \to M_2$ between symplectic manifolds (M_1, ω_1) and (M_2, ω_2) such that $f^*(\omega_2) = \omega_1$, then $f^*(H^k_{\omega_2\text{-}hr}(M)) = H^k_{\omega_1\text{-}hr}(M)$.

In [Ym] we observed that if M is a compact manifold and ω, ω' are symplectic forms on M such that $\omega - \omega' = d\gamma$, then $H^k_{\omega\text{-}hr}(M) = H^k_{\omega'\text{-}hr}(M)$, that is, the subspace $H^k_{\omega\text{-}hr}(M)$ of $H^k_{DR}(M)$ does not depend on the choice of symplectic form ω of the de Rham class $[\omega]$. In section 2 we give an outline of the proof. By this result, together with a theorem of Nomizu on compact nilmanifolds, we may assume that symplectic forms on compact nilmanifolds are left invariant and it is enough to consider subspaces of Lie algebra cohomology groups to investigate harmonic cohomology groups of compact nilmanifolds. Using this fact, Ibáñez, Rudyak, Tralle and Ugarte [IRTU2] studied 3-dimensional harmonic cohomology groups for compact 6-dimensional nilmanifolds and obtained more examples of flexible 6-dimensional nilmanifolds.

In section 3 we study harmonic forms on compact 2-step symplectic nilmanifolds G/Γ by investigating the star operator. In particular, we consider the difference of $\dim H^2_{DR}(G/\Gamma)$ and $\dim H^{2m-2}_{\omega\text{-}hr}(G/\Gamma)$. We also prove that invariant exact 3-forms are harmonic for compact 2-step symplectic nilmanifolds. In section 4 we give examples of compact 2-step symplectic nilmanifolds G/Γ such that the dimension of the harmonic cohomology group varies.

2. Operators on $\Omega^*(M)$ and harmonic cohomology groups

For a $2m$-dimensional symplectic manifold (M,ω) let $\mathbf{G}$ be a skew-symmetric bivector field dual to ω. Then $\mathbf{G}$ is a Poisson structure on M. By Darboux's theorem, in canonical coordinates $\{p_1, q_1, \cdots, p_m, q_m\}$, we can write the symplectic structure ω as

$$\omega = dp_1 \wedge dq_1 + \ldots + dp_m \wedge dq_m$$

and the Poisson structure $\mathbf{G}$ as

$$\mathbf{G} = \frac{\partial}{\partial q_1} \wedge \frac{\partial}{\partial p_1} + \ldots + \frac{\partial}{\partial q_m} \wedge \frac{\partial}{\partial p_m}.$$

Brylinski [B] defined the star operator

$$* : \Omega^k(M) \to \Omega^{2m-k}(M)$$

by requiring $\alpha \wedge * \beta = (\wedge^k(\mathbf{G}))(\alpha, \beta) v_M$ for k-forms α, β, where $v_M = \omega^m/m!$. We denote by $L_\omega = L : \Omega^k(M) \to \Omega^{k+2}(M)$ the operator defined by $L(\alpha) = \alpha \wedge \omega$. The star operator $*$ satisfies the identities

$$*^2 = id, \quad d^* = (-1)^k * d*,$$

that is, the Koszul differential d^* is the symplectic codifferential of the exterior differential d with respect to the star operator $*$.

Yan [Yn] introduced the operator $L^* : \Omega^k(M) \to \Omega^{k-2}(M)$ as the symplectic adjoint operator of L, that is, $L^* = *L*$. Then we see that $i(\mathbf{G}) = -L^*$. Furthermore, the operators L, d, d^* and L^* acting on the algebra $\Omega^*(M)$ satisfy the following commutator relations:

$$[L,\, d] = 0,\ [L,\, d^*] = d,\ [L^*,\, d^*] = 0 \ \text{ and } \ [L^*,\, d] = d^*. \tag{2.1}$$

Note that for a symplectic manifold, we have $dd^* + d^*d \equiv 0$ on $\Omega^*(M)$. Now, the operators L and L^* acting on the algebra $\Omega^*(M)$ also satisfy the relation $[L^*,\, L]\alpha = (m-k)\alpha$ for all k-forms α (see Yan [Yn]). We define an operator A on $\Omega^*(M)$ by

$$A = \sum_{k=0}^{2m} (m-k)\pi_k$$

where $\pi_k : \Omega^*(M) \to \Omega^k(M)$ is the projection. These operators satisfy the following relations:

$$[L^*,\, L] = A, \quad [A,\, L] = -2L, \quad [A,\, L^*] = 2L^*. \tag{2.2}$$

We consider the Lie algebra $\mathfrak{sl}(2)$ with the standard basis $\{H, X, Y\}$:

$$[X,\, Y] = H, \quad [H,\, Y] = -2Y, \quad [H,\, X] = 2X. \tag{2.3}$$

DEFINITION 2.1. Let V be a vector space (possibly infinite dimensional) and assume that the Lie algebra $\mathfrak{sl}(2)$ acts on V. We say that V is an $\mathfrak{sl}(2)$-module of finite H-spectrum if the following two conditions are satisfied:

(1) V can be decomposed as a direct sum of eigenspaces of H.

(2) H has only finitely many distinct eigenvalues.

Let V_k be the eigenspace of H with respect to the eigenvalue k. Then we have the following Lemma.

LEMMA 2.2 ([Yn]). *Let V be an $\mathfrak{sl}(2)$-module of finite H-spectrum. Then the linear mappings*

$$Y^k : V_k \to V_{-k} \quad \textit{and} \quad X^k : V_{-k} \to V_k$$

are isomorphisms. Furthermore, let

$$P_k = \{v \in V_k \mid Xv = 0\} = \{v \in V_k \mid Y^{k+1}v = 0\} \tag{2.4}$$

be the set of primitive elements in V_k. Then we have the following decompositions:

$$V_k = P_k \oplus \mathrm{Im}(Y), \quad V_{-k} = Y^k P_k \oplus Y^{k+1} P_{k+2} \oplus \cdots . \tag{2.5}$$

We have a representation of $\mathfrak{sl}(2)$ on $\Omega^*(M)$ by defining

$$X \longrightarrow *L*, \qquad Y \longrightarrow L, \qquad H \longrightarrow A.$$

It is easily seen that $\Omega^*(M)$ is an $\mathfrak{sl}(2)$-module of finite H-spectrum. Thus:

PROPOSITION 2.3 ([Yn]). (Duality on forms) *The linear mapping $L^k : \Omega^{m-k}(M) \to \Omega^{m+k}(M)$ is an isomorphism.*

PROPOSITION 2.4 ([Yn]). (Duality on harmonic forms) *The linear mapping $L^k : \mathcal{H}^{m-k}(M) \to \mathcal{H}^{m+k}(M)$ is an isomorphism. In particular, we have $H^{m+k}_{\omega\text{-}hr} = \mathrm{Im}\{L^k : H^{m-k}_{\omega\text{-}hr} \to H^{m+k}_{DR}(M)\}$.*

We remark that Yan [Yn] gave an alternative proof of Mathieu's Theorem in the Introduction by using Lemma 2.2 and Proposition 2.4. Note also that we have $H^i_{\omega\text{-}hr}(M) = H^i_{DR}(M)$ for $i = 0, 1, 2$. Thus we have the following consequence of Proposition 2.4.

COROLLARY 2.5. *We have*

$$H^{2m-1}_{\omega\text{-}hr} = \mathrm{Im}\{L^{m-1} : H^1_{DR}(M) \to H^{2m-1}_{DR}(M)\} \tag{2.6}$$

and

$$H^{2m-2}_{\omega\text{-}hr} = \mathrm{Im}\{L^{m-2} : H^2_{DR}(M) \to H^{2m-2}_{DR}(M)\}. \tag{2.7}$$

Using Corollary 2.5, Ibáñez, Rudyak, Tralle and Ugarte [IRTU1] computed the harmonic cohomology groups $H^4_{\omega\text{-}hr}(M)$ and $H^5_{\omega\text{-}hr}(M)$ for 6-dimensional compact symplectic nilmanifolds M.

The following observation due to Yan is the key to proving that the subspace $H^k_{\omega\text{-}hr}(M)$ of $H^k_{DR}(M)$ does not depend on the choice of symplectic form ω of the de Rham class $[\omega]$.

LEMMA 2.6 ([Yn]). *Let (M, ω) be a symplectic manifold. Put $P^{m-k}_\omega(M) = \{v \in H^{m-k}_{DR}(M) \mid L^{k+1}_{[\omega]} v = 0\}$. Then $P^{m-k}_\omega(M) \subset H^{m-k}_{\omega\text{-}hr}(M)$.*

From Lemma 2.6, Lemma 2.2 and Proposition 2.4 we have:

PROPOSITION 2.7 ([Ym]). *Let M be a compact manifold and ω, ω' symplectic forms on M such that $\omega - \omega' = d\gamma$. Then we have $H^q_{\omega\text{-}hr}(M) = H^q_{\omega'\text{-}hr}(M)$ for $q = 0, \dots, 2m$.*

3. Harmonic cohomology groups of nilmanifolds

Now we consider the case of compact nilmanifolds. Let $\mathfrak{g}$ be a Lie algebra and put $\mathfrak{g}^{(0)} = \mathfrak{g}$ and $\mathfrak{g}^{(i+1)} = [\mathfrak{g}, \mathfrak{g}^{(i)}]$ for $i \geq 0$. We say that a Lie algebra $\mathfrak{g}$ is $(r+1)$-step nilpotent if $\mathfrak{g}^{(r)} \neq (0)$ and $\mathfrak{g}^{(r+1)} = (0)$. We say a Lie group G is $(r+1)$-step nilpotent if the Lie algebra $\mathfrak{g}$ is $(r+1)$-step nilpotent. If G is a simply connected $(r+1)$-step nilpotent Lie group and Γ is a lattice of G, that is, a discrete subgroup of G such that G/Γ is compact, then we say G/Γ is an $(r+1)$-step compact nilmanifold. Note that, by Malcev's theorem, G admits a lattice if and only if $\mathfrak{g}$ admits a basis such that all the structural constants are rational. We denote by $\bigwedge^k \mathfrak{g}^*$ the space of all left G-invariant k-forms on G and regard it as a subspace of $\Omega^k(G/\Gamma)$. Then we have a subcomplex $(\bigwedge^* \mathfrak{g}^*, d)$ of the de Rham complex $(\Omega^*(G/\Gamma), d)$ of the compact nilmanifold G/Γ and denote by $H^k(\mathfrak{g})$ the k-th cohomology group of the complex $(\bigwedge^* \mathfrak{g}^*, d)$.

THEOREM 3.1 (Nomizu [N]). *For a compact nilmanifold G/Γ, the natural inclusion $\iota : (\bigwedge^* \mathfrak{g}^*, d) \to (\Omega^*(G/\Gamma), d)$ induces an isomorphism $\iota^* : H^k(\mathfrak{g}) \cong H^k_{DR}(G/\Gamma)$ on cohomology groups.*

For a symplectic form ω on a compact nilmanifold G/Γ, there exists a G-invariant closed 2-form ω_0 on G/Γ such that $\omega - \omega_0 = d\gamma$. Note that ω_0 is also a symplectic form on G/Γ. For a G-invariant symplectic form ω_0, we denote by $\mathcal{H}^k(\mathfrak{g})$ the space of all G-invariant harmonic k-forms on G/Γ and we put

$$H^k_{\omega_0\text{-}hr}(\mathfrak{g}) = \mathcal{H}^k(\mathfrak{g})/(B^k(\mathfrak{g}) \cap \mathcal{H}^k(\mathfrak{g})).$$

PROPOSITION 3.2 ([Ym]). *Let $(G/\Gamma, \omega)$ be a compact symplectic nilmanifold and let ω_0 be a G-invariant symplectic form such that $\omega - \omega_0 = d\gamma$ as above. Then we have*

$$H^k_{\omega\text{-}hr}(G/\Gamma) = H^k_{\omega_0\text{-}hr}(G/\Gamma) = H^k_{\omega_0\text{-}hr}(\mathfrak{g}) \tag{3.1}$$

for any k.

We consider the relation between $\dim H^1_{DR}(G/\Gamma)$ and $\dim H^{2m-1}_{\omega\text{-}hr}(G/\Gamma)$ for a compact nilmanifold G/Γ.

THEOREM 3.3 ([Ym]). *Let $(G/\Gamma, \omega)$ be an $(r+1)$-step compact symplectic nilmanifold. Then we have*

$$\dim H^1_{DR}(M) - \dim H^{2m-1}_{\omega\text{-}hr} \geq \dim \mathfrak{g}^{(r)}.$$

In particular, if M is a 2-step nilmanifold,

$$\dim H^1_{DR}(M) - \dim H^{2m-1}_{\omega\text{-}hr} = \dim[\mathfrak{g}, \mathfrak{g}]. \tag{3.2}$$

Let $\mathfrak{g}$ be an $(r+1)$-step nilpotent Lie algebra. Consider the descending central series $\{\mathfrak{g}^{(i)}\}$ of $\mathfrak{g}$, where $\mathfrak{g}^{(i+1)} = [\mathfrak{g}, \mathfrak{g}^{(i)}]$ and $\mathfrak{g}^{(0)} = \mathfrak{g}$. Let $\mathfrak{a}^{(i)}$ denote the complementary vector subspace to $\mathfrak{g}^{(i+1)}$ in $\mathfrak{g}^{(i)}$: $\mathfrak{g}^{(i)} = \mathfrak{g}^{(i+1)} + \mathfrak{a}^{(i)}$ (for $i = 0, 1, \ldots, r-1$). Let $n_i = \dim \mathfrak{a}^{(i)}$. For simplicity, put $\bigwedge^{i_0, \ldots, i_r} = \bigwedge^{i_0} \mathfrak{a}^{(0)*} \wedge \ldots \wedge \bigwedge^{i_r} \mathfrak{a}^{(r)*}$. Then we have

$$\bigwedge^s \mathfrak{g}^* = \sum_{i_0 + \ldots + i_r = s} \bigwedge^{i_0, \ldots, i_r}.$$

The following is due to Benson and Gordon [BG].

LEMMA 3.4 ([BG]). *Each closed 2-form $\theta \in \bigwedge^2 \mathfrak{g}^*$ belongs to*

$$\bigwedge^{1,0,\dots,0,1} + \sum \bigwedge^{i_0,\dots,i_{r-1},0}.$$

Let $\{\lambda_1, \dots, \lambda_{n_r}\}$ be a basis of $\bigwedge^{0,\dots,0,1}$. By Lemma 3.4, a G-invariant symplectic form ω can be written in the form

$$\omega = \beta_1 \wedge \lambda_1 + \dots + \beta_{n_r} \wedge \lambda_{n_r} \quad \text{modulo} \quad \sum \bigwedge^{i_0,\dots,i_{r-1},0}, \tag{3.3}$$

where $\beta_1, \dots, \beta_{n_r}$ are elements of $\bigwedge^{1,0,\dots,0}$. Since ω is non-degenerate, we see that $\beta_1, \dots, \beta_{n_r}$ are linearly independent. We extend these elements to a basis $\{\beta_1, \dots, \beta_{n_r}, \dots, \beta_{n_0}\}$ for $\bigwedge^{1,0,\dots,0}$. Put $\mathfrak{a}_1^{(0)^*} = \mathrm{span}\{\beta_1, \dots, \beta_{n_r}\}$ and $\mathfrak{a}_2^{(0)^*} = \mathrm{span}\{\beta_{n_r+1}, \dots, \beta_{n_0}\}$. Then $\mathfrak{a}^{(0)^*} = \mathfrak{a}_1^{(0)^*} + \mathfrak{a}_2^{(0)^*}$. Put $n_0^1 = \dim \mathfrak{a}_1^{(0)^*} = n_r$ and $n_0^2 = \dim \mathfrak{a}_2^{(0)^*} = n_0 - n_r$. For simplicity, we also put $\bigwedge^{(i_0^1,i_0^2),i_1,\dots,i_r} = \bigwedge^{i_0^1} \mathfrak{a}_1^{(0)^*} \wedge \bigwedge^{i_0^2} \mathfrak{a}_2^{(0)^*} \wedge \bigwedge^{i_1} \mathfrak{a}^{(1)^*} \wedge \dots \wedge \bigwedge^{i_r} \mathfrak{a}^{(r)^*}$. Moreover, let $\{\beta_1^{(k)}, \dots, \beta_{n_k}^{(k)}\}$ be a basis of $\mathfrak{a}^{(k)^*}$ and put

$$\{\omega_1, \dots, \omega_{2m}\} = \{\beta_1, \dots, \beta_{n_0}, \dots, \beta_1^{(k)}, \dots, \beta_{n_k}^{(k)}, \dots, \lambda_1, \dots, \lambda_{n_r}\}.$$

Let $\{X_1, \dots, X_{2m}\}$ be a basis of $\mathfrak{g}$ which is dual to the basis $\{\omega_1, \dots, \omega_{2m}\}$. If we write a symplectic form ω as

$$\omega = \sum a_{ij} \omega_i \wedge \omega_j \quad a_{ij} = -a_{ji} \in \mathbb{R},$$

then it is easy to see that the Poisson structure $\mathbf{G}$ dual to ω is given by

$$\mathbf{G} = -\sum c_{ij} X_i \wedge X_j, \tag{3.4}$$

where (c_{ij}) is the inverse of transpose matrix of (a_{ij}). Thus we have the following.

LEMMA 3.5. *With respect to the basis $\{X_1, \dots, X_{2m}\}$ above, $\mathbf{G}$ is of the form*

$$(c_{ij}) = \left(\begin{array}{c|c|c|c} 0_{n_r,n_r} & 0_{n_r,n_0-n_r} & 0_{n_r,2m-n_0-n_r} & E_{n_r} \\ \hline 0_{n_0-n_r,n_r} & * & * & * \\ \hline 0_{2m-n_0-n_r,n_r} & * & * & * \\ \hline -E_{n_r} & * & * & * \end{array}\right). \tag{3.5}$$

LEMMA 3.6. *Let $\omega = \beta_1 \wedge \lambda_1 + \dots + \beta_{n_1} \wedge \lambda_{n_1}$ modulo $\bigwedge^{2,0}$ be a G-invariant symplectic form on 2-step nilmanifold G/Γ, then we have*

$$d\lambda_k = \sum_{i<j\le n_1} b_{ij}^k \beta_i \wedge \beta_j + \sum_{i \le n_1 < j} b_{ij}^k \beta_i \wedge \beta_j \in \bigwedge^{(2,0),0} + \bigwedge^{(1,1),0}.$$

COROLLARY 3.7 ([SY]). *If $(G/\Gamma, \omega)$ is a 2-step symplectic nilmanifold, then we have $B^3(\mathfrak{g}) \subset \mathcal{H}^3(\mathfrak{g})$.*

PROOF. Since $d\bigwedge^{2,0} = 0$ and $d\bigwedge^{1,1} \subset \bigwedge^{3,0}$, we only consider the case of $\bigwedge^{0,2}$. From Lemma 3.6, we see that

$$\bigwedge^{0,2} \xrightarrow{d} \bigwedge^{(2,0),1} + \bigwedge^{(1,1),1}.$$

Thus, from (3.5) in Lemma 3.5, we obtain

$$i(\mathbf{G})(d\bigwedge^{0,2}) \subset \bigwedge^{1,0} \xrightarrow{d} 0,$$

which implies that $d\bigwedge^{0,2} \subset \mathcal{H}^3(\mathfrak{g})$. □

By a straightforward computation, we see that

$$\begin{aligned}(3.6)\quad &*(\omega_{i_1}\wedge\ldots\wedge\omega_{i_r})\\ &=\sum\nolimits_{j_1<\ldots<j_r}(-1)^r a\cdot\det(c^k_h)^{k=j_1,\ldots,j_r}_{h=i_1,\ldots,i_r}\omega_1\wedge\ldots\wedge\hat{\omega}_{j_1}\wedge\ldots\wedge\hat{\omega}_{j_r}\wedge\ldots\wedge\omega_{2m},\end{aligned}$$

where $\omega^m/m! = a\cdot(\omega_{j_1}\wedge\ldots\wedge\omega_{j_r})\wedge(\omega_1\wedge\ldots\wedge\hat{\omega}_{j_1}\wedge\ldots\wedge\hat{\omega}_{j_r}\wedge\ldots\wedge\omega_{2m})$.

PROPOSITION 3.8 ([SY]). *If G/Γ is a 2-step nilmanifold, then*

$$\bigwedge\nolimits^{(n_0^1,n_0^2-p),n_1-q}\subset\mathcal{H}^{2m-p-q}(\mathfrak{g}). \tag{3.7}$$

In particular,

$$\bigwedge\nolimits^{(n_0^1,n_0^2-1),n_1}+\bigwedge\nolimits^{(n_0^1,n_0^2),n_1-1}=\mathcal{H}^{2m-1}(\mathfrak{g}),$$

which implies $\dim H^{2m-1}_{hr}(\mathfrak{g}) = n_0^2 = \dim\mathfrak{g} - 2\dim[\mathfrak{g},\mathfrak{g}]$.

PROOF. From Lemma 3.6, it is obvious that $\bigwedge^{(n_0^1,n_0^2-p),n_1-q}\subset Z^{2m-p-q}(\mathfrak{g})$. Since $d^* = (-1)^k * d*$ on $\bigwedge^k(\mathfrak{g}^*)$, to see (3.7) it is enough to prove

$$*:\bigwedge\nolimits^{(n_0^1,n_0^2-p),n_1-q}\longrightarrow\bigwedge\nolimits^{p+q,0}.$$

Note that

$$\begin{aligned}&*(\omega_{i_1}\wedge\ldots\wedge\omega_{i_s})\\ &=\sum\nolimits_{j_1<\ldots<j_s}(-1)^s a\cdot\det(c^k_h)^{k=j_1,\ldots,j_s}_{h=i_1,\ldots,i_s}\omega_1\wedge\ldots\wedge\hat{\omega}_{j_1}\wedge\ldots\wedge\hat{\omega}_{j_s}\wedge\ldots\wedge\omega_{2m},\end{aligned}$$

where $s = n_0^1+n_0^2+n_1-p-q$. Thus if $\{j_1,\ldots,j_s\}\not\ni n_0+j$, then we see $\det(c^k_h)^{k=j_1,\ldots,j_s}_{h=i_1,\ldots,i_s}=0$ from Lemma 3.5.

In fact, noting that $n_0^1 = n_1$, we have

$$\det(c^k_h)^{k=j_1,\ldots,j_s}_{h=i_1,\ldots,i_s} = \begin{vmatrix} c^{j_1}_1 & \ldots & c^{j_1}_{n_0^1} & \ldots & c^{j_1}_{i_s}\\ \vdots & & \vdots & & \vdots\\ c^{j_s}_1 & \ldots & c^{j_s}_{n_0^1} & \ldots & c^{j_s}_{i_s}\end{vmatrix}$$

$$= \begin{vmatrix} c^{j_1}_1 & \ldots & \overset{j}{0} & \ldots & c^{j_1}_{n_0^1} & \ldots & c^{j_1}_{i_s}\\ \vdots & & \vdots & & \vdots & & \vdots\\ c^{j_s}_1 & \ldots & 0 & \ldots & c^{j_s}_{n_0^1} & \ldots & c^{j_s}_{i_s}\end{vmatrix} = 0.$$

Thus we see that if $\det(c^k_h)^{k=j_1,\ldots,j_s}_{h=i_1,\ldots,i_s}\neq 0$, then $\{j_1,\ldots,j_s\}\supset\{n_0+1,\ldots,n_0+n_1\}$. Therefore we have $*(\bigwedge^{(n_0^1,n_0^2-p),n_1-q})\subset\bigwedge^{p+q,0}$.

In particular, we see that $\bigwedge^{(n_0^1,n_0^2-1),n_1}+\bigwedge^{(n_0^1,n_0^2),n_1-1}\subset\mathcal{H}^{2m-1}(\mathfrak{g})$. Since $L^{m-1}:\mathcal{H}^1(\mathfrak{g})\to\mathcal{H}^{2m-1}(\mathfrak{g})$ is an isomorphism by Propositon 2.4, we see that $\dim\mathcal{H}^{2m-1}(\mathfrak{g}) = n_0$. On the other hand, $\dim(\bigwedge^{(n_0^1,n_0^2-1),n_1}+\bigwedge^{(n_0^1,n_0^2),n_1-1}) = (n_0-n_1)+n_1 = \dim\mathcal{H}^{2m-1}(\mathfrak{g})$. Thus we have established our second claim. □

COROLLARY 3.9 ([SY]). *If G/Γ is a 2-step nilmanifold, then*

$$\dim H^2_{DR}(G/\Gamma) - \dim H^{2m-2}_{\omega\text{-}hr}(G/\Gamma) = \tag{3.8}$$
$$\dim (d{\bigwedge}^{n_0-3,n_1} \cap \mathcal{H}^{2m-2}(\mathfrak{g})) + \dim d{\bigwedge}^{n_0-2,n_1-1} - n_1.$$

PROOF. Since $d{\bigwedge}^{n_0-2,n_1-1} \subset {\bigwedge}^{n_0,n_1-2}$, elements of $d{\bigwedge}^{n_0-2,n_1-1}$ are harmonic. Thus we see that $\dim H^2_{\omega\text{-}hr}(\mathfrak{g}) - \dim H^{2m-2}_{\omega\text{-}hr}(\mathfrak{g})$

$= \dim \mathcal{H}^2(\mathfrak{g}) - \dim(B^2(\mathfrak{g}) \cap \mathcal{H}^2(\mathfrak{g})) - \dim \mathcal{H}^{2m-2}(\mathfrak{g}) + \dim(B^{2m-2}(\mathfrak{g}) \cap \mathcal{H}^{2m-2}(\mathfrak{g}))$

$= \dim(B^{2m-2}(\mathfrak{g}) \cap \mathcal{H}^{2m-2}(\mathfrak{g})) - \dim B^2(\mathfrak{g})$

$= \dim(d{\bigwedge}^{n_0-3,n_1} \cap \mathcal{H}^{2m-2}(\mathfrak{g})) + \dim d{\bigwedge}^{n_0-2,n_1-1} - \dim B^2(\mathfrak{g})$

$= \dim(d{\bigwedge}^{n_0-3,n_1} \cap \mathcal{H}^{2m-2}(\mathfrak{g})) + \dim d{\bigwedge}^{n_0-2,n_1-1} - n_1.$ □

4. Examples

We give examples of compact 2-step symplectic nilmanifolds G/Γ with varying dimensions of harmonic cohomology groups.

EXAMPLE 1. For $p \geq 2$ let $\mathfrak{h}(1,p)$ be a 2-step nilpotent Lie algebra of dimension $2p+1$ spanned by $\{X_1, \dots, X_{2p+1}\}$ with $[X_1, X_i] = X_{p+i}$ for $i = 2, \dots, p$. We consider the Lie algebra $\mathfrak{g} = \mathfrak{h}(1,p) \oplus \mathbb{R}$ of dimension $2p+2$. Let X_{2p+2} denote a generator of the Lie algebra $\mathbb{R}$ and let $\{\omega_1, \dots, \omega_{2p+1}, \omega_{2p+2}\}$ be the dual basis of $\{X_1, \dots, X_{2p+1}, X_{2p+2}\}$. Then we have

$${\bigwedge}^{(0,1),0} \supset \operatorname{span}\{\omega_{2p+2}\}, \quad {\bigwedge}^{(0,0),1} = \operatorname{span}\{\omega_{p+2}, \dots, \omega_{2p+1}\}.$$

Consider a G-invariant symplectic structure ω. We write the Poisson structure $\mathbf{G}$ as $\mathbf{G} = -\sum c_{ij} X_i \wedge X_j$ with respect to the basis $\{X_1, \dots, X_{2p+1}, X_{2p+2}\}$ above. Then we have the following result.

PROPOSITION 4.1 ([SY]). *For a compact symplectic nilmanifold G/Γ with Lie algebra $\mathfrak{g}$, we have*

$$\dim H^2_{DR}(\mathfrak{g}) - \dim H^{2p}_{\omega\text{-}hr}(\mathfrak{g}) = \begin{cases} {}_pC_2 & \text{if } c_{12p+2} \neq 0, \\ {}_pC_2 + (p-1) & \text{if } c_{12p+2} = 0. \end{cases}$$

To prove Proposition 4.1, we use the relation (3.8) of Corollary 3.9, the expression of the Poisson structure $\mathbf{G}$ and the operator $d^* = [d, i(\mathbf{G})]$.

EXAMPLE 2 ([SY]). We consider a 2-step nilpotent Lie algebra of dimension $2m$ given by $\mathfrak{g} = \operatorname{span}\{X_1, \dots, X_m, Y_1, \dots, Y_m\}$ with $[X_i, X_{i+1}] = Y_i \quad i = 1, \dots, m$, where we set $X_{m+1} = X_1$. Then a simply connected nilpotent Lie group G with Lie algebra $\mathfrak{g}$ has a lattice Γ. Let $\{\beta_1, \dots, \beta_m, \lambda_1, \dots, \lambda_m\}$ be the dual basis of $\{X_1, \dots, X_m, Y_1, \dots, Y_m\}$. Note that $d{\bigwedge}^{k,0} = (0)$ and $d{\bigwedge}^{k,j} \subset {\bigwedge}^{k+2,j-1}$. It is easy to see that $d : {\bigwedge}^{0,j} \to {\bigwedge}^{2,j-1}$ is injective.

The space of invariant closed 2-forms $Z^2(\mathfrak{g})$ is given by

$$Z^2(\mathfrak{g}) = \Big\{\sum_{i=1}^m a_{ii+1}(\beta_i \wedge \lambda_{i+1} - \beta_{i+2}\wedge\lambda_i) + \sum_{i=1}^m a_{ii}\,\beta_i\wedge\lambda_i + \sum_{i=1}^m a_{i+1i}\,\beta_{i+1}\wedge\lambda_i \;\Big|\; a_{ii+1}, a_{ii}, a_{i+1i} \in \mathbb{R}\Big\} + \bigwedge^{2,0},$$

where we put $\beta_{m+1} = \beta_1, \beta_{m+2} = \beta_2, \lambda_{m+1} = \lambda_1$.

Now we consider a symplectic form ω given by an element of $\bigwedge^{1,1}$. Then the Poisson structure $\mathbf{G}$ is of the form $\mathbf{G} = -\sum_{i,j=1}^m c_{ij}\, X_i \wedge Y_j$ with respect to the basis $\{X_1, \cdots, X_m, Y_1, \cdots, Y_m\}$. We see that

$$i(\mathbf{G}) : \bigwedge^{3,0} \to (0), \quad i(\mathbf{G}) : \bigwedge^{2,1} \to \bigwedge^{1,0}, \quad i(\mathbf{G}) : \bigwedge^{1,2} \to \bigwedge^{0,1}$$

and the space of harmonic 3-forms $\mathcal{H}^3(\mathfrak{g})$ is given by

$$\mathcal{H}^3(\mathfrak{g}) = \bigwedge^{3,0} + Z^3(\mathfrak{g}) \cap \bigwedge^{2,1} + \mathcal{H}^3(\mathfrak{g}) \cap \bigwedge^{1,2}.$$

For $m \geq 6$, we see that

$$Z^3(\mathfrak{g}) \cap \bigwedge^{1,2} = \Big\{\sum_{j=1}^m b_j\, \beta_{j+1}\wedge\lambda_j\wedge\lambda_{j+1} \mid b_j \in \mathbb{R},\ j = 1, \cdots, m\Big\}$$

where we we put $\beta_{m+1} = \beta_1, \lambda_{m+1} = \lambda_1$.

Case 1. The case when $\omega = \sum_{j=1}^m a_{jj}\,\beta_j\wedge\lambda_j$ with $a_{jj} \neq 0$.

Note that $\mathbf{G} = -\sum_{j=1}^m \dfrac{1}{a_{jj}} X_j \wedge Y_j$ and hence we have $\mathcal{H}^3(\mathfrak{g}) \cap \bigwedge^{1,2} = (0)$ for $m \geq 6$.

Case 2. The case when $\omega = \sum_{j=1}^{m-1} a_{jj+1}(\beta_j \wedge \lambda_{j+1} - \beta_{j+2}\wedge\lambda_j)$.

Note that ω is given by a form $\begin{pmatrix} 0 & A \\ -{}^tA & 0 \end{pmatrix}$ for the basis above, where

$$A = \begin{pmatrix}
0 & a_{12} & 0 & 0 & 0 & \cdots & 0 & -a_{m-1\,m} & 0 \\
0 & 0 & a_{23} & 0 & 0 & \cdots & 0 & 0 & 0 \\
-a_{12} & 0 & 0 & a_{34} & 0 & \cdots & 0 & 0 & 0 \\
0 & -a_{23} & 0 & 0 & a_{45} & \cdots & 0 & 0 & 0 \\
0 & 0 & -a_{34} & 0 & 0 & \cdots & 0 & 0 & 0 \\
0 & 0 & 0 & -a_{45} & 0 & \cdots & 0 & 0 & 0 \\
\cdots & \cdots & \cdots & \cdots & \cdots & \cdots & \cdots & \cdots & \cdots \\
\cdots & \cdots & \cdots & \cdots & \cdots & \cdots & \cdots & \cdots & \cdots \\
0 & 0 & 0 & 0 & 0 & \cdots & a_{m-3\,m-2} & 0 & 0 \\
0 & 0 & 0 & 0 & 0 & \cdots & 0 & a_{m-2\,m-1} & 0 \\
0 & 0 & 0 & 0 & 0 & \cdots & 0 & 0 & a_{m-1\,m} \\
0 & 0 & 0 & 0 & 0 & \cdots & -a_{m-2\,m-1} & 0 & 0
\end{pmatrix}.$$

We can see that, for $m \geq 3$, the matrix A is non-degenerate if and only if $m = 3\ell$ for $\ell \in \mathbb{N}$. Put $A^{-1} = (c_{ij})$ for $m = 3\ell$. Then we see that the components of the matrix $A^{-1} = (c_{ij})$ satisfy the conditions

$$c_{jj} = 0 \text{ for } j = 1, \cdots, m, \quad c_{j+1j} = 0 \text{ for } j = 1, \cdots, m-1 \text{ and } c_{1m} = 0.$$

Since the poisson structure $\mathbf{G}$ is given by $\begin{pmatrix} 0 & -A^{-1} \\ {}^tA^{-1} & 0 \end{pmatrix}$, we see that, for $\alpha = \sum_{j=1}^m b_j\,\beta_{j+1}\wedge\lambda_j\wedge\lambda_{j+1}$, $i(\mathbf{G})\alpha = 0$, and hence α is harmonic. This implies that

$\mathcal{H}^3(\mathfrak{g}) \cap \bigwedge^{1,2} = Z^3(\mathfrak{g}) \cap \bigwedge^{1,2}$ and $\dim(\mathcal{H}^3(\mathfrak{g}) \cap \bigwedge^{1,2}) = 3\ell$ for $\ell \geq 2$. Thus, from Corollary 3.7, we see that the compact 2-step nilmanifolds G/Γ admit symplectic structures such that the dimension of the harmonic cohomology group $H^3_{\omega\text{-}hr}(G/\Gamma)$ varies.

References

[B] J.L. Brylinski, *A differential complex for Poisson Manifolds*, J. Differential Geom. **28** (1988), 93–114.

[BG] C. Benson and C. Gordon, *Kähler and Symplectic structures on nilmanifolds*, Topology **27** (1988), 513–518.

[IRTU1] R. Ibáñez, Yu. Rudak, A. Tralle, L. Ugarte, *On symplectically harmonic forms on six-dimensional nilmanifolds*, Comment. Math. Helv. **76** (2001), 89–109.

[IRTU2] —, *Symplectically harmonic cohomology of nilmanifolds*, arXiv:math.SG/0111074 7 Nov 2001.

[K] J.-L. Koszul, *Crochet de Schouten-Nijenhuis et cohomologie, in Élie Cartan et les mathématiques d'aujourd'hui*, Soc. Math. de France, Astérisque, hors serie (1985), 257–271.

[M] O. Mathieu, *Harmonic cohomology classes of symplectic manifolds*, Comment. Math. Helv. **70** (1995), 1–9.

[N] K. Nomizu, *On the cohomology of compact homogeneous spaces of nilpotent Lie groups*, Ann. of Math., **59** (1954), 531–538.

[SY] Y. Sakane and T. Yamada, *Harmonic forms on compact symplectic 2-step nilmanifolds*, preprint.

[TO] A. Tralle and J. Oprea, *Symplectic Manifolds with no Kähler Structure*, Lecture Notes in Mathematics 1661, Springer-Verlag, Berlin, 1997.

[YM] T. Yamada, *Harmonic cohomology groups on compact symplectic nilmanifolds*, to appear in Osaka Math. J.

[YN] D. Yan, *Hodge structure on symplectic manifolds*, Adv. in Math. **120** (1996), 143–154.

Department of Mathematics, Osaka University, Toyonaka, Osaka 560-0043, JAPAN

E-mail address: sakane@math.sci.osaka-u.ac.jp

Contemporary Mathematics
Volume **308**, 2002

Bonnet Pairs in the 3-Sphere

Boris A. Springborn

1. Introduction

Two non-congruent surfaces that are isometric and have the same mean curvature at corresponding points are called a *Bonnet pair of surfaces* or simply a *Bonnet pair.* The interest in such pairs arises from the following considerations: One of the fundamental problems of surface theory is to find invariants which characterize surfaces geometrically. The Bonnet theorem states that a surface is determined up to congruence by its first and second fundamental forms. However, this description has some redundancy since the first and second fundamental forms must satisfy the Gauss and Codazzi equations. So it may be that some pieces of information can be eliminated from this description, but which? It seems wise to retain the metric. But given the metric, we know the Gaussian curvature, i.e. the product of the two principal curvatures. The second fundamental form contains the following extra pieces of information: the mean curvature, i.e. half the sum of the two principal curvatures, and the principal curvature directions. It seems worthwhile to investigate what happens if we do not prescribe the curvature directions and ask the question: Is a surface already determined by its metric and mean curvature?

Bonnet himself [**6**] gave the following answer: In general, a surface is determined up to congruence by its metric and mean curvature. There are only three exceptions:

- *Constant mean curvature surfaces* can be continuously deformed while keeping their metric and mean curvature fixed.
- If a surface is not determined by its metric and mean curvature and the mean curvature is not constant, then there is either a one-parameter familiy of non-congruent isometric surfaces with the same mean curvature (*Bonnet surfaces*), or
- there is exactly one non-congruent isometric surface with the same mean curvature.

Constant mean curvature surfaces form their own field of study, and we will not go in this direction at all.

2000 *Mathematics Subject Classification.* Primary 53A05, 53C40.

Bonnet surfaces were studied by many mathematicians; an incomplete list includes Hazzidakis [**10**], Graustein [**9**], Cartan [**7**], Chern [**8**], Kenmotsu [**13**], Roussos [**16**], Bobenko and Eitner [**4, 5**]. Cartan develops a more or less complete local theory. Bobenko and Eitner describe Bonnet surfaces using Painlevé equations. This makes it possible to give a global classification.

Much less is known about the case in which there are exactly two non-congruent surfaces with the same mean curvature. However, there are trivial examples: helicoidal surfaces [**15**]. Suppose $F(x, y)$ is an immersion in $\mathbf{R}^3$ such that there is a non-compact one parameter subgroup $T(s)$ of the group of rigid motions with $F(x, y+s) = T(s)F(x, y)$. Then $-F(x, -y)$ is by symmetry an isometric immersion with the same mean curvature function. But there is in general no isometry S of $\mathbb{R}^3$, with $-F(x, -y) = SF(x, y)$. Note that if the helicoidal surface degenerates to a surface of rotation, the Bonnet pair degenerates to a single isothermic surface.

There is a relationship between Bonnet pairs in $\mathbb{R}^3$ and isothermic surfaces in S^3 that was discovered by Servant [**17**]. It is due to an equivalence of the Gauss and Codazzi equations specialized to either class of surfaces. This relationship was studied further by Bianchi [**1**] and Jonas [**11**], who consider Darboux transformations of Bonnet pairs. More recently, this relationship was rediscovered by Kamberov, Pedit and Pinkall [**12**], using their quaternionic calculus.

Lawson and Tribuzy [**14**] show among other things that there are no Bonnet pairs that are topologically spheres. It is an open question whether there are any compact Bonnet pairs. Bobenko [**3**] takes some steps towards attacking this problem by treating Bonnet pairs as integrable systems.

In this article, we study Bonnet pairs not in $\mathbf{R}^3$, but in the three dimensional sphere S^3. The author is not aware of any previous work done on Bonnet pairs in S^3. This theory is at first completely analogous to the $\mathbb{R}^3$-theory. However, it turns out that there are trivial compact examples. We present a duality transformation by which a Bonnet pair in S^3 is transformed into another such pair.

2. Surfaces in the 3-Sphere and Quaternionic Frames

In this section we outline the use of complex notation and quaternionic frames in the analytic description of conformally immersed surfaces in the 3-sphere. See also [**2**]. Throughout this paper, let D be a simply connected domain in $\mathbb{C}$ with coordinate $z = x + iy$.

Suppose $F : D \to S^3 \subset \mathbb{R}^4$ is a conformal immersion with normal $N : D \to S^3$. Conformality is equivalent to

$$\langle F_z, F_z \rangle = 0, \qquad \langle F_z, F_{\bar z} \rangle = \frac{1}{2}\, e^u, \qquad \langle F_{\bar z}, F_{\bar z} \rangle = 0,$$

for some real valued function u. (We use the Wirtinger operators $\partial_z = \frac{1}{2}(\partial_x - i\partial_y)$ and $\partial_{\bar z} = \frac{1}{2}(\partial_x + i\partial_y)$ and denote partial differentiation by subscripts. The scalar products are complexified bilinearly: $\langle x, y \rangle = \sum_1^n x_k y_k$ for $x, y \in \mathbb{R}^n$ or $\mathbb{C}^n$.)

The first fundamental form is therefore

$$\langle dF,\, dF \rangle = e^u\, dz\, d\bar z,$$

and the second fundamental form is

$$-\langle dF,\, dN\rangle = Q\,dz^2 + e^u H\,dz\,d\bar{z} + \overline{Q}\,d\bar{z}^2,$$

where

$$Q = \langle F_{zz}, N\rangle \quad \text{and} \quad H = 2e^{-u}\langle F_{z\bar{z}}, N\rangle.$$

The $(2,0)$-form $Q\,dz^2$ is the Hopf differential and H is the mean curvature function of F. The zeros of Q are the umbilics of F. If (and only if) F is parametrized by curvature-line coordinates, then Q takes either only real or only purely imaginary values. One obtains the following frame equations:

$$\begin{array}{rclcccccc} F_{zz} & = & & & u_z\,F_z & & & + & Q\,N \\ F_{z\bar{z}} & = & -\frac{1}{2}e^u\,F & & & & & + & \frac{1}{2}e^u H\,N \\ F_{\bar{z}\bar{z}} & = & & & & & u_{\bar{z}}\,F_{\bar{z}} & + & \overline{Q}\,N \\ N_z & = & & - & H\,F_z & - & 2e^{-u}Q\,F_{\bar{z}} & & \\ N_{\bar{z}} & = & & - & 2e^{-u}\overline{Q}\,F_z & - & H\,F_{\bar{z}} & & \end{array} \tag{2.1}$$

Now we introduce quaternionic frames. Identify $\mathbb{R}^4$ with the algebra $\mathbb{H}$ of quaternions, so that the canonical basis of $\mathbb{R}^4$ corresponds to the quaternions $\mathbf{1}, \mathbf{i}, \mathbf{j}, \mathbf{k}$. Further, use the following representation of the quaternions as complex 2×2-matrices:

$$\mathbf{1} = \begin{pmatrix} 1 & 0 \\ 0 & 1 \end{pmatrix},\ \mathbf{i} = \begin{pmatrix} 0 & -i \\ -i & 0 \end{pmatrix},\ \mathbf{j} = \begin{pmatrix} 0 & -1 \\ 1 & 0 \end{pmatrix},\ \mathbf{k} = \begin{pmatrix} -i & 0 \\ 0 & i \end{pmatrix}.$$

Then the special unitary group $SU(2)$ is just the unit sphere in $\mathbb{H}$. For $(\Phi_1, \Phi_2) \in SU(2) \times SU(2)$, the linear map $\mathbb{H} \to \mathbb{H}$, $X \mapsto \Phi_2^{-1} X \Phi_1$ is orientation preserving and orthogonal. One obtains thus a right action of $SU(2) \times SU(2)$ on $\mathbb{R}^4$. It defines a two-to-one Lie-group anti-homomorphism onto $SO(4)$, the kernel being $\{(\mathbf{1}, \mathbf{1}), (-\mathbf{1}, -\mathbf{1})\}$.

By a *quaternionic frame* of F we mean a map

$$(\Phi_1, \Phi_2) : D \to SU(2) \times SU(2)$$

with

$$\begin{array}{rcll} F & = & & {\Phi_2}^{-1}\mathbf{1}\Phi_1, \\ F_x & = & e^{u/2} & {\Phi_2}^{-1}\,\mathbf{i}\,\Phi_1, \\ F_y & = & e^{u/2} & {\Phi_2}^{-1}\,\mathbf{j}\,\Phi_1, \\ N & = & & {\Phi_2}^{-1}\mathbf{k}\Phi_1. \end{array}$$

The last equation could have been omitted in this definition, because it follows from the others. Instead of the second and third equation, it is sometimes helpful to use the equivalent

$$\begin{array}{rcl} F_z & = & e^{u/2}\ {\Phi_2}^{-1} \begin{pmatrix} 0 & 0 \\ -i & 0 \end{pmatrix} \Phi_1, \\ F_{\bar{z}} & = & e^{u/2}\ {\Phi_2}^{-1} \begin{pmatrix} 0 & -i \\ 0 & 0 \end{pmatrix} \Phi_1. \end{array}$$

This setup leads to the fundamental theorem of surface theory in the following guise.

PROPOSITION 1. *The metric $e^u\,dz\,d\bar{z}$, mean curvature function H, and Hopf differential $Q\,dz^2$ of a conformal immersion into S^3 satisfy the compatibility conditions*

$$\text{(Gauss)}\qquad u_{z\bar{z}} + \frac{1}{2}e^u(H^2+1) - 2e^{-u}|Q|^2 = 0,$$

$$\text{(Codazzi)}\qquad 2e^{-u/2}Q_{\bar{z}} = e^{u/2}H_z$$

and determine the immersion up to congruence.

Conversely, if the real valued functions u and H and the complex valued function Q satisfy these compatibility conditions, then the systems

$$(2.2)\qquad \begin{aligned}\Phi_{1z} &= U_1\Phi_1\\ \Phi_{1\bar{z}} &= V_1\Phi_1\end{aligned} \quad \text{and} \quad \begin{aligned}\Phi_{2z} &= U_2\Phi_2\\ \Phi_{2\bar{z}} &= V_2\Phi_2\end{aligned}$$

with

$$(2.3)\qquad \begin{aligned} U_1 &= \begin{pmatrix} u_z/4 & -e^{-u/2}Q \\ \frac{1}{2}e^{u/2}(H-i) & -u_z/4 \end{pmatrix}, \quad V_1 = \begin{pmatrix} -u_{\bar{z}}/4 & -\frac{1}{2}e^{u/2}(H+i) \\ e^{-u/2}\overline{Q} & u_{\bar{z}}/4 \end{pmatrix}, \\ U_2 &= \begin{pmatrix} u_z/4 & -e^{-u/2}Q \\ \frac{1}{2}e^{u/2}(H+i) & -u_z/4 \end{pmatrix}, \quad V_2 = \begin{pmatrix} -u_{\bar{z}}/4 & -\frac{1}{2}e^{u/2}(H-i) \\ e^{-u/2}\overline{Q} & u_{\bar{z}}/4 \end{pmatrix}, \end{aligned}$$

are compatible. Solutions $\Phi_1, \Phi_2 : D \to SU(2)$ yield a conformal immersion $F = {\Phi_2}^{-1}\Phi_1$ into S^3 with metric $e^u\,dz\,d\bar{z}$, mean curvature H and Hopf differential $Q\,dz^2$.

PROOF-OUTLINE. We omit the somewhat lengthy calculations needed to prove this. The main ingredient of the proof is the Maurer-Cartan lemma stating that a system $\Phi_z = U\Phi$, $\Phi_{\bar{z}} = V\Phi$ is compatible if and only if $U_{\bar{z}} - V_z + [U,V] = 0$. First, suppose (Φ_1, Φ_2) is a quaternionic frame of a conformal immersion in S^3 and define U and V by equations (2.2). Use the frame equations (2.1) to derive equations (2.3). Finally, show by straightforward calculation that the Gauss and Codazzi equations are equivalent to $U_{k\,\bar{z}} - V_{k\,z} + [U_k, V_k] = 0$. This also takes care of the converse. □

3. Bonnet Pairs

A Bonnet pair in S^3 is a pair of immersions that have the same metric and mean curvature, but are not congruent. Hence their Hopf differentials must be different. We specialize proposition 1 to Bonnet pairs.

PROPOSITION 2. *Suppose $F, \widetilde{F} : D \to S^3$ are two immersions with the same metric $e^u\,dz\,d\bar{z}$ and mean curvature function H, but non-identical Hopf differentials $Q\,dz^2$ and $\widetilde{Q}\,dz^2$. Let $Q_1 = i(Q-\widetilde{Q})/2$ and $Q_2 = (Q+\widetilde{Q})/2$, so that $Q = Q_2 - iQ_1$ and $\widetilde{Q} = Q_2 + iQ_1$. Then the following compatibility conditions are satisfied:*

$$(3.1)\qquad u_{z\bar{z}} + \frac{1}{2}e^u(H^2+1) - 2e^{-u}(|Q_1|^2 + |Q_2|^2) = 0,$$

$$(3.2)\qquad Q_1\overline{Q}_2 - \overline{Q}_1 Q_2 = 0,$$

$$(3.3)\qquad 2e^{-u/2}Q_{2\,\bar{z}} - e^{u/2}H_z = 0,$$

$$(3.4)\qquad Q_{1\,\bar{z}} = 0.$$

Conversely, suppose the functions $u, H : D \to \mathbb{R}$, $Q_1, Q_2 : D \to \mathbb{C}$ *satisfy the above equations. Then the systems*

$$\begin{Bmatrix} \Phi_{1z} = U_1\Phi_1 \\ \Phi_{1\bar z} = V_1\Phi_1 \end{Bmatrix}, \begin{Bmatrix} \Phi_{2z} = U_2\Phi_2 \\ \Phi_{2\bar z} = V_2\Phi_2 \end{Bmatrix}, \begin{Bmatrix} \widetilde\Phi_{1z} = \widetilde U_1\widetilde\Phi_1 \\ \widetilde\Phi_{1\bar z} = \widetilde V_1\widetilde\Phi_1 \end{Bmatrix}, \begin{Bmatrix} \widetilde\Phi_{2z} = \widetilde U_2\widetilde\Phi_2 \\ \widetilde\Phi_{2\bar z} = \widetilde V_2\widetilde\Phi_2 \end{Bmatrix},$$

with

(3.5)

$$U_1 = \begin{pmatrix} u_z/4 & -e^{-u/2}(Q_2 - iQ_1) \\ \frac{1}{2}e^{u/2}(H - i) & -u_z/4 \end{pmatrix}, \; V_1 = \begin{pmatrix} -u_{\bar z}/4 & -\frac{1}{2}e^{u/2}(H + i) \\ e^{-u/2}(\overline{Q}_2 + i\overline{Q}_1) & u_{\bar z}/4 \end{pmatrix},$$

$$U_2 = \begin{pmatrix} u_z/4 & -e^{-u/2}(Q_2 - iQ_1) \\ \frac{1}{2}e^{u/2}(H + i) & -u_z/4 \end{pmatrix}, \; V_2 = \begin{pmatrix} -u_{\bar z}/4 & -\frac{1}{2}e^{u/2}(H - i) \\ e^{-u/2}(\overline{Q}_2 + i\overline{Q}_1) & u_{\bar z}/4 \end{pmatrix},$$

$$\widetilde U_1 = \begin{pmatrix} u_z/4 & -e^{-u/2}(Q_2 + iQ_1) \\ \frac{1}{2}e^{u/2}(H - i) & -u_z/4 \end{pmatrix}, \; \widetilde V_1 = \begin{pmatrix} -u_{\bar z}/4 & -\frac{1}{2}e^{u/2}(H + i) \\ e^{-u/2}(\overline{Q}_2 - i\overline{Q}_1) & u_{\bar z}/4 \end{pmatrix},$$

$$\widetilde U_2 = \begin{pmatrix} u_z/4 & -e^{-u/2}(Q_2 + iQ_1) \\ \frac{1}{2}e^{u/2}(H + i) & -u_z/4 \end{pmatrix}, \; \widetilde V_2 = \begin{pmatrix} -u_{\bar z}/4 & -\frac{1}{2}e^{u/2}(H - i) \\ e^{-u/2}(\overline{Q}_2 - i\overline{Q}_1) & u_{\bar z}/4 \end{pmatrix}.$$

are all compatible. Solutions $\Phi_1, \Phi_2, \widetilde\Phi_1, \widetilde\Phi_2, : D \to SU(2)$ *yield conformal immersions* $F = {\Phi_2}^{-1}\Phi_1$, $\widetilde F = \widetilde\Phi_2^{-1}\widetilde\Phi_1$ *with metric* $e^u\, dz\, d\bar z$, *mean curvature* H, *and Hopf differentials* $(Q_2 - iQ_1)\, dz^2$ *and* $(Q_2 + iQ_1)\, dz^2$. *Hence, if* Q_1 *does not vanish identically, they form a Bonnet pair.*

PROOF. Subtract the Gauss equations of F and $\widetilde F$ to obtain $|Q| = |\widetilde Q|$ and hence equations (3.1) and (3.2). Adding and subtracting the Codazzi equations gives (3.3) and (3.4). The rest of the proposition follows immediately from proposition 1. □

The following proposition is obtained in exactly the same way as in the case of Bonnet pairs in $\mathbb{R}^3$; see [**3**].

PROPOSITION. *The immersions of a Bonnet pair in* S^3 *have umbilics at corresponding points. They are the zeros of a not identically vanishing holomorphic quadratic differential and therefore isolated.*

There are no Bonnet pairs that are spheres. If Bonnet pair tori exist, they have no umbilics. If Bonnet pairs of genus $g > 1$ *exist, they have* $4g - 4$ *umbilics, counting multiplicities.*

Standard arguments also imply:

LEMMA. *Let* $F, \widetilde F : D \to S^3$ *form a Bonnet pair with fundamental invariants as in proposition 2, and assume that they have no umbilics. Then the conformal parametrization can be chosen such that* $Q_1 \equiv \frac{1}{2}$ *and* Q_2 *is real valued.*

The development so far has been completely parallel to the case of Bonnet pairs in $\mathbb{R}^3$. With the following theorem, however, we depart from the common features and turn to the properties peculiar to Bonnet pairs in S^3. As in the $\mathbb{R}^3$ case, helicoidal surfaces provide trivial examples of Bonnet pairs. But in S^3, such surfaces can be compact. That is:

THEOREM. *There are compact Bonnet pairs in* S^3 *which are helicoidal immersed tori.*

Next we introduce a duality transformation for Bonnet pairs.

THEOREM. *Suppose* $F, \widetilde{F} : D \to S^3$ *form a Bonnet pair with fundamental invariants as in proposition 2. Assume the immersions have no umbilics and are parametrized such that* $Q = \frac{1}{2}(q - i)\, dz^2$ *and* $\widetilde{Q} = \frac{1}{2}(q + i)\, dz^2$ *with real-valued* q. *Let* (Φ_1, Φ_2) *and* $(\widetilde{\Phi}_1, \widetilde{\Phi}_2)$ *be quaternionic frames for* F *and* $\widetilde{F}$.

Then $G = \widetilde{\Phi}_1^{-1}\Phi_1$ *and* $\widetilde{G} = \widetilde{\Phi}_2^{-1}\Phi_2$ *are conformal immersions into* S^3 *forming another—the* dual*—Bonnet pair with metric* $e^{-u}\, dz\, d\bar{z}$, *mean curvature* q, *and Hopf differentials* $\frac{1}{2}(H - i)\, dz^2$ *and* $\frac{1}{2}(H + i)\, dz^2$. *The dual Bonnet pair to* $G, \widetilde{G}$ *is again* $F, \widetilde{F}$.

PROOF. The idea is to show that $(\mathbf{j}\Phi_1, \mathbf{j}\widetilde{\Phi}_1)$ is a quaternionic frame of an immersion G with the required fundamental invariants. Similarly, $(\mathbf{j}\Phi_2, \mathbf{j}\widetilde{\Phi}_2)$ is a quaternionic frame for $\widetilde{G}$. To this end, substitute $\frac{1}{2}$ for Q_1 and q for Q_2 in equations (3.5) and compare these matrices to the corresponding 'U' and 'V' matrices of $\mathbf{j}\Phi_1$, $\mathbf{j}\widetilde{\Phi}_1$, $\mathbf{j}\Phi_2$ and $\mathbf{j}\widetilde{\Phi}_2$. For example, for Φ_1, calculate $(\mathbf{j}\Phi_1)_z(\mathbf{j}\Phi_1)^{-1} = -\mathbf{j}U_1\mathbf{j}$ and $(\mathbf{j}\Phi_1)_{\bar{z}}(\mathbf{j}\Phi_1)^{-1} = -\mathbf{j}V_1\mathbf{j}$ and compare them to U_1 and V_1. □

COROLLARY. *The dual Bonnet pair satisfies the following equations.*

$$\begin{aligned} F &= \widetilde{G}^{-1}\widetilde{F}\, G, \\ F_x &= \widetilde{G}^{-1}\widetilde{F}_x\, G, \\ F_y &= \widetilde{G}^{-1}\widetilde{F}_y\, G. \end{aligned}$$

In other words, $(G, \widetilde{G})$, *interpreted as a map into* $SU(2) \times SU(2)$, *rotates the frame of* $\widetilde{F}$ *into the frame of* F.

The following interesting questions are left unanswered: Can the theory of Bonnet pairs in S^3 be generalized to encompass Bonnet pairs in $\mathbb{R}^3$? Is there an integrable system connected to Bonnet pairs in S^3? Are there compact examples apart from helicoidal tori?

References

[1] L. Bianchi. Sulle superficie a linee di curvatura isoterme. *Rend. Acc. Naz. dei Lincei*, 12:511–520, 1903.

[2] A.I. Bobenko. Surfaces in terms of 2 by 2 matrices. Old and new integrable cases. In A. P. Fordy and J. Wood, editors, *Harmonic Maps and Integrable Systems*, pages 83–127. Vieweg, Braunschweig, 1994.

[3] A.I. Bobenko. Exploring surfaces through methods from the theory of integrable systems: the Bonnet problem. These proceedings.

[4] A.I. Bobenko and U. Eitner. Bonnet surfaces and Painlevé equations. *J. Reine Angew. Math.*, 499:47–79, 1998.

[5] A.I. Bobenko and U. Eitner. *Painlevé Equations in the Differential Geometry of Surfaces*, volume 1753 of *Lecture Notes in Mathematics*. Springer, Berlin, 2000.

[6] O. Bonnet. Mémoire sur la théorie des surfaces applicables. *J. Éc. Polyt.*, 42, 1867.

[7] É. Cartan. Sur les couples de surfaces applicables avec conservation des courbures principales. *Bull. Sci. Math.*, II. Ser. 66:55–72 and 74–85, 1942. Also in Œvres Complètes, part. III.

[8] S.S. Chern. Deformation of surfaces preserving principal curvatures. In I. Chavel and H. M. Farkas, editors, *Differential Geometry and Complex Analysis*, pages 155–163. Springer, New York, 1985.

[9] J.N. Graustein. Applicability with preservation of both curvatures. *Bull. Amer. Math. Soc.*, 30:19–23, 1924.
[10] J.N. Hazzidakis. Biegung mit Erhaltung der Hauptkrümmungsradien. *J. Reine Angew. Math.*, 117:42–56, 1887.
[11] H. Jonas. Über ein die Verbiegung der Linienkongruenzen betreffendes Problem und über die Transformation der Bonnetschen Flächenpaare. *Mathematische Annalen*, 86:78–98, 1922.
[12] G. Kamberov, F. Pedit, and U. Pinkall. Bonnet pairs and isothermic surfaces. *Duke Math. J.*, 92:637–644, 1998.
[13] K. Kenmotsu. An intrinsic characterization of H-deformable surfaces. *J. London. Math. Soc.*, 49:555–568, 1994.
[14] H.B. Lawson Jr. and R. de A. Tribuzy. On the mean curvature function for compact surfaces. *J. Differ. Geom.*, 16:179–183, 1981.
[15] I.M. Roussos. The helicoidal surfaces as Bonnet surfaces. *Tohoku Math. J.*, 40:485–490, 1988.
[16] I.M. Roussos. Global results on Bonnet surfaces. *J. Geom.*, 65:151–168, 1999.
[17] M. Servant. Sur deux problèmes de géométrie. *C. R.*, 134:1291–1293, 1902.

TECHNISCHE UNIVERSITÄT BERLIN, SEKR. MA 8-5, STRASSE DES 17. JUNI 136, 10623 BERLIN, GERMANY

E-mail address: springb@math.tu-berlin.de

Contemporary Mathematics
Volume **308**, 2002

Subspaces in the Category of Symmetric Spaces

Makiko Sumi Tanaka

1. Introduction

Symmetric spaces are not only fundamental but elegant in their structures and properties. Historically, by applying some new theory to symmetric spaces, non-trivial concrete results were obtained. Conversely, investigation into symmetric spaces suggested problems which led to new theories. It is to be expected that such relations will persist in the future. So it seems worthwhile to study symmetric spaces in a more geometric way and make them easier to handle. This is the motivation for our theory developed in the series of papers [**N2**],[**N3**],[**NT1**] and [**NT2**].

The theory is based on pairs of subspaces which are called polars and meridians (§3). They were introduced by B. Y. Chen and T. Nagano in the 1970's ([**CN1**],[**CN2**]) and play an important role in the category of symmetric spaces (Theorem 2.1). On the other hand, symmetric R-spaces comprise a large class which includes Kähler symmetric spaces, the so called Hermitian symmetric spaces, in the category of compact symmetric spaces, and they have nice properties which have been studied by many mathematicians. In [**N1**] T. Nagano indicated that on a symmetric R-space M a noncompact Lie group L containing the automorphism group of M acts transitively. He chose a zonal spherical function f on M which turns out to be a Morse-Bott function on M. Making use of the Morse-Bott function f, M. Takeuchi studied the structure of M in detail in [**T1**]. The geometric properties of the noncompact Lie group L were studied by S. Peterson ([**P**]), M. Takeuchi ([**T2**]) and S. Gindikin–S. Kaneyuki ([**GK**]).

In the first part of this article we give new characterizations of symmetric R-spaces (Theorem 3.2 and Theorem 3.7), one of which is equivalent to that of O. Loos ([**L2**], [**L3**]) (Theorem 3.5).

In the next part we introduce the notion of regular triplet in a symmetric space M. A triplet $\{o, p, q\}$ of three points o, p and q in M is called a regular triplet if the composite of the point-symmetries at each point is the identity map. If M has no pole, M has a regular triplet $\{o, p, q\}$ if and only if M has a polar $M^+_{(p;o)}$ which

2000 *Mathematics Subject Classification.* Primary 53C35.

is isomorphic with the meridian to it (Theorem 4.10). Then o, p and q are vertices of "a right regular triangle" each of whose sides is isomorphic with $M^+_{(o;p)}$. The existence of a regular triplet is related to the principle of triality. We also prove that if an M without a pole has a regular triplet, there is a subspace of M which is isomorphic with $\boldsymbol{R}P^2$ and contains the regular triplet (Theorem 4.14).

This article is based on collaboration with Tadashi Nagano.

2. Preliminaries

We define the *category of symmetric spaces* as follows: (i) The category of symmetric spaces is a subcategory of the category of smooth manifolds of finite dimension. (ii) Every object M, called a *symmetric space*, admits the (point-) symmetry $s_x : M \to M$ such that s_x is involutive and x is an isolated fixed point of s_x. (iii) Every morphism $f : M \to N$ commutes with symmetries; that is, f satisfies $f \circ s_x = s_{f(x)} \circ f$ for every point x of M. (iv) Every symmetry s_x is a morphism.

When M is *connected*, M is an object if and only M is a symmetric space as defined in [**H**] and [**KN**]. Indeed there exists a unique affine connection on M (hence M is paracompact) such that $\mathrm{Aut}(M)$, the automorphism group of M, coincides with the affine transformation group. Moreover, the morphisms are exactly the totally geodesic maps (affine maps in [**KN**]). On the other hand, any discrete set can be an object. Thus we have significant concepts such as the 2-number ([**CN3**]). Our definition is somewhat similar to that of Loos [**L1**].

Every morphism $f : M \to N$ is a composite of an epimorphism and a monomorphism. In particular, the image $f(M)$ is also a symmetric space and we call it a *subspace* of N. For example, the fixed point set of an automorphism σ of M is a subspace of M.

From now on, we work in the category of Riemannian symmetric spaces unless otherwise stated, that is, we assume that every morphism satisfies the isometry condition, more precisely, it is isometric on the orthogonal complement of the kernel of the differential at each point. In particular, every symmetry preserves the Riemannian metric. Hereafter we study compact symmetric spaces only.

Let $M = G/K$ be a connected symmetric space, where G is the identity component of the automorphism group of M and K is the isotropy subgroup of G at a point o in M. Each connected component of the fixed point set of s_o is called a *polar* of o in M and denoted by $M^+_{(p)}$ or $M^+_{(p;o)}$ if it contains a point p. The connected component of the fixed point set of $s_p \circ s_o$ which contains p is called the *meridian* to $M^+_{(p)}$ at p, denoted by $M^-_{(p)}$ or $M^-_{(o;p)}$. By definition, we can easily see that $M^+_{(p)}$ and $M^-_{(p)}$ are *c-orthogonal* to each other, that is, the tangent spaces of $M^+_{(p)}$ and $M^-_{(p)}$ at p are the orthogonal complements to each other in that of M. Each polar is a K-orbit and the meridians to a polar are all G-congruent with each other. Polars and meridians are subspaces of M.

For example, if M is a compact Lie group with bi-invariant metric, the symmetry at $x \in M$ is defined by $s_x(y) = xy^{-1}x$ for every element y in M. Thus the polars of the identity element in M coincide with the conjugacy classes of involutive elements of M. In the case of $M = U(n)$, we have $M^+(p) = G_k(\boldsymbol{C}^n)$, the complex Grassmann manifold of k-dimensional subspaces in $\boldsymbol{C}^n$, and $M^-(p) = U(k) \times U(n-k)$ where $0 \le k \le n$.

Polars of M are deeply related to the topology of M, for instance, M is orientable if and only if every polar has even dimension: the Euler number $\chi(M)$ of M is equal to the sum of those of all the polars of M unless $\chi(M) = 0$. On the other hand, each meridian $M^-_{(p)}$ has the same rank as that of M. So we can see that the natural homomorphism $\pi_1(M^-_{(p)}) \to \pi_1(M)$ is surjective.

When a polar $M^+_{(p)}$ consists of a single point p, p is called a *pole* of o in M if p is not o. Such poles do not necessarily exist, and when they do, they are not necessarily unique. When there is a pole p of o in M, we have a double covering morphism $\pi : M \to N$ onto another symmetric space N, satisfying $\pi(o) = \pi(p)$. In this situation, let $C(o,p)$ denote the set of the midpoints of the geodesic arcs from o to p, which is a subspace of M. Each connected component of $C(o,p)$ is called a *centriole* for the pair (o,p) in M. The image of every centriole under π is a polar of $\pi(o)$ in N. Conversely, every polar of $\pi(o)$ in N is the image of a polar of o or a centriole for (o,p) in M under π. Moreover, M and N are locally isomorphic and there is a unique symmetric space $M^{\%}$ in the locally isomorphism class of M which is not a covering space of any other symmetric space if M is semisimple. This $M^{\%}$ is called the *bottom space* of M. The bottom space has no pole but the converse is not true; for example, $M = SU(n)$ with $n =$ odd has no pole and $M^{\%} = SU(n)/\boldsymbol{Z_n}$.

For example, the n-dimensional sphere $M = S^n$ has a pole and the centriole is the equator S^{n-1}. The bottom space of S^n is a projective space $\boldsymbol{R}P^n$ and its polar $\boldsymbol{R}P^{n-1}$ is the image of the centriole S^{n-1} under the double covering morphism.

By the classification of polars and meridians ([**CN2**],[**CN3**], for example) we find that every polar of M is a pole if and only if M is a Riemannian product of spheres and a torus.

The theoretical significance of the notion of polars and the meridians lies in the following fact.

THEOREM 2.1 (1.15 in [**N3**]). *Every compact connected symmetric space is globally determined by any pair consisting of a polar and the meridian to it.*

3. Symmetric R-spaces

A symmetric space M is called a *symmetric R-space* if it is a totally real subspace in a semisimple Kähler symmetric space M^c with $2\dim M = \dim M^c$. We call M^c the *complexification* of M; in fact a semisimple M^c is uniquely determined by M. Every Kähler symmetric space M^c contains a symmetric R-space M whose complexification is M^c. We exclude complex tori from being symmetric R-spaces following the conventional definition.

Table 1 is the list of symmetric R-spaces M such that the root system $R(M^c)$ is simple. For a Kähler symmetric space M with complex structure J, $\bar{M}$ denotes a Kähler symmetric space M with complex structure $-J$. In Table 2 we list the simple Kähler symmetric spaces $M = G/K$.

Notation. $UI(n) = U(n)/O(n)$, $UII(n) = U(2n)/Sp(n)$, $AII(n) = SU(2n)/Sp(n)$, $CI(n) = Sp(n)/U(n)$, $DIII(n) = SO(2n)/U(n)$, $EIII = E_6/T \cdot Spin(10)$, $EIV = E_6/F_4$, $EVII = E_7/T \cdot E_6$, $FII = F_4/Spin(9)$. $G_p(\boldsymbol{K}^n)(\boldsymbol{K} = \boldsymbol{R}, \boldsymbol{C}$ or $\boldsymbol{H})$ denotes the Grassmann manifold of p -dimensional subspaces in $\boldsymbol{K}^n$.

Table 1. Symmetric R-spaces and their complexifications.

R(M)	R(M^c)	M	M^c
T	C_1	$T = S^1$	S^2
$T \cdot A_{n-1}, n \geq 2$	C_n	$UI(n)$	$CI(n)$
$T \cdot A_{n-1}, n \geq 2$	C_n	$U(n)$	$G_n(\boldsymbol{C}^{2n})$
$T \cdot A_{n-1}, n \geq 2$	C_n	$UII(n)$	$DIII(2n)$
$T \cdot A_2$	C_3	$T \cdot EIV$	$EVII$
$T \cdot A_1$	C_2	$T \cdot S^m, m \geq 2$	$G_2^o(\boldsymbol{R}^{m+3})$
$D_n, n \geq 2$	C_n	$G_n(\boldsymbol{R}^{2n})$	$G_n(\boldsymbol{C}^{2n})$
$D_n, n \geq 2$	C_n	$SO(2n)$	$DIII(2n)$
$D_3 = A_3$	C_3	$AII(4)/\boldsymbol{Z}_2$	$EVII$
$D_2 = A_1 + A_1$	C_2	$S^p \cdot S^q,\ p, q \geq 2$	$G_2^o(\boldsymbol{R}^{p+q+2})$
$C_n, n \geq 1$	C_{2n}	$Sp(n)$	$CI(2n)$
$C_n, n \geq 1$	C_{2n}	$G_n(\boldsymbol{H}^{2n})$	$G_{2n}(\boldsymbol{C}^{4n})$
$C_1 = A_1$	C_2	$S^m,\ m \geq 2$	$G_2^o(\boldsymbol{R}^{m+2})$
$B_n, n \geq 2$	BC_n	$G_n(\boldsymbol{R}^{2n+m}),\ m \geq 1$	$G_n(\boldsymbol{C}^{2n+m})$
$B_n, n \geq 2$	BC_n	$SO(2n+1)$	$DIII(2n+1)$
B_2	BC_2	$G_2(\boldsymbol{H}^4)^{\%}$	$EIII$
$BC_n, n \geq 1$	BC_{2n}	$G_n(\boldsymbol{H}^{2n+m}),\ m \geq 1$	$G_{2n}(\boldsymbol{C}^{4n+2m})$
BC_1	BC_2	FII	$EIII$
C_n or BC_n		Kähler	$M \times \bar{M}$

Table 2. Simple Kähler symmetric spaces.

R(M)	G	$M = G/K$
C_n	$Sp(n)$	$CI(n)$
C_n	$U(2n)$	$G_n(\boldsymbol{C}^{2n})$
C_n	$SO(4n)$	$DIII(2n)$
C_2	$Spin(4+m)$	$G_2^o(\boldsymbol{R}^{4+m})$
C_3	E_7	$EVII$
BC_n	$U(2n+m)$	$G_n(\boldsymbol{C}^{2n+m})$
BC_n	$SO(4n+2)$	$DIII(2n+1)$
BC_2	$T \cdot E_6$	$EIII$

From the classification (see Table 1 and 2) we have the following facts:

Fact 1. A simple symmetric space is locally isomorphic with an R-space if and only if its root system $R(M)$ is of classical type.

Fact 2. The rank of R-space M, $\mathrm{r}(M)$, is equal to $\mathrm{r}(M^c)$ or $\frac{1}{2}\mathrm{r}(M^c)$, the latter if and only if M is simply connected.

Fact 3. The fundamental group $\pi_1(M)$ of an R-space M is isomorphic with $\boldsymbol{Z}$ if and only if $R(M) = T \cdot A_n$, with $\{0\}$ if and only if either $R(M) = C_n$ or BC_n, and, except for the cases $M = G_2(\boldsymbol{H}^4)^{\%}$ and $SO(5)$, with $\boldsymbol{Z_2}$ if and only if either $R(M) = B_n$ or D_n.

We refer to [**NT3**] for the following results.

LEMMA 3.1. *If a symmetric space M is a symmetric R-space, then all polars and meridians in M are symmetric R-spaces.*

THEOREM 3.2. *The following conditions* (i) *and* (ii) *are equivalent:* (i) *M is a symmetric R-space;* (ii) *some meridian M^- in M is a symmetric R-space, more precisely, M^- with the induced metric is isomorphic with a symmetric R-space N whose metric is induced from its complexification N^c.*

REMARK 3.3. We normalize the Riemannian metric of a symmetric R-space M as follows: normalize the Riemannian metric on each simple factor of the complexification M^c so that the maximum of the sectional curvature is equal to one, i.e. the sectional curvature of the Helgason sphere of M^c, which is isomorphic to S^2, is equal to one. Then we take the induced metric on M. For example, $M = GI = G_2/SO(4)$ is not a symmetric R-space and has a polar $M^+_{(p)} = S^2 \cdot S^2$ which is isomorphic with $M^-_{(p)}$. Here $S^2 \cdot S^2$ is not a symmetric R-space since these S^2 are not isomorphic with each other with respect to the induced metric. We emphasize that we work in the category of Riemmanian symmetric spaces.

Theorem 3.2 is a generalization of the following theorem for the case of Kähler symmetric spaces in [**NT2**].

THEOREM 3.4. *M is a Kähler symmetric space if and only if M has a meridian M^- which is Kählerian.*

The "only if" part of Theorem 3.2 is easily proved using the following theorem of O. Loos ([**L2**],[**L3**]), since the rank of M^- is equal to that of M.

THEOREM 3.5 (Loos). *M is a symmetric R-space if and only if M has a cubic unit lattice with respect to a Riemannian metric invariant under the identity component of its automorphism group, that is, a maximal torus A of M is isomorphic with a Riemannian product of circles which are isometric with each other.*

Conversely, we can prove Theorem 3.5 by using Theorem 3.2.

THEOREM 3.6. *Theorem 3.2 is equivalent to Theorem 3.5.*

We have another characterization of symmetric R-spaces.

THEOREM 3.7. *In the category of Riemannian symmetric spaces we have the following.*

(1) *S is a simple Kähler symmetric space or a non-Kählerian symmetric R-space whose complexification S^c is simple if and only if S is isomorphic with a centriole $C(o,p)$ in a symmetric space L_u/G which meets one of the shortest geodesic arcs joining o and p at its midpoint m, where L_u/G is either simple or a dot product $T \cdot M$ of T and a simple symmetric space M.*
(2) *The conditions in (1) imply that the c-orthogonal space to $C(o,p)$ is the dot product of the circle T and another space.*
(3) *Under the conditions in (1), S is Kählerian if and only if L_u/G is a Lie group.*

Table 3. Symmetric R-spaces $M = G/K = L/P$ and L_u/G.

M	L_u/G
$UI(n)$	$Sp(n)/U(n)$
$U(n)$	$SU(2n)/S(U(n) \times U(n))$
$UII(n)$	$SO(4n)/U(2n)$
$T \cdot EIV$	$E_7/(T \cdot E_6)$
$SO(n)$	$SO(2n)/S(O(n) \times O(n))$
$Sp(n)$	$Sp(2n)/(Sp(n) \times Sp(n))$
$G_n(\boldsymbol{H}^{n+m})$, $m \geq n$	$SU(2n+2m)/Sp(n+m)$
$G_n(\boldsymbol{R}^{n+m})$, $m \geq n$	$SU(n+m)/SO(n+m)$
$G_2(\boldsymbol{H}^4)^{\%}$	$E_6/Sp(4)^{\%}$
$S^p \cdot S^q$	$SO(p+q+2)/(SO(p+1) \times SO(q+1))$
FII	E_6/F_4
$AII(4)/\boldsymbol{Z}_2$	$E_7/(SU(8)/\boldsymbol{Z}_4)$

We list L_u/G in Table 3. It is folklore that a linear isotropy orbit of H_j, which satisfies $\langle H_j, \alpha_i \rangle = \delta_{ij}$ for every simple root α_i, and is such that $m_j = 1$ in the summation $\tilde{\alpha} = m_1\alpha_1 + \cdots + m_r\alpha_r$ of the highest root $\tilde{\alpha}$ of a simple L_u/G, is a symmetric R-space G/K whose complexification is simple and conversely.

REMARK 3.8. It is well-known that a noncompact Lie group L acts on a symmetric R-space M. In fact, L_u is a compact form of L ([**N1**]).

4. Regular triplets

Let M be a compact connected symmetric space, and o, p and q be three points in M. A triplet $\{o,p,q\}$ is called *a regular triplet* in M if $s_o \circ s_p \circ s_q = \mathrm{id}_M$. For example, the real, complex, quaternionic, or Cayley projective plane has a regular triplet. In fact, the polar $M^+_{(p)}$ of a point o in M is a projective line (hence a sphere or a circle) and by taking q, the pole of p in $M^+_{(p)}$, we have a regular triplet $\{o,p,q\}$. Then o, p and q are vertices of "a right regular triangle" each of whose sides is a projective line.

REMARK 4.1. As we will see later, every symmetric space which admits a regular triplet has such "a right regular triangle".

Since every point-symmetry is involutive, the next lemma follows immediately.

LEMMA 4.2. *If* $\{o, p, q\}$ *is a regular triplet, we have* $s_o \circ s_p = s_q$, $s_p \circ s_q = s_o$ *and* $s_q \circ s_o = s_p$.

Since the composite of two point-symmetries is an inner involution, we have the following consequence.

COROLLARY 4.3. *If* M *has a regular triplet,* M *is of inner type.*

REMARK 4.4. A simple symmetric space $M = G/K$ is of inner type if and only if one of the following conditions holds: (i) the rank of K is same as that of G and (ii) the Euler number $\chi(M)$ of M is positive.

The following lemmas describe some of fundamental properties of regular triplets and we use them to prove the main theorem (Theorem 4.10).

LEMMA 4.5. *Let* $\{o, p, q\}$ *be a regular triplet. Then the group generated by* $\{s_o, s_p, s_q\}$ *is isomorphic with that generated by* $\left\{\begin{pmatrix} \pm 1 & 0 \\ 0 & \pm 1 \end{pmatrix}\right\}$ *and its automorphism group is isomorphic with the symmetric group of degree 3.*

LEMMA 4.6. *If* $\{o, p, q\}$ *is a regular triplet,* s_o, s_p *and* s_q *commute.*

LEMMA 4.7. *Assume that* M *has no pole and let* $\{o, p, q\}$ *be a regular triplet in* M*. Then* $\{s_o, s_p, s_q\}$ *acts trivially on* $\{o, p, q\}$.

LEMMA 4.8. *Assume that* M *has no pole and let* $\{o, p, q\}$ *be a regular triplet in* M *and* m *the midpoint of a minimal geodesic arc between* p *and* q*. Then we have* $s_m(o) = o$.

LEMMA 4.9. *Assume that* M *has no pole and let* $\{o, p, q\}$ *be a regular triplet in* M*. Then the polar* $M^+_{(p;o)}$ *coincides with* $M^+_{(q;o)}$.

THEOREM 4.10. *Assume that* M *has no pole. Then the following three conditions are equivalent:* (i) *M has a regular triplet;* (ii) *there is a polar in M which is congruent with the meridian to it; and* (iii) *there is a polar in M which has a pole.*

REMARK 4.11. A simple M has no pole if and only if (i) M is the bottom space $M^{\%}$; or (ii) the root system of M is of type A_{2r} or type E_6. This condition is related with the fundamental group of $M^{\%}$.

Now we consider the case when M admits a pole. Let p be a pole of o in M and π the double covering morphism from M onto N which maps o and p onto the same point.

THEOREM 4.12. *Asume that* $N = \pi(M)$ *has no pole. Then* M *has a regular triplet if and only if* N *has a regular triplet.*

Since the complete list of (M^+, M^-) is known (see [**CN2**],[**CN3**], for example), we can find all M which have no pole but have a regular triplet by Theorem 4.10.

In [**NT1**] we investigated the relation between self-intersections and regular triplets on the sequence of symmetric spaces: $\boldsymbol{R}P^2 \subset \boldsymbol{C}P^2 \subset \boldsymbol{H}P^2 \subset FII \subset EIII \subset EVI \subset EVIII$. Here we use the notation of symmetric spaces in [**H**].

There is another sequence of quaternionic Kähler symmetric spaces of exceptional type: $GI \subset FI \subset EII \subset EVI \subset EIX$, where each of these spaces has a regular triplet. Thus we know that every inner symmetric space of exceptional type has a regular triplet. And every Kähler symmetric space with a pole (that is, whose root system is of type C_r) has a regular triplet. In fact, each of them includes a symmetric R-space as a centriole, which is isomorphic with its c-orthogonal space.

REMARK 4.13. K. Atsuyama studied the bottom spaces which have a regular triplet from the view point of "projective geometry in a wider sense" in [**A**]. B. Y. Chen studied regular triplets in [**C**], where he called them *antipodal* subsets and he proved the equivalence of conditions (i) and (ii) in Theorem 4.10 and Theorem 4.12.

If $M = G/K$ has a regular triplet $\{o, p, q\}$, the Lie algebra $\mathfrak{g}$ of G has an orthogonal direct sum decomposition $\mathfrak{g} = \mathfrak{k}_+ + \mathfrak{k}_- + \mathfrak{m}_+ + \mathfrak{m}_-$ with respect to s_o and s_p, and $\mathfrak{k}_- \cong \mathfrak{m}_+ \cong \mathfrak{m}_-$ as $\mathfrak{k}_+$-modules. Moreover, we have $[\mathfrak{k}_-, \mathfrak{m}_\pm] = \mathfrak{m}_\mp$ and $[\mathfrak{m}_+, \mathfrak{m}_-] = \mathfrak{k}_-$. There is an automorphism of $\mathfrak{g}$ which interchanges these three spaces cyclicly. This property is related with the triality of $Spin(8)$. In fact, if M is one of the spaces $FII \subset EIII \subset EVI \subset EVIII$, the isotropy subgroup contains a spinor group as a normal subgroup.

We have another observation on regular triplets.

THEOREM 4.14. *Let M have a regular triplet but no pole. Then there is a subspace in M which is isomorphic with $\boldsymbol{R}P^2$ and contains the regular triplet.*

References

[A] K. Atsuyama, *Projective spaces in a wider sense,* I, Kodai Math. J. **15** (1992), 324–340.

[C] B.Y. Chen, *A new approach to compact symmetric spaces and applications (a report on joint work with Professor T. Nagano),* Katholieke Universiteit Leuven, August 1987.

[CN1] B. Y. Chen and T. Nagano, *Totally geodesic submanifolds of symmetric spaces,* I, Duke Math. J., **44** (1977), 745-755.

[CN2] B.Y. Chen and T. Nagano, *Totally geodesic submanifolds of symmetric spaces,* II, Duke Math. J., **45** (1978) 405-425.

[CN3] B. Y. Chen and T. Nagano, *A Riemannian geometric invariant and its applications to a problem of Borel and Serre,* Trans. Amer. Math. Soc., **308**(1988), 273-297.

[GK] S. Gindikin and S. Kaneyuki, *On the automorphism group the generalized conformal structure of a symmetric R-space,* Diff. Geom. Appl. **8** (1998), 21-33.

[H] S. Helgason, *Differential Geometry, Lie Groups and Symmetric Spaces*, Academic Press, 1978.

[KN] S. Kobayashi and K. Nomizu, *Foundations of Differential Geometry I, II,* Wiley (Interscience), 1963, and 1969.

[L1] O. Loos, *Symmetric spaces I, II,* W. A. Benjamin, 1969.

[L2] O. Loos, *Jordan triple systems, R-spaces, and bounded symmetric domains,* Bulletin of AMS. **77** (1971), 558-561.

[L3] O. Loos, *Charakterisierung symmetrischer R-Räume durch ihre Einheitsgitter,* Math. Z. **189** (1985), 211-226.

[N1] T. Nagano, *Transformation groups on compact symmetric spaces,* Trans. Amer. Math. Soc. **118** (1965), 428-453.

[N2] T. Nagano, *The involutions of compact symmetric spaces,* Tokyo J. Math. **11** (1988), 57-79.

[N3] T. Nagano, *The involutions of compact symmetric spaces,* II, Tokyo J. Math. **15** (1992), 39-82.

[NT1] T. Nagano and M.S. Tanaka, *The involutions of compact symmetric spaces,* IV, Tokyo J. Math. **22** (1999), 193-211.

[NT2] T. Nagano and M.S. Tanaka, *The involutions of compact symmetric spaces,* V, Tokyo J. Math. **23** (2000), 403-416.

[NT3] T. Nagano and M.S. Tanaka, *The involutions of compact symmetric spaces,* VI, in preparation.
[P] S.P. Peterson, *Arithmetic distance on compact symmetric spaces,* Geom. Dedicata **23** (1987), 1-14.
[T1] M. Takeuchi, *Cell decompositions and Morse equalities on certain symmetric spaces,* J. Fac. Sci. Univ. of Tokyo, I, **12** (1965), 81-192.
[T2] M. Takeuchi, *Basic transformations of symmetric R-spaces,* Osaka J. Math. **25** (1988), 259-297.

DEPARTMENT OF MATHEMATICS, TOKYO UNIVERSITY OF SCIENCE, NODA-SHI, CHIBA 278-8510, JAPAN

E-mail address: tanaka_makiko@ma.noda.tus.ac.jp

Contemporary Mathematics
Volume **308**, 2002

Integral Geometry of Submanifolds of Real Dimension Two and Codimension Two in Complex Projective Spaces

Hiroyuki Tasaki

1. Introduction

Poincaré formulae are fundamental in integral geometry. Many mathematicians have obtained generalized Poincaré formulae for various homogeneous spaces. In particular, Howard [**2**] proved the following formula. For a Riemannian homogeneous space G/K with a unimodular Lie group G and submanifolds M and N in G/K which satisfy $\dim(G/K) \le \dim M + \dim N$,

$$(1.1) \qquad \int_G \mathrm{vol}(M \cap gN) d\mu_G(g) = \int_{M\times N} \sigma_K(T_x^\perp M, T_y^\perp N) d\mu_{M\times N}(x,y)$$

holds, where σ_K is defined by a certain integration over the isotropy group K. See (2.1) for its definition. We regard $\mathrm{vol}(M \cap gN)$ as a function defined on G. This function is measurable and we can integrate it. Although this formula holds in general, the integrand σ_K has been explicitly described only in a few cases. When G/K is a space of constant curvature, σ_K is constant and can be explicitly described.

Before the result of Howard, Santaló [**5**] established Poincaré formulae for complex submanifolds in complex projective spaces and Howard [**2**] reformulated them as follows. For any complex p-dimensional submanifold M and complex q-dimensional submanifold N in a complex projective space $\mathbf{C}P^n$ of dimension n which satisfy $n \le p + q$,

$$\begin{aligned} &\int_{U(n+1)} \mathrm{vol}(M \cap gN) d\mu_{U(n+1)}(g) \\ &= \frac{\mathrm{vol}(U(n+1))\mathrm{vol}(\mathbf{C}P^{p+q-n})}{\mathrm{vol}(\mathbf{C}P^p)\mathrm{vol}(\mathbf{C}P^q)} \mathrm{vol}(M)\mathrm{vol}(N) \end{aligned}$$

holds, where $\mathrm{vol}(\mathbf{C}P^0) = 1$.

2000 *Mathematics Subject Classification.* Primary 53C65.
Partly supported by a Grant-in-Aid for Scientific Research from the Japan Society for Promotion of Science.

Kang and the author considered submanifolds which are not necessarily complex in complex projective spaces in [**3**] and [**4**], and obtained the following Poincaré formulae. For any real 2-dimensional submanifold M and complex $(n-1)$-dimensional submanifold N in the complex projective space $\mathbf{C}P^n$, we have

$$\begin{aligned}(1.2)\qquad &\int_{U(n+1)} \#(M\cap gN)d\mu_{U(n+1)}(g)\\ &= \frac{\mathrm{vol}(U(n+1))\mathrm{vol}(N)}{2\mathrm{vol}(\mathbf{C}P^1)\mathrm{vol}(\mathbf{C}P^{n-1})}\int_M (1+\cos^2\theta_x)d\mu_M(x),\end{aligned}$$

where θ_x is the Kähler angle of M at x. Moreover, the above formula holds for any real $(2n-2)$-dimensional submanifold M and complex 1-dimensional submanifold N, where θ_x is the Kähler angle of $T_x^{\perp}M$.

For any real surfaces M and N in $\mathbf{C}P^2$, we have

$$\begin{aligned}(1.3)\qquad &\int_{U(3)} \#(M\cap gN)d\mu_{U(3)}(g)\\ &= \frac{\mathrm{vol}(U(3))}{\mathrm{vol}(\mathbf{R}P^2)^2}\int_{M\times N}(2+2\cos^2\theta_x\cos^2\tau_y+\sin^2\theta_x\sin^2\tau_y)d\mu_{M\times N}(x,y),\end{aligned}$$

where θ_x (τ_y) is the Kähler angle of M (N) at x (y).

In this paper we obtain a Poincaré formula for any submanifolds of real dimension 2 and codimension 2 in the complex projective space, which is a generalization of (1.2) and (1.3).

THEOREM 1.1. *For any real 2-dimensional submanifold M and real $(2n-2)$-dimensional submanifold N in the complex projective space $\mathbf{C}P^n$, we have*

$$\begin{aligned}&\int_{U(n+1)} \#(M\cap gN)d\mu_{U(n+1)}(g)\\ &= \frac{\mathrm{vol}(U(n+1))}{\mathrm{vol}(\mathbf{C}P^1)\mathrm{vol}(\mathbf{C}P^{n-1})}\\ &\quad\times\int_{M\times N}\left(\frac{1}{4}(1+\cos^2\theta_x)(1+\cos^2\tau_y)+\frac{n}{4(n-1)}\sin^2\theta_x\sin^2\tau_y\right)\\ &\quad\cdot d\mu_{M\times N}(x,y),\end{aligned}$$

where θ_x is the Kähler angle of T_xM and τ_y is the Kähler angle of $T_y^{\perp}N$.

If N is a complex submanifold, the Kähler angle τ_y is always equal to 0, so the formula (1.2) is a special case of that in Theorem 1.1. In the case of the complex projective plane, $\mathrm{vol}(\mathbf{R}P^2) = 2\mathrm{vol}(\mathbf{C}P^1)$ and

$$\begin{aligned}&2+2\cos^2\theta_x\cos^2\tau_y+\sin^2\theta_x\sin^2\tau_y\\ &=(1+\cos^2\theta_x)(1+\cos^2\tau_y)+2\sin^2\theta_x\sin^2\tau_y,\end{aligned}$$

so the formula (1.3) is a special case of that in Theorem 1.1.

For more general submanifolds in $\mathbf{C}P^n$, the author defined their *multiple Kähler angle* and formulated Poincaré formulae using this concept in [**6**].

2. Preliminaries

In this section we shall review the Kähler angle, the Poincaré formula obtained by Howard [**2**], and the coarea formula, which are needed for the formulation of Theorem 1.1 and its proof.

Let $\mathbf{C}^n$ be the n-dimensional complex vector space with the standard real inner product $\langle\cdot,\cdot\rangle$. For a real 2-dimensional vector subspace V in $\mathbf{C}^n$, we take an orthonormal basis v_1, v_2 of V. We set $\theta(V) = \cos^{-1}|\langle\sqrt{-1}v_1, v_2\rangle|$ and call $\theta(V)$ the *Kähler angle* of V. The definition of $\theta(V)$ is independent of the choice of the orthonormal basis v_1, v_2 of V. In this paper we do not consider orientations of subspaces and submanifolds, so the range of the Kähler angle is $[0, \pi/2]$. The subspace is complex if and only if its Kähler angle is equal to 0, and V and $\sqrt{-1}V$ are perpendicular if and only if the Kähler angle of V is equal to $\pi/2$.

The linear isotropy action of the complex projective space

$$\mathbf{C}P^n = U(n+1)/U(1)\times U(n)$$

on the tangent space at the origin is equivalent to the action of $U(1)\times U(n)$ on $\mathbf{C}^n$ defined by

$$(z, A)v = zvA^* \qquad ((z, A)\in U(1)\times U(n),\ v\in\mathbf{C}^n),$$

where we regard the elements of $\mathbf{C}^n$ as row vectors. We denote by $\mathbf{e}_1,\dots,\mathbf{e}_n$ the standard unitary basis of $\mathbf{C}^n$.

LEMMA 2.1 ([**3**]). *For $0\le\theta\le\pi/2$ we denote by G^n_θ the set of real 2-dimensional subspaces whose Kähler angle is equal to θ in $\mathbf{C}^n$. Then $U(n)$ acts transitively on G^n_θ. Moreover, we put $V_\theta = \mathrm{span}_{\mathbf{R}}\{\mathbf{e}_1, \cos\theta\sqrt{-1}\mathbf{e}_1 + \sin\theta\mathbf{e}_2\}$. Then $G^n_\theta = U(n)\cdot V_\theta$.*

Thus the Kähler angle is a complete invariant of real 2-dimensional subspaces in $\mathbf{C}^n$ with respect to the action of $U(n)$, that is, for real 2-dimensional subspaces V and W in $\mathbf{C}^n$, there exists $g\in U(n)$ which satisfies $W = g\cdot V$ if and only if $\theta(V) = \theta(W)$.

Next we recall the Poincaré formula obtained by Howard [**2**]. For two vector subspaces V and W in a real vector space with an inner product, we define $\sigma(V, W)$ by

$$\sigma(V, W) = |v_1\wedge\cdots\wedge v_p\wedge w_1\wedge\cdots\wedge w_q|,$$

where $\{v_i\}$ and $\{w_j\}$ are orthonormal bases of V and W respectively. The definition of $\sigma(V, W)$ is independent of the choice of these orthonormal bases. Let G/K be a Riemannian homogeneous space. We denote by o the origin of G/K. For any points x and y in G/K and vector subspaces V and W in $T_x(G/K)$ and $T_y(G/K)$ respectively, we define $\sigma_K(V, W)$ by

$$\sigma_K(V, W) = \int_K \sigma((dg_x)_o^{-1}V, dk_o^{-1}(dg_y)_o^{-1}W)d\mu_K(k), \tag{2.1}$$

where g_x and g_y are elements in G which satisfy $g_x o = x$ and $g_y o = y$. The definition of $\sigma_K(V, W)$ is independent of the choice of g_x and g_y, because $\sigma_K(V, W)$ is defined as an integral over K. By definition, for any g in G we have

$$\sigma_K(V, W) = \sigma_K(dgV, W) = \sigma_K(V, dgW) = \sigma_K(W, V).$$

Thus it is sufficient to consider σ_K only at the tangent space $T_o(G/K)$. In this case, we can regard σ_K as a $(K \times K)$-invariant function defined on the product of the Grassmann manifolds consisting of subspaces in $T_o(G/K)$. In the case of the complex projective space, if $\dim V = \mathrm{codim} W = 2$, in order to calculate $\sigma_K(V, W)$ we can suppose that without loss of generality $V = V_\theta$ and $W = V_\tau^\perp$.

Finally we recall the coarea formula. Let V and W be vector spaces with inner products and $\dim V = m \geq n = \dim W$. For a linear mapping $F : V \to W$ we define JF by

$$JF = \sup\{|F(u_1) \wedge \cdots \wedge F(u_n)| \mid u_1, \ldots, u_n \text{ is orthonormal in } V\}.$$

If F in not surjective, then $JF = 0$. If F is surjective, then we have

$$JF = \frac{|F(v_1) \wedge \cdots \wedge F(v_n)|}{|v_1 \wedge \cdots \wedge v_n|}$$

for any basis $v_1, \ldots, v_n$ of $(\ker F)^\perp$. Using this we can formulate the coarea formula as follows. Here for a Riemannian manifold X we denote by μ_X the measure on X naturally induced from the Riemannian metric of X.

THEOREM 2.2. *Let M and N be Riemannian manifolds and $\dim M = m \geq n = \dim N$. For a smooth mapping $f : M \to N$ and a μ_M measurable function ϕ on M, $f^{-1}(y)$ is a submanifold of M for almost all y in N with respect to μ_N and the function*

$$y \mapsto \int_{f^{-1}(y)} \phi(x) d\mu_{f^{-1}(y)}(x)$$

is a μ_N measurable function on N. Moreover, if $\phi J df$ is μ_M integrable or if $\phi \geq 0$, then we have

$$\int_N \left(\int_{f^{-1}(y)} \phi(x) d\mu_{f^{-1}(y)}(x) \right) d\mu_N(y) = \int_M \phi(x) J df_x d\mu_M(x).$$

We will use this formula to change the variable of integration.

3. Reduction of integration

As we mentioned in Section 2, it is sufficient to calculate the following integral on $U(1) \times U(n)$:

$$\sigma^n(\theta, \tau) = \int_{U(1)\times U(n)} \sigma(V_\theta, k^{-1} \cdot V_\tau^\perp) d\mu(k).$$

The Poincaré formula in the case $n = 2$ has been already established by (1.3), so we can assume that $n \geq 3$. In order to calculate this explicitly we write $e(\theta) = \cos\theta\sqrt{-1}\mathbf{e}_1 + \sin\theta\mathbf{e}_2$ and take the standard inner product $\langle \cdot, \cdot \rangle$ of the Lie algebra $\mathfrak{u}(n)$ of the unitary group $U(n)$ defined by

$$\langle X, Y \rangle = -\frac{1}{2}\mathrm{trace}(XY) \quad (X, Y \in \mathfrak{u}(n)).$$

By the definition of V_θ we have

$$\sigma^n(\theta, \tau) = \mathrm{vol}(U(1)) \int_{U(n)} |\langle \mathbf{e}_1 \wedge e(\theta), (\mathbf{e}_1 \wedge e(\tau))k \rangle| d\mu(k).$$

Here we note that $\mathrm{vol}(U(1)) = 2\pi/\sqrt{2}$. We use the compact symmetric pair $(U(n), U(2) \times U(n-2))$ for the calculation of the above integral on $U(n)$, and

$$\mathfrak{u}(n) = (\mathfrak{u}(2) + \mathfrak{u}(n-2)) + \mathfrak{m}, \qquad \mathfrak{m} = \left\{ \begin{bmatrix} 0 & X \\ -X^* & 0 \end{bmatrix} \,\middle|\, X \in M_{2,n-2}(\mathbf{C}) \right\}$$

is the canonical orthogonal direct sum decomposition of $\mathfrak{u}(n)$ associated with the compact symmetric pair $(U(n), U(2) \times U(n-2))$.

When $n = 3$, we take a maximal Abelian subspace $\mathfrak{a}$ defined by

$$\mathfrak{a} = \left\{ \begin{bmatrix} 0 & 0 & \phi_1 \\ 0 & 0 & 0 \\ -\phi_1 & 0 & 0 \end{bmatrix} \,\middle|\, \phi_1 \in \mathbf{R} \right\}.$$

In this case the set of positive restricted roots for a suitable order is $\{\phi_1, 2\phi_1\}$ and their multiplicities are $m(\phi_1) = 2$ and $m(2\phi_1) = 1$. So the fundamental cell in $\mathfrak{a}$ is defined by $\pi/2 \geq \phi_1 \geq 0$.

When $n = 4$, we take a maximal Abelian subspace $\mathfrak{a}$ defined by

$$\mathfrak{a} = \left\{ \begin{bmatrix} 0 & \Phi \\ -\Phi^* & 0 \end{bmatrix} \,\middle|\, \Phi = \begin{bmatrix} \phi_1 & 0 \\ 0 & \phi_2 \end{bmatrix}, \phi_1, \phi_2 \in \mathbf{R} \right\}.$$

In this case the set of positive restricted roots for a suitable order is

$$\{\phi_1 - \phi_2, \phi_1 + \phi_2, 2\phi_1, 2\phi_2\}$$

and their multiplicities are as follows.

$$m(\phi_1 - \phi_2) = 2, \quad m(\phi_1 + \phi_2) = 2, \quad m(2\phi_1) = 1, \quad m(2\phi_2) = 1.$$

So the fundamental cell in $\mathfrak{a}$ is defined by $\pi/2 \geq \phi_1 \geq \phi_2 \geq 0$.

When $n \geq 5$, we take a maximal Abelian subspace $\mathfrak{a}$ defined by

$$\mathfrak{a} = \left\{ \begin{bmatrix} 0 & \Phi \\ -\Phi^* & 0 \end{bmatrix} \,\middle|\, \Phi = \begin{bmatrix} \phi_1 & 0 & 0 & \cdots & 0 \\ 0 & \phi_2 & 0 & \cdots & 0 \end{bmatrix}, \phi_1, \phi_2 \in \mathbf{R} \right\}.$$

In this case the set of positive restricted roots for a suitable order is

$$\{\phi_1 - \phi_2, \phi_1 + \phi_2, 2\phi_1, 2\phi_2, \phi_1, \phi_2\}$$

and their multiplicities are as follows.

$$\begin{aligned} &m(\phi_1 - \phi_2) = 2, \quad m(\phi_1 + \phi_2) = 2, \quad m(2\phi_1) = 1, \quad m(2\phi_2) = 1, \\ &m(\phi_1) = 2(n-4), \quad m(\phi_2) = 2(n-4). \end{aligned}$$

So the fundamental cell in $\mathfrak{a}$ is defined by $\pi/2 \geq \phi_1 \geq \phi_2 \geq 0$.

In each case we denote by C the image under the exponential map of the fundamental cell and write $B = U(2) \times U(n-2)$. We define a mapping

$$\rho : B \times C \times B \to U(n) \; ; (s, a, t) \mapsto sat.$$

We will apply the coarea formula to the above mapping ρ. By Theorem 8.6 of Chapter VII in Helgason [**1**], ρ is surjective.

LEMMA 3.1. *For $(s, a, t) \in B \times C \times B$ such that a belongs to the interior of C, we have*

$$\rho^{-1}(sat) = \{(sz, a, z^{-1}t) \mid z \in Z_B(\mathfrak{a})\},$$

$$\mathrm{vol}(\rho^{-1}(sat)) = \sqrt{2}^{\dim Z_B(\mathfrak{a})} \mathrm{vol}(Z_B(\mathfrak{a})),$$

where $Z_B(\mathfrak{a})$ *is the centralizer of* $\mathfrak{a}$ *in* B.

Proof. We take (s', a', t') in $\rho^{-1}(sat)$. In the complex Grassmann manifold $U(n)/B$ we have $s^{-1}s'a'B = aB$, and aB is a regular point of a maximal torus of $U(n)/B$. So $a' = a$. We denote by A the analytic subgroup of $U(n)$ corresponding to $\mathfrak{a}$. Then A is a torus in $U(n)$. The centralizer $Z(A)$ of A in $U(n)$ is connected and is equal to the centralizer $Z(\mathfrak{a})$ of $\mathfrak{a}$ in $U(n)$. The centralizer $Z_B(A)$ of A in B is equal to $Z(A) \cap B$ and the centralizer $Z_B(\mathfrak{a})$ of $\mathfrak{a}$ in B is equal to $Z(\mathfrak{a}) \cap B$. Thus we have $Z_B(A) = Z_B(\mathfrak{a})$. Since $s^{-1}s'aB = aB$, we have $s^{-1}s' \in Z_B(A) = Z_B(\mathfrak{a})$. Hence there exists $z \in Z_B(\mathfrak{a})$ which satisfies $s' = sz$ and $sat = s'a't' = szat' = sazt'$, that is, $t' = z^{-1}t$. Therefore $(s', a', t') = (sz, a, z^{-1}t)$. Conversely, any element of this form belongs to $\rho^{-1}(sat)$ and we obtain $\rho^{-1}(sat) = \{(sz, a, z^{-1}t) \mid z \in Z_B(\mathfrak{a})\}$. From this description of $\rho^{-1}(sat)$ we have

$$\mathrm{vol}(\rho^{-1}(sat)) = \sqrt{2}^{\dim Z_B(\mathfrak{a})}\mathrm{vol}(Z_B(\mathfrak{a})).$$

The following lemma is proved by direct calculation.

LEMMA 3.2. *When* $n = 3$

$$Z_B(\mathfrak{a}) = \left\{ \begin{bmatrix} z_1 & 0 & 0 \\ 0 & z_2 & 0 \\ 0 & 0 & z_1 \end{bmatrix} \middle| \; z_1, z_2 \in U(1) \right\}$$

$$\mathrm{vol}(Z_B(\mathfrak{a})) = \sqrt{2}\mathrm{vol}(U(1))^2.$$

When $n = 4$

$$Z_B(\mathfrak{a}) = \left\{ \begin{bmatrix} z_1 & 0 & 0 & 0 \\ 0 & z_2 & 0 & 0 \\ 0 & 0 & z_1 & 0 \\ 0 & 0 & 0 & z_2 \end{bmatrix} \middle| \; z_1, z_2 \in U(1) \right\}$$

$$\mathrm{vol}(Z_B(\mathfrak{a})) = 2\mathrm{vol}(U(1))^2.$$

When $n \geq 5$

$$Z_B(\mathfrak{a}) = \left\{ \begin{bmatrix} z & 0 & 0 \\ 0 & z & 0 \\ 0 & 0 & w \end{bmatrix} \middle| \; z \in U(1) \times U(1),\ w \in U(n-4) \right\}$$

$$\mathrm{vol}(Z_B(\mathfrak{a})) = 2\mathrm{vol}(U(1))^2\mathrm{vol}(U(n-4)).$$

LEMMA 3.3. *We take* $(s, a, t) \in B \times C \times B$. *Then we have*

$$Jd\rho_{(s,a,t)} = \sqrt{2}^{\dim Z_B(\mathfrak{a})} \prod_{\lambda} \sin^{m(\lambda)} \lambda,$$

where λ *runs through the positive restricted roots.*

Proof. The mapping ρ is left and right B-equivariant, so it is sufficient to calculate $Jd\rho_{(e,a,e)}$. We denote by $\mathfrak{z}_B(\mathfrak{a})$ the Lie algebra of $Z_B(\mathfrak{a})$. Lemma 3.1 implies $dL^{-1}_{(e,a,e)}\mathrm{ker}d\rho_{(e,a,e)} = \{(X, 0, -X) \mid X \in \mathfrak{z}_B(\mathfrak{a})\}$. Thus we have

$$\begin{aligned} &dL^{-1}_{(e,a,e)}(\mathrm{ker}d\rho_{(e,a,e)})^{\perp} \\ &= \{(X, 0, X) \mid X \in \mathfrak{z}_B(\mathfrak{a})\} \oplus \{(Y, 0, Z) \mid Y, Z \in \mathfrak{z}_B(\mathfrak{a})^{\perp}\} \oplus \{(0, H, 0) \mid H \in \mathfrak{a}\}, \end{aligned}$$

which is an orthogonal direct sum decomposition. We calculate the action of $d\rho$ on each summand. For $X \in \mathfrak{z}_B(\mathfrak{a})$, $Y, Z \in \mathfrak{z}_B(\mathfrak{a})^\perp$ and $H \in \mathfrak{a}$ we have

$$\begin{aligned} df(X, 0, X) &= dL_a(2X), \\ df(Y, 0, Z) &= dL_a(\mathrm{Ad}(a)^{-1}Y + Z), \\ df(0, dL_aH, 0) &= dL_aH. \end{aligned}$$

Using these equations and the root space decomposition we can obtain

$$Jd\rho_{(s,a,t)} = \sqrt{2}^{\dim Z_B(\mathfrak{a})} \prod_\lambda \sin^{m(\lambda)} \lambda.$$

By the three lemmas mentioned above and the coarea formula, we have

$$\begin{aligned} &\int_{B\times C\times B} |\langle \mathbf{e}_1 \wedge e(\theta), (\mathbf{e}_1 \wedge e(\tau))sat\rangle| \sqrt{2}^{\dim Z_B(\mathfrak{a})} \prod_\lambda \sin^{m(\lambda)} \lambda d\mu(s, a, t) \\ &= \sqrt{2}^{\dim Z_B(\mathfrak{a})} \mathrm{vol}(Z_B(\mathfrak{a})) \int_{U(n)} |\langle \mathbf{e}_1 \wedge e(\theta), (\mathbf{e}_1 \wedge e(\tau))k\rangle| d\mu(k) \end{aligned}$$

and

$$\begin{aligned} &\int_{U(n)} |\langle \mathbf{e}_1 \wedge e(\theta), (\mathbf{e}_1 \wedge e(\tau))k\rangle| d\mu(k) \\ &= \frac{1}{\mathrm{vol}(Z_B(\mathfrak{a}))} \int_{B\times C\times B} |\langle \mathbf{e}_1 \wedge e(\theta), (\mathbf{e}_1 \wedge e(\tau))sat\rangle| \prod_\lambda \sin^{m(\lambda)} \lambda d\mu(s, a, t). \end{aligned}$$

Thus we have

$$\begin{aligned} &\sigma^n(\theta, \tau) \\ &= \mathrm{vol}(U(1)) \int_{U(n)} |\langle \mathbf{e}_1 \wedge e(\theta), (\mathbf{e}_1 \wedge e(\tau))k\rangle| d\mu(k) \\ &= \frac{\mathrm{vol}(U(1))}{\mathrm{vol}(Z_B(\mathfrak{a}))} \int_{B\times C\times B} |\langle \mathbf{e}_1 \wedge e(\theta), (\mathbf{e}_1 \wedge e(\tau))sat\rangle| \prod_\lambda \sin^{m(\lambda)} \lambda d\mu(s, a, t) \\ &= \frac{\mathrm{vol}(U(1))}{\mathrm{vol}(Z_B(\mathfrak{a}))} \\ &\quad \times \int_C \left(\int_{B\times B} |\langle \mathbf{e}_1 \wedge e(\theta), (\mathbf{e}_1 \wedge e(\tau))sat\rangle| d\mu(s, t) \right) \prod_\lambda \sin^{m(\lambda)} \lambda d\mu(a). \end{aligned}$$

Since the orthogonal projection $P : \mathbf{C}^n \to \mathbf{C}^2$ and the actions of elements of B preserve $\mathbf{C}^2$, we have

$$\begin{aligned} &\int_{B\times B} |\langle \mathbf{e}_1 \wedge e(\theta), (\mathbf{e}_1 \wedge e(\tau))sat\rangle| d\mu(s, t) \\ &= \int_B \left(\int_B |\langle (\mathbf{e}_1 \wedge e(\theta))t^{-1}, P((\mathbf{e}_1 \wedge e(\tau))sa)\rangle| d\mu(t) \right) d\mu(s). \end{aligned}$$

$P((\mathbf{e}_1 \wedge e(\tau))sa)$ represents the subspace $P(V_\tau sa)$. For almost all s and a the dimension of $P(V_\tau sa)$ is 2. We denote by ψ its Kähler angle. Of course ψ is a function of τ, s and a. Using the result in the case $n = 2$ mentioned in Section 1, we have

$$\begin{aligned}
&\int_B |\langle (\mathbf{e}_1 \wedge e(\theta))t^{-1}, P((\mathbf{e}_1 \wedge e(\tau))sa)\rangle| d\mu(t) \\
&= |P((\mathbf{e}_1 \wedge e(\tau))sa)| \int_B |\langle (\mathbf{e}_1 \wedge e(\theta))t^{-1}, \mathbf{e}_1 \wedge e(\psi)\rangle| d\mu(t) \\
&= \frac{\mathrm{vol}(U(n-2))}{\mathrm{vol}(U(1))} |P((\mathbf{e}_1 \wedge e(\tau))sa)| \sigma^2(\theta, \psi) \\
&= \frac{\mathrm{vol}(U(n-2))\mathrm{vol}(U(3))}{\mathrm{vol}(U(1))\mathrm{vol}(\mathbf{C}P^1)^2} |P((\mathbf{e}_1 \wedge e(\tau))sa)| \\
&\quad \times \left(\frac{1}{4}(1+\cos^2\theta)(1+\cos^2\psi) + \frac{1}{2}\sin^2\theta \sin^2\psi \right).
\end{aligned}$$

So the integral on $B \times B$ is given as follows.

$$\begin{aligned}
&\int_{B\times B} |\langle \mathbf{e}_1 \wedge e(\theta), (\mathbf{e}_1 \wedge e(\tau))sat\rangle| d\mu(s,t) \\
&= \frac{\mathrm{vol}(U(n-2))\mathrm{vol}(U(3))}{\mathrm{vol}(U(1))\mathrm{vol}(\mathbf{C}P^1)^2} \int_B |P((\mathbf{e}_1 \wedge e(\tau))sa)| \\
&\quad \times \left(\frac{1}{4}(1+\cos^2\theta)(1+\cos^2\psi) + \frac{1}{2}\sin^2\theta \sin^2\psi \right) d\mu(s) \\
&= \frac{\mathrm{vol}(U(n-2))^2\mathrm{vol}(U(3))}{\mathrm{vol}(U(1))\mathrm{vol}(\mathbf{C}P^1)^2} \int_{U(2)} |P((\mathbf{e}_1 \wedge e(\tau))sa)| \\
&\quad \times \left(\frac{1}{4}(1+\cos^2\theta)(1+\cos^2\psi) + \frac{1}{2}\sin^2\theta \sin^2\psi \right) d\mu(s).
\end{aligned}$$

In the case $\tau = 0$ we have already obtained a description of $\sigma^n(\theta, 0)$ in [**3**], which coincides with the result of Theorem 1.1. So we can assume $\tau \neq 0$. We define a mapping $p : U(2) \to (\mathbf{e}_1 \wedge e(\tau))U(2)\,;\, k \mapsto (\mathbf{e}_1 \wedge e(\tau))k$. As is shown in [**4**], we have $Jdp = 2\sqrt{2}\sin\tau$. By the coarea formula,

$$\begin{aligned}
&2\sqrt{2}\sin\tau \int_{U(2)} |P((\mathbf{e}_1 \wedge e(\tau))sa)| \\
&\quad \times \left(\frac{1}{4}(1+\cos^2\theta)(1+\cos^2\psi) + \frac{1}{2}\sin^2\theta \sin^2\psi \right) d\mu(s) \\
&= \mathrm{vol}(SO(2)) \int_{(\mathbf{e}_1 \wedge e(\tau))U(2)} |P(\xi a)| \\
&\quad \times \left(\frac{1}{4}(1+\cos^2\theta)(1+\cos^2\psi) + \frac{1}{2}\sin^2\theta \sin^2\psi \right) d\mu(\xi),
\end{aligned}$$

where ψ in the last integral is the Kähler angle of the subspace represented by $P(\xi a)$. So ψ is a function of ξ and a. In order to calculate the last integral we need explicit descriptions of $|P(\xi a)|$ and ψ. Since $\mathbf{e}_1$ and $e(\tau)$ belong to $\mathbf{C}^2$, we can regard $\mathbf{e}_1 \wedge e(\tau) \in \wedge_2^{\mathbf{R}}\mathbf{C}^2$ and $(\mathbf{e}_1 \wedge e(\tau))U(2) \subset \wedge_2^{\mathbf{R}}\mathbf{C}^2$. We define an orthonormal basis $\{A_i, B_i\}$ of $\wedge_2^{\mathbf{R}}\mathbf{C}^2$ by

$$\begin{aligned} A_1 &= \frac{1}{\sqrt{2}}(\mathbf{e}_1 \wedge \sqrt{-1}\mathbf{e}_1 + \mathbf{e}_2 \wedge \sqrt{-1}\mathbf{e}_2), \\ A_2 &= \frac{1}{\sqrt{2}}(\mathbf{e}_1 \wedge \mathbf{e}_2 - \sqrt{-1}\mathbf{e}_1 \wedge \sqrt{-1}\mathbf{e}_2), \\ A_3 &= \frac{1}{\sqrt{2}}(\mathbf{e}_1 \wedge \sqrt{-1}\mathbf{e}_2 + \sqrt{-1}\mathbf{e}_1 \wedge \mathbf{e}_2), \\ B_1 &= \frac{1}{\sqrt{2}}(\mathbf{e}_1 \wedge \sqrt{-1}\mathbf{e}_1 - \mathbf{e}_2 \wedge \sqrt{-1}\mathbf{e}_2), \\ B_2 &= \frac{1}{\sqrt{2}}(\mathbf{e}_1 \wedge \mathbf{e}_2 + \sqrt{-1}\mathbf{e}_1 \wedge \sqrt{-1}\mathbf{e}_2), \\ B_3 &= \frac{1}{\sqrt{2}}(\mathbf{e}_1 \wedge \sqrt{-1}\mathbf{e}_2 - \sqrt{-1}\mathbf{e}_1 \wedge \mathbf{e}_2). \end{aligned}$$

Using this orthonormal basis we can describe $(\mathbf{e}_1 \wedge e(\tau))U(2)$ as follows.

$$(\mathbf{e}_1 \wedge e(\tau))U(2) = \left(\frac{\cos\tau}{\sqrt{2}}A_1 + S^1\left(\frac{\sin\tau}{\sqrt{2}}\right)\right) \times S^2\left(\frac{1}{\sqrt{2}}\right),$$

where $S^1(\sin\tau/\sqrt{2})$ is the circle of radius $\sin\tau/\sqrt{2}$ in $\mathrm{span}_{\mathbf{R}}\{A_2, A_3\}$ and $S^2(1/\sqrt{2})$ is the 2-dimensional sphere of radius $1/\sqrt{2}$ in $\mathrm{span}_{\mathbf{R}}\{B_1, B_2, B_3\}$. The variable $\xi \in (\mathbf{e}_1 \wedge e(\tau))U(2)$ of the integral is described as follows.

$$\begin{aligned} \xi &= \frac{\cos\tau}{\sqrt{2}}A_1 + x_2A_2 + x_3A_3 + y_1B_1 + y_2B_2 + y_3B_3 \\ &\left(x_2^2 + x_3^2 = \frac{\sin^2\tau}{2}, y_1^2 + y_2^2 + y_3^2 = \frac{1}{2}\right). \end{aligned}$$

We take

$$a = \exp\begin{bmatrix} 0 & \Phi \\ -\Phi^* & 0 \end{bmatrix} \in C \qquad \left(\Phi = \begin{bmatrix} \phi_1 & 0 & 0 & \cdots & 0 \\ 0 & \phi_2 & 0 & \cdots & 0 \end{bmatrix}\right).$$

We have

$$\begin{aligned} P(A_1a) &= \frac{1}{2}(\cos^2\phi + \cos^2\phi_2)A_1 + \frac{1}{2}(\cos^2\phi - \cos^2\phi_2)B_1, \\ P(A_2a) &= \cos\phi_1\cos\phi_2A_2, \\ P(A_3a) &= \cos\phi_1\cos\phi_2A_3, \\ P(B_1a) &= \frac{1}{2}(\cos^2\phi_1 - \cos^2\phi_2)A_1 + \frac{1}{2}(\cos^2\phi_1 + \cos^2\phi_2)B_1, \\ P(B_2a) &= \cos\phi_1\cos\phi_2B_2, \\ P(B_3a) &= \cos\phi_1\cos\phi_2B_3 \end{aligned}$$

and

$$\begin{aligned} P(\xi a) = \frac{1}{2}&\left(\frac{\cos\tau}{\sqrt{2}}(\cos^2\phi_1 + \cos\phi_2) + (\cos^2\phi_1 - \cos\phi_2)y_1\right)A_1 \\ &+ \cos\phi_1\cos\phi_2x_2A_2 + \cos\phi_1\cos\phi_2x_3A_3 \\ &+ \frac{1}{2}\left(\frac{\cos\tau}{\sqrt{2}}(\cos^2\phi_1 - \cos\phi_2) + (\cos^2\phi_1 + \cos\phi_2)y_1\right)B_1 \\ &+ \cos\phi_1\cos\phi_2y_2B_2 + \cos\phi_1\cos\phi_2y_3B_3. \end{aligned}$$

In order to express $|P(\xi a)|$ simply we write

$$f(y_1) = \frac{\cos\tau}{2}(\cos^2\phi_1 + \cos^2\phi_2) + \frac{1}{\sqrt{2}}(\cos^2\phi_1 - \cos^2\phi_2)y_1.$$

Then we have $|P(\xi a)| = \sqrt{f(y_1)^2 + \sin^2\tau\cos^2\phi_1\cos^2\phi_2}$. Moreover, by the above expression for $P(\xi a)$ we have $\cos\psi = f(y_1)/|P(\xi a)|$. In the case from $n = 3$, we substitute $\phi_2 = 0$ in the above equations. The integrand of the above integral on $(\mathbf{e}_1 \wedge e(\tau))U(2)$ is

$$\begin{aligned}
&|P(\xi a)|\left(\frac{1}{4}(1+\cos^2\theta)(1+\cos^2\psi) + \frac{1}{2}\sin^2\theta\sin^2\psi\right)\\
&= \frac{1}{4}\left(2(1+\cos^2\theta)|P(\xi a)| - \frac{(2\cos^2\theta - \sin^2\theta)\sin^2\tau\cos^2\phi_1\cos^2\phi_2}{|P(\xi a)|}\right)
\end{aligned}$$

and we have

$$\begin{aligned}
&\int_{B\times B} |\langle \mathbf{e}_1 \wedge e(\theta), (\mathbf{e}_1 \wedge e(\tau))sat\rangle| d\mu(s,t)\\
&= \frac{\pi \mathrm{vol}(SO(2))\mathrm{vol}(U(n-2))^2\mathrm{vol}(U(3))}{8\mathrm{vol}(U(1))\mathrm{vol}(\mathbf{C}P^1)^2}\int_{S^2(1/\sqrt{2})}\\
&\left(2(1+\cos^2\theta)|P(\xi a)| - \frac{(2\cos^2\theta - \sin^2\theta)\sin^2\tau\cos^2\phi_1\cos^2\phi_2}{|P(\xi a)|}\right) d\mu(\xi).
\end{aligned}$$

For any $r > 0$ the mapping

$$(-r, r) \times S^1(r) \to S^2(r)\,;\,(t, x) \mapsto \left(t, \sqrt{1-(t/r)^2}x\right)$$

preserves area. Using this mapping to change variables, we have

$$\begin{aligned}
&\int_{B\times B} |\langle \mathbf{e}_1 \wedge e(\theta), (\mathbf{e}_1 \wedge e(\tau))sat\rangle| d\mu(s,t)\\
&= \frac{\pi^2 \mathrm{vol}(SO(2))\mathrm{vol}(U(n-2))^2\mathrm{vol}(U(3))}{4\sqrt{2}\mathrm{vol}(U(1))\mathrm{vol}(\mathbf{C}P^1)^2}\\
&\times \int_{-1/\sqrt{2}}^{1/\sqrt{2}} \left(2(1+\cos^2\theta)\sqrt{f(t)^2 + \sin^2\tau\cos^2\phi_1\cos^2\phi_2}\right.\\
&\left. - \frac{(2\cos^2\theta - \sin^2\theta)\sin^2\tau\cos^2\phi_1\cos^2\phi_2}{\sqrt{f(t)^2 + \sin^2\tau\cos^2\phi_1\cos^2\phi_2}}\right) dt.
\end{aligned}$$

Here we use $u = f(t)$ and the formulae $\int 1/\sqrt{x^2+c}dx = \log\left|x+\sqrt{x^2+c}\right|$, $\int \sqrt{x^2+c}dx = \frac{1}{2}\left(x\sqrt{x^2+c}+c\log\left|x+\sqrt{x^2+c}\right|\right)$, hence we obtain

$$\begin{aligned}
&\int_{-1/\sqrt{2}}^{1/\sqrt{2}}\left(2(1+\cos^2\theta)\sqrt{f(t)^2+\sin^2\tau\cos^2\phi_1\cos^2\phi_2}\right.\\
&\left.-\frac{(2\cos^2\theta-\sin^2\theta)\sin^2\tau\cos^2\phi_1\cos^2\phi_2}{\sqrt{f(t)^2+\sin^2\tau\cos^2\phi_1\cos^2\phi_2}}\right)dt\\
=&\frac{\sqrt{2}}{\cos^2\phi_2-\cos^2\phi_1}\int_{f(1/\sqrt{2})}^{f(-1/\sqrt{2})}\left(2(1+\cos^2\theta)\sqrt{u^2+\sin^2\tau\cos^2\phi_1\cos^2\phi_2}\right.\\
&\left.-\frac{(2\cos^2\theta-\sin^2\theta)\sin^2\tau\cos^2\phi_1\cos^2\phi_2}{\sqrt{u^2+\sin^2\tau\cos^2\phi_1\cos^2\phi_2}}\right)du\\
=&\frac{\sqrt{2}}{\cos^2\phi_2-\cos^2\phi_1}\left[(1+\cos^2\theta)u\sqrt{u^2+\sin^2\tau\cos^2\phi_1\cos^2\phi}\right.\\
&+2\sin^2\theta\sin^2\tau\cos^2\phi_1\cos^2\phi_2\\
&\left.\times\log\left|u+\sqrt{u^2+\sin^2\tau\cos^2\phi_1\cos^2\phi_2}\right|\right]_{f(1/\sqrt{2})}^{f(-1/\sqrt{2})}\\
=&\frac{\sqrt{2}}{\cos^2\phi_2-\cos^2\phi_1}\left(\frac{1}{2}(1+\cos^2\theta)(1+\cos^2\tau)(\cos^4\phi_2-\cos^4\phi_1)\right.\\
&\left.+2\sin^2\theta\sin^2\tau\cos^2\phi_1\cos^2\phi_2(\log\cos^2\phi_2-\log\cos^2\phi_1)\right).
\end{aligned}$$

Thus the integral on $B \times B$ is

$$\begin{aligned}
&\int_{B\times B}|\langle \mathbf{e}_1\wedge e(\theta),(\mathbf{e}_1\wedge e(\tau))sat\rangle|d\mu(s,t)\\
=&\frac{\pi^2\mathrm{vol}(SO(2))\mathrm{vol}(U(n-2))^2\mathrm{vol}(U(3))}{4\mathrm{vol}(U(1))\mathrm{vol}(\mathbf{C}P^1)^2(\cos^2\phi_2-\cos^2\phi_1)}\\
&\times\left(\frac{1}{2}(1+\cos^2\theta)(1+\cos^2\tau)(\cos^4\phi_2-\cos^4\phi_1)\right.\\
&\left.+2\sin^2\theta\sin^2\tau\cos^2\phi_1\cos^2\phi_2(\log\cos^2\phi_2-\log\cos^2\phi_1)\right)
\end{aligned}$$

and we obtain

$$\begin{aligned}
\sigma^n(\theta,\tau)=&\frac{\pi^2\mathrm{vol}(SO(2))\mathrm{vol}(U(n-2))^2\mathrm{vol}(U(3))}{4\mathrm{vol}(Z_B(\mathfrak{a}))\mathrm{vol}(\mathbf{C}P^1)^2}\\
&\times\int_C\left(\frac{1}{2}(1+\cos^2\theta)(1+\cos^2\tau)(\cos^4\phi_2-\cos^4\phi_1)\right.\\
&\left.+2\sin^2\theta\sin^2\tau\cos^2\phi_1\cos^2\phi_2(\log\cos^2\phi_2-\log\cos^2\phi_1)\right)\\
&\cdot\frac{\prod_\lambda\sin^{m(\lambda)}\lambda}{\cos^2\phi_2-\cos^2\phi_1}d\phi_1d\phi_2.
\end{aligned}$$

4. Calculation of σ^n

We derive an explicit description of $\sigma^n(\theta,\tau)$ in this section by using the last equation of Section 3. In the case $n \geq 4$, from the equation $\sin^2(\phi_1-\phi_2)\sin^2(\phi_1+$

$\phi_2) = (\cos^2\phi_2 - \cos^2\phi_1)^2$ and the results obtained in Section 3, we have

$$
\begin{aligned}
&\sigma^n(\theta,\tau)\\
&= \frac{\pi^2 \mathrm{vol}(SO(2))\mathrm{vol}(U(n-2))^2\mathrm{vol}(U(3))}{8\mathrm{vol}(U(1))^2\mathrm{vol}(U(n-4))\mathrm{vol}(\mathbf{C}P^1)^2}\\
&\times \int_C \Big(\frac{1}{2}(1+\cos^2\theta)(1+\cos^2\tau)(\cos^4\phi_2 - \cos^4\phi_1)\\
&+2\sin^2\theta\sin^2\tau\cos^2\phi_1\cos^2\phi_2(\log\cos^2\phi_2 - \log\cos^2\phi_1)\Big)\\
&\cdot(\cos^2\phi_2 - \cos^2\phi_1)\sin 2\phi_1 \sin 2\phi_2 \sin^{2(n-4)}\phi_1 \sin^{2(n-4)}\phi_2 d\phi_1 d\phi_2\\
&= \frac{\mathrm{vol}(U(n+1))}{\mathrm{vol}(\mathbf{C}P^1)\mathrm{vol}(\mathbf{C}P^{n-1})}\cdot\frac{n(n-1)(n-2)^2(n-3)}{8}\\
&\times \int_C \Big(\frac{1}{2}(1+\cos^2\theta)(1+\cos^2\tau)(\cos^4\phi_2 - \cos^4\phi_1)\\
&+2\sin^2\theta\sin^2\tau\cos^2\phi_1\cos^2\phi_2(\log\cos^2\phi_2 - \log\cos^2\phi_1)\Big)\\
&\cdot(\cos^2\phi_2 - \cos^2\phi_1)\sin 2\phi_1 \sin 2\phi_2 \sin^{2(n-4)}\phi_1 \sin^{2(n-4)}\phi_2 d\phi_1 d\phi_2\\
&= \frac{\mathrm{vol}(U(n+1))}{\mathrm{vol}(\mathbf{C}P^1)\mathrm{vol}(\mathbf{C}P^{n-1})}\cdot\frac{n(n-1)(n-2)^2(n-3)}{8}\\
&\times \int_0^1\int_0^{v_1} \Big(\frac{1}{2}(1+\cos^2\theta)(1+\cos^2\tau)((1-v_2)^2 - (1-v_1)^2)\\
&+2\sin^2\theta\sin^2\tau(1-v_1)(1-v_2)(\log(1-v_2) - \log(1-v_1))\Big)\\
&\cdot(v_1 - v_2)v_1^{n-4}v_2^{n-4}dv_2 dv_1,
\end{aligned}
$$

where we change the variables by $v_1 = \sin^2\phi_1$ and $v_2 = \sin^2\phi_2$. Calculations show $\int_0^1\int_0^{v_1}\frac{1}{2}((1-v_2)^2 - (1-v_1)^2)(v_1-v_2)v_1^{n-4}v_2^{n-4}dv_2dv_1 = \frac{2}{(n-3)(n-2)^2(n-1)n}$ and $\int_0^1\int_0^{v_1} 2(1-v_1)(1-v_2)(\log(1-v_2) - \log(1-v_1)) \times (v_1-v_2)v_1^{n-4}v_2^{n-4}dv_2dv_1 = \frac{2}{(n-3)(n-2)^2(n-1)^2}$. Therefore we obtain

$$
\begin{aligned}
&\sigma^n(\theta,\tau)\\
&= \frac{\mathrm{vol}(U(n+1))}{\mathrm{vol}(\mathbf{C}P^1)\mathrm{vol}(\mathbf{C}P^{n-1})}\cdot\frac{n(n-1)(n-2)^2(n-3)}{8}\\
&\times \left(\frac{2}{(n-3)(n-2)^2(n-1)n}(1+\cos^2\theta)(1+\cos^2\tau)\right.\\
&\left.+\frac{2}{(n-3)(n-2)^2(n-1)^2}\sin^2\theta\sin^2\tau\right)\\
&= \frac{\mathrm{vol}(U(n+1))}{\mathrm{vol}(\mathbf{C}P^1)\mathrm{vol}(\mathbf{C}P^{n-1})}\\
&\times\left(\frac{1}{4}(1+\cos^2\theta)(1+\cos^2\tau) + \frac{n}{4(n-1)}\sin^2\theta\sin^2\tau\right).
\end{aligned}
$$

In the case of $n = 3$, we can similarly calculate $\sigma^3(\theta, \tau)$ and obtain

$$\sigma^3(\theta,\tau) = \frac{\mathrm{vol}(U(4))}{\mathrm{vol}(\mathbf{C}P^1)\mathrm{vol}(\mathbf{C}P^2)} \times \left(\frac{1}{4}(1+\cos^2\theta)(1+\cos^2\tau) + \frac{3}{8}\sin^2\theta\sin^2\tau\right),$$

which completes the proof of Theorem 1.1.

References

[1] S. Helgason, Differential Geometry, Lie Groups, and Symmetric Spaces. Academic Press, New York London, 1978.
[2] R. Howard, The kinematic formula in Riemannian homogeneous spaces, Mem. Amer. Math. Soc., No.509, **106**, (1993).
[3] H.J. Kang and H. Tasaki, Integral geometry of real surfaces in complex projective spaces, Tsukuba J. Math., **25** (2001), 155–164.
[4] H.J. Kang and H. Tasaki, Integral geometry of real surfaces in the complex projective plane, Geometriae Dedicata, **90**, (2002), 99–106
[5] L.A. Santaló, Integral geometry in Hermitian spaces, Amer. J. Math., **74**, (1952), 423 – 434.
[6] H. Tasaki, Generalization of Kähler angle and integral geometry in complex projective spaces, "Steps in Differential Geometry", Proceedings of Colloquium on Differential Geometry, Debrecen, 2000 (edited by L. Kozma, P. T. Nagy and L. Tamássy) Published by the Institute of Mathematics and Informatics, University of Debrecen, 2001. Available electronically at: `http://www.emis.de/proceedings/CDGD2000/`

Institute of Mathematics, University of Tsukuba, Tsukuba, Ibaraki 305-8571, JAPAN

E-mail address: `tasaki@math.tsukuba.ac.jp`

Contemporary Mathematics
Volume **308**, 2002

Jacobi Fields along Harmonic Maps

John C. Wood

1. Introduction

A *Jacobi field* (for the energy) along a harmonic map φ is a vector field along φ which is in the kernel of the second variation of the energy. Equivalently, it is tangent to a variation of φ for which the tension field remains zero 'to first order'. We shall show that Jacobi fields preserve several properties of a harmonic map to first order, namely holomorphicity for maps between compact Kähler manifolds, conformality of maps from the 2-sphere, real isotropy of a harmonic map from the 2-sphere to a space form, and complex isotropy of a harmonic map from the 2-sphere to a complex space form. The main idea of the proofs is to define differentials depending on the map and the Jacobi field which are shown to be holomorphic and so vanish; thus generalizing the proof of isotropy of a harmonic map from the 2-sphere as given, for example, in [**22**].

Note that any harmonic map from the 2-sphere is weakly conformal and so is the same thing as a *minimal branched immersion* in the sense of [**12**].

Lastly, we mention that the Jacobi equation is preserved by *harmonic morphisms*; see [**19**] for this result, [**11**] for a bibliography and [**1**] for an account of the theory of harmonic morphisms.

The author thanks the organizers of the 9th MSJ-IRI, Tokyo 2000, for inviting him to present this paper, and Luc Lemaire for some useful comments on it.

2. Jacobi fields

Throughout this paper $M = (M, g)$ and $N = (N, h)$ will denote smooth (i.e. C^∞) compact Riemannian manifolds without boundary and $\varphi : M \to N$ will denote a smooth map. For any vector bundle $E \to M$, $\Gamma(E)$ will denote the space of smooth sections of E. Particularly important is the *pull-back bundle* $\varphi^{-1}TN \to M$; its smooth sections are called *vector fields along* φ.

2000 *Mathematics Subject Classification.* Primary 58E20, 53C43, 53A10.

We recall some basic definitions, see [**8**] for more details. The *energy* of φ is defined by the integral

$$E(\varphi) = \tfrac{1}{2}\int_M |\mathrm{d}\varphi|^2\,\omega_g$$

where $|\mathrm{d}\varphi|$ denotes the Hilbert–Schmidt norm of the differential of φ and ω_g denotes the volume form of the metric g. By a *(smooth) (1-parameter) variation* $\{\varphi_t\}$ *of* φ we mean a smooth map $M \times I \to N$, $(x,t) \mapsto \varphi_t(x)$, where I is an open interval of the real line containing 0, such that $\varphi_0 = \varphi$.

Given a smooth variation $\{\varphi_t\}$, we set

$$v = \left.\frac{\partial \varphi_t}{\partial t}\right|_0$$

where $\left.\frac{\partial}{\partial t}\right|_0$ denotes (partial) derivative with respect to t at $t = 0$; this defines a vector field along φ called the *variation vector field of* $\{\varphi_t\}$. Then, for any smooth map $\varphi : M \to N$ and any smooth variation $\{\varphi_t\}$ of it we have

$$\left.\frac{\mathrm{d}}{\mathrm{d}t}\right|_0 E(\varphi_t) = -\int_M \langle \tau(\varphi), v\rangle\,\omega_g \tag{1}$$

where $\tau(\varphi)$ denotes the vector field along φ called the *tension field* of φ given by $\tau(\varphi) = \operatorname{Trace}\nabla\mathrm{d}\varphi$.

Here and in the sequel, $\langle\ ,\ \rangle$ and ∇ denote the inner product and connection induced on the relevant bundle by the metrics and Levi-Civita connections on M and N, see, for example, [**8**] for this formalism.

The formula (1) shows that the number $\left.\frac{\mathrm{d}}{\mathrm{d}t}\right|_0 E(\varphi_t)$ depends only on the variation vector field v of $\{\varphi_t\}$. Given any $v \in \Gamma(\varphi^{-1}TN)$, there are infinitely many smooth variations $\{\varphi_t\}$ *tangent to* (i.e. *with variation vector field*) v. Writing the left-hand side of (1) as $\nabla_v E$ gives the *first variation formula for the energy*:

$$\nabla_v E = -\int_M \langle \tau(\varphi), v\rangle\,\omega_g\,. \tag{2}$$

A smooth map φ is called *harmonic* if the first variation (2) is zero for all $v \in \Gamma(\varphi^{-1}TN)$; equivalently, from (2), φ satisfies the equation

$$\tau(\varphi) = 0\,. \tag{3}$$

In local coordinates this is an elliptic semilinear system, not linear except when (N,h) is flat.

Now suppose that $\varphi : M \to N$ is harmonic. Then we can define its *second variation* as follows. By a *(smooth)* 2*-parameter variation* $\{\varphi_{t,s}\}$ of φ we mean a smooth map $M \times I^2 \to N$, $(x,t,s) \mapsto \varphi_{t,s}(x)$, where I^2 is an open connected subset of $\mathbb{R}^2$ containing $(0,0)$, such that $\varphi_{0,0} = \varphi$. The *Hessian of* φ is defined on a pair v, w of vector fields along φ by

$$H_\varphi(v,w) = \left.\frac{\partial^2 E(\varphi_{t,s})}{\partial t \partial s}\right|_{(0,0)} \tag{4}$$

where $\{\varphi_{t,s}\}$ is a smooth 2-parameter variation of φ with

$$\left.\frac{\partial \varphi_{t,s}}{\partial t}\right|_{(0,0)} = v \quad\text{and}\quad \left.\frac{\partial \varphi_{t,s}}{\partial s}\right|_{(0,0)} = w\,.$$

Then the left-hand side of (4) depends only on v and w and is given by

$$H_\varphi(v,w) = \int_M \langle J_\varphi(v), w\rangle\, \omega_g \tag{5}$$

where

$$J_\varphi(v) = \Delta v - \operatorname{Trace} R^N(\mathrm{d}\varphi, v)\mathrm{d}\varphi\,;$$

here Δ denotes the Laplacian on $\varphi^{-1}TN$ and R^N the curvature operator of N (conventions as in [**8**]). The mapping $J_\varphi : \Gamma(\varphi^{-1}TN) \to \Gamma(\varphi^{-1}TN)$ is called the *Jacobi operator* (for the energy); it is a self-adjoint elliptic *linear* operator. A vector field v along φ is called a *Jacobi field (along φ)* if it is in the kernel of the Jacobi operator, i.e., it satisfies the *Jacobi equation*

$$J_\varphi(v) \equiv \Delta v - \operatorname{Trace} R^N(\mathrm{d}\varphi, v)\mathrm{d}\varphi = 0\,. \tag{6}$$

By standard elliptic theory (see, for example, [**18**]) the set of Jacobi fields along a harmonic map is a finite dimensional vector subspace of $\Gamma(\varphi^{-1}TN)$.

We shall make use of the following interpretation of the Jacobi operator as the *linearization* of the tension field.

PROPOSITION 2.1. *Let $\varphi : M \to N$ be harmonic and let $v \in \Gamma(\varphi^{-1}TN)$. Let $\{\varphi_t\}$ be a smooth variation of φ tangent to v. Then*

$$J_\varphi(v) = -\left.\frac{\partial}{\partial t}\right|_0 \tau(\varphi_t)\,.$$

In particular, v is a Jacobi field along φ if and only if the tension field of φ_t is zero to first order, i.e.,

$$\left.\frac{\partial}{\partial t}\right|_0 \tau(\varphi_t) = \tau(\varphi) = 0\,. \tag{7}$$

PROOF. From (4) we have

$$\begin{aligned} H_\varphi(v,w) &= \left.\frac{\mathrm{d}}{\mathrm{d}t}\right|_0 \Big(\nabla_w E(\varphi_{t,0})\Big) \\ &= -\left.\frac{\mathrm{d}}{\mathrm{d}t}\right|_0 \int_M \big\langle \tau(\varphi_{t,0}), w\big\rangle\, \omega_g \\ &= \int_M \Big\langle -\left.\frac{\partial}{\partial t}\right|_0 \tau(\varphi_{t,0}), w\Big\rangle\, \omega_g\,. \end{aligned}$$

Comparing this with (5) gives the statement. □

We shall write the condition (7) succinctly as

$$\tau(\varphi_t) = o(t) \tag{8}$$

and we shall call a smooth variation $\{\varphi_t\}$ *harmonic to first order* if it satisfies this condition. Then the Proposition tells us that *a smooth variation $\{\varphi_t\}$ of a harmonic map φ is harmonic to first order if and only if it is tangent to a Jacobi field along φ.*

In particular, if $\{\varphi_t\}$ is a smooth variation of φ through harmonic maps, its variation vector field $v = \left.\dfrac{\partial\varphi_t}{\partial t}\right|_0$ is a Jacobi field. Conversely, a Jacobi field along a harmonic map is called *integrable* if it is tangent to a variation through harmonic maps. Not all Jacobi fields are integrable as well-known examples along geodesics show; see also [**20**] for two-dimensional examples. Integrability has applications to the study of spaces of harmonic maps and to the behaviour of a minimizing harmonic map near a singularity, see [**17**] and the references therein. It is thus an important question to determine for what pairs of Riemannian manifolds all Jacobi fields along harmonic maps are integrable. The known examples are very few in number; recently L. Lemaire and the author have added the case $M = S^2$, $N = \mathbb{C}P^2$ [**17**]. One main idea in the proof is that many properties of a harmonic map are 'preserved to first order' under a variation tangent to a Jacobi field. We shall explain this in the next two sections.

3. Preservation of conformality and isotropy to first order

We discuss two properties of a harmonic map from a 2-sphere to a Riemannian manifold which are preserved to first order under smooth variations tangent to a Jacobi field; the first, *conformality*, involves only the first order partial derivatives of the map, the second, *real isotropy* is a stronger condition involving higher order partial derivatives.

3.1. Preservation of conformality. Let $\varphi : (M,g) \to (N,h)$ be a smooth map between Riemannian manifolds. Then φ is called *weakly conformal* if, for each $x \in M$ there is a number $\lambda(x) \in [0,\infty)$ such that $|\mathrm{d}\varphi_x(X)| = \lambda(x)|X|$ for all $X \in T_xM$.

To formulate this in a way we can use, it is convenient to extend the inner product on TN to a complex-bilinear inner product $\langle\ ,\ \rangle^{\mathbf{C}}$ on the complexified tangent space $T^{\mathbf{c}}N = TN \otimes_{\mathbb{R}} \mathbb{C}$ (and so also on $\varphi^{-1}T^{\mathbf{c}}N = \varphi^{-1}TN \otimes_{\mathbb{R}} \mathbb{C}$). We also extend the differential of φ to a complex-linear map $\mathrm{d}^{\mathbf{c}}\varphi : T^{\mathbf{c}}M \to T^{\mathbf{c}}N$ between complexified tangent spaces.

Suppose now that M^2 is a Riemann surface, i.e. a 1-dimensional complex manifold. For any complex coordinate z on M^2 write

$$\frac{\partial^{\mathbf{c}}\varphi}{\partial z} = (\mathrm{d}^{\mathbf{c}}\varphi)\Big(\frac{\partial}{\partial z}\Big), \quad \frac{\partial^{\mathbf{c}}\varphi}{\partial \bar{z}} = (\mathrm{d}^{\mathbf{c}}\varphi)\Big(\frac{\partial}{\partial \bar{z}}\Big).$$

Then φ is weakly conformal if and only if

$$\eta^{\mathbf{R}}(\varphi) \equiv \Big\langle \frac{\partial^{\mathbf{c}}\varphi}{\partial z}, \frac{\partial^{\mathbf{c}}\varphi}{\partial z} \Big\rangle^{\mathbf{C}} = 0,$$

this condition being independent of the choice of complex coordinate; it is equivalent to the vanishing of the $(2,0)$-part of the pull-back $\varphi^{-1}h$ of the metric on N.

Extend the connection ∇ on $\varphi^{-1}TN$ by complex-bilinearity to $\varphi^{-1}T^{\mathbf{c}}N$ and write

$$\frac{\nabla}{\partial z} = \nabla_{\frac{\partial}{\partial z}}, \quad \frac{\nabla}{\partial \bar{z}} = \nabla_{\frac{\partial}{\partial \bar{z}}}.$$

Then, a smooth map $\varphi : M^2 \to N$ is harmonic if and only if

$$\frac{\nabla}{\partial \bar{z}}\frac{\partial^{\mathbf{c}}\varphi}{\partial z} = 0, \quad \text{equivalently,} \quad \frac{\nabla}{\partial z}\frac{\partial^{\mathbf{c}}\varphi}{\partial \bar{z}} = 0, \tag{9}$$

these expressions giving a non-zero multiple of the tension field with respect to any Hermitian metric on M^2; we thus recover the well-known result that harmonicity is independent of the choice of such a metric.

Now suppose $\varphi : M^2 \to N$ is harmonic. Then, by the first expression in (9), a smooth variation $\{\varphi_t\}$ of φ is harmonic to first order if and only if

$$\frac{\nabla}{\partial t}\Big|_0 \frac{\nabla}{\partial \bar{z}} \frac{\partial^{\mathbf{c}} \varphi_t}{\partial z} = 0\,; \tag{10}$$

swapping derivatives using the curvature of N shows that this is equivalent to

$$\frac{\nabla}{\partial \bar{z}} \frac{\nabla}{\partial z} \frac{\partial \varphi_t}{\partial t}\Big|_0 + R\Big(\frac{\partial^{\mathbf{c}} \varphi}{\partial \bar{z}}, \frac{\partial \varphi_t}{\partial t}\Big|_0\Big) \frac{\partial^{\mathbf{c}} \varphi}{\partial z} = 0\,.$$

It follows that $v \in \Gamma(\varphi^{-1}TN)$ is a Jacobi field along φ if and only if

$$\frac{\nabla}{\partial \bar{z}} \frac{\nabla}{\partial z} v + R\Big(\frac{\partial^{\mathbf{c}} \varphi}{\partial \bar{z}}, v\Big) \frac{\partial^{\mathbf{c}} \varphi}{\partial z} = 0 \tag{11}$$

or, equivalently (the conjugate of (11)),

$$\frac{\nabla}{\partial z} \frac{\nabla}{\partial \bar{z}} v + R\Big(\frac{\partial^{\mathbf{c}} \varphi}{\partial z}, v\Big) \frac{\partial^{\mathbf{c}} \varphi}{\partial \bar{z}} = 0\,.$$

We remark that these equations could be obtained directly from (6). Indeed, their sum is that equation and their difference is a consequence of the first Bianchi identity for the curvature.

Now let $\varphi : M^2 \to N$ be a smooth map and let $\{\varphi_t\}$ be a smooth variation of it. For any complex coordinate, set

$$\eta^{\mathbf{R}}(\varphi_t) = \Big\langle \frac{\partial^{\mathbf{c}} \varphi_t}{\partial z}, \frac{\partial^{\mathbf{c}} \varphi_t}{\partial z} \Big\rangle^{\mathbf{C}}.$$

Say that $\{\varphi_t\}$ is *conformal to first order* if $\eta^{\mathbf{R}}(\varphi_t) = o(t)$, i.e.

$$\frac{\partial}{\partial t}\Big|_0 \eta^{\mathbf{R}}(\varphi_t) = \eta^{\mathbf{R}}(\varphi) = 0\,.$$

Let $v = \dfrac{\partial \varphi_t}{\partial t}\Big|_0$ be the variation vector field of $\{\varphi_t\}$. Then, since ∇ is a Riemannian connection,

$$\frac{\partial}{\partial t}\Big|_0 \eta^{\mathbf{R}}(\varphi_t) = 2\Big\langle \frac{\nabla v}{\partial z}, \frac{\partial^{\mathbf{c}} \varphi}{\partial z} \Big\rangle^{\mathbf{C}}. \tag{12}$$

We call a vector field v along a smooth map $\varphi : M^2 \to N$ *conformal* if $\Big\langle \dfrac{\nabla v}{\partial z}, \dfrac{\partial^{\mathbf{c}} \varphi}{\partial z} \Big\rangle^{\mathbf{C}} = 0$; this condition is clearly independent of the choice of complex coordinate. Then from (12) we see that conformality to first order is equivalent to conformality of the variation vector field as follows:

LEMMA 3.1. *Let $\varphi : M^2 \to N$ be a weakly conformal map from a Riemann surface to a Riemannian manifold and let $v \in \Gamma(\varphi^{-1}TN)$. Then the following are equivalent:*

(i) *v is a conformal vector field along φ;*
(ii) there exists *a smooth variation of φ tangent to v which is conformal to first order;*

(iii) all *smooth variations of φ tangent to v are conformal to first order.* □

Any harmonic map from the 2-sphere S^2 to an arbitrary Riemannian manifold N is weakly conformal. This is proved by showing that, for any harmonic map $\varphi : M^2 \to N$ from a Riemann surface, $\eta^{\mathbf{R}}(\varphi)\, dz^2$ defines a *holomorphic quadratic differential*, i.e. a holomorphic section of the holomorphic line bundle $\otimes^2 T'_* M^2$; if $M^2 = S^2$, this bundle has negative degree and so any holomorphic section vanishes, see, for example, [**21**]. We generalize this in the next result; for later comparison we give two proofs.

PROPOSITION 3.2. *Any Jacobi field along a harmonic map $\varphi : S^2 \to N$ is conformal; equivalently, any smooth variation of φ tangent to a Jacobi field is conformal to first order.*

PROOF. *First proof:* Let $\varphi : M^2 \to N$ be a harmonic map from a Riemann surface and let v be a Jacobi field along it. Using the Jacobi equation (11), in any complex coordinate we have

$$\begin{aligned}\frac{\partial}{\partial \bar z}\Big\langle \frac{\nabla v}{\partial z}, \frac{\partial^{\mathbf{c}}\varphi}{\partial z}\Big\rangle^{\mathbf{C}} &= \Big\langle \frac{\nabla}{\partial \bar z}\frac{\nabla v}{\partial z}, \frac{\partial^{\mathbf{c}}\varphi}{\partial z}\Big\rangle^{\mathbf{C}} + \Big\langle \frac{\nabla v}{\partial z}, \frac{\nabla}{\partial \bar z}\frac{\partial^{\mathbf{c}}\varphi}{\partial z}\Big\rangle^{\mathbf{C}} \\ &= -\Big\langle R\Big(\frac{\partial^{\mathbf{c}}\varphi}{\partial \bar z}, v\Big)\frac{\partial^{\mathbf{c}}\varphi}{\partial z}, \frac{\partial^{\mathbf{c}}\varphi}{\partial z}\Big\rangle^{\mathbf{C}} + 0 \\ &= 0\,.\end{aligned}$$

Hence $\Big\langle \frac{\nabla v}{\partial z}, \frac{\partial^{\mathbf{c}}\varphi}{\partial z}\Big\rangle^{\mathbf{C}} dz^2$ defines a holomorphic differential on M^2; if $M^2 = S^2$ this must vanish. *Second proof:* Let $\varphi : M^2 \to N$ be a harmonic map from a Riemann surface and let $\{\varphi_t\}$ be a smooth variation of it tangent to a Jacobi field. Then, by Proposition 2.1,

$$\frac{\partial}{\partial \bar z}\eta^{\mathbf{R}}(\varphi_t) = 2\Big\langle \frac{\nabla}{\partial \bar z}\frac{\partial^{\mathbf{c}}\varphi_t}{\partial z}, \frac{\partial^{\mathbf{c}}\varphi_t}{\partial z}\Big\rangle^{\mathbf{C}} = o(t)\,.$$

It follows that

$$\frac{\partial}{\partial \bar z}\frac{\partial}{\partial t}\Big|_0 \eta^{\mathbf{R}}(\varphi_t) = \frac{\partial}{\partial t}\Big|_0 \frac{\partial}{\partial \bar z}\eta^{\mathbf{R}}(\varphi_t) = 0\,.$$

Hence, as well as $\eta^{\mathbf{R}}(\varphi)$, $\frac{\partial}{\partial t}\Big|_0 \eta^{\mathbf{R}}(\varphi_t)$ is a holomorphic quadratic differential and so must vanish if $M^2 = S^2$. □

3.2. Preservation of real isotropy. For any smooth map $\varphi : M^2 \to N$ from a Riemann surface to a Riemannian manifold, integers $r, s \geq 1$ and local complex coordinate z, set

$$\eta^{\mathbf{R}}_{r,s}(\varphi) = \Big\langle \frac{\nabla^{r-1}}{\partial z^{r-1}}\frac{\partial^{\mathbf{c}}\varphi}{\partial z}, \frac{\nabla^{s-1}}{\partial z^{s-1}}\frac{\partial^{\mathbf{c}}\varphi}{\partial z}\Big\rangle^{\mathbf{C}}.$$

Recall that φ is called *real isotropic* or *pseudoholomorphic* [**5**] if $\eta^{\mathbf{R}}_{r,s}(\varphi) = 0$ for all integers $r, s \geq 1$. Note that this condition is independent of the complex coordinate chosen and that any real isotropic map is weakly conformal.

Let $\varphi : S^2 \to N$ be a harmonic map from the 2-sphere to a *space form*, i.e. a Riemannian manifold of constant sectional curvature. Then [**4, 5**] φ is real isotropic. This is proved by showing, inductively that, for $k = 2, 3, \dots$ and $r + s = k$, $\eta^{\mathbf{R}}_{r,s}(\varphi) dz^k$ defines a *holomorphic k-differential*, i.e., a holomorphic section of

$\otimes^k T'_* S^2$, and so must vanish. It follows that all harmonic maps can be constructed explicitly from holomorphic maps, see [**5, 6, 2, 3, 15**]. We now generalize this argument. Let $\{\varphi_t\}$ be a 1-parameter variation of φ. Say that $\{\varphi_t\}$ *is real isotropic to first order* if $\eta^{\mathbf{R}}_{r,s}(\varphi_t) = o(t)$ for all r, s, i.e.,

$$\left.\frac{\partial}{\partial t}\right|_0 \eta^{\mathbf{R}}_{r,s}(\varphi_t) = \eta^{\mathbf{R}}_{r,s}(\varphi) = 0 \qquad (r, s \geq 1)\,.$$

Note that this condition is independent of the choice of complex coordinate. It can be formulated in terms of the variation vector field of $\{\varphi_t\}$ as follows. For any smooth map $\varphi : M^2 \to N$ from a Riemann surface to a Riemannian manifold, vector field v along φ, integers $r, s \geq 1$ and local complex coordinate z, set

$$j^{\mathbf{R}}_{r,s}(v) = \left\langle \frac{\nabla^r v}{\partial z^r}, \frac{\nabla^{s-1}}{\partial z^{s-1}} \frac{\partial^{\mathbf{c}} \varphi}{\partial z} \right\rangle^{\mathbf{C}} + \left\langle \frac{\nabla^{r-1}}{\partial z^{r-1}} \frac{\partial^{\mathbf{c}} \varphi}{\partial z}, \frac{\nabla^s v}{\partial z^s} \right\rangle^{\mathbf{C}}.$$

Then we have

LEMMA 3.3. *Let $\varphi : M^2 \to N$ be a real isotropic map from a Riemann surface into a space form and let v be a Jacobi field along it. Then, for any smooth variation $\{\varphi_t\}$ of φ tangent to v and any complex coordinate z on M^2 we have*

$$j^{\mathbf{R}}_{r,s}(v) = \left.\frac{\partial}{\partial t}\right|_0 \eta^{\mathbf{R}}_{r,s}(\varphi_t) \qquad (r, s \geq 1)\,.$$

PROOF. We show by induction that, for $k = 1, 2, \ldots$,

$$\left.\frac{\nabla}{\partial t}\right|_0 \frac{\nabla^{k-1}}{\partial z^{k-1}} \frac{\partial^{\mathbf{c}} \varphi_t}{\partial z} \in \theta_{k-1}(\varphi) + \frac{\nabla^k v}{\partial z^k} \tag{13}$$

where

$$\theta_{k-1}(\varphi) = \operatorname{span}\left\{ \frac{\partial^{\mathbf{c}} \varphi}{\partial z}, \ldots, \frac{\nabla^{k-2}}{\partial z^{k-2}} \frac{\partial^{\mathbf{c}} \varphi}{\partial z} \right\} \ (k \geq 2) \ \text{ and } \ \theta_0(\varphi) = \{\mathbf{0}\}\,;$$

note that this is independent of the choice of complex coordinate. For $k = 1$, (13) holds since $\left.\frac{\nabla}{\partial t}\right|_0 \frac{\partial^{\mathbf{c}} \varphi_t}{\partial z} = \left.\frac{\nabla}{\partial z} \frac{\partial \varphi_t}{\partial t}\right|_0 = \frac{\nabla v}{\partial z}$.

Suppose that (13) holds for $k-1$ for some $k \geq 2$. Then

$$\left.\frac{\nabla}{\partial t}\right|_0 \frac{\nabla^{k-1}}{\partial z^{k-1}} \frac{\partial^{\mathbf{c}} \varphi_t}{\partial z} = \frac{\nabla}{\partial z} \left.\frac{\nabla}{\partial t}\right|_0 \frac{\nabla^{k-2}}{\partial z^{k-2}} \frac{\partial^{\mathbf{c}} \varphi_t}{\partial z} + R\Big(\frac{\partial^{\mathbf{c}} \varphi}{\partial z}, v\Big) \frac{\nabla^{k-2}}{\partial z^{k-2}} \frac{\partial^{\mathbf{c}} \varphi}{\partial z}\,. \tag{14}$$

Now, by the induction hypothesis the first term on the right-hand side of (14) lies in $\theta_{k-1}(\varphi) + \dfrac{\nabla^k v}{\partial z^k}$. Also, by the well-known formula for the curvature of a space form (see, for example, [**13**, Chapter V]), the curvature term is a multiple of

$$\left\langle v, \frac{\nabla^{k-2}}{\partial z^{k-2}} \frac{\partial^{\mathbf{c}} \varphi}{\partial z} \right\rangle^{\mathbf{C}} \frac{\partial^{\mathbf{c}} \varphi}{\partial z} - \left\langle \frac{\partial^{\mathbf{c}} \varphi}{\partial z}, \frac{\nabla^{k-2}}{\partial z^{k-2}} \frac{\partial^{\mathbf{c}} \varphi}{\partial z} \right\rangle^{\mathbf{C}} v\,; \tag{15}$$

since the second term of this is zero by the isotropy of φ, (15) lies in $\theta_1(\varphi)$ and so in $\theta_{k-1}(\varphi)$. It follows that the right-hand side of (14) lies in $\theta_{k-1}(\varphi) + \dfrac{\nabla^k v}{\partial z^k}$ as required, completing the inductive step. Using (13) and the isotropy of φ we have, for any $r,s\geq 1$,

$$\begin{aligned}\frac{\partial}{\partial t}\Big|_0 \eta^{\mathbf{R}}_{r,s}(\varphi_t) &= \Big\langle \frac{\nabla}{\partial t}\Big|_0 \frac{\nabla^{r-1}}{\partial z^{r-1}}\frac{\partial^{\mathbf{c}}\varphi_t}{\partial z}, \frac{\nabla^{s-1}}{\partial z^{s-1}}\frac{\partial^{\mathbf{c}}\varphi}{\partial z}\Big\rangle^{\mathbf{C}} + \Big\langle \frac{\nabla^{r-1}}{\partial z^{r-1}}\frac{\partial^{\mathbf{c}}\varphi}{\partial z}, \frac{\nabla}{\partial t}\Big|_0 \frac{\nabla^{s-1}}{\partial z^{s-1}}\frac{\partial^{\mathbf{c}}\varphi_t}{\partial z}\Big\rangle^{\mathbf{C}} \\ &= \Big\langle \frac{\nabla^r v}{\partial z^r}, \frac{\nabla^{s-1}}{\partial z^{s-1}}\frac{\partial^{\mathbf{c}}\varphi}{\partial z}\Big\rangle^{\mathbf{C}} + \Big\langle \frac{\nabla^{r-1}}{\partial z^{r-1}}\frac{\partial^{\mathbf{c}}\varphi}{\partial z}, \frac{\nabla^s v}{\partial z^s}\Big\rangle^{\mathbf{C}} \\ &= j^{\mathbf{R}}_{r,s}(v)\end{aligned}$$

as required. □

Say that a vector field v along a smooth map $\varphi : M^2 \to N$ *preserves real isotropy to first order* if $j^{\mathbf{R}}_{r,s}(v) = 0$ for all $r,s \geq 1$; note that this condition is independent of the choice of local complex coordinate.

PROPOSITION 3.4. *Let $\varphi : S^2 \to N$ be a harmonic map from the 2-sphere to a space form. Then any Jacobi field preserves real isotropy to first order, equivalently, any smooth variation of φ tangent to a Jacobi field is real isotropic to first order.*

PROOF. Let $\{\varphi_t\}$ be any smooth variation of φ. We shall show that $\eta^{\mathbf{R}}_{r,s}(\varphi_t) = o(t)$ for all integers $r,s \geq 1$. In fact we show by induction that, for all $K \in \{1,2,\dots\}$,

$$\text{(16)} \qquad \text{(i)} \quad \frac{\nabla}{\partial \bar z}\frac{\nabla^{K-1}}{\partial z^{K-1}}\frac{\partial^{\mathbf{c}}\varphi_t}{\partial z} \in \theta_{K-1}(\varphi_t) + o(t)\,,$$

$$\text{(17)} \qquad \text{(ii)} \quad \text{for all } r,s \geq 1 \text{ with } r+s = K+1,\ \eta^{\mathbf{R}}_{r,s}(\varphi_t) = o(t)\,.$$

This is true for $K=1$ by (10) and Proposition 3.2.

Suppose that it is true for all $K < k$ for some $k \geq 2$. We shall show that it is true for $K = k$. Indeed,

$$\text{(18)} \qquad \frac{\nabla}{\partial \bar z}\frac{\nabla^{k-1}}{\partial z^{k-1}}\frac{\partial^{\mathbf{c}}\varphi_t}{\partial z} = R\Big(\frac{\partial^{\mathbf{c}}\varphi_t}{\partial z}, \frac{\partial^{\mathbf{c}}\varphi_t}{\partial \bar z}\Big)\frac{\nabla^{k-2}}{\partial z^{k-2}}\frac{\partial^{\mathbf{c}}\varphi_t}{\partial z} + \frac{\nabla}{\partial z}\frac{\nabla}{\partial \bar z}\frac{\nabla^{k-2}}{\partial z^{k-2}}\frac{\partial^{\mathbf{c}}\varphi_t}{\partial z}\,.$$

The curvature term is a multiple of

$$\Big\langle \frac{\partial^{\mathbf{c}}\varphi_t}{\partial \bar z}, \frac{\nabla^{k-2}}{\partial z^{k-2}}\frac{\partial^{\mathbf{c}}\varphi_t}{\partial z}\Big\rangle^{\mathbf{C}}\frac{\partial^{\mathbf{c}}\varphi_t}{\partial z} - \Big\langle \frac{\partial^{\mathbf{c}}\varphi_t}{\partial z}, \frac{\nabla^{k-2}}{\partial z^{k-2}}\frac{\partial^{\mathbf{c}}\varphi_t}{\partial z}\Big\rangle^{\mathbf{C}}\frac{\partial^{\mathbf{c}}\varphi_t}{\partial \bar z}\,.$$

The first term of this lies in $\theta_1(\varphi_t)$ and so in $\theta_{k-1}(\varphi_t)$; by the induction hypothesis, the second term is $o(t)$. Also, by the induction hypothesis, the last term of (18) lies in $\theta_{k-1}(\varphi_t)$, so that (16) holds for $K = k$. Next, for $r,s \geq 1$ with $r+s = k+1$ we have

$$\frac{\partial}{\partial \bar z}\eta^{\mathbf{R}}_{r,s}(\varphi_t) = \Big\langle \frac{\nabla}{\partial \bar z}\frac{\nabla^{r-1}}{\partial z^{r-1}}\frac{\partial^{\mathbf{c}}\varphi_t}{\partial z}, \frac{\nabla^{s-1}}{\partial z^{s-1}}\frac{\partial^{\mathbf{c}}\varphi_t}{\partial z}\Big\rangle^{\mathbf{C}} + \Big\langle \frac{\nabla^{r-1}}{\partial z^{r-1}}\frac{\partial^{\mathbf{c}}\varphi_t}{\partial z}, \frac{\nabla}{\partial \bar z}\frac{\nabla^{s-1}}{\partial z^{s-1}}\frac{\partial^{\mathbf{c}}\varphi_t}{\partial z}\Big\rangle^{\mathbf{C}}.$$

By the induction hypothesis, this is $o(1)$, hence $\dfrac{\partial}{\partial t}\Big|_0 \eta^{\mathbf{R}}_{r,s}(\varphi_t)dz^{k+1}$ defines a holomorphic $(k+1)$-differential on S^2. As before, this must vanish, so that $\eta^{\mathbf{R}}_{r,s}(\varphi_t) = o(t)$, thus completing the induction step. □

4. Preservation of holomorphicity and complex isotropy to first order

We give two properties of a harmonic map into a Kähler manifold which are preserved to first order under smooth variations tangent to a Jacobi field; the first is holomorphicity which involves only first order partial derivatives, the second is complex isotropy which involves higher order partial derivatives.

4.1. Preservation of holomorphicity. Let M and N be compact Kähler manifolds. Then we can decompose the complexified tangent spaces $T^{\mathbf{c}}M = TM \otimes_{\mathbb{R}} \mathbb{C}$ and $T^{\mathbf{c}}N = TN \otimes_{\mathbb{R}} \mathbb{C}$ into $(1,0)$ and $(0,1)$ parts, viz.,

$$T^{\mathbf{c}}M = T'M \oplus T''M \qquad T^{\mathbf{c}}N = T'N \oplus T''N\,. \tag{19}$$

We shall let $\langle\ ,\ \rangle^{\mathrm{Herm}}$ denote the *Hermitian extension* of the inner product on TN to $T^{\mathbf{c}}N$ given by

$$\langle v, w\rangle^{\mathrm{Herm}} = \langle v, \overline{w}\rangle^{\mathbf{C}}\,;$$

this restricts to a positive definite Hermitian inner product on $T'N$ and on its pull-back $\varphi^{-1}T'N$. Let $\varphi : M \to N$ be a smooth map. As before extend its differential to a complex-linear map $\mathrm{d}^{\mathbf{c}}\varphi : T^{\mathbf{c}}M \to T^{\mathbf{c}}N$; then we can consider the two components:

$$\partial\varphi : T'M \to T'N\,, \qquad \bar{\partial}\varphi : T''M \to T'N\,.$$

Recall that φ is called *holomorphic* if $\bar{\partial}\varphi = 0$ (and *antiholomorphic* if $\partial\varphi = 0$).

For simplicity we calculate in a local complex coordinate system (z^j) on M, writing

$$\frac{\partial\varphi}{\partial z^j} = (\partial\varphi)\big(\frac{\partial}{\partial z^j}\big)\,, \quad \frac{\partial\varphi}{\partial z^{\bar{j}}} = (\bar{\partial}\varphi)\big(\frac{\partial}{\partial z^{\bar{j}}}\big)\,;$$

note that

$$\frac{\partial^{\mathbf{c}}\varphi}{\partial z^j} = \frac{\partial\varphi}{\partial z^j} + \overline{\frac{\partial\varphi}{\partial z^{\bar{j}}}}\,.$$

Say that a smooth variation $\{\varphi_t\}$ is *holomorphic to first order* if $\bar{\partial}\varphi_t = o(t)$, i.e., for any local complex coordinate system (z^j),

$$\frac{\nabla}{\partial t}\Big|_0 \frac{\partial\varphi_t}{\partial z^{\bar{j}}} = \frac{\partial\varphi}{\partial z^{\bar{j}}} = 0 \quad \forall j\,.$$

For any $v \in \Gamma(\varphi^{-1}T^{\mathbf{c}}N)$, let v' denote its $(1,0)$ component under the decomposition (19). Note that, if v is real, i.e., lies in $\Gamma(\varphi^{-1}TN)$, v' determines v, indeed $v = v' + \overline{v'}$. Say that v is *holomorphic* if $\nabla_Z v' = 0$ for all $Z \in T''M$, equivalently,

$$\frac{\nabla v'}{\partial z^{\bar{j}}} = 0 \quad \forall j\,. \tag{20}$$

Holomorphicity to first order is then equivalent to holomorphicity of the variation vector field as follows (cf. Lemma 3.1):

LEMMA 4.1. *Let $\varphi : M \to N$ be a holomorphic map between Kähler manifolds and let $v \in \Gamma(\varphi^{-1}TN)$. Then the following are equivalent:*

(i) *v is a holomorphic vector field along φ;*
(ii) there exists *a smooth variation of φ tangent to v which is holomorphic to first order;*
(iii) all *smooth variations of φ tangent to v are holomorphic to first order.* □

PROOF. Noting that the Kähler condition on N implies that $(\nabla_Z v)' = \nabla_Z(v')$ for all $v \in \Gamma(\varphi^{-1}T^{\mathbf{c}}N)$, $Z \in T^{\mathbf{c}}M$, we have

$$\frac{\nabla}{\partial t}\Big|_0 \frac{\partial \varphi_t}{\partial z^{\bar{j}}} = \frac{\nabla}{\partial z^{\bar{j}}}\left(\frac{\partial \varphi_t}{\partial t}\Big|_0\right)' = \frac{\nabla v'}{\partial z^{\bar{j}}} \quad \forall j$$

from which the lemma follows. □

PROPOSITION 4.2. *Let $\varphi : M \to N$ be a holomorphic map between compact Kähler manifolds. Then any Jacobi field along φ is holomorphic; equivalently, any smooth variation of φ tangent to a Jacobi field is holomorphic to first order.*

PROOF. The harmonicity condition for a smooth map $\varphi : M \to N$ between Kähler manifolds is (with summation over i and j) $g^{i\bar{j}}\dfrac{\nabla}{\partial z^i}\dfrac{\partial \varphi}{\partial z^{\bar{j}}} = 0$. Hence a smooth variation $\{\varphi_t\}$ of a harmonic map $\varphi : M \to N$ is tangent to a Jacobi field v if and only if

$$\frac{\partial}{\partial t}\Big|_0 g^{i\bar{j}}\frac{\nabla}{\partial z^i}\frac{\partial \varphi_t}{\partial z^{\bar{j}}} = 0\,.$$

On using the curvature to commute the derivatives, this gives the Jacobi equation in the form

$$g^{i\bar{j}}\left\{\frac{\nabla^2 v'}{\partial z^i \partial z^{\bar{j}}} + R\Big(\frac{\partial^{\mathbf{c}}\varphi}{\partial z^i}, v\Big)\frac{\partial \varphi}{\partial z^{\bar{j}}}\right\} = 0\,.$$

If φ is holomorphic, $\partial\varphi/\partial z^{\bar{j}} = 0$ so that the curvature term vanishes; then integrating by parts gives

$$\int_M g^{i\bar{j}}\Big\langle \frac{\nabla v'}{\partial z^{\bar{j}}}, \frac{\nabla v'}{\partial z^{\bar{i}}}\Big\rangle^{\text{Herm}} \omega_g = 0\,.$$

Therefore $\dfrac{\nabla v'}{\partial z^{\bar{j}}} = 0$ for all j, so that v is holomorphic. □

For an alternative proof see [**17**].

4.2. Preservation of complex isotropy. Let N be a Kähler manifold. For any smooth map $\varphi : M^2 \to N$ from a Riemann surface, integers $r, s \geq 1$ and complex coordinate z, set

$$\eta^{\mathbf{C}}_{r,s}(\varphi) = \Big\langle \frac{\nabla^{r-1}}{\partial z^{r-1}}\frac{\partial \varphi}{\partial z}, \frac{\nabla^{s-1}}{\partial \bar{z}^{s-1}}\frac{\partial \varphi}{\partial \bar{z}}\Big\rangle^{\text{Herm}}\,.$$

Then φ is called *complex isotropic* if $\eta^{\mathbf{C}}_{r,s}(\varphi) = 0$ for all integers $r, s \geq 1$. Note that this condition is independent of the complex coordinate chosen and that it is stronger than real isotropy.

It is well-known that any harmonic map $\varphi : S^2 \to N$ from the 2-sphere to a complex space form (i.e. a Kähler manifold of constant holomorphic sectional curvature) is complex isotropic, see, for example, [**7, 10, 22**]. As in the real case, this is proved by showing inductively that, for $k = 2, 3, \ldots$ and $r+s = k$, $\eta^{\mathbf{C}}_{r,s}(\varphi)\mathrm{d}z^k$ defines a holomorphic k-differential on S^2 and so must vanish.

For any smooth map $\varphi : M^2 \to N$ from a Riemann surface to a Kähler manifold, vector field v along φ, integers $r, s \geq 1$ and local complex coordinate z, set

$$j^{\mathbf{C}}_{r,s}(v) = \Big\langle \frac{\nabla^r v}{\partial z^r}, \frac{\nabla^{s-1}}{\partial \bar{z}^{s-1}} \frac{\partial \varphi}{\partial \bar{z}} \Big\rangle^{\text{Herm}} + \Big\langle \frac{\nabla^{r-1}}{\partial z^{r-1}} \frac{\partial \varphi}{\partial z}, \frac{\nabla^s v}{\partial \bar{z}^s} \Big\rangle^{\text{Herm}} .$$

Then, analogously to Lemma 3.3, we have

LEMMA 4.3. *Let $\varphi : M^2 \to N$ be a complex isotropic map from a Riemann surface into a space form and let v be a Jacobi field along it. Then, for any smooth variation $\{\varphi_t\}$ of φ tangent to v and any complex coordinate z on M^2,*

$$j^{\mathbf{C}}_{r,s}(v) = \frac{\partial}{\partial t}\Big|_0 \eta^{\mathbf{C}}_{r,s}(\varphi_t) \qquad (r, s \geq 1) .$$

□

Say that v *preserves complex isotropy to first order* if $j^{\mathbf{C}}_{r,s}(v) = 0$ for all $r, s \geq 1$, and say that $\{\varphi_t\}$ *is complex isotropic to first order* if $\eta^{\mathbf{C}}_{r,s}(\varphi_t) = o(t)$, i.e.,

$$\frac{\partial}{\partial t}\Big|_0 \eta^{\mathbf{C}}_{r,s}(\varphi_t) = \eta^{\mathbf{C}}_{r,s}(\varphi) = 0 \qquad (r, s \geq 1) .$$

Note that these conditions are independent of the choice of local complex coordinate. Then we have the complex analogue of Proposition 3.4:

PROPOSITION 4.4. *Let $\varphi : S^2 \to N$ be a harmonic map from the 2-sphere to a complex space form. Then any Jacobi field preserves complex isotropy to first order, equivalently, any smooth variation of φ tangent to a Jacobi field is complex isotropic to first order.* □

The proofs proceed in the same way as for the real case, except that we must use the more complicated formula for the curvature of a complex space form (see, for example, [**14**, Chapter IX]). The calculations are omitted; they are simplified by the following easy consequence of that curvature formula:

LEMMA 4.5. *For any $X \in T^{\mathbf{C}}N$, $R\Big(X, \dfrac{\partial^{\mathbf{c}} \varphi}{\partial z}\Big) \dfrac{\nabla^{k-1}}{\partial z^{k-1}} \dfrac{\partial \varphi}{\partial z}$ is a linear combination of $\dfrac{\partial \varphi}{\partial z}$, $\dfrac{\nabla^{k-1}}{\partial z^{k-1}} \dfrac{\partial \varphi}{\partial z}$ and $\eta^{\mathbf{C}}_{k,1}(\varphi) X'$.* □

References

[1] P. Baird and J.C. Wood, *Harmonic morphisms between Riemannian manifolds*, London Math. Soc. Monogr. (N.S.), Oxford Univ. Press (to appear).

[2] J.L.M. Barbosa, On minimal immersions of S^2 into S^{2m}, *Trans. Amer. Math. Soc.* **210** (1975), 75–106.

[3] H.-J. Borchers and W.D Garber, Local theory of solutions for the O($2k+1$) σ-model, *Comm. Math. Phys.* **72** (1980), 77–102.

[4] E. Calabi, Minimal immersions of surfaces in Euclidean spheres, *J. Differential Geom.* **1** (1967), 111–125.

[5] E. Calabi, Quelques applications de l'analyse complexe aux surfaces d'aire minima, *Topics in complex manifolds* (Univ. de Montréal, 1967), pp. 59–81.

[6] S.S. Chern, On the minimal immersions of the two-sphere in a space of constant curvature, *Problems in analysis (Lectures at the Sympos. in honor of Salomon Bochner, Princeton Univ., Princeton, N.J., 1969)*, Princeton Univ. Press, Princeton, N.J., 1970, pp. 27–40.

[7] A.M. Din and W.J. Zakrzewski, General classical solutions in the CP^{n-1} model, *Nuclear Phys. B* **174** (1980), 397–406.
[8] J. Eells and L. Lemaire, *Selected topics in harmonic maps*, C.B.M.S. Regional Conf. Series 50, Amer. Math. Soc., 1983.
[9] J. Eells and L. Lemaire, Another report on harmonic maps, *Bull. London Math. Soc.* **20** (1988), 385–524.
[10] J. Eells and J.C. Wood, Harmonic maps from surfaces to complex projective spaces, *Adv. in Math.* **49** (1983), 217–263.
[11] S. Gudmundsson, *The bibliography of harmonic morphisms*, `http://www.maths.lth.se/matematiklu/personal/sigma/harmonic/bibliography.html`
[12] R.D. Gulliver, R. Osserman and H.L. Royden, A theory of branched immersions of surfaces, *Amer. J. Math.* **95** (1973), 750–812.
[13] S. Kobayashi and K. Nomizu, *Foundations of differential geometry, Vol. 1*, Interscience, New York, 1963 (reprinted by John Wiley & Sons, Inc., New York, 1996).
[14] S. Kobayashi and K. Nomizu, *Foundations of differential geometry, Vol. 2*, Interscience, New York, 1969 (reprinted by John Wiley & Sons, Inc., New York, 1996).
[15] H.B. Lawson, Surfaces minimales et la construction de Calabi–Penrose. *Seminar Bourbaki*, Vol. 1983/84. Astérisque No. 121–122 (1985), 197–211.
[16] L. Lemaire and J.C. Wood, On the space of harmonic 2-spheres in $\mathbf{C}P^2$, *Intern. J. Math.* **7** (1996), 211–225.
[17] L. Lemaire and J.C. Wood, Jacobi fields along harmonic 2-spheres in $\mathbf{C}P^2$ are integrable, *J. London Math. Soc.*, (to appear).
[18] E. Mazet, La formule de la variation seconde de l'énergie au voisinage d'une application harmonique, *J. Differential Geom.* **8** (1973), 279–296.
[19] S. Montaldo and J.C. Wood, Harmonic morphisms and the Jacobi operator, *Rend. Sem. Fac. Sci. Univ. Cagliari* (to appear).
[20] M. Mukai, The deformation of harmonic maps given by the Clifford tori, *Kodai Math. J.* **20** (1997), 252–268.
[21] J.C. Wood, Harmonic maps and complex analysis, *Proc. Summer Course in Complex Analysis, Trieste, 1975*, IAEA, Vienna, 1976, vol. III, pp. 289–308.
[22] J.C. Wood, Holomorphic differentials and classification theorems for harmonic maps and minimal immersions, *Global Riemannian Geometry*, ed. T.J. Willmore and N.J. Hitchin, Ellis Horwood, 1984, pp. 168–175.

Department of Pure Mathematics, University of Leeds, Leeds LS2 9JT, UK

E-mail address: `j.c.wood@leeds.ac.uk`

Contemporary Mathematics
Volume **308**, 2002

Denseness of Plain Constant Mean Curvature Surfaces in Dressing Orbits

Hongyou Wu

It is shown in [**3**] that any immersed constant mean curvature $H = \frac{1}{2}$ surface $\boldsymbol{f} : M \to \mathbb{R}^3$ (CMC surface) other than the sphere can be constructed from a meromorphic matrix valued 1-form

$$P = \begin{pmatrix} 0 & p \\ E/p & 0 \end{pmatrix} \mathrm{d}z,$$

the so called "normalized potential" for the surface, on the universal covering $\widetilde{M}$ of M. Here, p is a meromorphic function on $\widetilde{M}$, and the holomorphic function $E = -\frac{1}{4}\langle \boldsymbol{f}_{zz}, \boldsymbol{n}\rangle$ is the coefficient in the Hopf differential $E(z)(\mathrm{d}z)^2$ on the surface with $\boldsymbol{n}$ being the unit normal field along the surface such that $H = \frac{1}{2}$. Geometrically, up to a constant multiple, p is the holomorphic part of the conformal factor in the induced metric on $\widetilde{M}$ (see [**5**]). If $z_0 \in \widetilde{M}$ is the base point of the normalized potential, then $p(z_0) \neq 0$. In [**2**], Dorfmeister and Haak obtained an analytic characterization of the meromorphic functions p that can appear in P when E is given: P is the normalized potential for a (smooth) harmonic map into $\mathbb{S}^2 = \mathrm{SU}(2)/\mathrm{SU}(1)$ if and only if certain residue conditions on some functions formed by p and solutions to two ordinary differential equations are satisfied; and P is the normalized potential for (the Gauss map of) a CMC surface if and only if furthermore p has only zeros and poles of some special orders. For a simpler analytic characterization of the normalized potential P for a harmonic map into $\mathbb{S}^2$, see [**6**].

A direct consequence of these analytic characterizations of P is that when E is given, P can be very complicated: for example, P can have poles of very high orders. However, at high order poles of P, the residue conditions of [**2**] or [**6**] are not easy to verify in practice. So, it is helpful to have some classes of CMC surfaces with simple P's to approximate any given CMC surface, and we believe that such approximations are very desirable if one wants to visualize CMC surfaces such as the embedded ones with ends and the compact ones using the method of [**3**].

In this short paper, we will prove that the class of CMC surfaces with either holomorphic normalized potentials or normalized potentials having only poles of

2000 *Mathematics Subject Classification.* Primary 53A10, 53C42; Secondary 58E20, 35Q53.

order 2 away from the umbilic points is dense in each dressing orbit of CMC surfaces, where the topology on the dressing orbit is induced from the appropriate loop group (see Theorem 2.3). Note that for such normalized potentials, the residue conditions of [**2**] or [**6**] are equivalent to the simple requirement that the residue at any of their poles is zero, and that under this simple requirement, meromorphic functions p having only double zeros and double poles away from the umbilic points are always allowed in the normalized potential P for a CMC surface. In view of the fact [**4**] that, in general, there are (uncountably) many dressing orbits sharing the same Hopf differential, we conjecture that this class is one of the simplest classes that can be used to approximate all the CMC surfaces. Actually, it is not known whether every dressing orbit of CMC surfaces contains a CMC surface with a holomorphic normalized potential.

Even though the meromorphic functions p and E/p in a normalized potential can be approximated by simpler functions using results such as Runge's Theorem [**1**] in complex analysis, the difficulty in approximating a normalized potential comes from the topology on the dressing orbits.

We are grateful to Josef Dorfmeister for his interest in this work and for providing a proof of part of the property B2 in Section 1 that is shorter than the original proof. We would also like to thank the referee, whose suggestions led to several improvements in our presentation. The author was partially supported by NSF grant DMS-9973108 and by a grant from Northern Illinois University Graduate School.

1. Dressing Transformations

In this section, we first recall the dressing transformations $\mathcal{B}_t$ and $\mathcal{C}_t$ corresponding to

$$\begin{pmatrix} 1 & 0 \\ t\lambda & 1 \end{pmatrix} \quad \text{and} \quad \begin{pmatrix} 1 & t\lambda \\ 0 & 1 \end{pmatrix}, \tag{1.1}$$

respectively, where $t \in \mathbb{C}^* := \mathbb{C}\backslash\{0\}$, then present some basic properties of these transformations, and finally mention a property of dressing transformations in general. These results will be needed in the next section. For more details concerning dressing transformations, see [3], [4] and [2].

Let D be either the open unit disk D_1 centered at the origin or the whole $\mathbb{C}$, to one of which the universal covering of any surface other than the sphere is conformally equivalent. Consider the normalized potential

$$P = \begin{pmatrix} 0 & p \\ E/p & 0 \end{pmatrix} \mathrm{d}z \tag{1.2}$$

for a CMC surface $\boldsymbol{f} : D \to \mathbb{R}^3$ (with the origin as the base point, which will always be the case unless otherwise stated). For each $t \in \mathbb{C}^*$,

$$\begin{pmatrix} 0 & q \\ E/q & 0 \end{pmatrix} \mathrm{d}z = \mathcal{B}_t P := \begin{pmatrix} 1 & 0 \\ t\lambda & 1 \end{pmatrix} \bullet \begin{pmatrix} 0 & p \\ E/p & 0 \end{pmatrix} \mathrm{d}z \tag{1.3}$$

and

$$\begin{pmatrix} 0 & r \\ E/r & 0 \end{pmatrix} \mathrm{d}z = \mathcal{C}_t P := \begin{pmatrix} 1 & t\lambda \\ 0 & 1 \end{pmatrix} \bullet \begin{pmatrix} 0 & p \\ E/p & 0 \end{pmatrix} \mathrm{d}z \tag{1.4}$$

are also normalized potentials for CMC surfaces with the same Hopf differential. Moreover, by [**2**],

$$q(z) = \frac{p(z)}{\left(1 + t\int_0^z p(w)\,\mathrm{d}w\right)^2}, \qquad r(z) = \left(1 + t\int_0^z \frac{E(w)}{p(w)}\,\mathrm{d}w\right)^2 p(z). \tag{1.5}$$

Thus, the residue at any pole of p or E/p is zero. In particular, p and E/p have no simple poles.

For a meromorphic function h on D and an open subset $S \subseteq D$, we will denote the set of poles of h in S by $\mathrm{Pol}(h, S)$. If $\mathrm{Pol}(h, S)$ is finite, then the δ-neighborhood $\mathrm{Pol}_\delta(h, S)$ of $\mathrm{Pol}(h, S)$ in $\mathbb{C}$ is in S for all small $\delta > 0$. We will also use the function $m : D \to \{0, 1, 2, \cdots\}$ defined by

$$m(z) \text{ is the order of } z \text{ as a zero of } E. \tag{1.6}$$

Denote by D_ρ the open disk centered at the origin and with radius ρ, where $\rho \in (0, 1)$ if $D = D_1$ and $\rho > 0$ if D is the whole $\mathbb{C}$.

PROPOSITION 1.7. *Let P given by (1.2) be the normalized potential for a CMC surface without any pole on the boundary of D_ρ. Then, when $t \in \mathbb{C}^*$ is sufficiently close to 0, the normalized potential*

$$Q = \begin{pmatrix} 0 & q \\ E/q & 0 \end{pmatrix} \mathrm{d}z = \mathcal{B}_t P \tag{1.8}$$

has the following properties.

B1. *E/q carries all the poles of E/p in D_ρ with their orders unchanged; the other poles of E/q in D_ρ are the poles of p in D_ρ of orders $> m + 2$ with their orders reduced by $m + 2$.*

B2. *An order n pole z_0 of p in D_ρ yields exactly $n - 1$ double poles of q in D_ρ close to z_0 (but not at z_0) and away from the umbilic points; the only poles of q in D_ρ are such poles.*

B3. *Q does not have any pole on the boundary of D_ρ.*

PROOF. We restrict our attention to D_ρ to avoid repeating the words "in D_ρ". Note that p only has finitely many zeros and poles (in D_ρ). From Section 3.8 of [**2**], we have the following results:

i) a zero of p is a zero of q of the same order when $t \in \mathbb{C}^*$ is sufficiently close to 0;

ii) a pole of p of order 2 is neither a zero of q nor a pole of q; and

iii) a pole of p of order $n > 2$ is a zero of q of order $n - 2$.

Now, we prove B1. The poles of E/q are the zeros of q of orders $> m$. By the first formula in (1.5), any zero of q is either a zero of p or a pole of $1 + t\int_0^z p(w)\,\mathrm{d}w$, i.e., is either a zero of p or a pole of p. Thus, by i)—iii), the claims in B1 hold.

Next, we prove B2. Also by the first formula in (1.5), any pole of q is either a pole of p or a zero of $1 + t\int_0^z p(w)\,\mathrm{d}w$. So, in view of ii) and iii), the poles of q are among the zeros of $1 + t\int_0^z p(w)\,\mathrm{d}w$. Note that p is bounded on $D_\rho \backslash \mathrm{Pol}_\delta(p, D_\rho)$ for small $\delta > 0$ and D_ρ is also bounded. When $t \in \mathbb{C}^*$ is sufficiently close to 0, the zeros of $1+t\int_0^z p(w)\,dw$ are very close to (but not at) the poles of p and away from the zeros of p and E. Thus, when $t\in\mathbb{C}^*$ is sufficiently close to 0, the zeros of $1+t\int_0^z p(w)\,dw$ are the poles of q, and they are away from the umbilic points and very close to the poles of p. Let z_0 be a pole of p of order n. Then, $n\geq 2$, and

$$p(z)=\frac{A}{(z-z_0)^n}+\cdots \tag{1.9}$$

for some constant $A\neq 0$. Thus,

$$\int_0^z p(w)\,dw=-\frac{B}{(z-z_0)^{n-1}}+\cdots=-\frac{B}{(z-z_0)^{n-1}}[1+(z-z_0)h(z)] \tag{1.10}$$

for some constant $B\neq 0$ and function h holomorphic on a neighborhood of z_0 contained in D_ρ, say the disk $D_{2\tau}(z_0):=\{z\in\mathbb{C};\ |z-z_0|<2\tau\}$ with $\tau>0$. By using a smaller τ if necessary, we assume further that $1+(z-z_0)h(z)$ does not have a zero on $D_\tau(z_0)$ and $[1+(z-z_0)h(z)]^{\frac{1}{n-1}}$ is well defined on $D_\tau(z_0)$. When $t\in\mathbb{C}^*$ is sufficiently close to 0, the zeros of $1+t\int_0^z p(w)\,dw$ near z_0 live in $D_\tau(z_0)$ and are determined by

$$(z-z_0)^{n-1}-tB\big[1+(z-z_0)h(z)\big]=0, \tag{1.11}$$

i.e., by

$$\frac{(z-z_0)^{n-1}}{1+(z-z_0)h(z)}=tB. \tag{1.12}$$

Thus, these zeros of $1+t\int_0^z p(w)\,dw$ are preimages of the $(n-1)$-th roots of tB under the map

$$k(z)=\frac{z-z_0}{[1+(z-z_0)h(z)]^{\frac{1}{n-1}}}. \tag{1.13}$$

Since $k(z_0)=0$ and $k'(z_0)\neq 0$, we can even assume that k is one-to-one on $D_\tau(z_0)$ and, by the Open Mapping Theorem [**1**], $k(D_\tau(z_0))\ni 0$ is open. So, when $t\in\mathbb{C}^*$ is sufficiently close to 0, the $(n-1)$-th roots of tB are all in $k(D_\tau(z_0))$, and hence $1+t\int_0^z p(w)\,dw$ has at least $n-1$ distinct zeros in $D_\tau(z_0)$. On the other hand, let $C\geq 0$ be an upper bound for $|h|$ on the circle $\mathbb{S}_\tau(z_0):=\{z\in\mathbb{C};\ |z-z_0|=\tau\}$, then when

$$|t|<\frac{\tau^{n-1}}{|B|(\tau C+1)}\leq\frac{\tau^{n-1}}{|B|}, \tag{1.14}$$

there holds

$$\Big|\big\{(z-z_0)^{n-1}-tB[1+(z-z_0)h(z)]\big\}-\big[(z-z_0)^{n-1}-tB\big]\Big|<\big|(z-z_0)^{n-1}-tB\big| \tag{1.15}$$

on $\mathbb{S}_\tau(z_0)$. Note that we already assume that (1.11) does not have a zero on $\mathbb{S}_\tau(z_0)$. Thus, when t satisfies (1.14), (1.11) only has $n-1$ zeros in $D_\tau(z_0)$ by (1.15) and Rouché's Theorem [**1**]. Hence, when $t\in\mathbb{C}^*$ is sufficiently close to 0, (1.11) only has $n-1$ zeros in $D_\tau(z_0)$, and they are simple. Therefore, when $t\in\mathbb{C}^*$ is sufficiently close to 0, q has exactly $n-1$ poles near z_0, and each of them has order 2.

Finally, since P does not have a pole on the boundary of D_ρ, B3 is a consequence of B1 and B2 applied to an open disk slightly larger than D_ρ. □

REMARK 1.16. If p only has double poles in D_ρ, then q has the same number of double poles in D_ρ as p has in D_ρ when $t \in \mathbb{C}^*$ is sufficiently close to 0. So, in this case, the effect of $\mathcal{B}_t$ on the poles of p in D_ρ is relocation by small distances, and these distances approach 0 as t goes to 0.

Similar to Proposition 1.7, we have:

PROPOSITION 1.17. *Let*

$$Q = \begin{pmatrix} 0 & q \\ E/q & 0 \end{pmatrix} \mathrm{d}z \tag{1.18}$$

be the normalized potential for a CMC surface without any pole on the boundary of D_ρ. *Then, when* $s \in \mathbb{C}^*$ *is sufficiently close to* 0, *the normalized potential*

$$R = \begin{pmatrix} 0 & r \\ E/r & 0 \end{pmatrix} \mathrm{d}z = \mathcal{C}_s Q \tag{1.19}$$

has the following properties.

C1. r *carries all the poles of* q *in* D_ρ *with their orders unchanged; the other poles of* r *in* D_ρ *are the poles of* E/q *in* D_ρ *of orders* $> m+2$ *with their orders reduced by* $m+2$.

C2. *An order* n *pole* z_0 *of* E/q *in* D_ρ *yields exactly* $n-1$ *double poles of* E/r *in* D_ρ *close to* z_0 *(but not at* z_0*) and away from the umbilics; the only poles of* E/r *in* D_ρ *are such poles.*

C3. R *does not have any pole on the boundary of* D_ρ.

The proof of Proposition 1.17 is very similar to that of Proposition 1.7 and is omitted.

REMARK 1.20. If E/q only has double poles in D_ρ, then E/r has the same number of double poles in D_ρ as E/q has in D_ρ when $s \in \mathbb{C}^*$ is sufficiently close to 0. So, in this case, the effect of $\mathcal{C}_s$ on the poles of E/q in D_ρ is relocation by small distances, and these distances approach 0 as s goes to 0.

Combining Remarks 1.16 and 1.20 yields:

REMARK 1.21. If the normalized potential P given by (1.2) only has double poles in D_ρ, then when $t \in \mathbb{C}^*$ and $s \in \mathbb{C}^*$ are sufficiently close to 0, the effect of $\mathcal{C}_s\mathcal{B}_t$ on the poles of P in D_ρ is relocation by small distances, and these distances approach 0 as t and s go to 0.

The following result is a direct consequence of the definition of dressing transformations and the smoothness of the loop group splittings used in the definition. For notation, see [**3**].

PROPOSITION 1.22. *Let* P *given by (1.2) be the normalized potential for a CMC surface and* $K \subset D$ *a compact subset containing no poles of* P. *Then, for* *each $\epsilon > 0$, there exists a neighborhood $\mathcal{O}$ of the identity I in $\Lambda_B^+\mathrm{SL}(2,\mathbb{C})_\sigma$ such that for every element $g_+ \in \mathcal{O}$, the corresponding normalized potential*

$$\begin{pmatrix} 0 & q \\ E/q & 0 \end{pmatrix} \mathrm{d}z = (g_+)\bullet \begin{pmatrix} 0 & p \\ E/p & 0 \end{pmatrix} \mathrm{d}z \tag{1.23}$$

satisfies

$$\sup_K |p - q| < \epsilon, \qquad \sup_K \left| \frac{E}{p} - \frac{E}{q} \right| < \epsilon. \tag{1.24}$$

2. Denseness of Plain CMC Surfaces

A CMC surface $f : D \to \mathbb{R}^3$ is said to be *plain* if there is a point $z_0 \in D$ such that the normalized potential

$$P = \begin{pmatrix} 0 & p \\ E/p & 0 \end{pmatrix} \mathrm{d}z \tag{2.1}$$

for the surface with z_0 as the base point is holomorphic or has only poles of order 2 away from the umbilic points of the surface. Such a normalized potential for a plain CMC surface is called a plain one. In this section, we want to show that the plain CMC surfaces are dense in each dressing orbit of CMC surfaces.

First, we mention the following characterization of the plain normalized potentials for the plain CMC surfaces. This result is a direct consequence of the ones in [**2**] or [**6**].

PROPOSITION 2.2. *Let E be a holomorphic function on D and p a meromorphic function on D with $p(0) \neq 0$. Then, P given by (2.1) is the plain normalized potential for a CMC surface with 0 as the base point if and only if*

i) *all the poles and zeros of p are away from the zeros of E and have order 2,*

ii) *the residue at any pole of p or E/p is zero.*

Now, we proceed to prove the following main result of this paper.

THEOREM 2.3. *Fix a point $z_0 \in D$. Then, in each dressing orbit of CMC surfaces defined on D, the CMC surfaces having plain normalized potentials with z_0 as the base point are dense.*

PROOF. Without loss of generality, we assume that $z_0 = 0$. Let P given by (2.1) be the normalized potential for an arbitrary CMC surface.

First, pick a number ρ_1 in $(\frac{1}{2}, \frac{2}{3})$ if $D = D_1$ and in $(2, 3)$ if $D = \mathbb{C}$ such that there are no poles of P on the boundary of the disk D_{ρ_1}. Then, we can apply Proposition 1.7 to P on D_{ρ_1}. When $t_1 \in \mathbb{C}^*$ is sufficiently close to 0, q from

$$\begin{pmatrix} 0 & q \\ E/q & 0 \end{pmatrix} \mathrm{d}z = \mathcal{B}_{t_1} P \tag{2.4}$$

has in D_{ρ_1} only poles of order 2 away from the umbilic points, and E/q carries in D_{ρ_1} only the poles of E/p in D_{ρ_1} with their orders unchanged and the poles of p in

D_{ρ_1} of orders $> m+2$ with their orders reduced by $m+2$. Moreover, when $t_1 \in \mathbb{C}^*$ is sufficiently close to 0, there are no poles of q or E/q on the boundary of D_{ρ_1}. So, we can apply Proposition 1.17 to $\mathcal{B}_{t_1}P$ with $t_1 \in \mathbb{C}^*$ sufficiently close to 0. Let us fix a t_1 as needed above and look at $\mathcal{C}_{s_1}\mathcal{B}_{t_1}P$. When $s_1 \in \mathbb{C}^*$ is sufficiently close to 0, E/r from

$$\begin{pmatrix} 0 & r \\ E/r & 0 \end{pmatrix} \mathrm{d}z = \mathcal{C}_{s_1}\mathcal{B}_{t_1}P \tag{2.5}$$

has in D_{ρ_1} only poles of order 2 away from the umbilic points, and r carries in D_{ρ_1} only the poles of q with their order (=2) unchanged and the poles of E/q of orders $> m+2$ with their orders reduced by $m+2$. Also, when $s_1 \in \mathbb{C}^*$ is sufficiently close to 0, r and E/r still do not have any pole on the boundary of D_{ρ_1}. Note that a pole of p in D_{ρ_1} of order $n > 2m+4$ is a pole of r of order $n-2m-4$, a pole of E/p in D_{ρ_1} of order $n' > m+2$ is a pole of r of order $n'-m-2$, the other poles of P in D_{ρ_1} are not poles of $\mathcal{C}_{s_1}\mathcal{B}_{t_1}P$, and all the other poles of $\mathcal{C}_{s_1}\mathcal{B}_{t_1}P$ in D_{ρ_1} are of order 2 and away from the umbilic points. The above procedure can be repeated. Hence, there exists a positive integer n_1 such that for any nonzero complex numbers $t_1, s_1, t_2, s_2, \cdots, t_{n_1}$ and s_{n_1} sufficiently close to 0,

$$P_1 = \mathcal{C}_{s_{n_1}}\mathcal{B}_{t_{n_1}} \cdots \mathcal{C}_{s_1}\mathcal{B}_{t_1}P \tag{2.6}$$

is a normalized potential having in D_{ρ_1} only (finitely many) poles of order 2 away from the umbilic points. And the poles of P_1 in D_{ρ_1} are very close to (but not at) those of P in D_{ρ_1}. Moreover, the transformation $\mathcal{C}_{s_{n_1}}\mathcal{B}_{t_{n_1}} \cdots \mathcal{C}_{s_1}\mathcal{B}_{t_1}$ can be chosen as close to the identity as one wants, and applying a new $\mathcal{C}_s\mathcal{B}_t$ to P_1 only relocates slightly the poles of P_1 in D_{ρ_1} when $t \in \mathbb{C}^*$ and $s \in \mathbb{C}^*$ are sufficiently close to 0.

Next, let ρ_2 be a radius in $(\frac{2}{3}, \frac{3}{4})$ if $D = D_1$ and in $(3, 4)$ if $D = \mathbb{C}$ such that there are no poles of P_1 on the boundary of the disk D_{ρ_2}. Then, as above, there exists a positive integer $n_2 > n_1$ such that for any nonzero complex numbers t_{n_1+1}, $s_{n_1+1}, \cdots, t_{n_2}$ and s_{n_2} sufficiently close to 0,

$$P_2 = \Big(\prod_{j=n_1+1}^{n_2} \mathcal{C}_{s_j}\mathcal{B}_{t_j}\Big)P_1 = \Big(\prod_{j=1}^{n_2} \mathcal{C}_{s_j}\mathcal{B}_{t_j}\Big)P \tag{2.7}$$

has in D_{ρ_2} only (finitely many) poles of order 2 away from the umbilic points and that the poles of P_1 in D_{ρ_1} are only moved slightly.

Set $\rho_0 = 0$ (i.e., $D_{\rho_0} = \emptyset$). Apply the arguments as illustrated by the last two paragraphs infinitely many times. Then, we have a sequence of disks

$$D_{\rho_1} \subset D_{\rho_2} \subset D_{\rho_3} \subset \cdots \to D \tag{2.8}$$

and a sequence of positive integers

$$n_1 < n_2 < n_3 < \cdots \tag{2.9}$$

such that for any sequence of nonzero complex numbers

$$t_1,\ s_1,\ t_2,\ s_2,\ t_3,\ s_3,\ \cdots \tag{2.10}$$

sufficiently close to 0 and for each j, the normalized potential

$$P_j = \begin{pmatrix} 0 & p_j \\ E/p_j & 0 \end{pmatrix} \mathrm{d}z = \Big(\prod_{l=1}^{n_j} \mathcal{C}_{s_l}\mathcal{B}_{t_l}\Big)P \tag{2.11}$$

has in D_{ρ_j} only (finitely many) poles of order 2 away from the umbilic points; the poles of P_j, P_{j+1}, P_{j+2}, $\cdots$ in D_{ρ_j} form convergent sequences that approach distinct non-umbilic points in D_{ρ_j}; each such sequence z_j, z_{j+1}, z_{j+2}, $\cdots$ with $z_j \in D_{\rho_j}\backslash D_{\rho_{j-1}}$ stays in a small disk $D_{\tau_j}(z_j)$ with $\tau_j > 0$ such that $D_{2\tau_j}(z_j) \subset D_{\rho_j}$; and

$$\prod_{l=1}^{+\infty} \mathcal{C}_{s_l}\mathcal{B}_{t_l} \tag{2.12}$$

converges to a dressing transformation close to the identity I. Moreover, we can assume that all the disks $D_{2\tau_j}(z_j)$ in D are mutually disjoint. So, z_j is the only pole of P_j on $\overline{D_{2\tau_j}(z_j)} := \{z \in \mathbb{C};\ |z - z_j| \leq 2\tau_j\}$, and hence if z_j is a pole of p_j, p_j does not have any zero on $\overline{D_{2\tau_j}(z_j)}$ since its zeros are the poles of E/p_j [**2**] (if z_j is a pole of E/p_j, we may have to use a smaller τ_j to guarantee that E/p_j does not have any zero on $\overline{D_{2\tau_j}(z_j)}$: some zeros of E may not be poles of E/p_j). Then, for any $k > j$ and every pole z_j of p_j on D_{ρ_j}, the corresponding pole z_k of p_k is the only pole of p_k on $\overline{D_{2\tau_j}(z_j)}$, and p_k also does not have any zero on $\overline{D_{2\tau_j}(z_j)}$. Also, we have similar statements for E/p_k. Set

$$\widetilde{P} = \begin{pmatrix} 0 & \tilde{p} \\ E/\tilde{p} & 0 \end{pmatrix} \mathrm{d}z = \Big(\prod_{l=1}^{+\infty} \mathcal{C}_{s_l}\mathcal{B}_{t_l}\Big)P \tag{2.13}$$

and fix a j. Let z_j be a pole of p_j in $D_{\rho_j}\backslash D_{\rho_{j-1}}$ (the poles of E/p_j in $D_{\rho_j}\backslash D_{\rho_{j-1}}$ can be handled similarly) with z_j, z_{j+1}, z_{j+2}, $\cdots$ being the corresponding sequence mentioned above. Set

$$C_j = \inf\big\{|p_j(z)|;\ z \in D_{2\tau_j}(z_j), z \neq z_j\big\}, \tag{2.14}$$

then $C_j > 0$. When t_{n_j+1}, s_{n_j+1}, t_{n_j+2}, s_{n_j+2}, t_{n_j+3}, s_{n_j+3}, $\cdots$ are sufficiently close to 0, for each $k > j$,

$$\prod_{l=n_j+1}^{n_k} \mathcal{C}_{s_l}\mathcal{B}_{t_l} \quad \text{and} \quad \prod_{l=n_k+1}^{+\infty} \mathcal{C}_{s_l}\mathcal{B}_{t_l} \tag{2.15}$$

are sufficiently close to I and hence, by Proposition 1.22, we have

$$|p_j(z) - p_k(z)| < \frac{1}{2}C_j \tag{2.16}$$

on the circle $\mathbb{S}_{2\tau_j}(z_j)$ and

$$|p_k(z) - \tilde{p}(z)| < \frac{1}{4}C_j \tag{2.17}$$

on $\overline{D_{2\tau_j}(z_j)}\backslash D_{\tau_j/k}(z_k)$. Note that in total, on each $|t_m|$ or $|s_m|$, we have imposed only a finite number of positive upper bounds. Now, (2.14) and (2.16) imply that

$$|p_k(z)| \geq |p_j(z)| - |p_j(z) - p_k(z)| > C_j - \frac{1}{2}C_j = \frac{1}{2}C_j \tag{2.18}$$

on $\mathbb{S}_{2\tau_j}(z_j)$. Thus,

$$\left|\frac{1}{p_j(z)} - \frac{1}{p_k(z)}\right| = \left|\frac{p_j(z)-p_k(z)}{p_j(z)p_k(z)}\right| < \frac{C_j/2}{C_j \cdot C_j/2} = \frac{1}{C_j} \tag{2.19}$$

on $\mathbb{S}_{2\tau_j}(z_j)$ and hence, by the Maximum Modulus Theorem [**1**], on $\overline{D_{2\tau_j}(z_j)}$ since $1/p_j$ and $1/p_k$ are holomorphic on $D_{2\tau_j}(z_j)$ and continuous on $\overline{D_{2\tau_j}(z_j)}$. So,

$$\left|\frac{1}{p_k(z)}\right| \leq \left|\frac{1}{p_j(z)}\right| + \left|\frac{1}{p_j(z)} - \frac{1}{p_k(z)}\right| < \frac{1}{C_j} + \frac{1}{C_j} = \frac{2}{C_j} \tag{2.20}$$

on $\overline{D_{2\tau_j}(z_j)}$. From (2.17) and (2.20) we obtain that on $\overline{D_{2\tau_j}(z_j)} \backslash D_{\tau_j/k}(z_k)$,

$$|p_k(z) - \tilde{p}(z)| < |p_k(z)|, \tag{2.21}$$

$$|\tilde{p}(z)| \geq |p_k(z)| - |p_k(z) - \tilde{p}(z)| > \frac{1}{2}C_j - \frac{1}{4}C_j = \frac{1}{4}C_j. \tag{2.22}$$

By (2.21) and Rouché's Theorem, the sum of the orders of the poles of $\tilde{p}$ in $D_{\tau_j/k}(z_k)$ is at least 2. Note that $\tau_j/k \to 0$ as $k \to +\infty$. Let z_∞ be the limit of the sequence z_j, z_{j+1}, z_{j+2}, ... Then, z_∞ is a pole of $\tilde{p}$ and has an order of at least 2. A consequence of (2.22) is that $\tilde{p}$ does not have a zero on $\overline{D_{2\tau_j}(z_j)}$. Finally, (2.16) and (2.17) together yield that

$$|p_j(z) - \tilde{p}(z)| \leq |p_j(z) - p_k(z)| + |p_k(z) - \tilde{p}(z)| < \frac{3}{4}C_j < |p_j(z)| \tag{2.23}$$

on $\mathbb{S}_{2\tau_j}(z_j)$. By (2.23) and Rouché's Theorem again, z_∞ is the only pole of $\tilde{p}$ on $D_{2\tau_j}(z_j)$ and has order 2. Therefore, when t_1, s_1, t_2,s_2, t_3, s_3, $\cdots$ are sufficiently close to 0, $\widetilde{P}$ is a normalized potential close to P in the dressing orbit through P and has on the whole of D only poles of order 2 away from the umbilic points. □

References

[1] J. Conway, *Functions of One Complex Variable*, Second Edition, Springer-Verlag, 1986.

[2] J. Dorfmeister & G. Haak, *Meromorphic potentials and smooth CMC surfaces*, Math. Z. **224** (1997), 603–640.

[3] J. Dorfmeister, F. Pedit & H. Wu, *Weierstrass-type representation of harmonic maps into symmetric spaces*, Commun. Anal. & Geom. **6** (1998), 633–668.

[4] H. Wu, *On the dressing action of loop groups on constant mean curvature surfaces*, Tôhoku Math. J. **49** (1997), 599–621.

[5] H. Wu, *A simple way for determining the normalized potentials for harmonic maps*, Ann. Global Anal. Geom. **17** (1999), 189–199.

[6] H. Wu, *A new characterization of normalized potentials in dimension two*, Results in Math. **36** (1999), 184–194.

Department of Mathematics, Northern Illinois University, DeKalb, IL 60115, USA

E-mail address: wu@math.niu.edu

Titles in This Series